ASIAN AMERICANS AND THE LAW

Historical and Contemporary Perspectives

Series Editor

CHARLES McCLAIN

University of California, Berkeley

Series Contents

VOLUME

3

THE MASS INTERNMENT OF JAPANESE AMERICANS AND THE QUEST FOR LEGAL REDRESS

Edited with introductions by

CHARLES McCLAIN

Routledge
Taylor & Francis Group
LONDON AND NEW YORK

First published by 1994 Garland Publishing, Inc.

This edition published 2011 by Routledge
2 Park Square, Milton Park, Abingdon, Oxon OX14 4RN
52 Vanderbilt Avenue, New York, NY 10017

Routledge is an imprint of the Taylor & Francis Group, an informa business

Library of Congress Cataloging-in-Publication Data

The mass internment of Japanese Americans and the quest for legal redress / edited with introductions by Charles McClain.
p. cm. — (Asian Americans and the law ; v. 3)
Collection of previously published material.
Includes bibliographical references.
ISBN 0–8153–1866–9 (acid-free paper)
1. Japanese Americans—Evacuation and relocation, 1942–1945. 2. Japanese Americans—Civil rights. 3. World War, 1939–1945—Reparations. I. McClain, Charles, 1943– . II. Series.
D769.8.A6M33 1994
940.53'1503956073—dc20 94–22630
CIP

ISBN 13: 978-0-8153-1866-8 (pbk)

For Laurie

Contents

Series Introduction

According to the 1990 census there are now more than seven million persons of Asian ancestry living in the United States, representing close to three per cent of the total U.S. population. In California Asians represent almost ten per cent of the state's residents and that figure seems likely to increase in the years ahead. Asians (Chinese immigrants, in particular) have been present in significant numbers in the United States since the 1860s. What may not be known is that Asians have been deeply implicated in the legal history of this country since that time, as well. The law has had a profound, though often negative, impact on the experience of Asian communities in the United States. In turn the struggles of Asian Americans to vindicate their right to equal treatment under the law have contributed significantly to shaping the American legal order.

This Garland anthology gathers together some of the best writings on the large subject of Asian Americans and the law. As one might infer from the title, the orientation of this anthology is primarily historical. Most articles deal with past events, but they often make an effort to relate these events to contemporary developments. Furthermore, their authors come from a variety of disciplines besides history, including Asian American studies, law, and political science. Together they open up the possibility of a multidisciplinary understanding of the episodes and issues that are discussed. Finally, lest it be thought that the series focuses exclusively on the past, there are a number of articles in the collection that deal explicitly and in depth with contemporary themes, including themes that transcend individual Asian groups.

The series consists of four volumes and attempts to be comprehensive in its coverage. The first volume concerns Chinese immigrants and focuses on their struggles in the U.S. courts against discriminatory state and federal laws. Two volumes are dedicated to the Japanese, a group whose immigration dates from the 1890s, and who likewise encountered their share of discriminatory legislation. The first of these documents the rise of anti-Japanese agitation in the western states, the enactment of laws aimed at ousting the Japanese from their position in the agricultural economy, and the immigrant community's response. The second is devoted almost entirely to the forcible evacuation and mass internment of civilians of Japanese ancestry during World War II, a topic that has generated a torrent of literature and that constitutes a most extraordinary chapter in the legal history of Asians in this country. The last volume in the series is devoted to other Asian groups—Filipinos, Koreans, Asian Indians, Southeast Asians. These immigrant communities, whose size has increased dramatically in the last two decades, have had significant interactions with the American legal process. The volume also includes articles on the impact of the law on the indigenous population of Hawaii.

In the course of preparing this anthology I have incurred several debts of gratitude that I would like to acknowledge. I am very grateful to Roger Daniels and Neil Gotanda for their suggestions concerning articles to include in this collection. Naturally, the ultimate responsibility for the selection of articles is my own. I also want to express my thanks to Leo Balk, Carole Puccino, and the production staff of Garland Publishing for their patient assistance throughout this project.

INTRODUCTION

On February 19, 1942, two and a half months after the Japanese attack on Pearl Harbor, President Franklin Roosevelt issued Executive Order 9066, authorizing the Secretary of War or his military subordinates to designate certain sections of the country as "military areas" and to exclude any and all persons from them. A month later Congress enacted Public Law No. 503, making it a crime to disobey military orders concerning restrictions on residence in these areas. These two enactments would serve as the ultimate legal basis, first for military orders imposing a curfew on all persons of Japanese ancestry in coastal California, Washington and Oregon and parts of Arizona, and later for orders forcibly removing them from these areas to what the government euphemistically called "Relocation Centers" in the interior. More than 110,000 men, women and children, some 70,000 of them American citizens, were transported to these centers, none having been subjected to any individual adjudication as to loyalty or disloyalty, guilt or innocence of crime. The sites were surrounded by barbed wire and armed guards. Those inside were forbidden to leave without official permission. Places of this description, as the historian Roger Daniels reminds us, have been known to the world since the end of the nineteenth century, when the British used them to confine Boer civilians, as concentration camps.[1]

Several American citizens of Japanese descent chose to challenge the military decrees in court and their cases ultimately reached the Supreme Court of the United States. There they met largely with rebuffs. In 1943, in *Hirabayashi v. United States*[2], the Court held that the curfew imposed by Gen. J.L. DeWitt, head of the Western Defense Command, was within the joint war powers of Congress and the president and did not unconstitutionally discriminate against Japanese Americans (it reached the same result in the companion case of *Yasui v. United States*[3]).The following year, in *Korematsu v. United States*[4], the Court, in a 6-3 decision, declared constitutional the order excluding persons of Japanese ancestry from areas near the Pacific coast and affirmed the conviction of Fred Korematsu, a welder from San Leandro, California, for remaining in one of them.On the same day, in *Ex parte Endo*[5], the

Court held that neither the executive order nor the Congressional act authorized the continued detention in a Relocation Center of Mitsuye Endo, a young woman internee from Sacramento, whose loyalty the government conceded. It refused, however, to reexamine the constitutionality of the military orders that had led to her incarceration in the first place.

The net effect of this series of decisions was to legitimize the most extraordinary exercise of power by military authorities over civilians in American history. More importantly, it affirmed the principle that the military or any other arm of government could use membership in an ethnic group—and that alone—as the basis for depriving individuals of some of their most basic civil liberties. (Ironically, in *Korematsu* the Court had said that executive action curtailing the civil rights of racial minorities was to be viewed as suspect and was to be subjected to the most exacting scrutiny before being allowed to pass constitutional muster.)

The mass evacuation and internment[6] of Japanese Americans and the great legal cases this gave rise to have been the subject of several important books over the years, most recently Peter Irons' *Justice at War*. This work, in addition to providing a detailed and engaging narrative of the events, also brought to light the fact that the government withheld important information from the Supreme Court in connection with the internment cases, information which Irons suggests might have led to a different outcome had the Court known of it. There are a large number of articles on the subject as well, and I have selected some of the best for inclusion in this volume.

In 1943, before any of the Supreme Court decisions had come down, Harrop Freeman published an article in the *Cornell Law Quarterly* examining the evacuation and internment decisions. He put them in the context of past actions by the authorities in time of hostilities and revealed just how unprecedented in nature and scale they were. Reviewing the government's argument that "military necessity" lay behind these decisions, he found it to be notably unconvincing. Freeman noted the availability of other options and pointed to the evidence that racial stereotyping had clearly played a significant part in the government's actions. "When the final history of the Japanese evacuation is written," he wrote prophetically, "it will almost certainly appear that decisions were made on misinformation, assumptions, prejudices [and] half truths."[7]

Two years later Eugene Rostow, then on the Yale law faculty, published a scathing attack on the Supreme Court's decisions, entitled "The Japanese American Cases—A Disaster." It characterized the evacuation and internment as "*the worst blow our* liberties have sustained in many years," and took the Court to task for its handling of the cases that had grown out of them. Like Freeman, he concluded that the program had not been driven by military necessity but had been "conceived and put through by the organized minority whose business it has been for forty-five years to increase and exploit racial tensions on the West Coast."[8]

Another extremely critical article appeared at about the same time in the *Columbia Law Review* by Nanette Dembitz, the niece of the great Supreme Court Justice Louis D. Brandeis. It discussed in great and knowledgeable detail the full train of events leading to the evacuation decisions and provided a penetrating doctrinal analysis of the two Supreme Court opinions. She, too, called attention to the enormous role that politicians on the West Coast, nativist pressure groups, and the popular press had played in persuading the government to make its moves, to the government's own admission that it had the means readily at hand for distinguishing the loyal from the potentially disloyal, and finally to General DeWitt's obvious racial bias. She dissected with great skill the doctrinal inconsistencies implicit in the *Korematsu* opinion. Her conclusion was simple: the Justices of the Supreme Court had accepted uncritically the allegations of the government and abdicated their responsibility to stand as independent judges of the reasonableness of military controls over civilians. The Dembitz article takes on special significance—and a certain irony—when one realizes that Dembitz herself, while a Justice Department lawyer, had contributed to the preparation of the government's curfew and evacuation cases and indeed had prepared a memorandum pointing to the racial separateness of the Japanese American population.

The volume includes a range of other articles on the internment. Roger Daniels looks to the Canadian experience (Canada also interned its Japanese residents) for a comparative perspective. Sidney Fine seeks to explain why Justice Frank Murphy, author of the vigorous dissent in the *Korematsu* case (he had written that the internment was the product of racism pure and simple) went along with his brethren in the *Hirabayashi* curfew case. Philip Tajitsu Nash's *Yale Law Journal* review-essay brings to life the often wretched experiences of those confined to the internment camps. Sandra Takahata, the daughter of a Hawaii resident interned during the war, explores the interesting question of how a case like *Korematsu* might be decided now, given the major changes in constitutional doctrine that have occurred since the decision was handed down. I have also included several review-essays on *Justice at War,* each reflecting a slightly different perspective on the Irons' work.

Eugene Rostow observed in the conclusion of his 1945 article that the Japanese Americans had suffered great property losses as a result of their removal from their homes and argued that they were due generous financial redress. As Mark Iyeki makes clear in his article, Japanese Americans began to campaign for redress shortly after their release from the camps. They achieved a partial success in 1948 when Congress passed "The Japanese American Evacuation Claims Act." It gave persons of Japanese ancestry the right to compensation for real and personal property losses but not for loss of income or pain and suffering. Not only were the damages inadequate but the standard of proof was difficult to meet, and in the end the government paid out only $37 million dollars to claimants. A full-scale campaign to achieve redress and reparations began in the 1970s with the support of various Japanese American civil rights groups. One result was the creation in 1980 of a national commission to study the internment. This Commission on Wartime Relocation and Internment of Civilians concluded its work in 1982 with the publication of the following report: "The broad historical causes which shaped these decisions were race prejudice [it linked them to the long history of anti-Japanese agitation on the West Coast], war hysteria and a failure of political leadership." It recommended that $20,000 be paid every survivor of the internment camps, a recommendation that the U.S. Congress implemented in 1988 when it passed authorizing legislation (payments to survivors did not actually commence until 1990).

As the commission was concluding its work, Gordon Hirabayashi, Fred Korematsu, and Minoru Yasui with widespread community support launched a parallel effort to have their wartime convictions set aside. Invoking the little known writ of *coram nobis,* which allows courts to reexamine judgments on the basis of newly discovered facts, they argued that the government's misconduct (discovered after the war through the work accomplished by Irons) had prevented them from getting a fair hearing. They were all *ultimately* successful. Indeed, the federal courts that considered their petitions bent over backward to show their sympathy for the men and their revulsion at the government's wartime actions. Iyeki discusses these legal proceedings in his article.[9]

The concluding articles in the volume deal with other aspects of the Japanese American experience during World War II: the treason trial of Iva Ikuko Toguri d'Aquino (mislabelled "Tokyo Rose"), the successful campaign to have her pardoned and her citizenship restored, and the similarly successful litigation aimed at winning back American citizenship, renounced during internment, for Japanese American internees. Both Mrs. d'Aquino and the Japanese renunciants were represented by the San Francisco attorney, Wayne Collins. Collins, who represented the petitioner in *Korematsu* and who filed *amicus* briefs in the *Hirabayashi* and *Endo* cases, has come under severe criticism because of his strategy and tactics in those cases. Whatever one's evaluation of Collins' legal tactics, one cannot help but be impressed with the depth of his commitment to the Japanese American community, with his tenaciousness in defending its interests, and with his willingness to step forward and take on unpopular causes when few other lawyers seemed eager to do so.

Notes

1. Roger Daniels et al., eds., *Japanese Americans: From Relocation to Redress*, p. 6. I am grateful to Roger Daniels for his thoughts on the subject matter of this volume.
2. 320 U.S. 81 (1943).
3. 320 U.S. 115 (1943).
4. 323 U.S. 214 (1944).
5. 323 U.S. 283 (1944).
6. Although "internment" is the term most often used to describe the mass removal, under the purported authority of Executive Order 9066 and Public Law 503, of West Coast Japanese Americans to Relocation Centers inland, Roger Daniels argues that it is really a misnomer as applied to them. It ought properly to be reserved, he contends, for those specific Japanese nationals who were taken into custody by the FBI and other authorities as suspected security risks and removed to what were in fact called "internment camps." Daniels estimates that more than 8,000 Japanese were interned under this program, which employed very broad and vague criteria, as it turns out. Roger Daniels, "The Internment of Japanese Nationals" (Paper presented at the Annual Meeting, American Historical Association, San Francisco, Jan. 9, 1994). See also John J. Culley, "The Santa Fe Internment Camp and the Justice Department Program for Enemy Aliens," in R. Daniels et al., *Japanese Americans: From Relocation to Redress* (1991).
7. Not all articles were critical. A piece appearing in the same *issue*

of the *Law Quarterly* argued that the curfew and evacuation were amply justified and violated no one's constitutional rights. See Maurice Alexandre, "The Nisei—A Casualty of World War II," 28 *Cornell Law Quarterly* 385 (1943). Alexandre conceded that individual Japanese might be wronged by the government's actions, but everybody suffers in wartime, and they ought to see that as their contribution to the war effort. The most notable voice in the scholarly community arguing that the evacuation was justified was that of the Stanford political scientist, Charles Fairman. See "The Law of Martial Rule and the National Emergency," 55 *Harv. L. Rev.* 1251, at 1301–1302 (1942).

8. Eugene Rostow, "The Japanese American Cases—A Disaster," 54 *Yale Law Journal* 489, at 490, 496 (1945).
9. In Yasui's case the outcome was not so clear-cut. The government agreed that his conviction should be vacated but moved to dismiss his petition without any finding that his rights had been violated. The motion was granted. The story of the redress and reparations campaign and of the *coram nobis* case is also told at great length in Peter Irons, *Justice Delayed: The Record of the Japanese American Internment Cases* (1989). An attempt by camp survivors and their representatives to pursue tort and unjust takings of property claims against the federal government proved ultimately unsuccessful. See *United States v. Hohri*, 482 U.S. 64 (1987).

Further Reading

Roger Daniels, Sandra C. Taylor, Harry L. Kitano, eds., *Japanese Americans: From Relocation to Redress,* University of Washington Press, 1991.

Roger Daniels, *Concentration Camps, North America: Japanese in the United States and Canada During World War II,* R.E. Krieger Pub. Co., 1981.

Peter Irons, *Justice at War: The Story of the Japanese American Internment Cases,* Oxford University Press, 1983.

———, *Justice Delayed: The Record of the Japanese American Internment Cases,* Wesleyan University Press, 1989.

John Tateishi, *And Justice for All: An Oral History of the Japanese American Detention Camps,* Random House, 1984.

Jacobus tenBroek, Edward N. Barnhart, Floyd W. Matson, *Prejudice, War and the Constitution,* University of California Press, 1954.

Notes and Documents

The Decisions to Relocate the North American Japanese: Another Look

Roger Daniels

The author is professor of history in the University of Cincinnati.

IN A LITTLE over two months after the outbreak of the Great Pacific War in December 1941, both North American democracies—the United States and Canada—began the process of rounding up and incarcerating, in one way or another, persons of Japanese ancestry who happened to live in the coastal strip between the mountains and the Pacific Ocean in the province of British Columbia and in the states of Washington, Oregon, and California. The similarity in the actions of the two nations is not surprising. For over half a century each country had been dealing with small Asian minorities on the West Coast in quite similar ways, and, in each society, Asians had played the same kinds of economic roles. First Chinese and then Japanese arrived to become, figuratively, hewers of wood and drawers of water, or, if one prefers to be literal, builders of railroads, performers of stoop labor and domestic chores, and, particularly in British Columbia, fishermen. In each nation economic and racial hostility arose first among workingmen and their allies and then spread across most of western society.

The years 1940–1941 saw enclaves of Japanese clustered in coastal regions of each country. In the continental United States there were just over 125,000 ethnic Japanese, more than eighty-eight percent of whom lived on the Pacific slope. In Canada there were 23,000 who were even more highly concentrated: fewer than five percent lived outside of British Columbia. In both nations Japanese represented fewer than one-tenth of one percent of the total population, and even on the West Coast their proportion was not high: 1.2 percent of the population in the American Pacific states and 2.7 percent of the population of British Columbia.[1]

[1]See Roger Daniels, "The Japanese Experience in North America: An Essay in Comparative Racism," in *Canadian Ethnic Studies*, IX (1977), 91–100 and the references cited there.

Japanese were clearly second-class residents in each country. In Canada, Japanese immigrants could and did become naturalized citizens under color-blind federal statutes, but could not vote in the province of British Columbia, where almost all of them lived, because provincial law forbade Asians, regardless of citizenship, from voting. Only about a quarter of the Japanese Canadian population were Japanese nationals. In the United States, federal naturalization statutes, dating back to 1870, prohibited Asians from being naturalized, but, since under the Fourteenth Amendment to the Constitution "all persons" born in the U.S. were citizens legally entitled to suffrage and other rights, native born Japanese American adults had the franchise and were beginning to use it. About two-thirds of the American Japanese were citizens. In each country the overwhelming majority of the native-born generation, children of marriages first consummated in the years around 1920, were minors. In 1942, as is well known, these differentials in citizenship status were of little significance, as, in each nation, citizen and alien, adult and child, male and female, were all herded out of the coastal zones and into the interior. The only criterion was ethnicity: all Japanese had to go.[2]

In the United States the instrument of evacuation was the United States Army since President Franklin Roosevelt had delegated authority, on February 19, 1942, in Executive Order 9066, to the Secretary of War and his subordinates. In Canada, Prime Minister William Lyon Mackenzie King used an Order in Council, P. C. 1486, issued five days later, which delegated power to the Minister of Justice who utilized the Royal Canadian Mounted Police as his instrument. In each nation, a special civilian administrative body was created to warehouse and manage the evacuees: the American War Relocation Authority and the Canadian British Columbia Security Commission. Also in each country, the places of evacuation were similar in their isolation, but the techniques of restraint used were different. Canadian Japanese were initially segregated largely by sex, with able-bodied men sent to various work camps, and women, children, and older men sent to "housing centres" in the interior valleys of British Columbia where the physical restraints were imposed largely by geography—the same kind of exile that had been used by the Czarist Russian government. In the United States, incarceration was overt: Japanese family units were shipped off to ten isolated sites where flimsy barrack-like buildings were erected and then surrounded by barbed-wire, watch towers, and armed guards—the kind of restraint usually called "concentration camps" but which the federal authorities insisted were "relocation centers."

Thus the two governments in early 1942 treated Japanese Americans and Canadians in quite similar ways. A mistrusted yet innocent ethnic minority clustered geographically in enclaves on the Pacific Coast of each nation was uprooted from its homes and placed in alien and largely hostile environments on the basis of a wholly fictitious military necessity that was a mere fig leaf

[2]The comparison is made more fully in my *Concentration Camps, North America: Japanese in the United States and Canada during World War II* (Malabar, Fla., 1981).

for a racism that had been endemic against Asians for nearly a century. The means used—the military in one nation, the national police in the other—were closely related, while the places of incarceration were almost equally isolated and unpleasant even though the conditions of confinement were clearly more restrictive in the United States. And, it must be noted, the Canadian government, unlike the American, did not kill any of the people it incarcerated. Both countries, in spite of a show of fairness, simply cheated their Japanese residents out of much of their real and personal property, often to the advantage of Caucasian citizens who were allowed to buy at prices well below the market value.

Faced with these similarities the historian of North America is forced to ask if all these similarities were simply fortuitous. Was there joint preplanning? Were the similarities due to mimesis? Or were basic attitudes and institutions on either side of the forty-ninth parallel so related that similar reactions naturally ensued?

None of those who have written about the evacuation, including me, has previously found any evidence of preplanning.[3] It is now clear, as I shall shortly demonstrate, that there were discussions between the two governments before the war about "doing something" with West Coast Japanese if war came. These discussions, which might be called planning, were not unrelated to the actions eventually taken against the Japanese in both countries. The real question is, perhaps, whether these discussions had any demonstrable impact on the decisions eventually taken, or whether they are merely further evidence of a similar set of mind, a kind of international conditioned reflex.

The major evidence for preplanning comes from the records of the Permanent Joint Board of Defense, a mixed Canadian-American group established as one of the results of the Ogdensburg, New York, meetings between President Roosevelt and Prime Minister King on August 17–18, 1940. The board had a predominantly military membership with a civilian chairman and secretary from each country. The surviving member, Canadian civilian Hugh L. Keenleyside, remembered in 1979:

---

[3]The major book-length accounts of the American experience are Morton Grodzins, *Americans Betrayed: Politics and the Japanese Evacuation* (Chicago, 1949); Jacobus ten Broek, Edward N. Barnhart, and Floyd W. Matson, *Prejudice, War and the Constitution* (Berkeley and Los Angeles, 1954); Allan R. Bosworth, *America's Concentration Camps* (New York, 1967); Audrie Girdner and Anne Loftis, *The Great Betrayal: The Evacuation of the Japanese-Americans during World War II* (New York, 1969); Roger Daniels, *Concentration Camps, USA: Japanese Americans and World War II* (New York, 1971); Daniels, *The Decision to Relocate the Japanese Americans* (Philadelphia, 1975); and Michi Weglyn, *Years of Infamy: The Untold Story of America's Concentration Camps* (New York, 1976). On the Canadian experience, the major book-length accounts are Forrest La Violette, *The Canadian Japanese and World War II: A Sociological and Psychological Account* (Toronto, 1948); Ken Adachi, *The Enemy That Never Was: A History of the Japanese Canadians* (Toronto, 1976); Barry Broadfoot, *Years of Sorrow, Years of Shame* (Toronto, 1977); and W. Peter Ward, *White Canada Forever: Popular Attitudes and Public Policy toward Orientals in British Columbia* (Montreal, 1978). Also see Ann Gomer Sunahara, *The Politics of Racism: The Uprooting of Japanese Canadians during the Second World War* (Toronto, 1981).

informal discussion about Canadian and American plans for dealing with persons of Japanese racial origin who were living on the Pacific Coast took place from time to time between members of the Joint Board on Defence. There was no formal debate on the subject, and certainly no prior agreement between the Governments on what should be done, nor I believe was there any kind of informal agreement reached.[4]

Some twenty years earlier, Keenleyside had written:

a vigorous effort was made by certain members of the Board to put through a resolution urging the Governments to deport or to place in custody those "elements in the population of Japanese racial origin." More moderate counsels prevailed and the Board merely expressed the belief that watchfulness was essential and that "a coincidence of policy" between Washington and Ottawa would be useful. The eventual action was similar in the two countries and reflects credit on neither. Canadian behaviour was particularly discreditable because the numbers involved were small and the individuals, unlike the Germans or Italians resident in Canada, easily identifiable, their previous behaviour had been above reproach, and the way in which the Japanese-Canadians' property was liquidated was both callous and mean. The whole operation was a cheap and needless capitulation to popular prejudice fanned by political bigotry or ambition or both.[5]

The leading American authority on the Joint Board, Colonel Stanley W. Dziuban, author of the volume on U.S.-Canadian military cooperation in the U.S. Army's official history of World War II, says absolutely nothing about any Joint Board discussions of possible actions against resident Japanese even though he had a manuscript version of Keenleyside's article at his disposal. He prints an appendix of thirty-three formal recommendations which the board made between August 26, 1940, and September 1, 1945, none of which deals with resident Japanese.[6]

Happily, papers once classified are now open so that we can look behind these secondary accounts. After the Joint Board's twenty-third meeting, held in Montreal on November 10–11, 1941, the following was the fifth item entered into its journal:

---

[4]Letter to Roger Daniels, two pages, typed with ink corrections, Sept. 19, 1979, in my possession. Keenleyside, born on July 7, 1898, took a B.A. at the University of British Columbia and an M.A. and Ph.D. at Clark University. After a brief academic career teaching history, he entered government service in 1928. By the time of his appointment to the Joint Board he was also Assistant Under Secretary of State for External Affairs. He was secretary of the Canadian Section of the Joint Board from 1940 to 1944 and its acting chairman in 1944–1945.

[5]H. L. Keenleyside, "The Canada-United States Permanent Joint Board on Defence, 1940–1945," *International Journal*, XVI (1960–1961), 63.

[6]Col. Stanley W. Dziuban, *Military Relations between the United States and Canada, 1939–1945* (Washington, D.C., 1959), 347–365.

In the course of its review of the defense position on the Pacific Coast, the Board gave attention to the problem caused by the existence in Canada and in the United States of elements in the population of Japanese racial origins. The opinion was expressed that an effort should be made to see that, in the event of hostilities with Japan, the two governments should follow policies of a similar character in relation to those racial groups. It was agreed that it would be desirable that steps be taken to arrange for consultation between the appropriate officers of the two governments, with a view to bringing about a practicable coincidence of policies in this matter in Canada and in the continental United States.[7]

Clearly, Keenleyside's account was more accurate than Dziuban's, and I, for one, can find no excuse for his total suppression of the topic.[8]

What effect, if any, did the board's policy statement have on relocation policies? As near as I have been able to discover, there was not, in the weeks between the Montreal meeting and Pearl Harbor, any consultation "between appropriate officers of the two governments," although it is difficult to believe that some of the officers involved did not discuss it with some of their colleagues or even opposite numbers. By mid-January 1942, there is some correspondence about the Joint Board statement between the Ministry of External Affairs and the State Department.[9] I very much doubt that those persons in the American government who were most responsible for the decision to relocate the Japanese Americans—Secretary of War Henry L. Stimson and his subordinates—were influenced by or even had an awareness of the Joint Board's statement. There is certainly no reference to it in any logical place in their papers. As for the head of the American government, it is difficult to say. Roosevelt had direct contact with Fiorello La Guardia, the impetuous mayor of New York who wore many hats, including one as a member of the Joint Board and another as director of the federal Office of

[7]File 3370, MG 26J4, vol. 320, W. L. Mackenzie King Papers, Public Archives of Canada. This is a fruit of Ann Sunahara's research which she generously shared with me.

[8]Dziuban, *Military Relations*, contains two references to the relocation of Japanese. He mentions that Japanese were evacuated from Prince Rupert (p. 240), and he states, without documentation (p. 284): "In Canada as well as in the United States, an evacuation of west coast Japanese took place after Pearl Harbor. Beginning in the spring of 1942 U.S. authorities raised the question of the security of the Canadian National Railways, along which internee camps were located at a number of points. To minimize the threat of sabotage, Canada closed certain of the camps and took additional police measures at others."

In Keenleyside's article of 1960–1961, he indicates that the board statement emanated from the "meeting in New York in November" of 1941. This is an obvious slip, as that meeting took place in Montreal. The preceding meeting had been in New York (Sept. 9–10, 1941). I suspect, but cannot now demonstrate, that the discussion about what to do with Japanese took place in more than one meeting.

[9]Secretary of State for External Affairs to Canadian Minister in Washington, Jan. 14, 1942, and George A. Gordon, U.S. Dept. of State, to Hume Wrong, Minister Counselor of Canada, Feb. 5, 1942, both in Historical Division, Dept. of External Affairs, Ottawa, file 3464-B-40C. Again, I owe these references to the generosity of Ann Sunahara.

Civilian Defense. La Guardia may have mentioned it to the President. It is likely that La Guardia's Joint Board service conditioned his views about Japanese. In 1924 he had been one of just three members of the House of Representatives who had been willing to continue Japanese immigration; in the days after Pearl Harbor, La Guardia, in his capacity as civilian defense director, pointedly omitted mention of Japanese in two public statements calling for decent treatment of enemy aliens and suggesting that Germans and Italians be presumed loyal until proven otherwise. By implication, at least, Japanese were presumed to be disloyal.

On the Canadian side, with a much smaller government bureaucracy, there seems to have been a somewhat wider awareness of what had taken place at the Montreal meeting. There are records of the Joint Board statement in the Mackenzie King Papers, and King, of course, could have mentioned the matter to FDR. The Joint Board statement, however, seems not to have been an overt factor in the Canadian decision to relocate, which, after all, came after the American decision had been made.

I have encountered one other piece of evidence that indicates some propensity toward unity of action between the two governments in the diplomatic journals of Jay Pierrepont Moffat, which are in the Houghton Library at Harvard. Moffat, who served as American Minister to Canada from June 4, 1940, to January 24, 1943, was in Ottawa at the time of the Pearl Harbor attack, according to his journal entry for December 8, 1941:

> Yesterday afternoon, while returning from a walk I turned on the radio . . . in time to hear the announcement of the Japanese attack on Pearl Harbor. . . . I got in touch with [Norman] Robertson [Under Secretary of State for External Affairs] by telephone who told me that Canada was preparing its declaration of war. . . . Today I called on Mr. Robertson. . . . [He expressed the hope] that the two Governments would be able to concert as to their policies in regard to interning Japanese citizens. There would, of course, be a number of individual internments here, but the Government had hoped not to have to intern all Japanese. However that might be difficult in view of the treacherous nature of the Japanese attack, the evidence of premeditation, the news reports coming through from Latin American countries that they had interned all Japanese residents, etc.[10]

The significant aspects of this conversation are, first of all, that it took place. It illustrates nicely, I think, the kinds of cooperation that could exist

[10]Jay Pierrepont Moffat, "Diplomatic Journals," vol. 47, June 13, 1941–June 11, 1942, Houghton Library, Harvard. Professor Blanche Linden read and transcribed material from the diaries for me. None of the matters quoted in this essay were considered important enough to print by the editor of the diary, although a glancing reference, misdated by one day, mentions "concerting policy about interning Japanese citizens." Nancy Harvison Hooker, ed., *The Moffat Papers: Selections from the Diplomatic Journals of Jay Pierrepont Moffat, 1919–1943* (Cambridge, Mass., 1956), 357.

between the two governments. Second, it argues that apparently neither Robertson nor Moffat had any awareness of the Joint Board statement of the previous month, although evidence exists elsewhere to show that Robertson was aware of it by mid-January 1942. If this matter ever came up again, it was not recorded in Moffat's rather full journal. The only other indication in Moffat's journal of an awareness of a Japanese problem in Canada is in a December 11, 1941, memorandum of a conversation with Admiral Nelles, chief of the Naval Staff, in which Moffat noted the admiral's concern about "what amounted to a minor panic in British Columbia. For the first time the people in that province who had been apathetic in the extreme had woken up to the fact that a war was raging." [11] And although it is not entirely apposite here, I should report a memorandum of a third conversation, this one held with Jack Pickersgill, secretary to Prime Minister King, on November 26, 1941, in which Moffat reports that "in Pickersgill's opinion a war with Japan is about the only thing he could visualize which would solve Canada's domestic difficulties." [12]

In the weeks after Pearl Harbor, there were, of course, numerous instances of Canadian-American conversations about the menace of the Japanese, external and internal. There were daily contacts among diplomats, generals, admirals, and civilian officials, but none that I have examined indicates either policy formulation or preplanning. Of course, given the swollen nature of modern bureaucracies, one can never be sure that one has seen all the relevant surviving papers, and additional such material may be unearthed.

* * *

How significant is the material presented here? It certainly will not involve a major revision of anything now in print and it does not change appreciably the sorry story of what happened to Japanese in North America in the aftermath of Pearl Harbor. To return to the questions posed earlier, I think that it now must be agreed that at least a step toward preplanning had been taken. I have not, however, been able to discover, as yet, any causal links between either the Joint Board's statement or the diplomatic consultation and what both governments did to their Japanese in February 1942 and after. Therefore, the major significance of this material, I believe, is to demonstrate further the propensity that then existed in North American society to treat "enemies" of another race quite differently than "enemies" who happened to be white. It also demonstrates again that, when leaders gather to talk about defense from foreign dangers, they often, if not usually, talk about violating the civil liberties of their own citizens as well.

---

[11] Moffat, "Diplomatic Journals," vol. 47.
[12] *Ibid.*

RACIAL DISCRIMINATION AND THE MILITARY JUDGMENT:

The Supreme Court's Korematsu and Endo Decisions

NANETTE DEMBITZ*

INTRODUCTION

While the bulk of the controls exercised over the civil liberties of the civilian population during the present war closely resemble those prevailing in World War I, significant departures from social and legal precedent have been made in the program of restraints over American citizens of Japanese descent. This program, which was initiated in the early months of the war, and is partially continuing in effect, dealt with all persons of Japanese descent residing in the State of California, the western portions of Washington and Oregon, and in southern Arizona, including approximately 70,000 native-born American citizens.[1]

The close of some aspects of the program was written by the Supreme Court's recent decisions in *Korematsu v. United States*[2] and in *Ex parte Mitsuye Endo.*[3] In the *Korematsu* case the Court upheld the orders issued by the Commanding General of the Western Defense Command from March through July, 1942, which excluded from their homes and from the area all persons of Japanese descent residing in the area described above: *i.e.,* the entire State of California and substantial portions of Oregon, Washington and Arizona. The orders applied regardless of the individual's American citizenship and regardless of his loyalty to this country. The Court also sustained the initial phase of the detention of these persons, which was coupled with their exclusion and was likewise effected regardless of citizenship or loyalty.

---

* This article expresses the personal views of the author and in no way represents the official views of any governmental agency.

1. The total number of persons of Japanese descent, both alien and citizen, residing in this area was approximately 112,000; they constituted about ninety per cent of the total population of the United States of such descent but less than two per cent of the general population in this area. 16TH CENSUS OF THE UNITED STATES, POPULATION, 2D SERIES, CHARACTERISTICS OF POPULATION (Dept. of Commerce, 1940), for California, p. 10; for Oregon, p. 10; for Washington, p. 10; for Arizona, p. 10; H. R. REP. No. 2124, 77th Cong., 2d Sess. (1942) 94, 96. While no significant distinction was made in the operation of the program between aliens and citizens, the litigation and the more important social and legal problems concern the latter. The various measures and issues will therefore be discussed herein with reference to the citizen group. All the citizens are native-born, since naturalization is not open to persons of Japanese descent. Ozawa v. United States, 260 U. S. 178 (1922).

2. 65 Sup. Ct. 193 (1944), *reargument denied,* Feb. 12, 1945.

3. 65 Sup. Ct. 208 (1944).

The issue in the *Endo* case was the validity of one of the latter phases of the detention program: the validity of the detention of admittedly loyal citizens and of the refusal to release them except upon their acceptance of certain conditions, when the justification offered for these restraints was that hardship and disorder might result because of racial prejudice against these loyal citizens if they were released unconditionally to proceed to destinations of their own choice. The Court invalidated this part of the program. Coincident with the issuance of the *Korematsu* and *Endo* decisions, the Commanding General of the Western Defense Command withdrew orders which had barred the return of these citizens, regardless of their loyalty, to California and the evacuated portions of the other West Coast states, ever since their original exclusion therefrom almost three years before.[4]

The significance of these decisions arises from the fact that the measures with which they dealt combined a number of diverse features of social and political consequence. In the first place, the Japanese ancestry program brought to our law the first Federal measure of racial discrimination applicable to citizens; that is, the first instance in which the applicability of a deprivation or restraint imposed by the Federal Government depended solely upon the citizen's race or ancestry. Furthermore, the *Korematsu* case, with the earlier decision in *Hirabayashi v. United States*,[5] upholding a more minor restraint—a curfew, on all persons of Japanese ancestry—are the first instances of Supreme Court approval of such a discrimination, whether State or Federal.[6]

---

4. The decisions were issued December 18, 1944. The withdrawal of the orders was announced December 17, to be effective January 2, 1945. N. Y. Times, Dec. 18, 1944, p. 1, col. 2. The interrelationship of the withdrawal and the decisions is discussed *infra* pp. 220-223.

5. 320 U. S. 81 (1943).

6. In the cases upholding segregation measures, which comprise the largest group dealing with race, the Court has not countenanced racial discrimination since these measures have been upheld only insofar as equal, though separate accommodations were provided for all races. Cases upholding segregation: Gong Lum v. Rice, 275 U. S. 78 (1927) (schools); McCabe v. Atchison, T. & S. F. Ry., 235 U. S. 151 (1914) (railroads); Cumming v. Richmond County Board of Education, 175 U. S. 528 (1899) (schools); Plessy v. Ferguson, 163 U. S. 537 (1896) (all public conveyances). Cases disapproving: Missouri v. Canada, 305 U. S. 337 (1938) (schools); *see* Mitchell v. United States, 313 U. S. 80, 97 (1941) (railroads). And where the segregation would result in a deprivation, the fact that it was claimed to "promote the public peace by preventing race conflicts" did not bring it within the police powers of the State. Buchanan v. Warley, 245 U. S. 60, 81 (1917). There appears to have been no federal case prior to the *Hirabayashi* case in which either a State or Federal racial discrimination against citizens was upheld. See Hill v. Texas, 316 U. S. 400 (1942), and cases cited (disapproving discrimination with respect to jury duty); Yu Cong Eng. v. Trinidad, 271 U. S. 500 (1926) (disapproving discrimination with respect to use of Chinese language by Chinese merchants); Yick Wo v. Hopkins, 118 U. S. 356 (1886) (disapproving discrimination with respect to Chinese launderers); Wong Wai v. Williamson, 103 Fed. 1 (N. D. Cal. 1900) (disapproving discrimination against Chinese with respect to

But the program marks a development in the law not only because of its use of a racial classification but also because of the types of controls employed. Protective custody, which was in issue in the *Endo* case, and which was, as will be pointed out below, likewise involved in the *Korematsu* case, had never before been practised in this country on a mass basis.[7] Aside from this phase of the program, however, which was not contemplated at its commencement,[8] the very inception of the program was based on an approach to protection of this country's internal security which had not been used during World War I. For the primary purpose in initiating the program was to restrain citizens who, it was thought, might commit acts helpful to the enemy but who had not in fact committed any proscribed act.

Except in the case of alien enemies,[9] no measures had been used during the first World War, or existed at the commencement of this war,[10] to deal with persons who were merely suspected of being danger-

---

vaccination for the bubonic plague where there was no evidence of particular susceptibility to the disease on the part of this race). See also Ah Sin v. Wittman, 198 U. S. 500 (1905) and Soon Hing v. Crowley, 113 U. S. 703 (1885), where the Court found there was insufficient proof that legislation was being discriminatorily enforced against Chinese.

The so-called Japanese land laws do not involve racial discrimination against citizens, since they prohibit ownership only by aliens ineligible for citizenship, Frick v. Webb, 263 U. S. 326 (1923); Webb v. O'Brien, 263 U. S. 313 (1923); Porterfield v. Webb, 263 U. S. 225 (1923); Terrace v. Thompson, 263 U. S. 197 (1923).

7. The statement in the text is limited to protective custody on a mass basis since detention in a case such as that of a material witness [United States v. Von Bonim, 24 F. Supp. 867 (S. D. N. Y. 1938)] may be thought to represent an instance of individual protective custody.

8. See *infra,* pp. 200-202.

9. Power to intern or otherwise restrict nationals of enemy countries was granted to the President by the Act of July 6, 1798 [1 STAT. 577 (1798), 50 U. S. C. §§ 21-23 (1940)], as amended. During the last war and the present one, enemy aliens considered dangerous to the security of this country have been interned. See Presidential Proclamations of April 6, 1917, 40 STAT. 1650; November 16, 1917, 40 STAT. 1716; December 7, 1941, 6 FED. REG. 6321 (1941); and of December 8, 1941, 6 FED. REG. 6323, 6324 (1941); REP. ATT'Y GEN. (1917) 59-61; Wilson, *Treatment of Civilian Alien Enemies* (1943) 37 AM. J. INT. L. 42; H. R. REP. No. 2124, 77th Cong., 2d Sess. (1942) 27-28, 157-160; Hoover, *Alien Enemy Control* (1944) 29 IOWA L. REV. 396.

10. Though Section 2 of the Alien Registration Act of 1940 [54 STAT. 671, 18 U. S. C. § 10 (1940)] was not a "dangerous persons" measure because it proscribed definite and specified activity, this Act, making it a crime to be affiliated with, knowing its purpose, an organization or group encouraging violent overthrow of the government, is pertinent to a consideration of the development of measures against suspected persons. For this Act penalized behavior at a level farther removed from the overt act the statute was designed to prevent than did the preexisting anti-subversive legislation. And it has been pointed out that under this Act the concept of guilt by association was re-introduced into federal law for the first time since the repeal in 1921 of a similar section of the Espionage Act of 1917, which in turn was the first federal legislation of this type since the Alien and Sedition Law of 1798. *See* CHAFEE, FREE SPEECH IN THE UNITED STATES (1941) 446, 465-483; Note, *The Legal Limits of Personal Liberties in Time of War* (1941) 90 U. OF PA. L. REV. 598. As to the previous federal legislation and state legislation,

ous. And prior to the last war the only major attempts in this country to restrain such persons were those employed during the Civil War through the device of the suspension of the privilege of the writ of *habeas corpus,*[11] which served to deprive the person affected of the ordinary means for securing release from imprisonment. During the present war, however, the fear that the country could not be sufficiently protected without the restraint of "suspicious" citizens gave rise not only to the West Coast program but also to the Individual Civilian Exclusion program. The latter, which was executed under the same Congressional and Executive authorizations as the West Coast program,[12] consisted of the issuance by military commanders of orders prohibiting the continuance of individual citizens in their places of residence and their entrance into large areas of the United States.[13] In addition, the Civil War device was used in Hawaii, where the suspension of the privilege of the writ of *habeas corpus* was employed by the military authorities in that Territory in part to accomplish the detention of "suspicious" citizens.[14]

An analogy in English practice, so often of comfort to Americans

---

see Chafee, *Freedom of Speech in War Time* (1919) 32 HARV. L. REV. 932; Note, *Federal Sedition Bills: Speech Restriction in Theory and Practice* (1935) 35 COLUMBIA LAW REV. 917.

11. In addition to the Civil War suspensions, the privilege of the writ was suspended for a short time in the Philippine Islands by order of the Governor General of the Philippines [see Barcelon v. Baker, 5 Phil. Rep. 87 (1905), *appeal dismissed,* 203 U. S. 174 (1906)], and for a period of eight months in Massachusetts on the occasion of Shay's Rebellion in 1786 by order of the Massachusetts legislature. *See* MINOT, THE HISTORY OF THE INSURRECTION IN MASSACHUSETTS IN THE YEAR 1786 AND THE REBELLION SUBSEQUENT THEREON (2d ed., 1810) 64-65. There have also been several suspensions of short duration ordered by State Governors in connection with labor disputes. *See* RANKIN, WHEN CIVIL LAW FAILS (1939) *passim.*

The broadest suspension invoked by Lincoln was that in Proclamation No. 1 of Sept. 24, 1862, 13 STAT. 730 (1862), which effected a suspension throughout the United States as to persons "guilty of any disloyal practice." For similar language in Lincoln's proclamation and orders authorizing his generals to suspend the writ in specified areas, *see* 6 RICHARDSON, MESSAGES AND PAPERS OF THE PRESIDENTS (1896) 16-19. Subsequently, under an Act of Congress granting him the authority [12 STAT. 755 (1863)], Lincoln provided for a suspension where "by authority of the President," officers of the United States "hold persons . . . as prisoners of war, spies, or aiders or abettors of the enemy" or as members of the armed forces, deserters, or draft evaders, or for offenses against the armed services. [Proclamation No. 7 of Sept. 15, 1863, 13 STAT. 734 (1863)]. A similar suspension was effected by the Confederacy. *See In Re* Rafter, 2 Jones L. 537 (N. C. 1864).

12. Act of March 21, 1942, c. 191, 56 STAT. 173 (1942), 18 U. S. C. § 97a (Supp. 1943) set forth *infra* p. 181; E. O. 9066, issued Feb. 19, 1942, 7 FED. REG. 1407 (1942), discussed *infra* pp. 181.

13. *See infra* p. 229.

14. Anthony, *Martial Law in Hawaii* (1942) 30 CALIF. L. REV. 371 and (1943) 31 CALIF. L. REV. 477; Armstrong, *Martial Law in Hawaii* (1943) 29 A. B. A. J. 698; Frank, *Ex Parte Milligan v. The Five Companies: Martial Law in Hawaii* (1944) 44 COLUMBIA LAW REV. 639, and see *infra* p. 234.

when novel measures are undertaken in this country, exists with respect to our control of suspicious citizens, in that in both the last and the present war England has interned such citizens. The British approach to the problem, however, differed in many respects from that adopted in this country; in particular, the measures are dissimilar in that the power to deal with dangerous citizens was not in Britain placed in the hands of the military.[15] In fact there would appear to have been a studied effort in Britain during the present war to prevent power over civilian liberties falling into military hands.[16] In this country, on the other hand, the military authorities were connected with all the novel measures of wartime control over the personal liberty of civilians in that they were the progenitors of the Japanese ancestry program, the Individual Civilian Exclusion program, and of numerous controls in Hawaii over an extended period of time, and in that they were also the administrators of these measures except for one of the latter phases of the Japanese program. By virtue of these measures the military authorities assumed during this war an unprecedented position in respect of the liberties of the American civilian populace. The period in American history when the authority of the military was most nearly comparable to that during this war was the Civil War years,[17] but even then the scope of the military

15. As to the British detention measures during the two Wars, see Carr, *Crisis Legislation in Britain* (1940) 40 COLUMBIA LAW REV. 1309; Carr, *A Regulated Liberty* (1942) 42 COLUMBIA LAW REV. 339; Laski, *Civil Liberties in Great Britain in Wartime* (1942) 2 BILL OF RIGHTS REV. 243; Cohn, *Legal Aspects of Internment* (1941) 4 MOD. L. REV. 200; Keith, *The War and the Constitution* (1941) 4 MOD. L. REV. 82; Note (1917) 33 L. Q. REV. 205. Britain also used the device in the last War of excluding suspected individuals from strategic areas, though this power likewise did not rest in the hands of the military authorities. See Ronnfeldt v. Phillips, 35 L. T. R. 46 (K. B. 1918). The present regulation and statute on evacuation [Defense (General) Regulations (S. R. & O. 1939, No. 927), Part 2, section 21; Civil Defense Act 1939, 2 & 3 GEO. VI, c. 31] are directed primarily at clearing areas subject to actual or apprehended attack of persons and property which might be injured or might encumber military operations, rather than at removing suspected persons therefrom.

16. Special war-zone courts, to function in the event that extraordinary judicial remedies were required because of apprehended or actual enemy action, were established to preclude the necessity or opportunity for military courts to perform this role. Carr, *supra* note 15, at 1317-1318; Keith, *supra* note 15, at 99.

17. See measures discussed in note 11 *supra*.

During the last War, an unsuccessful attempt to extend Army jurisdiction took the form of a bill which provided that

> ". . . owing to changes in the conditions of modern warfare whereby the enemy now attempts to attack and injure the successful prosecution of the war . . . by the United States by means of civilians behind the lines spreading false statements and propaganda, injuring and destroying the . . . utilities prepared . . . for the use of the land and naval forces of the United States, . . .",

any person in the United States who published or printed any false matter designed to interfere with the successful prosecution of the war was to be deemed a spy and subject to trial by a military tribunal. H. R. REP. No. 11556, 65th Cong., 2d Sess. (1918); SEN. REP. No. 4364, 65th Cong. 2d Sess. (1918). The bill was killed in Committee because of the strong disapproval of the then Attorney General and the President. [56 CONG. REC. 307 (1918); 56 CONG. REC. 5471 (1918)].

controls and their impact upon the population in loyal territory[18] were more confined than during the present period. Moreover, the occasion for the much debated exercise of authority by the military during the Civil War[19] differs notably from that during the present war with foreign enemies, because of the greater fluidity of the line between enemy and loyal territory in a civil war and the greater basis for suspicion that civilian officials themselves may be hostile to the prosecution of the war.

In view of this background, the legal problems arising from the Japanese ancestry program which are of major significance are the scope of judicial review of action by military authorities respecting civilians and the scope of judicial review of race discrimination, whether imposed by military or civilian authorities; and it is on these two problems, with respect to which the recent Supreme Court decisions are significant precedents, that the major emphasis of this article will be placed.

I. THE KOREMATSU DECISION

A. The Court's Review of Factual Justification for Racial Exclusion

Korematsu, an American citizen of Japanese ancestry, was convicted under an information charging him with having remained in San Leandro, California after the date on which all persons of such ancestry had been ordered excluded therefrom by an order of Lieutenant General John L. De Witt, Commanding General of the Western Defense Command.[20] The prosecution was brought under the Act of March 21, 1942, which provided:

---

18. The authority exercised by the occupying armies in conquered Confederate territory is irrelevant to this article since such territory had the same status as foreign enemy territory. *See* The Grapeshot, 9 Wall. 129, 131-133 (U. S. 1869).

19. The display of interest by the Bar in the present situation has been exceedingly slight compared to the fervor with which the Civil War measures were attacked and defended. Some of the better examples of the many tracts on the subject are: Binney, The Privilege of the Writ of Habeas Corpus Under the Constitution (2d ed. 1862); Gross, A Reply to Horace Binney's Pamphlet on Habeas Corpus (1862); Wharton, Answer to Mr. Binney's Reply to "Remarks" on His Treatise on the Habeas Corpus (1862); Jackson, Authorities Cited Antagonistic to Horace Binney's Conclusions on the Writ of Habeas Corpus (1862); Anonymous, *Article VIII (respecting the Merryman Decision)* (1861) 93 North American Review 471, 490-491.

20. The Circuit Court upheld the conviction on the basis of the Supreme Court's *Hirabayashi* decision with scant discussion. Korematsu v. United States, 140 F.(2d) 289 (C. C. A. 9th, 1943). Prior to the rendition of this opinion, the Supreme Court had held [Korematsu v. United States, 319 U. S. 432 (1943)] on a certification of the question from the circuit court, that the order of the district court that "pronouncing of judgment be suspended" and Korematsu placed on probation was an appealable order. The Supreme Court thus eradicated the distinction which had been drawn by two circuits between such an order and an order imposing judgment but suspending its execution, which had been held by the Supreme Court in Berman v. United States, 302 U. S. 211 (1937) to be appealable.

"That whoever shall enter, remain in, leave, or commit any act in any military area or military zone prescribed, under the authority of an Executive order of the President, by the Secretary of War, or by any military commander designated by the Secretary of War, contrary to the restrictions applicable to any such area or zone or contrary to the order of the Secretary of War or any such military commander, shall, if it appears that he knew or should have known of the existence and extent of the restrictions or order and that his act was in violation thereof, be guilty of a misdemeanor and upon conviction shall be liable to a fine of not to exceed $5,000 or to imprisonment for not more than one year or both for each offense."[21]

The Executive Order to which the Act referred was issued by the President on February 19, 1942 and provided that "Whereas the successful prosecution of the war requires every possible protection against espionage and sabotage," the Secretary of War and Military Commanders designated by him were authorized "to prescribe military areas in such places and of such extent as he or the appropriate Military Commander may determine, from which any or all persons may be excluded, and with respect to which the right of any person to enter, remain in, or leave shall be subject to whatever restrictions the Secretary of War or the appropriate Military Commander may impose in his discretion."[22] The developments leading to the promulgation of the Order and the Act will be discussed *infra*;[23] they show *inter alia* that though the power of formulation and execution of the exclusion program was conferred upon the military authorities, the proposal for racial exclusion did not in fact stem exclusively from these authorities but had been backed by numerous West Coast groups and organizations.

Pursuant to this authorization General De Witt, the duly designated military commander of the Western Defense Command, created military areas throughout his command.[24] He then effected the exclusion of all persons of Japanese ancestry from California, western Washington, western Oregon, and southern Arizona (hereinafter termed the exclusion area) by dividing this area into 108 subdivisions. So-called Civilian

---

21. 56 STAT. 173 (1942), 18 U. S. C. § 97a (Supp. 1943). This Act was the prototype for a statute conferring authority upon the Secretary of Navy to control ingress to, and egress from, vessels and waterfront facilities. Act of July 9, 1943, 157 STAT. 391 (1943), 18 U. S. C. § 96 (Supp. 1943). However, the scope of the power granted by this statute and by its parallel Executive Order [No. 9074, 7 FED. REG. 1587 (1942)], and the activity undertaken thereunder is insignificant compared to that granted and undertaken under the Act of March 21 and its parallel Order. As to exclusions from waterfront facilities and vessels, see §§ 9.1-9.4, Coast Guard Regulations, 8 FED. REG. 16867 (1943).

22. 7 FED. REG. 1407 (1942). The Order authorized the use of troops to enforce the restrictions imposed thereunder.

23. Pp. 206-207.

24. Public Proclamation No. 1, issued March 2, 1942 [7 FED. REG. 2320 (1942)]; Public Proclamation No. 2, issued March 16, 1942 [7 FED. REG. 2405 (1942)].

Exclusion Orders were issued at intervals from March 24 to July 22 dealing with the subdivisions *seriatim;* each ordered that all persons of Japanese ancestry "be excluded" from a specified subdivision on, and not be "found" therein after, a specified date six days following the date of publication of the order.[25]

The record in Korematsu's case showed by stipulation and uncontroverted evidence that he had violated the Exclusion Order applicable to San Leandro, which was the place of his residence. It further showed without contradiction that Korematsu had been born in California; that he had never departed from the continental limits of the United States nor renounced his American citizenship nor did he have any form of allegiance to any country other than the United States; that he was willing to render any service requested of him in the war against Japan and had no sympathy for Japan in the war; that he was a registered voter and was in various respects assimilated into the American community; and that he had remained in his place of residence in violation of the Exclusion Order because of friendly relations with the residents of the locality, particularly with a girl not of Japanese ancestry, and because he considered himself an American.[26] As the Supreme Court stated, no question was raised as to his loyalty.[27]

The majority opinion, written by Mr. Justice Black, upheld Korematsu's conviction on the basis of *Hirabayashi v. United States, supra,* in which the Court had upheld an order issued by General De Witt on March 24, 1943 directing all persons of Japanese ancestry to remain in their homes from 8 P.M. to 6 A.M. Vigorous dissenting opinions were written by Mr. Justices Roberts, Murphy and Jackson. In the immediately following pages, the issues and doctrine of the *Hirabayashi* and *Korematsu* cases will be compared. It will be shown that the latter opinion, though involving a much more drastic measure of racial discrimination than the former, employed a far less careful standard of review, accompanied by a peculiar disregard of the realities of the action under review; and, rather than remedying the weaknesses of the prior opinion, it abandoned the safeguards against arbitrary official ac-

25. The Orders applicable to Military Area No. 1, in which lay the whole exclusion area except eastern California, were issued from March 24 [Order No. 1, 7 Fed. Reg. 2581 (1942)] to May 30, 1942 [Order No. 99, 7 Fed. Reg. 4436 (1942)] and those for the remainder of California, which lay in Military Area No. 2, from June 30 [Order No. 100, 7 Fed. Reg. 5369 (1942)] to July 22, 1942 [Order No. 108, 7 Fed. Reg. 5916 (1942)]. As to the arrangement of the subdivisions, see Final Report, Japanese Evacuation from the West Coast, 1942, submitted by Lt. Gen. J. L. De Witt to the Chief of Staff, dated June 5, 1943 (hereinafter cited as Final Report) Map Insert I, foll. p. 289.

26. Brief for the United States, Korematsu v. United States, Supreme Court, No. 22, Oct. Term 1944, pp. 4-5.

27. 65 Sup. Ct. 193, 194 (1944).

tion which that decision had created. A later section of this article will touch on the possible justifications, *dehors* the issues of the case, for the *Korematsu* opinion; but it will be demonstrated that this dangerous opinion, particularly surprising since it is from the pen of Mr. Justice Black,[28] is not in fact justified by considerations of sound policy lying either within or without the issues of the case.

1. *The Standard of Review in the Hirabayshi Case.*

At the commencement of its opinion in the *Hirabayashi* case, the Court, by Mr. Chief Justice Stone, expressed the test of whether a measure is within the war power in language which had not previously been used in considerations of the scope of this power,[29] and which smacked of the Court's discussions of due process of law with respect to peace-time regulations.[30] The curfew order, the opinion declared,

28. See comment in Chamberlain, *Nine Young Men*, Life Magazine, Jan. 22, 1945, p. 76, that Justice Black generally emphasized the importance of personal rights and civil liberties, "however much he may have seemed to backslide in the Korematsu Japanese detention case" (at p. 84).

29. In the bulk of the previous cases involving war measures, it had been sufficient for the Court to refer in general terms to the great breadth of the war power and the wide discretion allowed in its exercise, either because (1) the major question had concerned the interpretation or effect of the measure rather than its basic validity; or (2) because the major question had concerned the effect of various specific Constitutional limitations and guaranties; or (3) merely because the Court regarded the fact that the measure was comprehended by the war power as too clear for debate. Of the cases cited on the war power in the *Hirabayashi* opinion, the Prize Cases, 2 Black 635, 671 (U. S. 1862); Miller v. United States, 11 Wall. 268, 303-314 (U. S. 1870); and Martin v. Mott, 12 Wheat. 19, 29 (U. S. 1827) are in the first category; Selective Draft Law Cases, 245 U. S. 366 (1917) and *Ex parte* Quirin, 317 U. S. 1, 28-29 (1942) are in the second; and McKinley v. United States, 249 U. S. 397 (1919), is in the third. For references to the scope of the war power in cases of the first type, *see* Hamilton v. Dillin, 21 Wall. 73, 88 (U. S. 1874); United States v. Chemical Foundation, 272 U. S. 1, 12 (1926); and Raymond v. Thomas, 91 U. S. 712, 714-715 (1875); of the second, *Ex parte* Milligan, 4 Wall. 2, 21 (U. S. 1866); and of the third, Dakota Central Co. v. South Dakota, 250 U. S. 163, 183 (1919); Highland v. Russell Car Co., 279 U. S. 253, 262 (1929). In a few previous cases in which it was necessary for the Court to particularize with respect to the scope of the war power, the issue had concerned measures regulating the civilian economy disconnected from, or subsequent to, hostilities. *See* Stewart v. Kahn, 11 Wall. 493, 506-7 (U. S. 1870), in which the Court upheld the power of Congress to toll state statutes of limitation during and after the Civil War, stating that "the [war] power is not limited to victories in the field and the dispersion of the insurgent forces. It carries with it inherently the power to guard against the immediate renewal of the conflict, and to remedy the evils which have arisen from its rise and progress." *See* Hamilton v. Kentucky Distilleries, 251 U. S. 146, 161 (1919) citing the *Stewart* case, and upholding the power of Congress over intoxicating liquor during the demobilization period.

30. Aside from the cases in which war measures had been held invalid under the due process clause because of the denial of procedural rights [United States v. Cohen Grocery Co., 255 U. S. 81 (1921) (excessively vague and indefinite definition of criminal offense); *Ex parte* Milligan, 4 Wall. 2 (U. S. 1866) (criminal trial of civilian by military court)], the war power cases in which due process had been discussed had been few and the due process issue of a relatively minor nature. See Hamilton v. Kentucky Distilleries, 251 U. S. 146 (1919), and Highland v. Russell Car Co., 279 U. S. 253 (1929), cited *supra* note 29.

was within the scope of the combined war power of Congress and the Executive[31] if "in the light of all the facts and circumstances there was any substantial basis for the conclusion . . . that the curfew as applied was a protective measure necessary to meet the threat of sabotage and espionage which would substantially affect the war effort and which might reasonably be expected to aid a threatened enemy invasion," and if there was "reasonable ground for believing that the threat . . . is [was] real."[32] The narrowing of the issue to justifiability as a protection against espionage and sabotage in the event of threatened invasion resulted from the statement of purpose in the preamble to Executive Order 9066, under which all the orders here under discussion were promulgated, the statement of findings in General De Witt's proclamations,[33] and the emphasis in the Government's brief on these factors as justification for the measure.[34] As to the factual element of threatened invasion, the Court, after reviewing the military events subsequent to the attack on Pearl Harbor and the significance of that attack concluded that "reasonably prudent men charged with the responsibility of our national defense had ample ground for concluding that they must face the danger of invasion."[35] But the more difficult factual question related to the basis for the conclusion that the curfew should be applied to all American citizens of Japanese ancestry in the area.

On this question the Court departed from the relatively high standard of review which it had previously enunciated. The premises supporting the conclusion as to the application of the curfew were, according to the Government's argument, the potential disloyalty of at least a minority of the citizens of Japanese ancestry and the difficulty of identifying and restricting the potentially disloyal as a separate group from the loyal.[36] At that time General De Witt's Report on the evacuation program had not been published, so there was no question of whether the

31. The war power of the Executive was delegated by Executive Order 9066 [7 FED. REG. 1407 (1942)] and that of Congress by the Act of March 21. The Court solved the difficult question of the constitutionality of the latter delegation largely by reliance on the Congressional knowledge, and therefore ratification, of General De Witt's proclamations of March 2 and 16 and of his intention to promulgate the curfew [Hirabayashi v. United States, 320 U. S. 81, 91, 102-104 (1943)].

32. 320 U. S. 81, 95 (1943).

33. Both the proclamations of March 2 and March 16 recited that

"the entire Pacific Coast of the United States, . . . by its geographical location is particularly subject to attack, to attempted invasion by the armed forces of nations with which the United States is now at war, and, in connection therewith, is subject to espionage and acts of sabotage, thereby requiring the adoption of military measures necessary to establish safeguards against such enemy operations." [7 FED. REG. 2320, 2406 (1942)].

34. Brief for the United States, Hirabayashi v. United States, Supreme Court, Oct. Term 1942, No. 870, pp. 12-16.

35. 320 U. S. 81, 94 (1943).

36. Brief for the United States, cited *supra* note 34, pp. 61-62.

Government had to try to support General De Witt's generalizations on the subject of racial characteristics (see *infra*).[37] Rather than proving in the trial court the existence of the "Constitutional facts",[38] the Government asked the Court to judicially notice the bases for a belief in the existence of various circumstances encouraging an attachment to Japan on the part of at least some of the second and third generations of the population of Japanese ancestry.[39] It is to be noted that none of the assertions by the Government as to such bases were denied by Hirabayashi. That there was "some basis" for the belief, however, was the maximum that could be established beyond controversy with respect to the existence of most of these alleged circumstances by the documentary authorities presented.[40] A "reasonable" man could not and would not have come to

37. There is good ground for stating that if the reasons for the promulgation of the measure had been known, the Court should have considered its justifiability only on the basis of such reasons. See *infra* note 107.

38. That is, the facts relevant to the determination of the Constitutional question. Such facts are for the Court to "ascertain as it sees fit," rather than for the jury. Chastleton Corp. v. Sinclair, 264 U. S. 543, 548 (1924). They are "matters of fact that are merely premises to a rule of law. . . . As the judge is bound to declare the law, he must know or discover the facts that establish the law." Prentis v. Atlantic Coast Line, 211 U. S. 210, 227 (1908). *See* 1, 9 WIGMORE, EVIDENCE (3rd ed. 1940) §§ 4L, 2555; Barnett, *External Evidence of the Constitutionality of Statutes,* (1924) 58 AM. L. SCHOOL REV. 88; Bikle, *Judicial Determination of Questions of Fact Affecting the Constitutional Validity of Legislative Action* (1924) 38 HARV. L. REV. 6; Note, *Consideration of Facts in "Due Process" Cases* (1930) 30 COLUMBIA LAW REV. 360; Denman, *Comment on Trials of Fact in Constitutional Cases* (1935) 21 A. B. A. J. 805; Note, *Presentation of Facts Underlying the Constitutionality of Statutes* (1936) 49 HARV. L. REV. 631.

39. The Court will notice facts if it believes, either on the basis of its preexisting knowledge, documentary authority supplied by counsel, or documentary authority discovered by its own research, that the truth of the fact is generally accepted by the public or by persons well acquainted with the field of knowledge in issue. If, however, there is a bona fide dispute about the existence of the fact and the Court believes that the truth is not or cannot be established to a convincing degree of probability by extra-record sources, it will remand the case for the taking of evidence on the Constitutional facts. Compare cases in which Constitutional facts were judicially noticed: Adkins v. Children's Hospital, 261 U. S. 525, 560 (1923) (Court relied on a "notorious fact" known to it to disprove the inference from a "mass of reports" offered by counsel); Ashwander v. Valley Authority, 297 U. S. 288 (1936); Mugler v. Kansas, 123 U. S. 623 (1887); and West Coast Hotel v. Parrish, 300 U. S. 379 (1937) (no citations of documentary authority); Phelps Dodge Corp. v. Labor Board, 313 U. S. 177 (1941); Nebbia v. New York, 291 U. S. 502 (1934); United States v. Carolene Products Company, 304 U. S. 144 (1938); Schoellenberger v. Pennsylvania, 171 U. S. 1 (1898) (citing documentary authority); with cases of remands for the taking of evidence on constitutional facts: Borden's Co. v. Baldwin, 293 U. S. 194 (1934); Chastleton Corp. v. Sinclair, 264 U. S. 543 (1924); United States v. Rio Grande Irrigation Co., 184 U. S. 416 (1902), and Hammond v. Schiappi Bus Line, 275 U. S. 164 (1927). If the Court believes the fact is sufficiently established from knowledge outside the record, the Court will reject an offer of evidence to prove the contrary. Jacobson v. Massachusetts, 197 U. S. 11, 31-34 (1905).

40. For example, on the one hand it was stated that the teaching in the Japanese language schools was influenced by the Japanese government, and on the other hand reputable authority stated the schools were innocuous, that pupils attended them to be able to converse with their parents and because of increased employment opportunities with Japanese speaking employers, and that they did not

a positive conclusion, on the basis of the available documentary data, that most of the supposed influences toward disloyalty did in fact exist; a belief in their existence could not be said to rest on "reasonable or substantial grounds" insofar as the phrase connotes that a fact is established by a preponderance of evidence after the weighing of an adequate amount of data on both sides.

Thus, with the exception of reference to the education of about 10,000 American-born children in Japan, which was correctly treated as an established fact, the Court pointed only to facts which were "generally believed" or which have "been deemed" to exist or to conditions which "may well" have existed, which would have encouraged an attachment to Japan.[41] On the basis of this data and without explicitly considering any facts pertaining to the difficulty or impossibility of separating the loyal from the disloyal, the Court concluded that the imposition of the curfew on a racial basis was valid because:

> "Whatever views we may entertain regarding the loyalty to this country of the citizens of Japanese ancestry, we cannot reject as unfounded the judgment of the military authorities and of Congress that there were disloyal members of that population, whose number and strength could not be precisely and quickly ascertained. We cannot say that the war-making branches of the Government did not have ground for believing that in a critical hour such persons could not readily be isolated and separately dealt with, and constituted a menace to the national defense and safety, which demanded that prompt and adequate measures be taken to guard against it."[42]

Thus while the Court's stated general criterion for determining whether an act is within the war power was whether the belief in its necessity is supported by reasonable and substantial grounds, it departed from this standard in reviewing the racial facts, if the terms "reasonable" and "substantial" have any meaning.[42a] Though it is of course cor-

differ from other language schools in this country. Similarly, compare statements that there was a marked cleavage between the viewpoints and associations of the first and second generation with the view that a strong parental influence existed. Brief for the United States, cited *supra* note 34, pp. 30-31, 22-23.

It seems doubtful that the truth as to the existence and degree of the alleged influences could have been established any better by evidence than by the available documentary authorities.

41. 320 U. S. 81, 97-98 (1943).

42. 320 U. S. 81, 99 (1943).

42a. That is, it is inconsistent to state that there was "reasonable ground for believing that the threat [of espionage and sabotage by citizens of Japanese ancestry] . . . is real" [320 U. S. at 95 (1943)] if there were less than reasonable grounds for believing that conditions encouraging disloyalty existed, since disloyalty was one of the factors necessary to create the threat. While it was true, as the Court stated, that had the conditions discussed by the Government in fact existed, "Congress and the Executive could reasonably have concluded that these conditions have encouraged the attachment of members of this group to Japan and Japanese institutions" (*id.* at 98), the Court could not point to reasonable grounds for the conclusion that the requisite conditions existed.

rect that governmental authorities "have constitutional power to appraise the danger in the light of facts of public notoriety,"[43] here the opinion itself shows that the danger was appraised not in the light of "facts" reasonably established by consideration of an adequate amount of data but of widely held suspicions, such as may be possessed by every group of society with respect to every other group. A typical instance is the statement, frequently and positively made, that the children of persons of Japanese ancestry have close filial ties and are thus easily dominated by their parents,[43a] as contrasted with the findings by reputable sociologists that the second-generation generally strive to disassociate themselves from the ways of their parents even more than in the usual immigrant families.[44]

The Court has in other decisions stated that especial attention must be given to the basis for a measure affecting fundamental liberties, and insisted on a strong affirmative showing in support of such a measure rather than merely the absence of a showing against it.[44a] The need for this principle is particularly sharp in the field of racial discrimination. For there is no belief as to a racial characteristic, no matter how prejudiced it may seem to those who do not share it, which could be rejected as "unfounded" or groundless.[45]

Nor could the application to measures of racial discrimination of the standard of reasonableness and of the principle of a strong affirmative supporting showing be rejected as too inflexible to meet the needs of the country; for it may be reasonable to take a measure for the protection

---

43. 320 U. S. 81, 102 (1943).

43a. *See* Brief for the United States, cited *supra* note 34, p. 22; FINAL REPORT, *supra* note 25, at 9.

44. SCHREIKE, ALIEN AMERICANS, A STUDY OF RACE RELATIONS (1936) 36-39; ICHIHASHI, JAPANESE IN THE UNITED STATES (1932) 348; *Hearings before Select Committee of the House of Representatives Investigating National Defense Migration on H. R. 113,* 77th Cong., 1st Sess. (1941), Part 29, 11148, 11223. See also SECOND QUARTERLY REPORT, WAR RELOCATION AUTHORITY (June 1 to Sept. 30, 1942) 55-58.

44a. "Mere legislative preferences or beliefs respecting matters of public convenience may well support regulation directed at other personal activities, but be insufficient to justify such as diminishes the exercise of rights so vital to the maintenance of democratic institutions [referring to the rights to freedom of speech and of the press]." Schneider v. State, 308 U. S. 147, 161 (1939). To the same effect, *see* Thomas v. Collins, 65 Sup. Ct. 315, 322-323 (1945); West Virginia State Bd. of Education v. Barnette, 319 U. S. 624, 639 (1943); Martin v. Struthers, 319 U. S. 141, 144 (1943); *cf.* Note, *Presumption of Constitutionality* (1931) 31 COLUMBIA LAW REV. 1136, 1145.

45. For example, while the bald statement that members of racial Group A are generally avaricious would be viewed as bigoted and prejudiced, the statement can easily be intellectualized by selecting some environmental factor which appears to form a logical premise for the conclusion, as that members of Group A were generally poverty-stricken in their youth. The fallacy lies in the failure to consider all the diverse factors in the lives of the individual members of Group A, which cause the singled-out environmental factor of poverty to have a different effect in the lives of the various members of the group.

of the country if the detriment to which the country might otherwise be exposed is sufficiently great, even though there are less than reasonable grounds for believing that all the circumstances which might cause the detriment in fact exist. Thus, for example, the determination to take *some* action against the citizens of Japanese ancestry might have been justifiable on the basis of the degree of danger in the situation as a whole, even if the belief that there was more potential disloyalty among them than the rest of the population was based on less than reasonable grounds.

But the Court's failure to require a showing that the method used to meet the danger was reasonable, rather than that there was merely ground for believing it should be adopted, seems unjustifiable. The Court's insistence on an affirmative showing in support of measures affecting fundamental civil liberties, referred to above, required a showing not only that the measure might be reasonably deemed to advance a permissible objective, but also a showing that it was reasonable to select the particular measure, with the restrictions it involved, to secure that objective.[46] In view of the ease with which a plausible foundation for a belief in racial distinctions can be advanced, and the ease with which measures on the basis thereof can be advocated on the plea of emergency, and in view of the fact that human equality is fully "as vital to the maintenance of democratic institutions"[47] as any of the other Constitutional freedoms, the principle of these civil liberties cases must be fully and vigorously applied to measures of racial discrimination. It is in no way overstringent or impractical to establish as a judicial standard of review that: When the method chosen to meet the danger is one of racial discrimination, it should not be deemed reasonable unless the Government sustains the burden of demonstrating that available less stringent and more limited alternatives could not reasonably have been considered adequate. In the *Hirabayashi* case this would, of course, have required the Government to demonstrate that the less drastic measure of segregating the loyal from the disloyal and acting only against the latter would not have afforded a reasonable degree of protection.

In the present era when the ease with which racial prejudice can be engendered, and its effects, are so well-known and when the possibility of its flourishing in the stresses of the post-war period are evident, this

46. Schneider v. State, 308 U. S. 147, 161-162 (1939); Martin v. Struthers, 319 U. S. 141, 147 (1943), and Mr. Justice Murphy, concurring, at 151; Thornhill v. Alabama, 310 U. S. 88, 105-106 (1940). Compare Miller v. Schoene, 276 U. S. 272, 278-280 (1928).

47. See quoted portion of Schneider v. State, 308 U. S. 147 (1937), *supra* note 44[a].

decision, the first in which the Court upheld racial discrimination[48] is of especial importance.[49] It is fortunate that the Court tempered its approval by emphasizing that "in most circumstances racial distinctions are irrelevant,"[50] but it is most unfortunate that the Court neglected to issue a caveat with respect to the requisite basis for establishing the existence of racial distinctions, and that it neglected to lay down a rule requiring a stronger showing than a mere "not unfounded belief" to justify dealing with a situation by race discrimination.

2. *The Standard of Review in the Korematsu Opinion.*

Since the curtailment involved in the curfew was a relatively minor one, compared by the Court to the "establishment of fire lines during a fire, or the confinement of people to their houses during an air raid alarm,"[51] it was to be hoped that the Court would, when confronted with the far more drastic measure of exclusion, correct any doctrinal weakness of the *Hirabayashi* opinion. The *Korematsu* opinion is a grave disappointment to any such hopes. To be sure, the majority opinion tends to equate the exclusion with the curfew, by terming the exclusion "temporary," which, as Mr. Justice Roberts stated in his dissent, "is a substitution of a hypothetical case for the case actually before the court."[51a]

It must be kept uppermost in mind that the determination which led to the imposition of measures of racial discrimination rather than other forms of control was not that there were disloyal persons in the population of Japanese ancestry but that such disloyal persons could not be separately dealt with from the other members of the group. And as to this determination the Court had not in the *Hirabayashi* opinion made any real examination of the factual basis. The casual treatment of this factor in the *Hirabayashi* case may be to a large extent explicable in view of the patent difficulty of segregation for purposes of the curfew, since it was put into effect over the whole area simultaneously prior to

48. *Supra* note 6.

49. It might be stated that the foregoing discussion attaches too much importance to what could be regarded merely as a question of the Court's emphasis. However, it is believed that the lack of stress in the *Hirabayashi* opinion on the matters discussed above opened the way for the unmistakably dangerous doctrine of the *Korematsu* opinion, discussed *infra*.

50. 320 U. S. 81, 101 (1943).

51. *Id.* at 99.

51[a]. 65 Sup. Ct. 193, 200 (1944). The evacuees were not allowed to return for a period of approximately three years (*infra*, pp. 220-221). At the time of the evacuation there was no announced intention to permit a return at any time; employment connections were severed and property was disposed of. (See FINAL REPORT, *supra* note 25, at 55-64 as to the latter aspect of the evacuation.) The great majority of the evacuees are not now in a position to return though the ban on their residence has now been revoked (*infra*, p. 223).

the commencement of the exclusions.[52] However, the exclusion, unlike the curfew, was effected over a period of four months and was not completed for approximately five months after its authorization by Executive Order 9066 and approximately eight months after Pearl Harbor; the exclusion in the particular subdivision in which Korematsu resided was not ordered until nearly three months after the authorization.[53] No attempt will be made here to discuss all of the facts pertaining to the possibility of distinguishing over this period of time at least those individuals who could most easily be cleared of suspicion, and of exempting them from the exclusion. Some of the pertinent considerations are stated in Justice Murphy's penetrating and painstaking dissent. He points out that

> "the fact that conditions were not such as to warrant a declaration of martial law adds strength to the belief that the factors of time and military necessity were not as urgent as they have been represented to be.
>
> Moreover, there was no adequate proof that the Federal Bureau of Investigation and the military and naval intelligence services did not have the espionage and sabotage situation well in hand during this long period. Nor is there any denial of the fact that not one person of Japanese ancestry was accused or convicted of espionage or sabotage after Pearl Harbor while they were still free, a fact which is some evidence of the loyalty of the vast majority of these individuals and of the effectiveness of the established methods of combatting these evils. It seems incredible that under these circumstances it would have been impossible to hold loyalty hearings for the mere 112,000 persons involved—or at least for the 70,000 American citizens—especially when a large part of this number represented children and elderly men and women." (Footnotes omitted).[54]

In support of the latter statement Justice Murphy cites the experience of the British government in examining approximately 74,000 German and Austrian aliens over a period of six months at the commencement of the war.

In addition to the facts reviewed by Justice Murphy, it is to be observed that published material, of which the Court had notice, indicated that the intelligence agencies possessed, on the basis of pre-war investigations, compilations of information as to suspicious members of the

52. The curfew was established by Public Proclamation No. 3, issued March 24, 1942, effective March 27; it applied to all alien Italians, alien Germans, and all persons of Japanese ancestry in Military Area No. 1, and in various zones in Military Areas 2, 3, 4, 5 and 6 [7 FED. REG. 2543 (1942)].

53. As to the time of issuance of the 108 exclusion orders by which the exclusion was effected, see note 25 *supra*. The order applicable to Korematsu's place of residence was No. 34, issued May 3, 1942, effective May 9 [7 FED. REG. 3967 (1942)].

54. 65 Sup. Ct. 193, 205 (1944).

West Coast population of Japanese ancestry;[55] that the identity of the group most open to suspicion: American citizens who had lived and been educated in Japan, could have been definitely established from steamship records;[56] and that the justification sometimes suggested for the mass measure that even named individuals of the Japanese race cannot be apprehended because "they all look alike"[57] is clearly false in view of the fact that dangerous Japanese aliens have been apprehended by the thousands all over the country, including the West Coast, under the Alien Enemy Control program.[58]

Nevertheless, despite the differences as to the time factor between the curfew and the exclusion, despite such considerations as those pointed out above with respect to the possibility of separation, and despite the difference in the degree of deprivation involved in the exclusion as compared to the curfew, Justice Black found it possible to dispose of the whole question of separation of the loyal from the disloyal—basic to whether the measure was a justifiable security measure or was an overly facile disposition of a problem by racial discrimination and disregard of minority rights—by relying on the *Hirabayashi* opinion. He quoted the passage thereof set forth above in which the Court had stated that the belief in disloyalty and in the impossibility of a rapid separation were not unfounded, and concluded that in the case of the exclusion, like the curfew, "we could not reject the finding of the military authorities that it was impossible to bring about an immediate segregation of the disloyal from the loyal."[59]

Justice Black's reference to a "finding" by the military authorities tends to make the Court's acceptance of the military action more palatable, in that it connotes the military's deliberate consideration of the facts respecting segregation. It is reminiscent of the *Hirabayashi* opinion where considerable reliance was placed upon the circumstance that formal and

55. Consider Anonymous (An Intelligence Officer), *The Japanese in America, The Problem and the Solution,* HARPER'S MAGAZINE, Oct. 1942, p. 489. See p. 564 for biographical note respecting the author, an Intelligence Officer whose function over a period of years had been to study the West Coast Japanese population; he has been identified as Lt. Comdr. Ringle of the Office of Naval Intelligence. *See* McWILLIAMS, PREJUDICE (1944) 114-115. The article is cited in Brief for the United States, cited *supra* note 26, p. 12. It appears that the Intelligence agencies did not recommend the mass measure (*Ibid.*); consider statement by the Attorney General, whose Department includes the F.B.I., cited *infra* note 207, and the unanswered challenge on this point in Brief for American Civil Liberties Union, *amicus curiae,* Supreme Court, No. 22, Oct. Term, 1944.

56. See FINAL REPORT, *supra* note 25, at 14; Anonymous, *supra* note 55, at 490-493, 496-7.

57. See Judge Denman, concurring in Korematsu v. United States, 140 F.(2d) 289, 300 (C. C. A. 9th, 1943).

58. *Infra* note 124.

59. 65 Sup. Ct. 193, 195 (1944).

explicit findings had been made as to the necessity for the curfew.[60] In the *Korematsu* opinion, however, the reference to findings is a peculiar euphemism, and, in fact, misleading. No findings whatever were made by the military authorities as to the possibility of segregating the disloyal from the loyal. The only documentary exploration of the considerations motivating the exclusion is given in General De Witt's official description and justification of the evacuation, submitted to the Chief of Staff in June 1943; General De Witt indicates clearly his belief that Japanese ancestry (rather than education, associations, previous expressions of anti-Japanese sentiment or any other factors) determined loyalty and that no person of such ancestry could be trusted not to aid the enemy. There is no indication whatsoever in this Report that he had weighed, and particularly that he had weighed with the care that the making of a "finding" implies, the possibility of separating the disloyal from the loyal. As Justice Murphy points out:

> "In it he refers to all individuals of Japanese descent as 'subversive,' as belonging to 'an enemy race' whose 'racial strains are undiluted,' and as constituting 'over 112,000 potential enemies . . . at large today' along the Pacific Coast." (Footnotes omitted.)[61]

Further, disregarding even the fact that some citizens of Japanese ancestry had openly opposed Japanese imperialism long prior to the war[62] and that such citizens at least might be presumed loyal, if only for the reason that they would be *personae non grata* to Japan, General De Witt

---

60. 320 U. S. 81, 103-105 (1943). The discussion of findings is in connection with the question of the delegation of legislative power. The Court's emphasis in that opinion on findings is of interest since it had to search the series of orders to spell out their existence; there was no statutory direction that findings be made; and the administrative order involved was of a general legislative nature, rather than of a quasi-judicial nature relating to an individual situation. The requirement of findings is more rare, and its rationale not as clear, in the case of orders of the former than of the latter nature. *See* Feller, *Prospectus for the Further Study of Federal Administrative Law* (1937) 47 YALE L. J. 647, 665-669. The Court indicates in one passage of the opinion that the mere recitation of findings is itself important although the final conclusion refers to the fact that the findings were adequately supported by the facts [320 U. S. 81, 104-105 (1943)].

While the requirement of findings may be of some value *per se* in that the findings presumptively indicate the agent's awareness of the standard [see City of Yonkers v. United States, 320 U. S. 685 (1944); *cf.* United States v. Carolina Carriers Corp., 315 U. S. 475, 489 (1942)], it would seem that this requirement is of real value only if coupled with the requirement that the finding be supported by evidence so that the Court may determine if the presumption is true and the findings not a mere "talismanic phrase" [Mr. Justice Frankfurter, dissenting in the *Yonkers* case, at 692]. Some of the orders connected with the evacuation program demonstrate that the recitation may be meaningless. Compare, for example, the recitation in General De Witt's Public Proclamation No. 8 with the reasons he announced for the measure prescribed therein (*infra* pp. 200-202).

61. 65 Sup. Ct. 193, 203 (1944).

62. See *Hearings* cited *supra* note 44, at pp. 11220-11229.

stated there was "no ground for assuming any Japanese . . . will not turn against this nation."[63]

The Court did not, of course, endorse these views; it could not without complete abandonment of the Constitution and basic American tenets subscribe to the concept that no type of environment, education, or association could produce a loyal citizen of Japanese ancestry or that with respect to this one class of citizens no individual investigation could be relied upon. But Justice Black's reliance upon a mythical finding, which would not in any event have been a sufficient one to support the exclusion, does belie his introductory rhetorical statement that "courts must subject them [racial restrictions] to the most rigid scrutiny."[64]

a. *Special treatment of action by military, as distinguished from civilian, authorities*

In the *Korematsu* opinion the Court completely disregards and abandons the *Hirabayashi* language as to reasonableness and substantiality; further, in its language other than that quoted from the *Hirabayashi* opinion, it is not even concerned with whether the beliefs of the military authorities were "not unfounded."[65] Rather, the *Korematsu* opinion indicates that there is no basis for invalidating war-time action by military authorities, save perhaps by a showing of malice and a lack of good faith on the part of the military. The Court points to views which it assumed the military authorities possessed, which it then regards as rendering the exclusion constitutional. It states as follows:

> "In the light of the principles we announced in the *Hirabayashi* case, we are unable to conclude that it was beyond the war power of Congress and the Executive to exclude those of Japanese ancestry from the West Coast war area at the time they did.[66] True, exclusion from the area in which one's home is located is a far greater deprivation than constant confinement to the home from 8 p.m. to 6 a.m. Nothing short of apprehension by the proper military authorities of the gravest imminent danger to the public safety can constitutionally justify either. But exclusion from a threatened area, no less than curfew, has a definite and close relationship to the prevention of espionage

63. FINAL REPORT, *supra* note 25, at 34.
64. 65 Sup. Ct. 193, 194 (1944).
65. It is to be observed that the Court's total Constitutional discussion is comprehended in its consideration of whether the exclusion was "beyond the war power" [65 Sup. Ct. 193, 194 (1944)], and it makes no reference to the due process clause. While the Court's major discussion in the *Hirabayashi* opinion of the substantive validity of the curfew was included in its consideration of the war power rather than of the due process clause, the criterion it there stated for determining whether a measure was within that power would necessitate the type of consideration frequently associated with the due process clause.
66. The term "war area" was invented by the Court; the actual scope of the evacuated territory is not considered or mentioned.

and sabotage. The military authorities, charged with the primary responsibility of defending our shores, concluded that curfew provided inadequate protection and ordered exclusion.

* * * *

Like curfew, exclusion of those of Japanese origin was deemed necessary because of the presence of an unascertained number of disloyal members of the group, most of whom we have no doubt were loyal to this country. It was because we could not reject the finding of the military authorities that it was impossible to bring about an immediate segregation of the disloyal from the loyal that we sustained the validity of the curfew order as applying to the whole group.[67] In the instant case, temporary exclusion of the entire group was rested by the military on the same ground. The judgment that exclusion of the whole group was for the same reason a military imperative answers the contention that the exclusion was in the nature of group punishment based on antagonism to those of Japanese origin. That there were members of the group who retained loyalties to Japan has been confirmed by investigations made subsequent to the exclusion. Approximately five thousand American citizens of Japanese ancestry refused to swear unqualified allegiance to the United States and to renounce allegiance to the Japanese Emperor, and several thousand evacuees requested repatriation to Japan.[68]

* * * *

He [Korematsu] *was* excluded because we are at war with the Japanese Empire, because the properly constituted military authorities feared an invasion of our West Coast and felt constrained to take proper security measures, because they decided that the military urgency of the situation demanded that all citizens of Japanese ancestry be segregated from the West Coast temporarily,[69] and finally, because Congress, reposing its confidence in this time of war in our military leaders—as inevitably it must—determined that they should have the power to do just this.[70] There was evidence of disloyalty on the part of some, the military authorities considered that the need for action was great, and time was short." (Court's footnote omitted.)[71]

---

67. See discussion of this mythical finding, *supra* p. 192.

68. The danger of reliance on events occurring subsequent to a measure to justify it is well-illustrated in this case, since officials conducting these investigations stated that the measures themselves probably caused in some cases the attitudes to which the Court refers (*infra* p. 212, note 155). In any event the circumstances cited by the Court do not of course bear on the crucial factor of the possibility of segregation of the disloyal from the remainder of the 70,000 citizens, except in that they give rise to a slight inference of its feasibility.

69. As to whether the segregation was temporary, see *supra* p. 189. The Court's statement has the implication that an entirely different plan which might have been followed was in fact followed: that of taking the citizens in each subdivision to a point outside the area for the purpose of investigation, and allowing those found loyal immediately to return.

70. The Court did not need to consider the question of the constitutionality of the delegation of legislative power; since the intention to enforce exclusion, like the intention to impose the curfew, was known to Congress, this question was controlled by the *Hirabayashi* case. See *supra* note 31, and *United States v. Hirabayashi*, 320 U. S. 81, 90-91 (1943).

71. 65 Sup. Ct. 193, 194-197 (1944).

The *Korematsu* opinion is thus pervaded by an acceptance of supposed reasons for action by the military authorities, with, at the most, only casual consideration of the reality of these reasons or their factual basis. In the special confidence and special status it thus accords to judgments of military authorities as compared to those of civilian agencies it is again a marked contrast to the *Hirabayashi* opinion. There the fact that the program was being executed by the military not only did not appear to influence the Court's approach to the question of review; but in fact the Court appeared to minimize the very existence of this circumstance by repeatedly mentioning that the curfew measure expressed the will of Congress and the Executive, as well as the military authorities.[72] It is significant that Mr. Justice Jackson, dissenting, ascribes the majority opinion in the *Korematsu* case wholly to the circumstance that the action under review was undertaken by military authorities. He states that "what the Court appears to be doing, whether consciously or not," is to "distort the Constitution to approve all that the military may deem expedient" and he regards the review accorded to the military action as "wholly delusive."[73] It may be noted that the majority opinion does not mention this contention, basic in Justice Jackson's dissent, though it does attempt to answer Justice Roberts' dissent and may be considered to have referred obliquely to Justice Murphy's as well.[74]

Whether any justification exists for this special treatment of the judgment of military authorities with respect to such a program as the instant one will be discussed after a consideration of the Court's treatment of detention in Assembly Centers, since this is characterized by the same approach as that to exclusion.

b. *Circumstances dictating special need for rigid scrutiny of exclusion*

It is pertinent to note briefly the background of prejudice on the West Coast against persons of Japanese descent, to which Justice Black adverted. Even the most active opponents of the exclusion have agreed that the possibility of invasion of the West Coast required special vigilance to protect the country against espionage or sabotage by disloyal Americans, including those of Japanese ancestry. But the pertinence of the well-known background of prejudice is that it should have compelled a particularly "rigid scrutiny" of whether the type of measure adopted

72. 320 U. S. 81, 93, 94, 95, 97, 98-99, 101, 102 (1943).
73. 65 Sup. Ct. 193, 207, 208 (1944).
74. The Court attempts to deal with Justice Roberts' dissent in its denial that the case involves imprisonment in a concentration camp [65 Sup. Ct. 193, 197 (1944)]; and its denial that the exclusion was based on "antagonism to those of Japanese origin" [*Id.* at 195] is probably directed at Justice Murphy's dissent.

was reasonably required for the safety of the country, especially since mass racial exclusion was the perfect answer to the recurrent search by West Coast pressure groups for a method of eliminating economic competition from the minority of Japanese ancestry.[75] And it has been suggested that the difference in the economic status of persons of Japanese ancestry on the West Coast and in Hawaii, and the resulting difference in public sentiment towards their continuance of residence, may have contributed largely to the difference in their treatment.[76] In Hawaii only those citizens of Japanese ancestry who were suspected of disloyalty, totalling a very small per cent of the total number of such citizens, have been interned, together with suspected citizens of other ancestries; those who were not suspected were not subjected to exclusion or restraint.[77] But it would seem that if the presence of the racial group as a whole was a danger in California, it presented a greater danger in Hawaii. In Hawaii persons of Japanese ancestry totalled thirty-seven per cent of the total population as compared to less than two per cent of the total population on the West Coast.[78] In Hawaii the factors encouraging an attachment to Japan were no less; and another attack on Pearl Harbor was constantly apprehended.[79] Certainly, even if exclusion by transportation to the mainland would have been a difficult problem, mass detention, which was in fact, as appears *infra,* the "exclusion" method used on the West Coast, would not have been an insuperable task if the danger had been deemed sufficiently great.

As to the urgencies of the military situation, while there were, as found in the *Hirabayashi* case, reasonable grounds for a belief that invasion was threatened, the "leisure and deliberation"[80] with which the exclusion was accomplished might well have given the Court reason to be wary as to the degree of the threat and as to whether it was the sole motive for, or justified, such a drastic move as mass racial exclusion. Another military factor strongly emphasized by General De Witt also gives rise to doubt: numerous reports as to the occurrence of shore-to-ship signalling.[81] Aside from any question as to the reliability of the reports and as to whether the exclusion was the reason for the decrease

75. See HOUSE REPORT cited *supra* note 9, at 72-90 as to this search.
76. MCWILLIAMS, PREJUDICE (1944) 143.
77. *Id.* at 142. As to internment of citizens regardless of ancestry, see authorities cited *supra* note 14.
78. 16TH CENSUS OF THE UNITED STATES, *supra* note 1, for Hawaii, p. 21; and *supra* note 1.
79. See statement of Secretary of the Navy, N. Y. Times, Dec. 6, 1942, p. 71, col. 2; p. 69, col. 2.
80. Mr. Justice Murphy, dissenting in the *Korematsu* case, 65 Sup. Ct. 193, 205 (1944).
81. FINAL REPORT, *supra* note 25, at 4, 8.

therein,[82] it would seem that if such signalling had been a grave military problem it would have been met by clearing a narrow coastal strip of all inhabitants or allowing only investigated and accredited persons in such a strip, rather than removing citizens of Japanese ancestry from a region extending more than 200 miles back from the Coast. Likewise, of course, it should be borne in mind that the event of actual invasion might have inspired disloyalty on the part of members of other races as well,[83] and that for this and other reasons relating to military efficiency, it was likely that a coastal strip would be completely evacuated of civilians in case the threatened invasion became more imminent.[84]

However, there can be no doubt that the sweeping measure was an accurate response to the more voluble West Coast sentiment[85] and thus eliminated dissension and discord about the program, among others than the minority group affected by it. And from the viewpoint of a military commander, the complete elimination of a problem and of all agitation about it by one broadside move would no doubt appear appropriate. These considerations, however, should only have again stimulated the Court's "rigid scrutiny."

B. The Court's Treatment of Detention

1. *Necessity for Consideration of Detention in the Korematsu Case*

While the order which Korematsu was charged with disobeying was termed an "Exclusion Order," its effect was to subject him not only to exclusion but also to detention—or, it might be said, to detention and also to exclusion, insofar as a person detained elsewhere is necessarily excluded from his residence. The first Exclusion Order in the series of

---

82. General De Witt does not give any details as to the reports, except that they decreased after the exclusion. They would have of course decreased after the exclusion if they had been prompted by hysteria and suspicion as well as if there had actually been signalling by the excludees. The General does not attempt to account for their continuance after the exclusion. Distrust of General De Witt's justifications for the mass exclusion seems indicated by the Solicitor General's disclaimer of reliance on the General's report except for very limited purposes. See Brief for the United States, cited *supra* note 26, at p. 11.

83. It may be noted in this connection that the principal defendants in pre-war cases involving charges of communicating, or attempting to communicate, information relative to the national defense of the United States to Japan were Caucasians. See Annual Rep. Att'y Gen. (1937) 76; McWilliams, Prejudice (1944) 110-111.

84. See British regulation providing for such a general evacuation, *supra* note 15.

85. See House Report, cited *supra* note 9, pp. 139-156; *e.g.*, statement of the American Legion of California, Los Angeles Evening Herald Express, Jan. 27, 1942; resolution of Ventura County Board of Supervisors, Los Angeles Times, Feb. 4, 1942; proposal of Los Angeles County Defense Council, Los Angeles Examiner, Feb. 12, 1942; resolution of American Legion University Post No. 11, Seattle Times, Feb. 18, 1942.

108, issued on March 24, 1942, had not had this effect;[86] but as stated by Mr. Justice Black:

> ". . . on May 9, the effective date of the exclusion order (applicable to Korematsu), the military authorities had already determined that the evacuation should be effected by assembling together and placing under guard all those of Japanese ancestry, at central points, designated as 'assembly centers,' in order 'to insure the orderly evacuation and resettlement of Japanese voluntarily migrating from military area No. 1 to restrict and regulate such migration.' Public Proclamation No. 4, 7 Fed. Reg. 2601. And on May 19, 1942, eleven days before the time petitioner was charged with unlawfully remaining in the area, Civilian Restrictive Order No. 1, 8 Fed. Reg. 982, provided for detention of those of Japanese ancestry in assembly or relocation centers."[87]

And as explained by Mr. Justice Frankfurter in his concurring opinion:

> ". . . the various orders issued by General De Witt . . . put upon Korematsu the obligation to leave Military Area No. 1, but only by the method prescribed in the instructions, *i.e.*, by reporting to the Assembly Center."[88]

In view of these circumstances, as to the true nature of exclusion, the question was presented to the Court of whether Korematsu could contend that his conviction was unconstitutional on the basis of the invalidity of the detention in Assembly Centers or was limited to the issue of the validity of an order to leave his place of residence. The question was complicated by the circumstance that the order to report for detention to an Assembly Center was not, either by design or accident, explicitly incorporated in any written document.[89] It is not known whether the evacuees were so instructed orally when they reported for instructions on evacuation, which they were ordered to do by the Exclusion Orders under pain of criminal prosecution, or whether they were never so informed, but were merely taken to the Centers when they reported for evacuation.[90] Once in the Centers they found themselves detained by military police;[91] and except for a few who were released therefrom

86. Described *infra* p. 201.
87. 65 Sup. Ct. 193, 196 (1944).
88. 65 Sup. Ct. 193, 197 (1944).
89. The Exclusion Orders merely stated that "persons within the bounds of an established Assembly Center pursuant to instructions from this Headquarters are excepted from the [exclusion] provisions of this order while those persons are in such Assembly Centers"; and the instructions accompanying each order provided that the affected persons "will be evacuated" on the exclusion date, and would be furnished transportation to an Assembly Center.
90. The FINAL REPORT, *supra* note 25, does not make any statement on this matter though it describes the evacuation process in detail. See FINAL REPORT, c. X. As to the mechanics of the transportation under military control to the Centers, *see* HOUSE REPORT, cited *supra* note 9, at 40.
91. FINAL REPORT, *supra* note 25, at 183. Somewhat after the inauguration of detention, provision therefor was embodied in Civilian Restrictive Order No. 1, issued on May 19, 1942 [8 FED. REG. 982 (1942)].

under various restrictions, the entire group of evacuees were transferred after varying periods of time, lasting from a few weeks to several months, from the Assembly Centers to Relocation Centers,[92] where they were subject to further detention of a type discussed *infra*.

Justice Black based his view that Korematsu could only raise the issue of the validity of an order to depart from the area largely on the ground that there were two separable provisions of the orders: namely, to depart from the area and to report to an Assembly Center, and that there was a separate offense of failing to depart for which Korematsu was prosecuted. In addition to this artificial breakdown of requirements,[93] Justice Black also argued as a makeweight that the issue of detention was not within the framework of the pleadings or evidence, and that it "will be time enough" to decide this issue when an "assembly . . . order is applied or is certain to be applied."[94] The implication in this statement that the proper way for Korematsu to have challenged the detention was by *habeas corpus* after he had been subjected to it was vigorously rejected by Mr. Justice Roberts in his dissent.[95] While the Court had held in respect to two instances of war-time regulations (*Yakus v. United States*[96] and *Falbo v. United States*[97]—both cited by

92. 1740 persons were released as members of agricultural work groups to work in specified counties; a few were released under regulations respecting parties to mixed Caucasian-Japanese marriages and persons whose family members had left the area to establish homes in the interior prior to the inauguration of detention. FINAL REPORT, *supra* note 25, at 243-246, 278. With a total of 110,219 evacuees, 108,503 persons entered Relocation Centers. *Id.* at 84-346, 279. The evacuation to Assembly Centers was completed in July, 1942 (*supra* note 25) and the transfer to Relocation Centers in November of that year. *Id.* at 282.

93. While Justice Black stated that each requirement imposed "distinct duties," there was in fact a duty not to depart except by the method of being transported to an Assembly Center. Under Public Proclamation No. 4, issued March 27, 1942 [7 FED. REG. 2601 (1942)] persons of Japanese ancestry were forbidden to leave Military Area No. 1 except as permitted by future orders; and all parts of that Area were thereafter affected by exclusion orders. A similar prohibition with respect to the California portion of Military Area No. 2 was contained in Public Proclamation No. 6 [7 FED. REG. 4436 (1942)], issued on June 2, 1942 before the issuance of the Exclusion Orders applicable to that region.

94. 65 Sup. Ct. 193, 197 (1944).

95. 65 Sup. Ct. 193, 201 (1944).

96. 321 U. S. 414 (1944). The Court held that the validity of a maximum price regulation of the O. P. A. could not be raised as a defense in a criminal prosecution for its violation, in view of the fact that the defendant had not availed himself of a special administrative and civil procedure for challenging the regulation, provided by Congress as the exclusive means for challenge, which the Court considered an adequate means.

97. 320 U. S. 549 (1943). There it was held that an individual ordered to report for assignment to work of national importance under the Selective Training and Service Act may not, in a prosecution for his failure to report, raise the defense that he had been erroneously classified by his Local Board as the result of a proceeding that was not fairly conducted. See (1944) 57 HARV. L. REV. 577 with respect to lower court decisions in accord during the last War; the question did not then reach the Supreme Court. Compare Keith, *supra* note 15, at 85, with respect to prosecution for violation of British billeting orders despite the pendency of an action to test the validity of the order.

the Government)[98] that obedience to an order might be exacted prior to its challenge, no special circumstances such as existed in those cases applied to the detention in Assembly Centers;[99] and the depreciation of the value of personal freedom involved in the extension of the holdings in those cases to such a detention would therefore be unjustified.

This question will not be further elaborated in view of the fact that the majority opinion upholds Korematsu's conviction both on the basis that exclusion and detention are separable and on the view that if inseparable the detention was valid. However, the fact that the majority was split on this question may be in part the explanation for its disposition of the validity of the detention in three brief sentences. To appraise the cogency of these three sentences it is necessary to describe the nature of the Assembly Center detention and the Constitutional problem it raises, neither of which are explained in the majority opinion.

2. *Validity of Assembly Center Detention*

The detention in Assembly Centers was of all evacuees, without distinction on the basis of loyalty, for the stated purpose, primarily, of preventing disturbances due to hostility toward the evacuees in the com-

98. Brief for the United States, *supra* note 26, pp. 33-37.

99. In the *Falbo* case, the Court reasoned that the order to report did not necessarily mean that the subject of the order would be forced to serve in view of the possibility of his rejection on physical grounds; that even if it was constitutionally necessary for Congress to provide for judicial review of the classification made by the Draft Board, it did not have to provide for such review before final acceptance for service; and that the circumstances of the Act's adoption show that Congress did not intend to permit the "litigious interruption of the process of selection which Congress created . . . for mobilizing national manpower in the shortest practicable period" [320 U. S. 549, 554 (1943)]. The decision was referred to in the *Yakus* case as a holding on failure to exhaust an administrative remedy [321 U. S. 414, 446 (1944)]. In Korematsu's case there was not, as in Falbo's case, any administrative machinery for relief from the effects of the disobeyed order.

In the *Yakus* case, in determining whether the statutory procedure was adequate to meet the demands of due process so that no hearing on the validity of a regulation need be afforded in a prosecution, the Court considered the contention that the statutory procedure did not provide for the injunction of the operation of the regulation while its validity was being determined. The Court's conclusion that the denial of injunctive relief would not be unconstitutional was based on its view that: "If the alternatives, as Congress could have concluded, were war-time inflation or the imposition on individuals of the burden of complying with a price regulation while its validity is being determined, Congress could constitutionally make the choice in favor of the protection of the public interest from the dangers of inflation." [321 U. S. 414, 439 (1944)].

It is also to be noted that the *Falbo* case only involved the validity of an individual ruling, and that the *Yakus* ruling was stated in the opinion not necessarily to be applicable in the case of a regulation alleged to be void on its face. On the latter point, *cf.* Smith v. Cahoon, 283 U. S. 553, 562 (1931); Anniston Mfg. Co. v. Davis, 87 F.(2d) 773 (C. C. A. 5th, 1937), *aff'd,* 301 U. S. 337 (1937), *rehearing denied,* 302 U. S. 772 (1937); Barrow v. Owen, 89 F.(2d) 476 (C. C. A. 5th, 1937).

munities to which they might have migrated.[100] Prior to the inauguration of detention, there were two successive evacuation methods. The first was simply to make it fairly clear that some sort of evacuation would be forced at some time in the future and to encourage "voluntary" migration.[101] The second was embodied in Civilian Exclusion Order No. 1 of March 24, 1942, the only one of this type issued.[102] It provided that the persons in the affected subdivision could migrate before the exclusion date to destinations of their own choice, if approved by the military authorities, and that if they did not so proceed, they would be evacuated to Army reception centers on the specified exclusion date. At the time that evacuation to, and detention in, Assembly Centers was put into effect, General De Witt's Assistant Chief of Staff, Civil Affairs Division, who recommended this procedure in place of the continued use of orders similar to Order No. 1,[103] explained the change on the basis that it was necessary in order "to insure an orderly evacuation and protect the Japanese" and that "State officials had said 'We won't take them'; and there was a great possibility of some untoward incident."[104] Similarly General De Witt stated in his Final Report that "essentially, military necessity required only that the Japanese population be removed from the coastal area and dispersed in the interior" and that the detention was "due primarily to the fact that the interior states would not accept an uncontrolled Japanese migration."[105]

It may be startling to observe that the apprehension of disloyalty among citizens of Japanese ancestry was viewed as necessitating their mass racial exclusion but was not viewed as sufficiently grave to require the segregation of the disloyal from the loyal or any special precautionary measures with respect to the activities of any of the evacuees in any other part of the country. However, regardless of one's appraisal of the wisdom of the program, it is clear from the above statements of the military authorities, from the attempts to secure migration prior to the initiation of detention, and from the fact that no program for the segregation of the loyal from the disloyal was commenced until established by the

100. It may be noted that in Canada where, as in the United States, the majority of the residents of Japanese ancestry lived on the West Coast and a similar state of hostility towards them prevailed, a similar evacuation and detention program was adopted. See *Removal of Japanese from Protected Areas,* REPORT OF THE BRITISH COLUMBIA SECURITY COMMISSION (March 4, 1942 to October 31, 1942) *passim.* The fact that the measures were used in Canada, however, hardly seems of much pertinence in determining whether they were desirable from the standpoint either of the prosecution of the war or the protection of civil liberties.

101. FINAL REPORT, *supra* note 25, at 41-48, 107, 109.

102. 7 FED. REG. 2543 (1942).

103. FINAL REPORT, *supra* note 25, at 65.

104. H. R. REP. No. 1911, 77th Cong., 2d Sess. (1942) 6, 8.

105. FINAL REPORT, *supra* note 25, at 43-44, 105.

War Relocation Authority approximately four months after the detentions in Assembly Centers began,[106] that such segregation was not the purpose of the Assembly Center detention at issue in the *Korematsu* case. It seems clear that a purpose which demonstrably was not a motivation for the detention cannot be urged as a justification for it, any more than in the case of any other confinement.[107]

To consider, then, the constitutionality of this detention on the basis of its announced purposes, it seems evident beyond question that one of the stated purposes of the detentions—that of assisting and protecting the evacuees—does not, without more, furnish a justification which has, or should have, any place in constitutional law. That is, it would be a totally anomalous doctrine that the Federal Government can impose restrictions on an individual solely for what is thought to be his own individual good, without a justification in terms of the relation of the control of that individual to the interests of society; or that it can force the acceptance of benefits unless the imposition of the benefits is justified by the social good. Under our theory of government the individual's freedom of action may be restricted only when his freedom impinges on the welfare of others. Thus it must be shown that the "protection" of the evacuees from violence was related to the successful prosecution of the war, or, under the theory that the war power includes the power to take measures for the public welfare necessitated by the war's prosecution,[108] that such protection was otherwise related to the public welfare. While the theme of benefaction which runs through the utterances of the

---

106. See *infra* pp. 210-212.

107. That is, a jail sentence could not be upheld on the basis that the evidence showed the prisoner committed a crime other than that for which he was sentenced. And the test of validity in the *Hirabayashi* opinion,—of whether under the circumstances a particular measure could reasonably have been thought necessary,—appears to be directed at the ascertainment, as the ultimate fact, of whether the authorities were motivated by reasonable purposes; it employs an objective rather than a subjective standard but does not appear open to the interpretation that purposes which clearly were not in fact part of the motivation may be accorded weight. It should be noted that such an interpretation would be inconsistent with the Court's emphasis on the findings by the authority promulgating the order with respect to the basis therefor (*supra*, p. 191). Furthermore such an interpretation would permit the validation of a measure without any official ever determining that it was desirable on the basis of a Constitutional objective, since the Court only determines at the most whether it might reasonably have been thought to serve such an objective. *Cf.* Securities and Exchange Commission v. Chenery Corp., 318 U. S. 80, 87 (1942) with respect to the Court's consideration of the validity of an administrative order only on the grounds of support found by the administrator.

108. *Stewart v. Kahn, Raymond v. Thomas, Hamilton v. Kentucky Distilleries*, cited *supra* note 29.

Such relationship to the public welfare would also have to be shown if the detention were viewed, by a considerable stretch of reasoning, as an exercise of a power possessed by the Federal government, apart from its other powers, to preserve the public peace and the sovereignty of the United States [*In re* Neagle, 135 U. S. 1 (1890); *Ex parte* Siebold, 100 U. S. 371, 395 (1879); *In re* Debs, 158 U. S. 564 (1894)].

military[109] as well as, subsequently, of the War Relocation Authority,[110] may have given the officials involved a feeling of satisfaction, it does not make the deprivations and restraints imposed on the donees any more constitutional.

If the constitutionality of the protective custody in Assembly Centers were to receive serious analysis, and the danger of violence affecting the war effort or the general public welfare were to be evaluated, it would first be necessary to consider that the maximum degree of migration which could have reasonably been anticipated from a continuance of optional migration to the interior or to Army shelters would have been small. For in addition to the inference from common sense that many would not have had the means or wished to take the risk of such a venture, the figures on "voluntary" migration in General De Witt's report prove the truth of this inference; *inter alia,* they show that the entire group who were in fact given the option under Exclusion Order No. 1 of migration or entrance into Army reception centers chose the latter alternative.[111]

To take the side most favorable to the military, it is true, nevertheless, that in view of the hostility expressed in the Middle Western states to the evacuees,[112] even a small migration might have resulted in some acts of violence and a substantial amount of agitation. But a reasonable apprehension of this result would not in itself be enough to justify the measure. In the command and discipline of troops, in the waging of a battle, or in the government of an occupied country while it is the scene of warfare, no doubt the absence of all elements of dissension and clamor is a desirable objective; however, in the domestic control of civilians under the Constitution, a measure of drastic racial discrimination can scarcely be justified merely on the basis that it will serve this objective. It is noteworthy, but perhaps natural if the whole program is viewed as a military maneuver, that General De Witt's report nowhere analyzes the scope of the possible disturbances or their effect on the prosecution of the war, but regards merely the alleviation of "tension" and the prevention of "incidents involving violence" as a justification for the detention.[113] Assuming that protective custody can be accepted as a rea-

109. See Public Proclamation No. 4, 7 FED. REG. 2601 (1942), which in part had the effect of imposing detention (*supra* p. 198) and which recited as one of the grounds for its promulgation that it was "necessary in order to provide for the welfare . . . of [the] Japanese." See also FINAL REPORT, *supra* note 25, at 8-9, 105.

110. *E.g.,* SECOND QUARTERLY REPORT OF WAR RELOCATION AUTHORITY (July 1 to Sept. 30, 1942) 16-17.

111. FINAL REPORT, *supra* note 25, at 107, 109, 49, 363.

112. H. R. REP. No. 1911, 77th Cong., 2d Sess. (1942) 27-30. See H. R. REP., *supra* note 9, at 17.

113. FINAL REPORT, *supra* note 25, at 105.

sonable method of preventing disturbances of such a type or scale that they might affect the prosecution of the war, it does not appear that such disturbances could reasonably have been anticipated from a continuance of the program of voluntary shelter and guidance, particularly if the Federal government had indicated that lawlessness against such migrants as there were would be punished.[114] And from the standpoint of the general welfare aside from the prosecution of the war, it would hardly seem that an intention to better race relations by removing the possibility of agitation could be seriously considered as a justification for the detention; the commencement of such betterment by the incarceration of the racial minority would not seem a very promising approach.

The Assembly Center detention was of importance not only in itself as a progressive step in the increasingly drastic program of racial discrimination, but also because, though of relatively short duration, it laid the groundwork for the subsequent detention in Relocation Centers which is continuing to date (discussed *infra*). In view of these circumstances and in view of the stated purposes of the detention in Assembly Centers, of which the Court showed itself fully aware in the *Endo* opinion,[115] the adequacy of the following sentences of the opinion, which disposed of detention, can be judged:

> "The Assembly Center was conceived as a part of the machinery for group evacuation. The power to exclude includes the power to do it by force if necessary. And any forcible measure must necessarily entail some degree of detention or restraint whatever method of removal is selected."[116]

The easy disposition of the question by implying that detention was necessary in order to get the evacuees to move, and thus was on a parity with putting them into trains if they would not get aboard voluntarily, is somewhat startling. Justice Roberts, it should be noted, viewed the detention somewhat more seriously, as presenting

> "the case of convicting a citizen as a punishment for not submitting to imprisonment in a concentration camp, based on his ancestry, and solely because of his ancestry, without evidence or inquiry concerning his loyalty and good disposition towards the United States."[117]

114. It may be noted that General De Witt mentions reports as to the occurrence of "incidents" in California prior to the evacuation (FINAL REPORT, *supra* note 25, at 43, 105-106); however, he does not suggest, and it could hardly be thought, that these "incidents," though in a locality of unusual hostility, caused an interference with the prosecution of the war.
115. 65 Sup. Ct. 208, 212, 215 (1944).
116. 65 Sup. Ct. 193, 197 (1944).
117. *Id.* at 198.

and was of the opinion that on the basis of this formulation of the question, the violation of Constitutional rights was obvious.[118]

C. Justifiability of the Court's Treatment of Military Judgment

The doctrine of review expressed in the *Korematsu* opinion[119] was not only not in conformity with that of the *Hirabayashi* opinion, but also was a departure from opinions reviewing military judgments with respect to more customary subjects for action by the military, including even judgments made in the course of actual deployments against enemy troops.[120] For the Court to accord in the instant case a degree of acceptance to the military judgment beyond that accorded in the latter type of case, where a special standard of review may be justifiable, is particularly surprising; for from the standpoint of the *expertise* involved, it is highly questionable that the military judgment should here even be accorded the weight customarily given to determinations of expert administrative agencies. It hardly seems that the determination of the necessity for evacuation and of the policies controlling it could properly be considered an essentially "military matter." It is true that the measure related to the preparation of a possible site of combat for successful military operations. On the other hand, the evacuation encompassed a large territory which was not at the time and might never be the scene of combat and which far exceeded the size of the areas traditionally protected from espionage and sabotage by military guards—those directly contiguous to military installations; it involved broad interests of civilians; and the knowledge upon which the desirability of the evacuation was to be determined was not secret military information nor ex-

118. *Ibid.* See Justice Jackson's dissent, at 206. Justice Murphy dissented on the basis of the unconstitutionality of the exclusion alone, regardless of the detention.

119. Described *supra* pp. 193-195.

120. *See* Mitchell v. Harmony, 13 How. 115, 134 (U. S. 1851), where the Court considering whether the taking of property by an officer was justified, stated: "And, if with such information as he had a right to rely upon, there is reasonable ground for believing that the peril is immediate and menacing, or the necessity urgent, he is justified in acting upon it. . . ." *See* United States v. Russell, 13 Wall. 623, 627-629 (U. S. 1871); Dooley v. United States, 182 U. S. 222, 234 (1901); United States v. Diekelman, 92 U. S. 520, 527 (1875); Raymond v. Thomas, 91 U. S. 712, 716 (1875); consider Luther v. Borden, 7 How. 1, 45-46 (U. S. 1849); Sterling v. Constantin, 287 U. S. 378, 399-400 (1932); Moyer v. Peabody, 212 U. S. 78, 85 (1909). While the Court held in Martin v. Mott, 12 Wheat. 19 (U. S. 1827) and The Prize Cases, 2 Black 635 (U. S. 1862), that the military judgments there in issue were entirely irreviewable, they both concerned a wholly distinguishable and unique type of judgment. The first case related to the judgment of the President himself to call out the militia in the event of invasion or threatened invasion, and the second concerned the judgment of the President that domestic foes should be treated as belligerents and that their suppression necessitated the use of force.

clusively in the hands of the military. The only circumstance as to which the military would have had information which could not be disclosed—the danger of invasion—was the one fact on which there was also ample public knowledge, as shown by the Court's review of the military situation in the *Hirabayashi* case.[121] As to the other facts on which the justifiability of the exclusion and detention rested, the words of Justice Murphy are unquestionably correct:

> "Justification for the exclusion is sought, . . . mainly upon questionable racial and sociological grounds not ordinarily within the realm of expert military judgment, supplemented by certain semi-military conclusions drawn from . . . [the] use of circumstantial evidence."[122]

The manner in which the military acquired the authority to determine the measures to be taken and to execute them is of importance in considering the military *expertise* with respect to the whole program. The initial measures taken on the West Coast with respect to persons of Japanese ancestry were taken immediately after Pearl Harbor by the Attorney General as a part of Alien Enemy Control program, under the Alien Enemy Act of 1798.[123] In addition to minor restrictions, all Japanese aliens who were suspected of being in any way dangerous, on the basis of the information theretofore collected by the F.B.I. and by the intelligence divisions of the Army and Navy, were interned; the Attorney General took the further measure of prohibiting the residence of all Japanese aliens in areas along the Coast surrounding facilities of extraordinary importance to the prosecution of the War.[124] In the designation of these areas and in general in the handling of the aliens, the Attorney General had received, and in many cases accepted, the recommendations of General De Witt.[125]

But beginning about the middle of January, 1942, West Coast organizations and newspapers, which had not until then been active in this matter, commenced to agitate for sweeping restrictions to be applied alike to the Japanese aliens and to the 70,000 native-born American citizens of Japanese ancestry.[126] Military authorities at the same time indicated that they too favored more drastic measures than those enforced by the Attorney General. An irreconcilable difference of opinion arose between General De Witt and the Attorney General when the

121. *Supra* p. 184.
122. 65 Sup. Ct. 193, 203 (1944).
123. Cited *supra* note 9.
124. H. R. Rep. No. 1911, *supra* note 112, at 2; H. R. Rep. No. 2124, cited *supra* note 9, at 160-161, 302-314.
125. Final Report, *supra* note 25, at 1-6.
126. See authorities cited *supra* note 85.

former recommended that the program of establishing prohibited zones for Japanese aliens in the immediate vicinity of important installations be transformed into a prohibition on their residence in large areas and a prohibition on the residence of American citizens of Japanese ancestry as well. The Attorney General replied that as to the aliens he would not take such a step without convincing evidence of the military necessity therefor; that American citizens could only be evacuated as a matter of military necessity; and that if there were such a necessity for citizen evacuation, it should be undertaken by the War Department and not by the Department of Justice.[127]

On February 13th, West Coast sentiment, which had been demanding action similar to that proposed by General De Witt, was expressed in the recommendation sent to the President by the entire West Coast Congressional delegation, urging action against all persons of Japanese ancestry, regardless of citizenry, who might be "disloyal and subversive." The delegation suggested that "by utilizing the military authority of the Army to effect the partial or complete evacuation of strategic areas, . . . we feel that the Army or the Department of Justice may rightfully remove any or all persons whom they may select from such areas and prohibit their return. This might require the principles of martial law. . . ."[128] Following General De Witt's formulation of an exclusion plan substantially the same as that thereafter effected,[129] Executive Order No. 9066 of February 19th, drafted by the War Department with the concurrence of the Department of Justice for the purpose of authorizing such an evacuation,[130] was promulgated by the President, granting the military the power both to determine the scope of the program and to execute it.

Thus, it may be inferred that the deposit of the authority in the hands of the military was not because the proposal to exclude was one which originated in military minds, nor because the military were peculiarly competent to determine the proper manner of executing it. One reason rather clearly seems to have been that it was thought such deposit would insure the constitutionality of the exclusion. The fact that the Court responded in the anticipated manner is not to say that the grant of authority to the military with respect to any subject is sufficient to insure its constitutionality; but the danger of the Court's doctrine as to review, if combined with a failure to scrutinize the "military" character of the action, is apparent.

---

127. FINAL REPORT, *supra* note 25, at 6-7.
128. H. R. REP. No. 1911, *supra* note 112, at 3.
129. FINAL REPORT, *supra* note 25, at 33 *et seq.*
130. *Id.* at 25.

The language of the *Korematsu* opinion represents such a marked departure from that used in *Hirabayashi* and even in preceding cases that it stimulates conjecture as to the ends which the Court may have expected it to serve.[131] There is no indication that the change of approach was thought necessary on the ground the military authorities were being unduly impeded in taking action by the prospect of the more rigorous court review to which they might be subjected under the previous decisions; and since under the latter the justifiability of the action would in any event be appraised in view of the degree of danger which evoked it, it does not seem that the opinion can be explained on this basis. A somewhat more likely possibility is that the Court thought that a doctrine of subjecting activity of military authorities to close scrutiny in war-time would have a deleterious effect on war morale. It would seem, however, that any effect which scrutiny or reversal of military control of civilians could have on confidence in the military conduct of combat against enemy troops, would be slight at best, and could be largely overcome by the Court's emphasizing the distinction between the two spheres of activity. As to the effect on confidence in military judgment in non-combat matters, it would appear that nothing is gained by the fostering of confidence in the general omniscience of the military other than the encouragement of an irrational adherence to a species of demagoguery and of the leadership principle. However, under the special circumstances of this case, inter-related as it is with the *Endo* decision and the social problem of the return of loyal persons of Japanese descent to the West Coast, the Court's abstention from close scrutiny of the initial measures of the military authorities may find some explanation not ascertainable from a mere study of the opinions themselves. This possibility will receive consideration following the discussion of the *Endo* opinion.

Some justification for the Court's approach may be thought to lie in an entirely different area: that it is justified because the validation of the measures on a special basis inapplicable to civilian agencies would prevent the decision from serving as a precedent for upholding peace-time attempts at racial discrimination. It is obvious, however, that though counsel could argue that racial discrimination was upheld in the

---

131. It is to be observed that the Court could hardly have been fearful of employing a more stringent standard of review because of any effect that a possible resulting invalidation of Korematsu's conviction might have had on measures at present in effect. The only surviving measure: the detention of the potentially disloyal (*infra* p. 219), would not have been affected thereby. If a *habeas corpus* case were brought by one of these detainees, his case would be governed by the present need for his detention (see *infra* p. 214), rather than the validity of the initial exclusion and detention.

Korematsu case only because the war power and the authority of the military were given special treatment, the case could easily be cited as a precedent for a peace-time measure on the basis that other emergencies may be as threatening as military ones.[132] And in evaluating the importance of restricting the decision's effect as a peace-time precedent as against the establishment of a doctrine of acceptance of military judgment, it must be borne in mind that the war power is considered to extend beyond the period of hostilities,[133] which itself may be of long duration, and that, aside from judicial precedents, measures undertaken by the military for the duration of the war may cause social changes and set social precedents which outlive the war.[134] In view of these considerations, the degree to which the Court's approach will limit the effect of the decision as a peace-time precedent does not compensate for the dangerous consequences of the doctrine of acceptance of military war-time judgments respecting the control of civilian liberties. Illustrations from war time litigation other than the Japanese ancestry cases of the value of a more meaningful standard of review, are given in the last section of this article.

II. DETENTION IN WAR RELOCATION AUTHORITY CENTERS AND THE ENDO DECISION

A. The War Relocation Authority Program

All of the evacuees with the exception of a small minority were transferred from the Assembly Centers to Centers constructed by the War Relocation Authority as the latter were completed.[135] In the W. R. A. Centers the detention of the evacuee continued, enforced by military police,[136] unless and until he was released pursuant to the requirements imposed by W. R. A. regulations.

The W. R. A. had been created in the Executive Office of the President by Executive Order No. 9102 issued March 18th, 1942.[137] Its functions were to provide for the "relocation, maintenance, supervision" and "insofar as feasible and desirable, for the employment" of the evacuees. The Centers erected by the War Relocation Authority

---

132. *See* Home Bldg. & L. Assn. v. Blaisdell, 290 U. S. 398, 426, 447 (1934); Chastleton Corp. v. Sinclair, 264 U. S. 543, 548 (1924); Pound, *Civil Rights During and After the War* (1943) 17 Tenn. L. Rev. 706, 722; Corwin, *American Government in War Time* (1943) 37 Am. Pol. Sci. Rev. 18, 25.

133. See authorities *supra* note 108.

134. See Chafee, *supra* note 10, with respect to the renewal after the last War of a wartime anti-civil liberties behavior.

135. See *supra* note 92.

136. First Quarterly Report of the War Relocation Authority (March 18 to June 30, 1942) 32; McWilliams, Prejudice (1944) 158-159.

137. 7 Fed. Reg. 2165 (1942).

were designed to afford better living facilities than the Army Reception and Assembly Centers;[138] and the removal of the evacuees from the latter to the former was completed by November, 1942.[139] The detention in the W. R. A. Centers was effected pursuant to General De Witt's Public Proclamation No. 8, which designated each War Relocation Authority Center within the Western Defense Command as a "War Relocation Project Area," and prohibited persons of Japanese ancestry residing in such areas from leaving them except upon a "written authorization executed by or pursuant to the express authority of this headquarters setting forth the effective period of said authorization and the terms upon and purposes for which it had been granted."[140] General De Witt thereafter gave the War Relocation Authority written authority to issue authorizations pursuant to Public Proclamation No. 8 for persons to leave these areas.[141]

The first measure for the conditional release of the evacuees was taken in response to the demand for agricultural labor, in which many of the evacuees had previously been engaged; beginning in May, 1942, evacuees were released from the Assembly Centers, under rigid restrictions, in agricultural work groups.[142] A program for conditional releases on an individual basis was not put into effect until July, 1942. The individual leave program then established by the War Relocation Authority, with refinement and improvement in many of its details, continued until the *Endo* decision.[143] Under the latest regulations[144] a leave application is to be granted "where the applicant agrees to make the reports required of him [with respect to changes in place of employment and residence after leaving the center], where there is no reasonable cause to believe that he will not have employment or other means of support or that he cannot otherwise successfully maintain residence at the proposed destination, and where there is no reasonable

138. FIRST QUARTERLY REPORT OF THE WAR RELOCATION AUTHORITY (March 18 to June 30, 1942) at 6.

139. *Supra* note 92.

140. 7 FED. REG. 8346 (1942). A similar prohibition, applicable both to Assembly and Relocation Centers, had been provided by Civilian Restrictive Order No. 1, 8 FED. REG. 982 (1943).

141. FINAL REPORT, *supra* note 25, at 530. Proclamation W D-1 of the Secretary of War, dated August 13, 1942, prohibited persons leaving the Relocation Centers outside of the Western Defense Command without permission of the Director of the W.R.A. or the Secretary of War. 7 FED. REG. 6593 (1942).

142. FINAL REPORT, *supra* note 25, at 243-246. The Orders under which these conditional releases were granted are printed at 8 FED. REG. 982 to 986 (1943).

143. The program was first established by the W.R.A.'s Administrative Instruction No. 22, dated July 20, 1942. This was followed by the Regulations issued Sept. 26, 1942, 7 FED. REG. 7656 (1942), and a Handbook of July 20, 1943, issued as a supplement to these Regulations.

144. Dated Jan. 1, 1944, 9 FED. REG. 154 (1944). The Handbook of July 20, 1943, continued in effect as a supplement to the revised Regulations.

cause to believe that the issuance of leave in the particular case will interfere with the war program or otherwise endanger the public peace and security." All leave permits are revocable in the event that the War Relocation Authority believes changed circumstances dictate the revocation. The determination of whether the applicant can "otherwise successfully maintain residence at the proposed destination" depends on whether his presence in the community is likely to cause disturbance because of racial hostility,[145] and the determination of the last named condition depends upon whether on the basis of investigation of the applicant there is ground for suspicion that he is sympathetic with the enemy.[146] This investigation includes consideration of all data obtainable from the Federal Bureau of Investigation, and it also included until May, 1944, by which date the cases of most of the evacuees had been considered, consultation between the War Relocation Authority and representatives of the Office of Naval Intelligence and of the Military Intelligence Service.[147]

Under the War Relocation Authority procedure the loyalty determination was made in a preliminary step, termed leave clearance, prior to consideration of a leave application;[148] thus the evacuees who received such clearance were deemed by the War Relocation Authority to be free of any taint or suspicion of disloyalty, but were nevertheless detained pending their demonstration of ability and willingness to fulfill the other requirements for release. Furthermore, again despite the loyalty clearance, detainees were barred by General De Witt's orders from returning to their former places of residence or to any part of the evacuated area: the State of California, western Washington and Oregon, and southern Arizona;[149] and the W. R. A. regulations prohibited releases for destinations barred by the military authorities.[150]

In the Relocation Centers the evacuees deemed ineligible for leave because of possible disloyalty were segregated and separately interned from those who had been cleared respecting loyalty;[151] the Tule Lake Center, where the much-publicized rioting took place in November, 1943, was then and is now being operated as a segregation center for the

---

145. See W.R.A. Handbook of July 20, 1943, § 60.4.3.
146. *Id.*, § 60.6.6.
147. Regulations, *supra* note 144, § 5.3(b); W.R.A. Handbook of July 20, 1943, § 60.10.2, and SEN. Doc. No. 96, 78th Cong., 1st Sess. (1943) 18.
148. Regulations, *supra* note 144, § 5.3(b).
149. Public Proclamation No. 7, 7 FED. REG. 4498 (1942); Public Proclamation No. 11, 7 FED. REG. 6703 (1942).
150. Regulations, *supra* note 144, § 5.7(c).
151. *Id.*, § 5.3(b); SEN. Doc. No. 96, *supra* note 147, *passim*.

former group.[152] On the basis of the segregation operation, the War Relocation Authority estimated that a maximum of approximately 6,000 out of the total 70,000 citizen-evacuees might be considered potentially disloyal;[153] it is to be observed, however, that this does not mean there is evidence establishing by a high degree of probability that all these persons would aid the enemy but merely that there is some danger they would do so.[154] It also does not mean that this whole group presented such a danger at the time of the evacuation; the attitude that the evacuee prefers to live in Japan to the United States, which is sufficient for him to be classed with the potentially "disloyal", has in some instances been stimulated by the evacuation and detention themselves.[155] As of July 29, 1944 the figures with respect to leave showed that including minor children, whose status depends in general on that of their parents,[156] and including both aliens and citizens, there were 28,911 persons on leave; 61,002 persons detained in Centers other than Tule Lake, that is, persons who had either received leave clearance, or were deemed eligible for it but had not formally obtained it; and 18,684 persons at Tule Lake.[157]

The detention of persons who were suspected of no disloyalty but were detained on the basis of what may be termed the social requirements, was invalidated by the Supreme Court in the *Endo* case on December 18, 1944, after the detention program had been in effect for approximately two and one-half years.[157a]

B. The Holding in the Endo Case

Mitsuye Endo filed a petition for a writ of *habeas corpus* in the District Court for the Northern District of California, in July, 1942 praying for a writ of *habeas corpus,* directed to the military and

---

152. Report on the Tule Lake Riot, and Minority Views on Tule Lake Segregation Center presented by Herman P. Eberharter, Subcommittee of the Special Committee on Un-American Activities, 78th Cong., 2d Sess. (1944) (committee print).

153. Statement of Acting Director of W.R.A., *Hearings before House of Representatives Committee on Immigration and Naturalization,* 78th Cong., 2nd Sess. (1944) 57-58.

154. See Sen. Doc. No. 96, *supra* note 147, for a description of the basis of segregation.

155. Statement of Acting Director of W.R.A. *Hearings, supra* note 153, at 39; Second Quarterly Report of W.R.A. (July 1 to Sept. 30, 1942) 51-52; Semi-Annual Report of W.R.A. (Jan. 1 to June 30, 1943) 11-14; McWilliams, Prejudice (1944) 180-185.

156. Regulations, *supra* note 144, § 5.3(b).

157. *Ex parte* Mitsuye Endo, 65 Sup. Ct. 208, 215-216, n. 19.

157a. Such detention was also in part terminated, effective Jan. 20, 1945, by Public Proclamation No. 21, issued by the Commanding General of the Western Defense Command, Dec. 17, 1944 [10 Fed. Reg. 53 (1944)].

W. R. A. authorities, to secure her release from a War Relocation Authority Center. Her petition alleged *inter alia* that she was born in California in 1920 and was a loyal citizen of the United States, with allegiance to no other country; that she was law-abiding and no charge had been made against her; and that she had become, after investigation by the officials of the State of California as to her efficiency, fitness, and moral responsibility, a permanent Civil Service employee of that State, but was unable to perform her duties because of her imprisonment in the Center. She had been evacuated from her place of residence in Sacramento, California pursuant to the applicable Exclusion Order. On the basis of the petition, an affidavit setting forth the regulations of the War Relocation Authority and the fact that Endo had not applied for leave, and an affidavit by Endo's attorney that she had not applied for leave because she could not secure through the War Relocation Authority permission to return to Sacramento, California, the District Court denied the petition in July, 1943.[158]

The court did not specify its grounds in detail, the order merely stating that "it appearing upon the face of the petition that petitioner is not entitled to a writ of *habeas corpus,* and it further appearing that she has not exhausted her administrative remedies," the petition was denied.[159] The exhaustion doctrine, however, could not be correctly applied to foreclose consideration of Endo's case on the merits, and it was not relied upon in the Circuit or Supreme Courts.[160]

158. See Statement of facts in *Endo* opinion, at 214-215; and in Briefs for the United States, and for the American Civil Liberties Union, *amicus curiae,* Supreme Court, Oct. Term 1944, No. 70, at 5-8 and 1-3, respectively.

159. *Ibid.*

160. In general where the doctrine is applied, the pursuit of the administrative remedy will not itself subject the litigant to the injury which he is attempting to escape or to any irreparable damage. For instances of the non-application of the doctrine of exhaustion where the litigant complains that substantial and irreparable damage will occur in the very course of his subjection to the administrative process, consider Utah Fuel Co. v. National Bituminous Coal Commission, 306 U. S. 561 (1939); Smith v. Illinois Bell Telephone Co., 270 U. S. 587 (1926); *see* Euclid v. Ambler Realty Co., 272 U. S. 365, 386 (1926); Pennsylvania v. West Virginia, 262 U. S. 553, 592 (1923); Oklahoma Natural Gas Co. v. Russell, 261 U. S. 290, 293 (1923). Compare Myers v. Bethlehem Corp., 303 U. S. 41 (1938) where the Court stated that the rule of exhaustion could not be circumvented by the mere allegation that the holding of a hearing in itself would cause irreparable injury. In rare cases where the pursuit of the administrative process has involved the detention of the pursuer the Court has, again giving expression to the rule of the above cases, not rested on the doctrine of exhaustion except with respect to the necessity for its determining the validity of the possible ultimate administrative action, and has considered whether the detention was constitutional. *See* United States v. Sing Tuck, 194 U. S. 161, 168-169 (1904) where the Court held that detention of persons denied admission to the country as aliens pending their administrative appeal to have their citizenship determined, was lawful. *Cf.* Tod v. Waldman, 266 U. S. 113 (1924). The Court therefore had to consider the validity of the detention to which Endo was subjected by the W.R.A.'s administrative procedure: that is, the detention she had to undergo in order to pursue her administrative remedy, as distinguished from any detention which might have resulted from a failure in such a pursuit.

In February, 1943, while the case was pending on appeal in the Circuit Court of Appeals, Endo applied for, and was granted leave clearance.[161] In view of the rule that a *habeas corpus* case is to be determined on the basis of the facts existing at the time of the final decision therein,[162] the Circuit Court pointed out, in certifying questions of law to the Supreme Court, that the case therefore involved the validity of the detention of a loyal citizen for failure to satisfy the requirements of the War Relocation Authority with respect to satisfactory prospective employment and place of residence.[163]

The Supreme Court, by unanimous decision, with the Court's opinion delivered by Justice Douglas, and concurring opinions by Justices Roberts and Murphy, held that "Endo should be given her liberty . . . [because] the War Relocation Authority . . . has no authority to subject citizens who are concededly loyal to its leave procedure," nor to detain such citizens, nor to release them only conditionally.[164] This holding is based on the following reasoning: Any power of detention possessed by the War Relocation Authority, which was holding Endo, is derived from Executive Order 9066, and its ratifying Act of March 21st, 1942, through a re-delegation by General De Witt of the authority conferred thereby. Detention is not mentioned explicitly in this Order nor in the Act, nor in the latter's legislative history, nor in Executive Order 9102, establishing the War Relocation Authority, which in any event was only issued to implement the measures already authorized by 9066. Accordingly, any power to detain must be implied. But the purpose of Executive Order 9066 and the ratifying Act limits the scope of powers implied therein; such purpose was protection against espionage and sabotage. Insofar as "the power to detain is derived from the power to protect the war effort against sabotage and espionage," the detention of a concededly loyal citizen, "which has no relationship to that objective is unauthorized."[165] Insofar as the power to detain or to grant a conditional release might "be implied as a useful or convenient step in the evacuation program," which program was authorized by the Order and the Act, it likewise follows that the detention of concededly loyal citizens is unauthorized.[166] For the lawful character of the original evacuation "was derived from the fact that it was an espionage and sabotage

---

161. See authorities *supra* note 158.

162. Mensevich v. Tod, 264 U. S. 134, 136 (1924); Storti v. Massachusetts, 183 U. S. 138, 143 (1901).

163. See Certificate to the Supreme Court in Endo v. Eisenhower, Case No. 10605, Circuit Court of Appeals for the Ninth Circuit.

164. 65 Sup. Ct. 208, 216, 218-219 (1944).

165. *Id.* at 218.

166. *Ibid.*

measure" and any implied power to detain in connection therewith "must be narrowly confined to the precise purpose of the evacuation program."[167] As to the point that the detention of loyal evacuees was authorized, aside from these Orders and the Act, by virtue of appropriations to the War Relocation Authority, the Court determined that there had been no ratification of such detention through these appropriations.[168]

The Court also decided an important procedural point by its determination that the case had not become moot by virtue of the fact that Endo was no longer within the territorial jurisdiction of the District Court. Though she had been within such jurisdiction at the time of the filing of her petition, she had been transferred to a Relocation Center outside of that jurisdiction during the pendency of her appeal to the Circuit Court. However, the Supreme Court was informed that if the District Court issued the writ and directed it to certain officials within its jurisdiction Endo's corpus would be produced and the writ would be fully honored. Rejecting indications in some lower court opinions that the place of confinement must be within the court's territorial jurisdiction at the time of the issuance of the writ,[169] the Court held, as a matter of first impression, that the District Court did not lose its jurisdiction in this case by virtue of Endo's removal since "a person in whose custody she is remains within the district."[170] This practical test of whether the writ can be effectuated seems to go as far as possible, in the troublesome removal situation, to preserve the expeditious character[171] of *habeas corpus;* the benefit of relieving the petitioner so far as possible from the necessity for filing another petition in the new District is particularly apparent in the instant case, where the petition had already been pending more than a year at the time of Endo's removal.

The holding as to the scope of the Executive Orders and the ratifying Act of March 21, 1942 is clearly correct; a contrary holding would have imputed to Congress and the Executive a too-casual approach to "the sensitive area of rights specifically guaranteed by the Constitution."[172] But it is with respect to authorization for the detention outside of these measures that the decision might have gone the other way.

---

167. *Ibid.*
168. *Id.* at 219.
169. See cases cited in opinion, at 220.
170. 65 Sup. Ct. 208, 220 (1944).
171. See statutes governing *habeas corpus,* REV. STAT. §§ 755-761 (1875), 28 U. S. C. §§ 455-461 (1940); Storti v. Massachusetts, 183 U. S. 138, 143 (1901); *Ex parte* Baez, 177 U. S. 378, 388-389 (1900); Holiday v. Johnston, 313 U. S. 342, 352-353 (1940) with respect to the intended expedition of this remedy.
172. 65 Sup. Ct. 208, 217 (1944).

As to Congress, however, the Court decided that the W. R. A. appropriation acts could not be considered a ratification of the detention of loyal evacuees, despite the circumstances that the War Relocation Authority's leave program was mentioned "in the committee hearings on the 1944 Appropriation Act . . . and on the floor of the House prior to the passage of the 1944 Act" as well as in other preceding "reports to Congress and in Congressional hearings," because "no sums were earmarked for the single phase of the total program which is here involved."[173] Justice Roberts, concurring in the result on the basis that the detention was authorized but unconstitutional, maintained that "the decision now adds an element never before thought essential to congressional ratification."[174]

The Court's limiting of the doctrine of ratification in this opinion may be compared to its very extensive reliance thereon in the *Hirabayashi* opinion. There the Court strained to find expressions of policy extrinsic to the terms of the Act of March 21, 1942, which expressions that statute was regarded as ratifying, in order to hold the delegation of legislative power involved in the curfew order to the narrowest possible scope.[175] But while refraining from countenancing a relinquishment of Congressional function through a broad delegation, despite the Government's argument *inter alia* that delegations of the war power may constitutionally be broader than that of other powers,[176] the Court's extensive reliance on the doctrine of ratification allowed Congress to perform its function by acting in a supervisory rather than a policy-making role.[177] The restriction of the application of the doctrine in the *Korematsu* case seems sound from a theoretical standpoint in that it ascribes a more deliberative character to legislative action. From a practical aspect, however, Congress is thus permitted to acquiesce and assist in the Executive's execution of a program over a period of years without

173. *Id.* at 219.
174. *Id.* at 222.
175. 320 U. S. at 102-103.
176. See Brief for the United States, *supra* note 158, at 67-70, stressing the similarity between delegations of the war power and delegations of power in the field of foreign relations, which, under United States v. Curtiss-Wright Corp. [299 U. S. 304, 318, 320 (1936)] may be broader than that of other powers. As to the demand for, and effect of, broad delegations in times of crisis, *cf.* Lowenstein, *Balance Between Legislative and Executive Power* (1938) 5 U. of Chi. L. Rev. 566; Sharp, *Separation of Powers* (1935) 2 U. of Chi. L. Rev. 385; Note, *Democratic Process in Time of Emergency* (1941) 30 Geo. L. J. 63; Note, *The War Relocation Authority and Japanese-Americans: Constitutional Aspects* (1943) 11 Geo. Wash. L. Rev. 482, 491.
177. The significance of this difference in function is heightened if it is true, as has been maintained, that the Executive branch is more susceptible to pressure than the Legislative. See R. B. Pound, *Civil Rights During and After the War* (1943) 17 Tenn. L. Rev. 706.

assuming any responsibility therefor, a result which does not seem salutary unless the program, differing from that here involved, is of such a nature that its undertaking can be viewed as only a matter for Executive concern.[178]

If the Court had found that Endo's detention was authorized, it is very highly probable that it would have felt compelled to declare it unconstitutional. In addition to the considerations pointed out[179] in connection with the constitutionality of Assembly Center detention, it must be assumed that hostility towards a citizen of Japanese ancestry would be somewhat reduced by the knowledge that he had been administratively determined to be loyal, and the possibility of the Government's utilizing this fact to diminish the hostility cannot be overlooked. And the experience of the W. R. A. in administering the leave program irrefutably confirmed the truth of the inference[180] that if the centers had been operated as shelters rather than as internment centers the detainees would have voluntarily refrained from going to communities where they would have been likely to be destitute or to meet violence.[181] Further, it appeared at the oral argument that the Court would have been strongly influenced by the consideration that the detention of citizens of Japanese ancestry to prevent lawless action against them by persons of other races had continued for two and one-half years, and that the circumstances offered in justification for such detention might well continue for an indefinite additional period, even after the cessation of hostilities.[182]

Consideration of the *Endo* opinion confirms the view that the Court

---

178. The Court did not consider whether the detention was authorized by the President as Chief Executive and Commander-in-Chief, acting apart from the two Executive Orders and from Congress. Though the requirements of publicity and specificity might well be salutary with respect to delegations of the sweeping powers of the Commander-in-Chief, there is no rule that the President's delegations to Executive officials must be by Executive order; and that the President authorized the detention is strongly indicated by a letter of the President, quoted by the Court in another connection, transmitting to Congress a report on the W.R.A. [65 Sup. Ct. 208, 219 (1944); SEN. Doc. No. 96, *supra* note 147]. While it would seem that a negative answer to the question of Presidential authorization, apart from Congressional, would have been necessary in order to complete the Court's justification for non-consideration of the Constitutional issue, it may have felt encouraged not to deal with this question by the fact that the Government only attempted to support the detention as an exercise of the joint power of Congress and the President.

179. See *supra*, pp. 203-204.

180. See *supra*, p. 203.

181. The War Relocation Authority found that the evacuees on the whole were reluctant to leave the protection of the centers, even though the Authority on occasion furnished free transportation and limited cash grants to encourage them to leave. See *Hearings before the Sub-committee of the Committee on Appropriations of the House of Representatives*, 78th Cong., 1st Sess. Part 2 (1944) (Testimony of Director of War Relocation Authority), at 710.

182. The war power outlasts the period of actual hostilities. See cases *supra* note 108. Detention of this type could of course also be attempted as an exercise of a State's police power.

resolutely half-closed its eyes with respect to the facts involved in the *Korematsu* case. The summary given above of the facts respecting Assembly Center detention[183] shows that it was not related to the purpose of preventing espionage and sabotage any more than was Endo's detention, both being related only in that they were undertaken for another purpose but grew out of the exclusion, which was directly related to such prevention. In the *Endo* decision the Court quotes, in tracing the evolution of the program, two of the three passages from General De Witt's Report which make clear the purpose of the Assembly Center detention.[184] And under the *Endo* decision the purpose of the detention is obviously the determinant of whether it is authorized. Yet, while the Court states that Endo's detention cannot be sustained on the ground of power implied "as a useful or convenient step in the evacuation program" because "its [the evacuation's] lawful character was derived from the fact that it was an espionage and sabotage measure, not that there was community hostility to this group of American citizens," it nevertheless states, obviously with an eye to the *Korematsu* case, that it does not mean that no power to detain was "necessary to the successful operation of the evacuation program."[185] And Assembly Center detention was in fact, as we have seen, upheld in the *Korematsu* case as "a part of the machinery for group evacuation," without the Court's even addressing itself explicitly to the question of authority.[186]

1. *Effect of Holding on Functioning of W. R. A. Centers*

The effect of the *Endo* decision was not, and will not be, to cause a sudden exodus from the W. R. A. camps of all those persons who are cleared of any suspicion of disloyalty and who are free to go under that decision. The W. R. A. has announced that it will be a considerable period of time before all these individuals leave the Centers, which will be operated as voluntary shelters.[187] Their reluctance to leave, which has already been adverted to in connection with the probable unconstitutionality of their detention, is due in part to emotional factors and apprehension with respect to their reception in strange places, and in large

183. Pp. 200-202.
184. 65 Sup. Ct. 208, 212, 215 (1944).
185. *Id.* at 218.
186. Not only was there in the *Korematsu* case a question of Congressional intention but also a wholly unconsidered but important Constitutional question as to whether the detention involved an excessive delegation of legislative power. Since, as the Court states in the *Endo* opinion, the legislative history of the Act of March 21 does not show that detention was within the Congressional contemplation, this Constitutional question was not controlled by the *Hirabayashi* decision, in which the holding rested on the legislative history (*supra* note 31).
187. N. Y. Times, Dec. 18, 1944, p. 1, col. 2; p. 10, col. 3.

part to their difficulty in finding employment. For many, the disruption of the West Coast communities in which they lived may mean permanent unemployment.[188]

But the fact that these persons voluntarily remain in the camps does not lessen the significance of the fact that they were heretofore detained therein or that the detention has been invalidatd by the *Endo* decision. Only the greatest cynic as to the value of liberty can believe that the distinction between voluntary residence and forced confinement is unimportant to the resident or as a matter of principle and precedent. True, a residence which is suffered due to economic and other pressures may not be viewed as wholly volitional. Yet the difference is vast between such compulsions, which are experienced to varying degrees by all sections of the population, and compulsion imposed by positive Governmental action directed against persons of a particular ancestry, based in part on Governmental acceptance of the existence of racial prejudice against such persons.

2. *Status of Persons Segregated as Potentially Disloyal*

A word must be said about the present status of those citizens segregated by the W. R. A. as possibly disloyal, whose detention was not invalidated by the *Endo* decision. While the treatment of such citizens also involves a question of discrimination in that suspicious citizens of other ancestries have not been subjected to detention,[189] there is no pending litigation on this question nor is there likely to be any. The post-war problem that would be created by the fact that these citizens, whether or not actually disloyal, would have been in internment for the duration, was decreased by the recent passage of an amendment to the Nationality Act of 1940.[190] By this amendment a citizen may, subject to various restrictions, renounce his nationality while in this country, rather than only when outside the country as theretofore; this provision makes it possible to treat those citizens who wish to take advantage of it as aliens and for them to be deported or repatriated to Japan.

---

188. Particularly in the case of those who sold goods or services to other persons of Japanese ancestry, and would not be able to secure a sufficient number of patrons without a concentration of such persons because of a lack of patronage by Caucasians. And many evacuees were removed from state civil service lists. For a thorough presentation of the facts respecting evacuation, see Brief of Japanese American Citizens League, *amicus curiae,* Supreme Court No. 22, Oct. Term, 1944, p. 81.

189. See *infra* p. 229, as to the exclusion program for such citizens.

190. §§ 401(i) and 403(a), NATIONALITY ACT OF 1940, 54 STAT. 1168, 8 U. S. C. § 801, as amended by Act of July 1, 1944, PUB. L. No. 405, 78th Cong., 2d Sess. See *Hearings Before House Committee on Immigration and Naturalization on H. R. 270* (1944) 78th Cong., 2d Sess. (Bills to Expatriate Certain Nationals of the United States, etc.)

C. Return to the West Coast

In the *Endo* opinion the Court regards Endo as detained solely by the W. R. A., and the opinion does not in any way reflect on the scope of authority assumed by, or the activities of, the Commanding General of the Western Defense Command. This view seems correct from the standpoint that the Commanding General left the W. R. A. free to release anyone it desired, at least unless he proposed to return to the West Coast.[191] However, the conclusion that Endo was in the custody of both the Commanding General and the W. R. A. and that the W. R. A. exercised authority with respect to her release in part for and on behalf of the Commanding General, has some basis in the facts that the Authority functioned under a re-delegation of General De Witt's power to prohibit persons from leaving military areas and that the detention was in fact enforced by military police.[192] In any event the result of the Court's determination that she was detained exclusively by the W. R. A. was to put definitively outside of the issues of the case the prohibition imposed by the Commanding General on the return of persons of Japanese ancestry to the evacuated area.[193] This prohibition was, judging from the petition and other papers in the case, the condition which was the chief source of Endo's unwillingness to attempt to leave the camp by following the W. R. A.'s leave application procedure, since such procedure did not provide for release to proceed to a destination contrary to the military prohibition.[194]

On December 17, 1944, the day preceding the issuance of the decisions in the *Korematsu* and *Endo* cases, the Commanding General of the Western Defense Command announced the revocation, effective January 2, 1945, of the ban on the residence of all persons of Japanese ancestry, whether or not loyal, in the States of California, western Oregon,

---

191. Final Report, *supra* note 25, at 242. Since the W.R.A. did not release anyone to return in violation of the ban (*supra* p. 211), the issue of whether the Commanding General would have taken action to prevent the W.R.A. from doing so was never posed.

192. See *supra,* note 91.

193. If the Commanding General of the Western Defense Command and the military authorities had been viewed as in whole or in part Endo's custodian, the opinion might well have included a determination that the West Coast ban could not be imposed upon her. For it probably would not have been a fulfillment of the Court's requirement that her custodian release her unconditionally if the custodian had permitted her to leave but restricted her subsequent freedom of action, even though the latter restriction was embodied in a separate set of regulations from that providing for her detention. And, regardless of the identity of the custodian, if the Court had gotten to a consideration of the Constitutionality of the leave regulations, it would probably have had to consider the validity of the prohibition, since one of the purposes in requiring a leave application was to give the Authority opportunity to appraise the proposed destination and to continue the detention if the applicant proposed to go to the prohibited area.

194. See *supra,* note 150.

western Washington, and southern Arizona, from which they had been evacuated approximately three years before.[195] Instead, only those deemed as individuals to be potentially dangerous, *i.e.*, those still subject to detention under the *Endo* decision, are to be excluded. The withdrawal of the prohibition had some appearance of disconnection from the *Endo* decision because of its temporal precedence, the fact that the ban was not directly involved in the case, and the fact that it overshadowed in news-value in many newspapers the Supreme Court decisions.[196] It seems evident *inter alia*, however, from the circumstances set forth in the following discussion of the revocation of the ban that the coincidence of events was not fortuitous.

While the invalidity of the ban as applied to the thousands of citizens who had been investigated and cleared as loyal by the W. R. A. would not have followed directly from the Endo decision, it would seem that the Army would necessarily have been involved in unsuccessful litigation following the *Endo* decision if it had continued the general ban.[197]

Constitutional justification for the blanket prohibition on the return of citizens of Japanese ancestry to their homes disappeared after the elapse of sufficient time for individual investigations and for determinations that individuals were not open to suspicion of a tendency to aid the enemy.[198] Such determinations were made as early as November,

---

195. Public Proclamation No. 21, issued Dec. 17, 1944, 10 FED. REG. 53 (1944); N. Y. Times, Dec. 18, 1944, p. 1, col. 2.

196. See for example San Francisco Chronicle, Dec. 19, 1944, p. 1; Baltimore Evening Sun, Dec. 18, 1944, p. 3.

197. While the *Endo* holding as to the lack of authority to restrain a concededly loyal citizen would apply to a prohibition of movement as well as to detention, the loyalty of citizens cleared by the W.R.A. would not necessarily have to be conceded by the military authorities who were imposing the ban. However, it would have been impossible for the Army to establish to a Court's satisfaction in the great majority, if not all of the cases, that there were grounds for refusing to concede the loyalty of the citizens cleared by the W.R.A. Accordingly, at least under the authorities preceding the *Korematsu* opinion, the ban would have been held invalid as to such citizens. (The cases would have presented the same issue as the individual exclusion cases, discussed *infra* p. 232.) Thus, in view of the great number who had been found by the W.R.A. to be loyal, the general ban on all persons of Japanese ancestry could not well be continued. The procedure which the Army adopted in this dilemma was announced together with the revocation of the ban; the Army apparently was not ready to concede completely the correctness of the W.R.A. determinations, and stated that officers would, by the effective date of January 2, tour the W.R.A. camps, examine the W.R.A. records, and compile an exclusion list. N. Y. Times, Dec. 18, 1944, p. 1, col. 3 and p. 10, col. 3.

See comment on the ban in Justice Murphy's concurring opinion, 65 Sup. Ct. 208, 221 (1944).

198. After such determinations the only circumstance that could have been offered as a justification for barring them from this area would have been the hostility against them in the area; this circumstance would not render the prohibition Constitutional for the reasons discussed *supra* pp. 203-204 in connection with the protective custody issue. It is to be noted that exclusion on the ground of such community hostility would also be invalid under the *Endo* decision.

1942.[199] And the lack of justification became more dramatic with public announcements by the military authorities in January, 1944 indicating that the danger of invasion of the West Coast had wholly or largely abated.[200] Nevertheless, the ban continued for another year. Meanwhile, attention was necessarily focused on the problem by the litigation commenced in August, 1944 by citizens attempting to return to the coast[201] and by the appeals of some West Coast elements to bar forever the return of persons of Japanese descent.[202] These appeals were in many cases based on specious economic reasoning, rather than on grounds of military necessity;[203] and the continuance of the ban afforded further time for attempts to arouse such sentiment.

At the argument of the *Endo* case on October 10th and 11th, 1944, the questioning and comments of several of the Justices clearly demonstrated an adverse reaction to the Government's position.[204] The interim between the argument and the decision, which was surprisingly long considering that the case was *habeas corpus,* allowed time for official consideration of the method and effect of revocation of the prohibition.[205]

When or how the prohibition would have been revoked if the *Endo* litigation had not occurred cannot be predicted. The military authorities did not point to any reason—which can receive consideration in consistence with the principle of racial equality—for their change of view as to the necessity for mass exclusion except that there had now been time for individual investigations;[206] as pointed out above, such investiga-

199. See SECOND QUARTERLY REPORT OF WAR RELOCATION AUTHORITY (July 1 to Sept. 30, 1942) 15, 18.

200. The military authorities indicated in their public statements that attack and invasion were no longer to be viewed as real dangers. See joint statement by the War and Navy Departments, N. Y. Times, Jan. 14, 1944, p. 7, col. 5. See similar statement by Canadian authorities, N. Y. Times, Jan. 16, 1944, § 1, p. 16, col. 1.

201. San Francisco Chronicle, Aug. 23, 1944.

202. MCWILLIAMS, PREJUDICE (1944) c. 7. And see p. 251, quoting General De Witt's notorious public "a Jap's a Jap" statement, which gave encouragement to this campaign.

203. *Id.* at 233-234.

204. Several Justices seemed particularly disturbed by the lack of an indication in the Government's argument of any definite limitation on the Government's general power to remedy conditions arising from the prosecution of the war and to continue the detention. (See argument in Brief for United States, Supreme Court, Oct. Term 1944, No. 70, pp. 80-82.)

205. The general knowledge that the revocation was intimately connected with the cases seems indicated by the statement in one news report on the revocation that "The Army's action, incidentally, anticipated any decision by the Supreme Court on pending appeals over the exclusion issue." N. Y. Times, Dec. 18, 1944, p. 10, col. 3. Consider also account of various investigations, inquiries, and statements regarding the ban, indicating that the Army was considering its revocation, in the period between the argument of the case and Dec. 17, Los Angeles Times, Dec. 18, 1944, p. 8.

206. See authorities cited *supra* note 195. While the military authorities also

tions were being completed more than two years earlier. Nevertheless, whether or not belated, the revocation was applauded by leaders of organizations of American citizens of Japanese ancestry,[207] and must be universally applauded, as a signal victory in the recognition of their rights as citizens and in the maintenance of democratic freedom.

It was suggested above[208] that the problem of the return to the West Coast might have influenced the Court's approach to the question of review in the *Korematsu* opinion and might offer some justification for it. If closer scrutiny had been given the factual basis for the actions at issue in that case, it would have been very difficult for the Court to avoid consideration of General De Witt's highly vulnerable report. It may have been hoped that the Court's non-critical acceptance of the military judgment would aid in securing the acceptance by the West Coast of the judgment of the military authorities that the return of the loyal had become feasible. The degree to which confidence in the military judgment could affect the situation was and is, however, limited by the fact that the campaigners for "keeping the Japs out of California" made little attempt to ground their position in terms of military necessity.[209]

Due to their economic dislocation occurring by reason of the evacuation itself and the approximately three-year interim since then, and due also in part apparently to fear of community hostility, relatively few of the loyal citizens have in fact returned.[210] With exhortations by Army and local officials, accompanying the revocation of the ban, for the maintenance of law and order, there was an initial mobilization and display of West Coast sentiment in favor of fairness and tolerance.[211] Subsequently, however, a campaign of opposition to the return of the evacuees based openly, because of the impossibility of a plea of military necessity, on economic motives, was launched.[211a] Despite the Federal Government's and the Army's implication in the situation, the problems of public sentiment and of the suppression of violence have been left

---

referred to the improved military situation, military danger on the Coast would not justify exclusion on a mass basis after the lapse of sufficient time for individual investigations except on the "a Jap's a Jap" theory.

207. N. Y. Times, Dec. 18, 1944, p. 10, col. 3.

208. See *supra* p. 208.

209. McWILLIAMS, *loc. cit. supra* note 202.

210. N. Y. Times, Jan. 4, 1945, p. 8, col. 4; N. Y. Times, March 18, 1945, § 1, p. 17, col. 1.

211. See authorities cited *supra* note 195; San Francisco Chronicle, Dec. 18, 1944, p. 6; Los Angeles Times, Dec. 18, 1944, p. 1; Arizona Daily Star, Dec. 18, 1944, p. 1; N. Y. Times, Dec. 31, 1944, § 1, p. 15, col. 3; N. Y. Times, Jan. 21, 1945, § 1, p. 4, col. 3.

211a. See, *e.g.* P. M., Feb. 25, 1945, § 1, p. 11, col. 1, reproducing advertisement in Hood River (Ore.) Sun for Feb. 2, 1945, which stated that good farm land is limited and should be restricted to white Americans.

to local officials; and probably as a result of the Federal withdrawal from the situation, accompanied by insufficient diligence on the part of local sheriffs, some attempts at violence have occurred.[212] The situation, however, is not extreme, and latest reports indicate that at least in the State of California, the sheriffs will in the future perform their duties with greater vigilance.[212a]

III. VALUE OF JUDICIAL REVIEW OF MILITARY JUDGMENTS RESPECTING CONTROL OF CIVILIANS

A. Feasibility of Review

It has already been indicated that, in the writer's opinion, judgments by military authorities respecting the control of civilians should be reviewed on the same basis as judgments by civilian authorities, substantially in accordance with the principle expressed but not completely followed in the *Hirabayashi* case. In view of the disregard of this principle in the *Korematsu* opinion, it is especially important to demonstrate the feasibility and value of its application. Justice Jackson's dissenting opinion in the *Korematsu* case is an appropriate springboard for such a demonstration, for it points to most of the objections raised to review of the type here advocated.

Justice Jackson's thesis is, in summary, that a court is unable to determine whether a military judgment rests on reasonable military grounds and that even if it could and did determine that the judgment rested on such grounds, it should not therefore hold the military action Constitutional. It should instead determine Constitutionality irrespective of the alleged military grounds. Accordingly, while the Court should not have interfered with the Army's action, it should not approve and assist in the execution of the exclusion order by upholding Korematsu's conviction since a peace-time order of this type would be deemed unconstitutional.[213]

Justice Jackson fully recognizes that his doctrine that the Court should not attempt to determine whether or not military acts are supported by reasonable military necessity seems dangerous. For "the

---

212. N. Y. Times, March 18, 1945, § 1, p. 17, col. 1.

212[a]. *Ibid.* (report of address to California sheriffs by Robert Kenny, Attorney-General of California, warning them that they would be replaced unless they upheld the law).

213. It would be unconstitutional because "if any fundamental assumption underlies our system, it is that guilt is personal and not inheritable. . . . But here is an attempt to make an otherwise innocent act a crime merely because this prisoner is the son of parents as to whom he had no choice, and belongs to a race from which there is no way to resign." 65 Sup. Ct. 193, 206 (1944).

existence of a military power resting on force, so vagrant, so centralized, so necessarily heedless of the individual, is an inherent threat to liberty."[214] But he "would not lead people to rely on this Court for a review that seems . . . wholly delusive." Their protection is only in the sense of responsibility and self-restraint of the military authorities for "if the people ever let command of the war power fall into irresponsible and unscrupulous hands, the courts wield no power equal to its restraint."[215]

Justice Jackson's view that the Court in the *Korematsu* opinion "distort[s] the Constitution to approve all that the military may deem expedient . . . , whether consciously or not" and that it "may as well say that any military order will be constitutional and have done with it"[216] seems a sound explanation of the effect of the majority's doctrine of review. And it seems equally sound to conclude that if this type of review is all that can be accorded, protection lies only in the self-restraint of the military. But Justice Jackson's ultimate conclusion as to the proper scope of review, with the "threat to liberty" which his conclusion entails, does not seem valid. For the Justice's implication that the majority tried, but failed, to determine whether the order rested on reasonable grounds does not seem a true reflection of the opinion. Rather, it seems apparent in the Court's opinion that it did not attempt to make such a determination,[217] and it is the purpose of this section to show that it could and should have done so.

As to the Court's capability of making a determination of reasonableness, Justice Jackson states that the Court had

> "no choice but to accept General De Witt's own unsworn, self-serving statement, untested by any cross-examination that what he did was reasonable. . . . In the very nature of things military decisions are not susceptible of intelligent judicial appraisal. They do not pretend to rest on evidence, but are made on information that often would not be admissible and on assumptions that could not be proved. Information in support of an order could not be disclosed to courts without danger that it would reach the enemy. Neither can courts act on communications made in confidence."[218]

Bearing in mind constantly that we are here concerned only with one distinct category of military decisions (those respecting civilian controls), the difficulties suggested by Justice Jackson with respect to

214. As to the especial danger which Justice Jackson believes results in the *Korematsu* case from review of the military grounds for the order, and which outweighs this danger, see *infra* p. 237.
215. 65 Sup. Ct. 193, 208 (1944).
216. *Id.* at 207.
217. See *supra* pp. 193-195.
218. 65 Sup. Ct. 193, 207 (1944).

review of military decisions do not appear substantial. While he emphasizes as a hindrance to review the fact that such decisions do not rest on evidence, military decisions of the type here under consideration need not be regarded as in a special class from the standpoint of the feasibility of their basis on evidence. The customary rules as to the necessity for notice and hearing and as to review on the basis of evidence of record or on other bases, can be applied to such decisions and have been so applied without difficulty.

Orders such as the exclusion order, being of general rather than individual applicability, and thus of a so-called quasi-legislative nature, are not in general Constitutionally required to be made on the basis of notice and hearing or evidence of record.[219] And as in the case of any other quasi-legislative measure, the facts necessary to a determination of the constitutionality of a quasi-legislative measure adopted by the military authorities with respect to civilians may be reviewed by the Court on the basis of judicial notice or, where that method is the more appropriate, on the basis of evidence adduced by trial.[220] The former method, as has already been observed, was utilized in the *Hirabayashi* case;[221] as to the latter, in the recent cases involving the validity of martial law in Hawaii, both Admiral Nimitz and General Richardson testified as to the danger of invasion of Hawaii, which was the purported justification for the continuance of martial law.[222] Similarly, in the individual civilian exclusion cases Army officers testified as to the general military situation and as to the special dangers of espionage and sabotage in the areas from which the subjects were excluded.[223] The conduct of the trials did not differ from that of any other civil case, and the officers were of course subject to cross-examination. And in the *Korematsu* case, if, for

---

219. United States v. Bush, 310 U. S. 371 (1940); *see* Norwegian Nitrogen Products Co. v. United States, 288 U. S. 294, 303, 307, 308 (1933); WILLOUGHBY, THE CONSTITUTION (2d ed. 1929) 653-654; Davis, *The Requirement of Opportunity to be Heard in the Administrative Process* (1942) 51 YALE L. J. 1093. It may be noted, however, that just as civilian authorities may as a matter of good practice hold a public hearing prior to the promulgation of such an order when the situation allows of the attendant delay in its issuance, the similar feasibility of a hearing with respect to non-combat military action is demonstrated by the hearings held by a Congressional investigating committee prior to the exclusion.

220. See notes 38 and 39 *supra* as to the use of these two methods of presenting constitutional facts.

221. See *supra* p. 185.

222. In the Matter of the Application of Lloyd C. Duncan for a Writ of Habeas Corpus, and In the Matter of the Application of Harry E. White for a Writ of Habeas Corpus (Nos. 10763 and 10774, combined for argument and decision) 146 F.(2d) 576 (C. C. A. 9th, 1944), *cert. granted,* Feb. 12, 1945.

223. Alexander v. De Witt, 141 F.(2d) 573 (C. C. A. 9th, 1944); Scherzberg v. Madeira, 57 F. Supp. 42 (E. D. Pa. 1944); Ebel v. Drum, 52 F. Supp. 189 (D. Mass. 1943); Schueller v. Drum, 51 F. Supp. 383 (E. D. Pa. 1943); Wilcox v. De Witt, decided Sept. 4, 1943 (S. D. Cal.).

example, the Court had sought to consider factual questions bearing on the segregation of loyal and disloyal such as the amount of information possessed by the intelligence agencies on the population of Japanese ancestry and the length of time reasonably required to hold hearings, and had thought it could not secure sufficient enlightenment by judicial notice of documentary sources, there was no reason why a trial could not have been directed on such issues.[224] Finally, if the decision is of a quasi-judicial nature so that notice and hearing would ordinarily be required, the feasibility of imposing this requirement with respect to a military decision is demonstrated by the individual civilian exclusion litigation described below.

The misleading qualities of the lump-concepts of "military decisions," "military orders," or "military judgments" should likewise be avoided in considering the other point mentioned by Justice Jackson with respect to the feasibility of review: the possibility that the submission to the Court of those facts necessary for judicial determination of the reasonableness of the orders would result in disclosures helpful to the enemy. The use of these phrases should not cause a mental assimilation of the decisions here in question, respecting control of civilians, to those respecting the conduct of an armed encounter. In the case of all the new wartime military controls over civilian liberties—the Japanese-ancestry program, the individual exclusion program, and martial law in Hawaii—the reasonableness of the controls depended upon general sociological facts,[225] facts respecting an individual's activities and associations,[226] or facts dealing with the progress of the war but in no case involving technical or detailed combat considerations.[227] The only instance in which the military authorities maintained that there was even a problem of possible disclosures of aid to the enemy was in the individual civilian ex-

224. It may be noted that the Government suggested in the *Endo* case the advisability of a remand for the taking of evidence as to the validity of the continued exclusion from the West Coast, if the Court were to determine that this question was in issue. Brief for the United States, *supra* note 204, pp. 60-61.

225. See *Hirabayashi* decision and Hawaiian cases cited *supra* note 222.

226. See civilian exclusion cases cited *supra* note 223.

227. For example, the testimony of General Richardson and Admiral Nimitz in the cases cited *supra* note 222 related to the importance of Hawaii in the defense of the United States, the atmospheric conditions surrounding Hawaii which would facilitate the surprise approach of enemy aircraft, the activity by Japan in the area since Pearl Harbor, the advantage to Japan of hit-and-run raids or the landing of espionage agents, the general probability of such efforts, and the general potentiality of defense against them. See Record in cases cited, pp. 1006-1038, 1072-1078.

The opponent of the military authorities may be at a greater disadvantage than other litigants against the Government because of the difficulty of securing opposing expert witnesses and of obtaining information from these authorities. However, these difficulties are no argument against judicial review, though they may furnish reason for placing the burden of proving reasonableness on the authorities, rather than of proving unreasonableness upon their opponents.

clusion cases, where a question arose with respect to the disclosure of the sources of information about the excludees. However, this problem, described a little more fully below, never became an issue in any of the litigation, and would in any event only force a modification rather than an abandonment of review.[228]

It is to be emphasized that the thesis here expressed—that there is no basis for differentiation in the review procedure between orders by military authorities respecting civilians and those by other authorities, does not mean that the military authorities are to be held to an unduly rigid standard of conduct. Whether the official be military or civilian, it may be the very uncertainties of the situation which make his order reasonable; and the reasonableness of the restriction on the individual, regardless of the imposing authority, of course depends upon the degree of danger the restriction is intended to prevent.[229]

Military control over civilians is now in issue before the Supreme Court in pending cases concerning the validity of two phases of martial law in Hawaii. The cases test the validity of trial and sentence of civilians by military courts, one in August, 1942 for the offense of embezzlement under the Territorial laws of Hawaii,[230] and the other in March, 1944 for an offense created by the Commanding General of assault on a military sentry.[231] It is believed that the declaration of martial law, which preceded these measures of military control, should be treated as any other finding by an Executive as to the degree and type of danger on which his action is grounded,[232] and furnishes no basis for an immunity from, at the least,[233] a review of the reasonableness of these con-

228. See *infra* p. 231.

229. See discussion of standard, *supra* pp. 187-188.

230. The *White* case, *supra* note 222.

231. The *Duncan* case, *supra* note 222.

232. The executive finding is entitled to weight but must be supported by the facts to be valid. See treatment of findings in *Hirabayashi* decision, discussed *supra* note 60. *Cf.* Block v. Hirsh, 256 U. S. 135, 154 (1921); Chastleton Corp. v. Sinclair, 264 U. S. 543, 547 (1924). In the instant cases, the declaration of martial law should not perhaps be accorded any weight as an indication of the executive judgment, since the Governor of Hawaii, who issued the declaration, testified that he did not consider the continuance of military trials necessary but allowed them to continue as a compromise with the military. See decision issued April 13, 1944, United States District Court for Territory of Hawaii, In the Matter of the Application of Lloyd C. Duncan for a Writ of Habeas Corpus, April Term 1944, *Habeas Corpus* No. 298, and Record therein, pp. 825-889.

It is to be noted that the question of whether a given act by the military authorities should be viewed as an exercise of martial law is in many cases a close one. See Corwin, *op. cit. supra* note, at 20-21, suggesting, prior to the *Hirabayashi* decision, that the whole Japanese-ancestry program should be appraised as an exercise of martial law.

233. In *Ex parte* Milligan, 4 Wall. 2 (U. S. 1866), the Court went further. It established a rigid rule, rather than the general rule of reasonableness, that criminal trials of civilians by military courts are unconstitutional except when the civil courts are unable to function. See Frank, *supra* note 14, at 639.

trols. Again, as with all controls imposed by military authorities, measures taken under a declaration of martial law may be categorized on the basis of whether they are in the nature of combat operations by or against an armed force; the test formulated in *Sterling v. Constantin* of "good faith and honest judgment," which is a somewhat looser standard than that of reasonableness, was directed at, and is appropriate to a measure of that nature.[234] In the pending cases, however, the question is merely whether the military situation in Hawaii, with respect to which the military authorities testified, justified a change in judicial procedure from civil to military. Whether or not it did so may be estimated by weighing the validity of the supporting argument in the Circuit Court opinion which upheld the military trials.[235]

B. The Value of Review of The Reasonableness of Military Orders

1. *Individual Civilian Exclusion Program*

The individual civilian exclusion program furnishes an example of the benefit of judicial review of the reasonableness of military orders to the general polity of the country, as well as the benefit of the military authorities' being held to customary procedural requirements. These orders, which were issued under the authority of Executive Order 9066 and the Act of March 21st, were directed at the prevention of espionage and sabotage and applied to citizens who had shown some interest or sympathy for the enemy countries by their pre-war activities. The order in each case directed that the subject remove himself from his place of residence in the exclusion area and remain outside the various exclusion areas, which consisted of the East and West coastal states plus Vermont, Pennsylvania, Arizona, and parts of other states.[236]

234. 287 U. S. 378 (1932). The Court was there considering the use of the militia to maintain order in insurrectionary or similar situations.

235. Cited *supra* note 222. The only relevance of military trials to the defense of Hawaii against attack, pointed to by the Court, were that (1) "citizens of Japanese extraction could not lawfully be excluded from jury panels on the score of race," an intimation that they should be treated as disloyal regardless of the fact that those suspected of disloyalty were interned (*supra* p. 196); (2) "Indeed the mere assembling of juries and the carrying on of protracted criminal trials might well constitute an invitation to disorder as well as an interference with the vital business of the moment"; (3) "And the summary punishment of criminal offenders of every sort might conceivably serve to discourage the commission of offenses immediately endangering the general security." This reasoning must be evaluated in the light of the facts that though Hawaii "continued in imminent danger of attack from the air, of submarine forays and commando raids from the sea" (Circuit Court's opinion), there was no threat that any attack would succeed to a degree at all comparable to that at Pearl Harbor, nor was there any threat at all of a land invasion (Testimony of Adm. Nimitz and Gen. Richardson, cited *supra* note 227).

236. See cases cited *supra* note 223. The East Coast exclusion area was narrowed in Nov. 1943. Ebel v. Drum, 52 F. Supp. 189 (D. Mass. 1943).

a. Procedure

The standard procedure used in the issuance of the orders was that the subject of the order

> "was served with a notice which in substance advised him that a Board of Officers had been appointed by the Commanding General [of the area in which he resided] . . . to consider the question whether military necessity required the issuance of an order excluding him from certain Military Areas. . . . This document further stated that said Board would be convened at a certain place . . . [at a specified time] . . . and that at his option he might appear before the Board for the purpose of being informed of the general nature and scope of the inquiry and afforded an opportunity to present evidence in his own behalf, also to answer questions or make a statement under oath or affirmation; likewise that material in the hands of the Board would not be made available for his inspection, that all matters pertaining to the inquiry were confidential, and no publicity would be given them by the Board. In addition this notice advised him that he might be accompanied by counsel who would be allowed to act only as his personal advisor, but would not be heard by the Board nor permitted to examine witnesses. . . ."[237]
>
> "No opportunity for cross-examination of witnesses was afforded him, nor was he informed of the nature and cause of the accusations against him. . . . The subject matters discussed at the hearing were the facts and circumstances reflected in the various confidential reports made by the investigative agencies acting for the military authorities, among which were the Military Intelligence Division, Federal Bureau of Investigation, Naval Intelligence, and Immigration and Naturalization authorities. It was upon these reports that the military authorities . . . relied in reaching the conclusion that military necessity demanded the exclusion of the [subject] . . . on the ground he was found to be 'dangerous to the national defense.' "[238]

Thus the facts upon which the order was based were not disclosed to the subject, and the basis for the ultimate conclusion to exclude him was only specified to the extent that he was stated to be "dangerous to the national defense." The facts upon which the order was predicated were only made known to the subject if and when he litigated its validity. At the trial of the civil actions for declaratory judgments or injunctions in these cases

> "the defendants introduced in evidence the original record in the exclusion proceedings . . . from which the names of confidential informants, the discussion of the source of their information, and other reports revealing information and investigation technique, such as reports of the Federal Bureau of Investigation, were deleted."[239]

---

237. Wilcox v. De Witt, decided Sept. 4, 1943 (S. D. Cal.).
238. Ebel v. Drum, 52 F. Supp. 189, 192-193 (D. Mass. 1943).
239. *Id.* at 192.

The Army's withholding of the information, subsequently revealed in the course of litigation, from the subject at the time of the Army Board hearing seems to have been an arbitrary and unjustified hindrance to the subject's ability to prepare his defense; and the inaccuracies in the Army's information which were likely to arise as a result of the subject's curtailed opportunity for rebuttal were hardly remedied by the review by Army officers of the case against the subject in "twenty-four separate steps."[240] The Army's failure to make an attempt to bring its procedure into conformity with the usual requirements of a fair hearing does not seem to have been due to a conviction of military urgency, but merely to its theory that a fair hearing was not a Constitutional prerequisite in the issuance of this type of order; and its subsequent disclosure of the information on which the order was based, excepting disclosures damaging to the internal security, was clearly due to its desire to have the litigated orders upheld.[241] The question of what type of hearing, if any, was Constitutionally required was not considered in any of the decisions. However, the practical importance of holding the military authorities to the requirements of due process seems to the writer amply demonstrated by the fact that it was mainly in order to avert a judicial decision that the order violated due process that the military authorities ultimately revealed to the subject information which in ordinary fairness he should have known and which could be revealed in complete consistency with security considerations. And since the requirements of a fair hearing in all events vary with the necessities of the case,[242] it would seem that the military authorities could[243] and should be held to such requirements if the question were judicially considered.

240. *Ibid.* More than fifteen officers engaged in a review of the file. Schueller v. Drum, 51 F. Supp. 383, 385 (E. D. Pa. 1943).

241. See Defendants' Memorandum of Law, Scherzberg v. Madeira, 57 F. Supp. 42 (E. D. Pa. 1944), pp. 41-43.

242. *See* Morgan v. United States, 304 U. S. 1, 19 (1937); Morgan v. United States, 298 U. S. 468, 481 (1936); Hurwitz v. North, 271 U. S. 40, 42 (1926); Tisi v. Tod, 264 U. S. 131, 134 (1924).

243. The question of whether the defects in the Army's procedure could be corrected by a hearing at the trial, after the issuance of the order when its validity was litigated, would depend on the established criterion of whether there was such a temporal emergency as to justify the elimination of a hearing before the issuance of the order. *See* Bourjois v. Chapman, 301 U. S. 183, 189 (1937); Phillips v. Commissioner, 283 U. S. 589, 597 (1930); Bragg v. Weaver, 251 U. S. 57, 59 (1919); North American Cold Storage Co. v. Chicago, 211 U. S. 306 (1908) with respect to the validity of an after-hearing. The only extraordinary procedural factor was that some of the sources of information about the subjects of the orders had to be kept confidential so that their further utility would not be destroyed; and it is clear that under otherwise appropriate circumstances, the fact that the subject's right to rebuttal was curtailed by such non-disclosure should not render the order invalid. But Court approval of the non-disclosure of confidential sources, without abandonment of due process and judicial review, is not unheard of. Low Wah Suey v. Backus, 225 U. S. 460 (1912); Kwock Jan Fat v. White, 253 U. S.

b. *Substantive Validity*

Despite Justice Jackson's rather defeatist view on the possibility of judicial restraint of military orders, in three out of the five civil actions which tested the validity of individual civilian exclusion orders and went to a decision on the merits,[244] District Court decisions were rendered against the military authorities. In one a declaratory judgment was rendered and in two injunctions were granted.[245] In these three cases the courts employed criteria indicated in the *Hirabayashi* opinion and based on the language in previous cases such as *Mitchell v. Harmony*[246] and *Sterling v. Constantin*.[247] They held the orders invalid because they were not necessary to prevent a "danger to the Government [which] is real, impending, and imminent"[248] and because there was not "present a reasonable and substantial basis for the judgment the military authorities made, i.e., that the threat of espionage and sabotage to our military resources was real and imminent."[249] The basis for these conclusions was in essence that in view of the fact that no danger of invasion was shown or even alleged, and in view of the fact that all the normal agencies for security and protection were able to function, the public safety did not require the employment of this extraordinary measure—the uprooting of citizens guilty of no wrong-doing—in the case of such persons as the subjects of the orders. In one of the two cases in which an order was upheld, the court stated that it could not see reason for believing the subject's "presence, in the present state of the war with Germany, or with

---

454, 459 (1920). *But see* Chew Hoy Quong v. White, 249 Fed. 869 (C. C. A. 9th, 1918). Compare *In re* Quarles and Butler, 158 U. S. 532 (1892); Vogel v. Gruaz, 110 U. S. 311 (1884); Segurola v. United States, 16 F.(2d) 563 (C. C. A. 1st, 1926). And the need for non-disclosure merely calls for application of the doctrine illustrated by the cases cited *supra* note 242 that the hearing must give as adequate opportunity for defense as is possible consistently with other demands of public policy.

For a somewhat contrary view to that here expressed *see* Liversidge v. Anderson, [1942] A. C. 206, 221, where Viscount Maughan, speaking for the House of Lords, gives as one ground for the ruling that the reasons for an individual's detention are not subject to judicial review the fact that a claim of privilege on the grounds of the public interest could be maintained respecting non-disclosure of the reasons for the internment. *Cf.* Martin v. Mott, 12 Wheat. 19, 30 (U. S. 1827).

244. A number of suits were instituted which did not go to final decision because the orders were suspended or withdrawn prior to trial. Alexander v. De Witt, 141 F.(2d) 573 (C. C. A. 9th, 1944), involved only a question of equity pleading.

245. Schueller v. Drum, 51 F. Supp. 383 (E. D. Pa. 1943) (declaratory judgment); Ebel v. Drum, 52 F. Supp. 189 (D. Mass. 1943) and Scherzberg v. Madeira, 57 F. Supp. 42 (E. D. Pa. 1944) (injunctions).

246. 13 How. 115 (U. S. 1851).

247. 287 U. S. 378 (1932).

248. Schueller v. Drum, 51 F. Supp. 383, 387 (E. D. Pa. 1943), *substantially followed in* Scherzberg v. Madeira, 57 F. Supp. 42 (E. D. Pa. 1944).

249. Ebel v. Drum, 52 F. Supp. 149, 197 (D. Mass. 1943).

Japan, either, [as] being a danger to the military security of the country or of this Coast"; however, due to a misconstruction of the *Falbo* decision,[250] the judge held that he should not "interfere with the military judgment."[251] It is important to observe that the same result would have been likely in all of these cases if they had succeeded and followed the *Korematsu* opinion.

Since the military authorities have not made public such statistical or descriptive information on the program as a whole as would be necessary for an evaluation, no attempt will be made herein to consider all the possible criticisms that could be expressed as to the general utility of this individual civilian exclusion program. In general, however, it may be observed that the individual exclusion technique, which apparently was inspired by the West Coast exclusion[252] and by British precedent,[253] would seem to be a totally inadequate weapon against citizens presenting any real danger and at the same time an encouragement, because of its mildness as compared to imprisonment, to the casual undertaking of a substantial interference with the liberty of citizens who presented no such threat as to justify it. But it is unnecessary to debate the merits of this position, for the history of the program seems to indicate the realization by the military authorities themselves of the lack of utility of the program, and, in addition, the significance of judicial review in encouraging a reconsideration of this question by them. For the military authorities have appealed none of the adverse

---

250. 320 U. S. 549 (1943).

251. Labedz v. Kramer, 55 F. Supp. 25 (D. Ore. 1944). The facts with respect to the potential dangerousness of the excludee in the other case in which the order was upheld [Wilcox v. De Witt, decided Sept. 4, 1943 (S. D. Cal.)] seem the weakest of any of the litigated cases. But there the Court relied on the concurring opinion of Justice Douglas in Hirabayashi v. United States, 320 U. S. 81, 106 (1942), in which he stated that it had been justifiable to treat West Coast citizens of Japanese ancestry on a mass rather than an individual basis and that therefore "the orders as applied to [each individual] are not to be tested by the substantial evidence rule." It was of course erroneous to apply this language in the *Wilcox* case since the exclusion order therein was an individual order.

The Wilcox exclusion order was revoked after the filing of a notice of appeal.

252. The purpose of the drafting of Executive Order 9066, under which these orders were promulgated, was to secure authorization for the Japanese-ancestry exclusion (see *supra* p. 207) and the Act of March 21 was requested of Congress by the War Department for the same purpose. [Hirabayashi v. United States, 320 U. S. 86, 89 (1942).] It may be noted, however, that these individual exclusion orders can probably be deemed within the intended scope of the Act since the Senate Committee reporting the Bill referred to the possible creation of military areas other than those on the Pacific Coast and a similar reference was made on the floor of the Senate [SEN. REP. No. 1711, 77th Cong., 2d Sess. (1942) 2; 88 CONG. REC. 2724, 2726 (1942)].

253. See *supra* note 15. Such individual exclusion orders might have been justifiable in Britain due to a threat of invasion, the absence of which in this country at the time of issuance of the litigated orders was a considerable factor in the decisions holding them unconstitutional (*supra* p. 232).

District Court decisions, and the issuance of exclusion orders, excepting in the case of persons of Japanese ancestry, has stopped. Yet this inaction could not be said to be attributable to an improving military situation; danger of invasion was at no time emphasized as the justification for the orders, and in fact one of the military experts in the trial of the most recent case, conceding that such danger could not be relied upon to justify the order, maintained that the more the United States took the offensive, the greater the danger of espionage and sabotage.[254] And since the military could hardly have concluded on the basis of these three District Court decisions that it was hopeless to carry the litigation further, it seems that the military authorities must have engaged in a reconsideration of the necessity for these orders and for the program as a whole. While it is not maintained that this program would have been continued indefinitely in the absence of judicial review, the re-scrutiny must be attributed at least in part to the fact that the orders were being subjected to judicial scrutiny and might eventually have been subject to condemnation by the higher courts.

2. *Other Litigation*

That the existence of judicial review may of itself, apart from final adjudications of invalidity, have the effect of encouraging a desirable self-review by military authorities of their orders is indicated by the history of martial law in Hawaii. As noted above, one of the important aspects of martial law was the suspension on December 7th, 1941 of the privilege of the writ of *habeas corpus,* which *inter alia* enabled the military authorities to arrest and imprison American citizens who were suspected of sympathy for the enemy.[255] Their imprisonment was of indefinite duration, with no provision for their release after any stated interval or on any stated occasion. In all three cases in which petitions for writs of *habeas corpus* were filed, the cases were mooted by the release of the petitioning prisoners, after periods of incarceration lasting from ten to twenty-one months.[256] While it is not of course maintained

254. Testimony of Colonels McDonnell and Forney, in *Scherzberg v. Madeira,* typed transcript of record, pp. 111, 1942, quoted in Defendants' Memorandum of Law, *supra* note 241, pp. 14, 16.

255. See authorities cited *supra* note 14.

256. A petition for a writ of *habeas corpus* was filed on Feb. 19, 1942 in the case of Zimmerman, a civilian citizen who had been interned by order of the Commanding General of the Hawaiian Department. The District Court stated that the writ should issue but that it must be denied because the court was under duress by virtue of the Commanding General's prohibition of the issuance of such a writ. The Circuit Court affirmed on other grounds [Zimmerman v. Walker, 132 F.(2d) 442 (C. C. A. 9th, 1942)]. While the petition for certiorari was pending,

that these prisoners would never have been released but for the courts, nevertheless the influence of the institution of judicial review is clear from the fact that in three cases, which were the only ones of this type brought,[257] prisoners incarcerated at varying times were released at varying times after varying periods of incarceration. The course of these three cases indicates that one of the great benefits of *habeas corpus* may continue to be the prevention of a prisoner's remaining in jail not so much because of malice as because his "sufferings are unknown or forgotten."[258]

The most important instance of official self-correction under the stimulus of judicial action is that of the revocation of the ban on residence of loyal citizens of Japanese ancestry on the West Coast; as pointed out *supra*, pp. 221-222, the continuance of this prohibition had been unjustified for an extended period of time but it was lifted only under the impetus of the *Endo* case and the prospect of consequent unsuccessful litigation with respect to it.

C. Conclusions as to the Need for Review of the Reasonableness of Military Controls Over Civilians

The stress has been laid in the foregoing discussion on the influence of judicial review in stimulating military self-correction because instances of such self-correction, in place of an argument as to the merits

---

the case became moot by Zimmerman's release, in the spring of 1943 on the Mainland [*cert. denied*, Zimmerman v. Walker, 319 U. S. 744 (1943)].

One Glockner and one Seifert were similarly interned in December, 1941 and December, 1942 respectively. They were released on the Mainland in October, 1943 (Orders issued by "Office of Military Governor, Territory of Hawaii," dated Oct. 14, 1943), after a district court decision granting writs in both cases (In the Matter of Applications of Glockner and Seifert for Writs of Habeas Corpus, decided by the United States District Court for the Territory of Hawaii, Aug. 24, 1943).

257. Without any intimation as to whether the military authorities have any desire similarly to moot the *Duncan* and *White* cases now pending in the Supreme Court (*supra* note 222), it is to be noted that the release of these men is an entirely different problem from that posed in the above cases. Duncan and White are under definite sentences for alleged criminal conduct [Duncan, though tried for a military offense, also committed the crime of assault under Hawaii Rev. Laws (1935) c. 166, § 5659] and their re-trial for their alleged acts might be impossible because of the Constitutional guarantee against double jeopardy.

The remaining *habeas corpus* case from Hawaii [Steer v. Spurlock, 146 F.(2d) 656 (C. C. A. 9th, 1944)] differs somewhat from any of the other cases. There the Commanding General remitted the unexecuted portion of Spurlock's sentence after the latter filed a petition for *certiorari* in the Supreme Court (No. 841, Oct. Term, 1944). The Government thereupon filed a Suggestion of Mootness, conceding, however, that in the event the case were not moot, it was possible that the District Court's finding as to the military procedure by which Spurlock had been convicted might furnish a basis for sustaining the issuance of the *habeas corpus* writ. This finding had not been considered by the Circuit Court which had reversed the District Court's grant of the writ.

258. I Bl. Comm. *136.

of the various orders, seem to demonstrate beyond possibility of debate that some of the military orders were unwise. While some of the questions of law which were subject to judicial consideration in the instances given above involved matters other than the basic reasonableness of the order, the instances bear out the general thesis that official self-restraint on which Justice Jackson believes we must place reliance,[259] should not be our sole protection against arbitrary action.

The fact that the courts have been a stimulus to the exercise of statesmanlike self-restraint by the military authorities may be thought in part attributable to wariness of unfavorable court decisions which would establish precedents against future attempts by the military to exercise authority, and would create unfavorable public opinion with respect to military controls over civilians. To as great an extent, however, it is also attributable to the simple fact that review or imminent review by a public body, be it judicial or legislative, tends to focus general official attention on a problem to a much greater extent, say, than the routine reception of letters of protest from affected parties.

The possibility that official deliberation and self-restraint would be encouraged by the absence of judicial review on substantive grounds deserves some consideration.[260] The British practice points up this question, especially with respect to review of such orders as the individual civilian exclusion orders; for the House of Lords held that it was not within its power to consider whether an administrative official acted upon reasonable grounds in issuing a similar type of order.[261] But the British practice with respect to such orders seems to emphasize by contrast the need for judicial review in this country. It is not true in the British situation that the self-restraint of the officials administering the program is the sole source of protection against arbitrary action, overzealousness, or negligence. For in Britain Parliament assumed responsibility for a close scrutiny of the orders and of the regulations under which they were made; furthermore the regulations were in the first instance subject to consideration by officials other than those who administered them.[262] Thus without judicial review, the benefit of care-

---

259. See quotation from Justice Jackson's dissent, *supra* p. 225. As pointed out above, the writer agrees with Justice Jackson that under the opinion of the majority in the *Korematsu* case there is little recourse but to rely on such self-restraint, and only disagrees with his view that the function of review must be as limited as that expressed by the majority opinion.

260. Relying on the fact that the judiciary will determine constitutionality, other branches of the government may not make diligent attempts to confine their measures to constitutional limits. As to the legislature, see CHAFEE, FREE SPEECH IN THE UNITED STATES (1941) 469.

261. Liversidge v. Anderson, [1942] A. C. 206.

262. See Carr, *supra* note 15; British Library of Information, New York City Release, April, 1942, *Detention of British Subjects Under the Emergency Powers Act.*

ful and continuing review and an objective appraisal by an outside party was obtained. In this country, however, there is no tradition for Congress to perform such a role, especially in the field of personal liberties. To take the programs here under consideration as an example, there was at no time Congressional inquiry into the individual civilian exclusion program or martial law in Hawaii, despite their long continuance. With respect to the West Coast program, the view indicated by the Supreme Court in the *Endo* opinion that despite the reports concerning, and appropriations for, the W. R. A., the majority of Congressmen were not fully aware of its program is probably factually correct. And to a large extent the Congressional interest that was displayed with respect to the W. R. A. was in response to reports that persons of Japanese ancestry were being too well-treated rather than because of a desire to see that their liberties were protected against bureaucratic zeal, neglect or error.[263]

To do full credit to Justice Jackson's thesis that the existence of reasonable grounds for a military order should not be judicially reviewed, one point which is strongly influential in his thinking deserves attention. He emphasizes that the concept that action based on reasonable military grounds is Constitutional resulted in the establishment by virtue of the *Korematsu* case of a precedent that a drastic racial discrimination is Constitutional.[264] It must be agreed that the establishment of some precedents detrimental to civil liberties might be deemed a logical consequence of review of the grounds for military orders. For it seems somewhat more likely that new and greater restrictions on liberty will be imposed in war than in peace; such restrictions might well be upheld even if the Court applied the standard of reasonableness rather than the *Korematsu* standard, thus creating judicial precedents susceptible of

263. See Reports by House Committee Investigating Un-American Activities, *supra* note 152.

264. Justice Jackson states:

"Much is said of the danger to liberty from the Army program for deporting and detaining these citizens of Japanese extraction. But a judicial construction of the due process clause that will sustain this order is a far more subtle blow to liberty than the promulgation of the order itself. A military order, however unconstitutional, is not apt to last longer than the military emergency. Even during that period a succeeding commander may revoke it all. But once a judicial opinion rationalizes such an order to show that it conforms to the Constitution, or rather rationalizes the Constitution to show that the Constitution sanctions such an order, the Court for all time has validated the principle of racial discrimination in criminal procedure and of transplanting American citizens. The principle then lies about like a loaded weapon ready for the hand of any authority that can bring forward a plausible claim of an urgent need. . . . But if we review and approve, . . . [the] passing incident becomes the doctrine of the Constitution. There it has a generative power of its own, and all that it creates will be in its own image. Nothing better illustrates this danger than does the Court's opinion in this case." 65 Sup. Ct. 193, 207 (1944).

adaptation to peace-time emergencies. But this consequence seems less to be feared than two of the major consequences of Justice Jackson's doctrine: (1) that insofar as the military authorities enforced their orders by force rather than by civil process there could be no judicial interference with their activities; and (2) that since the reasonableness of the ground for their orders would never be in issue, neither would there be any judicial protection against arbitrary action nor would there be the encouragement to official self-restraint which stems from the existence of judicial review of reasonableness.[265] Further, periods of war and peace are not disconnected eras;[266] the peace-time social scene emerges from the war-time, and that which might emerge as a result of the war-time existence of such unrestrained military force is a subject for apprehension.

Summary

The statements of theoreticians on the subject of separation of powers and judicial review that these cognate principles are essential protections against dictatorship,[267] finds empirical support in the situation described in this article. The effect of judicial review with respect to the measures herein considered has been to force reconsideration of ill-considered measures and of exercises of authority which, once adopted, were continued regardless of their continued necessity. Without debating whether military authorities are more prone of themselves to these evils than civilians, at least it is clear that, concomitant with the escape into authoritarianism expressed in the very handing over to military authorities of control over civilian liberties, the military authorities enjoy a greater immunity from inquiry and publicity. This immunity seems stronger in the field of personal than of property rights, and it is upon such immunity that arbitrary exercises of power flourish. Furthermore, while civilian officials are ordinarily appointed because of a particular competence in the field of activity to which they are assigned, the "expertise" of a Commanding General with respect to the con-

---

265. And the situation would necessarily cause the military orders to be enforced in all instances by force rather than by civil process. In evaluating Justice Jackson's theory, it must of course be borne in mind that he does not believe judicial review of the reasonableness of military orders can be effective (discussed *supra* pp. 224-225).

266. See *supra* p. 209 and Lowenstein, *supra* note 176, at 582.

267. GELLHORN AND KERN, ADMINISTRATIVE LAW CASES AND COMMENTS (1939) 37; Jacoby, *Delegation of Powers and Judicial Review* (1936) 36 COLUMBIA LAW REV. 871, 905; Lowenstein, *Legislative Control of Political Extremism in European Democracies,* II (1938) 38 COLUMBIA LAW REV. 725, 767-774; Lowenstein, *loc. cit. supra* note 176; Sharp, *Separation of Powers* (1935) 2 U. OF CHI. L. REV. 385; Wilson, *Separation of Powers under Democracy and Fascism* (1937) 52 POL. SCI. Q. 481.

trols over civilians which he executes is at best fortuitous. For these reasons judicial review of the controls by military authorities over civilians is a necessary protection, and it can and must be equally rigid as the review of similar acts by civilian authorities. The acceptance of military judgment with respect to such controls, as in the *Korematsu* opinion, does not represent a fulfillment of judicial responsibility.

Such acceptance of military judgment in the *Korematsu* opinion resulted in the establishment of an extremely weak standard of review of the justifiability of measures of racial discrimination, which it is believed will stand as an insidious precedent, unless corrected, for the emergencies of peace as well as of war. The standard here urged is that a measure cannot validly be based on a belief in the existence of a racial characteristic unless such belief is supported by reasonable grounds, except in circumstances where the danger threatened in the absence of some action on the basis of the belief is of extreme and overwhelming nature. In that event and in all events, however, a measure of racial discrimination cannot validly be used as the basis for meeting the danger unless the Government demonstrates that available less stringent and more limited alternative methods could not reasonably have been deemed adequate.

Mr. Justice Murphy and the Hirabayashi Case

SIDNEY FINE

The author is professor of history in the University of Michigan.

To QUOTE one of Mr. Justice Murphy's devoted law clerks, "The one discordant note in [Frank] Murphy's opposition to racism was struck in his concurring vote and opinion in *Hirabayashi* v. *United States*."[1] Recognizing that he concurred with considerable reluctance, commentators on the case have attempted "to emphasize how limited was his acquiescence."[2] The recently opened Frank Murphy Papers[3] reveal that he really intended to dissent in the case, wrote a dissenting opinion, and circulated it among his brethren on the United States Supreme Court, but in the end was dissuaded from issuing it.

The legal troubles of George Hirabayashi, a native-born American of Japanese descent and a resident of Seattle, Washington, began on February 19, 1942, when President Franklin D. Roosevelt, by executive order, authorized the Secretary of War and such military commanders as he might designate to prescribe military areas from which any or all persons might be excluded and to impose such restrictions as these officials saw fit on the right of any person to enter, remain

The preparation of this article was facilitated by a grant to the author from the Horace H. Rackham School of Graduate Studies of the University of Michigan.

[1] Eugene Gressman, "Mr. Justice Murphy—A Preliminary Appraisal," *Columbia Law Review*, L (Jan., 1950), 36; 320 U.S. 81 (1943).

[2] John P. Frank, "Justice Murphy: The Goals Attempted," *Yale Law Journal*, LIX (Dec., 1949), 11; Alfred L. Scanlan, "The Passing of Justice Murphy—The Conscience of a Court," *Notre Dame Lawyer*, XXV (Fall, 1949), 23; Gressman, "Mr. Justice Murphy," 36; C. Herman Pritchett, *The Roosevelt Court: A Study in Judicial Politics and Values, 1937-1947* (New York, 1948), 121-122; Byron F. Lindsley, "Constitutional Liberties of Japanese-Americans as Affected by the War," *Georgetown Law Journal*, XXXII (Jan., 1944), 191.

[3] The Frank Murphy Papers are located in the Michigan Historical Collections, Ann Arbor, Michigan.

in, or leave such areas.[4] The stated purpose of the executive order was the protection of national defense resources from espionage and sabotage. On March 18, 1942, the President issued an executive order establishing the War Relocation Authority and authorizing its director to formulate and carry out a program for the removal and relocation of persons designated under the February 19 order. Three days later, Congress, in a statute described by Edward S. Corwin as "just about the most heartless measure ever enacted" by the national legislature,[5] provided that whoever knowingly entered, remained in, or left a prescribed military zone or committed therein an act contrary to the restrictions applicable to the area was to be deemed guilty of a misdemeanor.

On March 2, 1942, pursuant to the executive order of February 19, 1942, Lt. General J. L. DeWitt, Military Commander of the Western Defense Command, designated two military areas within his command and then on March 24 proclaimed a curfew applicable to all alien Japanese, Germans, and Italians and all persons of Japanese ancestry within Military Area No. 1, which included the city of Seattle. Subsequently, DeWitt issued a series of Civilian Exclusion Orders and a Public Proclamation excluding all persons of Japanese descent from portions of Military Area No. 1, instructing them to report to Civil Control Stations, and stating that it would be necessary to provide for their evacuation and resettlement. Hirabayashi, then a senior at the University of Washington, did not comply with the military orders applicable to him as an American of Japanese descent. He was indicted and tried for having failed both to observe the curfew order and to report to the Civil Control Station in his area, was found guilty by jury trial of having violated the act of March 21, 1942, and was sentenced to three months' imprisonment for each of the two counts of the indictment, the sentences to run concurrently. His case was appealed to the United States Supreme Court, which heard arguments on May 10 and 11, 1943, and handed down its opinion on June 21, 1943. It was "the first important test of civil liberties" to reach the Supreme Court in World War II.[6]

Speaking for the Court, Chief Justice Harlan Fiske Stone argued that since Hirabayashi's sentences were to run concurrently, it was

[4] The factual background of the case is conveniently summarized in *Hirabayashi* v. *United States*, 320 U.S. 81, pp. 85-89.

[5] Corwin, *Total War and the Constitution* (New York, 1947), 98.

[6] Washington *Star*, June 24, 1943.

unnecessary for the Court to consider the questions raised by his conviction for failure to report to a Civil Control Station, an obvious preliminary to his exclusion from the Pacific Coast, provided that his conviction for violation of the curfew order could be sustained.[7] Although it would have been perfectly proper for the Supreme Court to have addressed itself to the validity of the exclusion order,[8] the Chief Justice chose to focus on the curfew order alone, which permitted the Court to "present a front of rather wary unanimity to the world."[9] The questions for the Court to decide, Stone contended, were whether the curfew order was promulgated by General DeWitt in the exercise of an unconstitutional delegation by Congress of its legislative powers and whether it unconstitutionally discriminated between persons of Japanese ancestry and of other ancestry in violation of the Fifth Amendment.

Since Stone thought the evidence indisputable that the act of March 21, 1942, had confirmed and ratified the executive order of February 19, 1942, he addressed himself to the issue of whether Congress and the Executive had the constitutional authority to impose the curfew restriction and to leave it to designated military commanders to appraise the relevant conditions and to decide that the order before the Court was an appropriate means of carrying out the executive order. Broadly defining the war power, in the words of Charles Evans Hughes, as "the power to wage war successfully" and as extending "to every matter and activity so related to war as substantially to affect its conduct and progress," Stone found both the executive order and the statute at issue to be legitimate exercises of the power to wage war conferred on the President and the Congress by the Constitution. He thought that it could not be doubted that "reasonably prudent men" had ample grounds for concluding that the United States at the time the curfew order was issued faced the danger of a Japanese invasion, that they had to take measures to cope with this possibility, and that in the selection of the appropriate means they had to take the internal situation of the country into account. Where conditions called for the exercise of judgment and discretion and the choice of means by those branches of the

[7] For Stone's opinion, see *Hirabayashi* v. *United States*, 320 U.S. 81, pp. 83-105.

[8] On this point, see Eugene V. Rostow, "The Japanese-American Cases—A Disaster," *Yale Law Journal*, LIV (June, 1945), 504.

[9] Jacobus ten Broeck *et al.*, *Prejudice, War and the Constitution* (Berkeley, 1954), 212.

government on which the Constitution had conferred the war-making power, it was "not for any court to sit in review of the wisdom of this action or substitute its judgment for theirs."

Since the curfew order, however, applied only to those citizens who were of Japanese descent, the Court had to inquire "whether in the light of all the facts and circumstances there was any substantial basis for the conclusion" that the particular curfew order before the Court was necessary to protect the Pacific Coast against espionage and sabotage. Arguing that espionage by persons sympathetic with the Japanese government had played a part in the attack on Pearl Harbor, Stone contended that at the time of the threatened invasion, "the nature of our inhabitants' attachment to the Japanese enemy" was understandably "a matter of grave concern." He noted that most of the approximately 112,000 persons of Japanese descent residing on the Pacific Coast, about two-thirds of whom were American citizens, were concentrated in or near Seattle, Portland, and Los Angeles, all in Military Area No. 1. He maintained that there was "support" for the view that social, political, and economic conditions prevailing in the United States in the twentieth century had intensified the feeling of "solidarity" among the Japanese in America and had in large measure "prevented their assimilation as an integral part of the white population." He called attention to the large number of Japanese-American children sent outside regular public-school hours to Japanese-language schools, some of which were believed to be sources of "Japanese nationalistic propaganda," the ten thousand *Nisei* children sent to Japan for all or part of their education, the maintenance by the Japanese government of a system of dual citizenship, particularly with regard to persons of Japanese descent born abroad before December 1, 1924, the limited "social intercourse" between Japanese and whites on the Pacific Coast, and the irritating restrictions imposed on Japanese-Americans that had enhanced their sense of isolation and attachment to Japan. "Whatever views we may entertain regarding the loyalty to this country of the citizens of Japanese ancestry," Stone wrote, "we cannot reject as unfounded the judgment of the military authorities and of Congress that there were disloyal members of the population, whose number and strength could not be precisely and quickly ascertained" and who could not, in a critical hour, be dealt with separately.

Stone rejected the argument that the curfew order discriminated against citizens of Japanese ancestry in violation of the Fifth Amendment. He noted that although the amendment did not contain an equal protection clause, distinctions between citizens solely because of their ancestry were "odious" to a free people and legislative classification based on race alone had often been held to constitute a denial of equal protection. The danger of espionage and sabotage in time of war and of threatened invasion, however, required the military to consider all the facts bearing on the population in the danger zone. "Because racial discriminations are in most circumstances irrelevant and therefore prohibited, it by no means follows that, in dealing with the perils of war, Congress and the Executive are wholly precluded from taking into account those facts and circumstances which are relevant to measures for our national defense and for the successful prosecution of the war, and which may in fact place citizens of one ancestry in a different category from others."

Stone thought that the Court had presented "facts and circumstances" with respect to Japanese-Americans residing on the Pacific Coast that afforded grounds for the military to distinguish citizens of Japanese ancestry from other groups in the United States. The fact that the invasion was threatened by Japan automatically set Japanese-Americans apart from other citizens since it had been "demonstrated by experience" that residents having an ethnic affiliation with an invading enemy may be "a greater source of danger" to the imperiled nation than those of different ancestry. There was thus a "rational basis" for the decision that had been made, and it was irrelevant to consider whether the Court would have made a similar decision. Hirabayashi's conviction for violating the curfew order, Stone concluded, was "without constitutional infirmity."

It is agreed by most commentators on the Hirabayashi case that Stone reviewed in an altogether perfunctory manner the judgment of the military that the curfew order was necessary to effect the purposes of the executive order of February 19, 1942.[10] The Chief

[10] Rostow, "The Japanese-American Cases," 503, 505, 507, 508, 531; ten Broeck *et al.*, *Prejudice, War and the Constitution*, 215-216, 308; Milton R. Konvitz, *The Alien and the Asiatic in American Law* (Ithaca, 1946), 253; Morton Grodzins, *Americans Betrayed: Politics and the Japanese Evacuation* (Chicago, 1949), 352-354; Nanette Dembitz, "Racial Discrimination and the Military Judgment: The Supreme Court's Korematsu and Endo Decisions," *Columbia Law Review*, XLV (March, 1945), 186-188. "At the suggestion of one of the war hawks [on the Court]," Stone, declares Alpheus Thomas Mason, "appears to have sanctioned shrinking judicial review of the war power almost to the vanishing point." Mason, *Harlan Fiske Stone* (New York, 1956), 675. It is not always possible in the works noted above to separate out criticism of the Court's standards of review in the Hirabayashi case from the standards later applied in *Korematsu* v. *U.S.*, 323 U.S. 214 (1944).

Justice, in effect, answered in the negative the question he had raised in the famous footnote 4 of his opinion in the Carolene Products Company case as to whether statutes directed at "national . . . or racial minorities" were "to be subjected to more exacting judicial scrutiny under the general prohibitions of the Fourteenth Amendment than are most other types of legislation" and whether "prejudice against discrete and insular minorities" might call for a "more searching judicial inquiry" than was normally required.[11] The test set forth in *Hirabayashi* was whether "in the light of all the facts and circumstances there was any substantial basis" for the judgment of the military, but the Court did not actually apply even this test in determining whether the curfew order before it was a necessary and appropriate means to prevent espionage and sabotage. The "facts" which it claimed it had presented with regard to Japanese-Americans and which it offered as affording a "rational basis" for the curfew order were not facts at all but simply the prejudiced racial opinions of a vocal minority on the Pacific Coast.[12] That General DeWitt shared these prejudices is revealed in his Final Recommendation of February 14, 1942, in which he asserted that "the Japanese race" was "an enemy race" and that although many citizens of Japanese descent had been " 'Americanized,' " the "racial strains" were "undiluted," and it therefore followed that "along the vital Pacific Coast over 112,000 potential enemies of Japanese extraction are at large today." To be sure, not a single act of sabotage had been committed up to that time by these "potential enemies," but this very fact, DeWitt incomprehensibly argued, was "a disturbing and confirming indication that such action will be taken."[13]

The Supreme Court also assumed that time had not been available to investigate the loyalty of Americans of Japanese descent on an individual basis, but this too was a questionable conclusion. The curfew order had not been issued until three-and-one-half months after the attack on Pearl Harbor, and in the interim no effort had been made to weed out the disloyal from the loyal. Hawaii, under

---

[11] *U.S.* v. *Carolene Products Co.*, 304 U.S. 144 (1938), pp. 152-153. For Stone's rejection of the standards of review considered in the footnote, see Wallace Mendelson, "A Note on a Famous Legal Footnote," *Journal of Politics*, XXV (May, 1963), 373-376.

[12] ten Breck *et al.*, *Prejudice, War and the Constitution*, 96, 268-287, 309-310, 326; Rostow, "The Japanese-American Cases," 496-497, 505-507, 519-520; Grodzins, *Americans Betrayed*, 356; Harrop A. Freeman, "Genesis, Exodus, and Leviticus: Genealogy, Evacuation, and Law," *Cornell Law Quarterly*, XXVIII (June, 1943), 440-450.

[13] Cited in ten Broeck *et al.*, *Prejudice, War and the Constitution*, 110.

martial law, facing a greater threat than did the Pacific Coast and with a proportionately larger population of Japanese descent, had nevertheless treated both Japanese aliens and Hawaiians of Japanese descent on an individual basis, and this was the procedure Great Britain had followed in dealing with the enemy aliens in its midst.[14]

Considering the important place occupied by the Bill of Rights in his value system and his confidence in the nation's ability to assimilate peoples of varying backgrounds, colors, and creeds, it is not surprising that Frank Murphy should almost instinctively have taken exception to the failure of the Chief Justice to scrutinize the judgment of the military in a more searching manner and to his unsupported observations concerning the Japanese in America. Throughout his long public career, Murphy had been dedicated to the cause of civil liberties, and it was he who as Attorney General of the United States had created the Civil Liberties Unit in the Department of Justice in 1939 and had pledged the federal government "to protect civil liberties for . . . the people of all racial extractions in our midst."[15] As lawyer, judge, and mayor in Detroit and as governor of Michigan, Murphy had developed very close and affectionate ties with the ethnic groups in the city and the state, and he had come to look upon ethnic and cultural differences in the United States as a source of national strength rather than of national weakness.[16]

At the Supreme Court conference on the Hirabayashi case, on May 16, 1943, there seemed to be general agreement that the decision should deal only with Hirabayashi's violation of the curfew order. "I want it done on [the] narrowest possible points," declared Justice Hugo L. Black. In outlining the case, the Chief Justice concentrated on the two principal issues to which he later devoted his decision, namely, whether there had been an unconstitutional delegation of power and whether citizens of Japanese descent had been discriminated against in violation of the Constitution. "It is jarring to me that U.S. citizens were subjected to this treatment," Stone stated with regard to the discrimination question, but he did not regard the action taken as unreasonable under the circumstances. Justices

[14] Rostow, "The Japanese-American Cases," 494-495, 507-508.

[15] Murphy, "Civil Liberties," Speech of March 27, 1939, Box 67, Murphy Papers. The name of the Civil Liberties Unit was later changed to "Civil Liberties Section" and, more recently, to "Civil Liberties Division."

[16] See, for example, Detroit *News*, March 22, 1931.

Stanley F. Reed and Owen J. Roberts seem to have had at least some doubts concerning the application of the curfew order to "a certain type of citizen," but if this point was debated at great length, it is not evident in the notes that Murphy made of the conference deliberations.[17] Since Murphy's conference notes for this as for nearly all the other cases in which he participated contain no reference to what he himself might have contributed to the discussion, we do not know whether he objected on this occasion to the Chief Justice's line of reasoning.

Since Murphy, however, believed that the record disclosed neither "reasonable grounds" for the military's action nor evidence of the " 'general disloyalty' " of Japanese-Americans,[18] he decided to dissent from Stone's opinion. Early in June at least some of his brethren had in their hands the draft of a powerul opinion in which Murphy argued that discrimination among citizens on the basis of their ancestry was "so utterly inconsistent with our ideas and traditions, and in my judgment so contrary to constitutional requirements, that I cannot lend my assent.'"[19]

Despite the extensive scope of the war power, it did not follow, Murphy wrote, that the broad guaranties of the Bill of Rights and other provisions of the Constitution protecting essential liberties were suspended by "the mere existence of a state of war." Had the progress of the war and the possibility of invasion justified the declaration of martial law and had martial law been proclaimed, Hirabayashi would have had "no legal ground for objection." But since the government had decided not to resort to martial law, it was necessarily subject to the constitutional limitations governing the exercise of the war power. The act of Congress of March 21, 1942, in Murphy's view, transgressed these limitations.

> To vest the military authorities with uncontrolled discretion to establish military zones outside the actual theatre of military operations, in which constitutional guaranties may be suspended or disregarded and the civil authorities be subject to the order of military commanders, utterly

[17] Conference Notes on *Hirabayashi* v. *U.S.*, May 16, 1943, Box 132, Murphy Papers. Anticipating his *Korematsu* dissent, Justice Robert H. Jackson declared at the conference that he did not believe that military commanders were bound by due process of law.

[18] Murphy to John [Pickering], June 8, 1943, Box 132, Murphy Papers.

[19] The drafts of Murphy's proposed dissent are in *ibid.* The printed drafts are virtually identical. There is one typewritten draft that differs somewhat from the printed versions.

subverts the security afforded to individuals by the Bill of Rights and amounts to a pure and unqualified delegation of legislative authority which is not justified by any previous decision of this Court.

Murphy found the statute not only "unconstitutional in its broad aspects," but he thought the particular action taken under it was also "defective." "I would protect rights on the basis of ancestry," he wrote to Felix Frankfurter on June 5. "But I would never deny them."[20] The curfew order was "discriminatory" and hence did not

conform to the requirement of due process of law contained in the Fifth Amendment. It was true that the amendment did not contain an equal protection clause, but it did not follow that "there may not be discrimination of such an injurious character in the application of laws as to amount to a denial of due process as that term is used in the Fifth Amendment. I think that point is reached when we have one law for the majority of our citizens and another for those of a particular racial heritage." Under the curfew order, at least seventy thousand American citizens had been

placed under a special ban and deprived of their liberty because of a particular racial inheritance. This is the first time, so far as I am aware, that a substantial restriction of the personal liberty of citizens based solely on the accident of race or ancestry has been upheld by this Court. The result is to permit the creation in this country of two classes of citizens for the purposes of the war—to sanction discrimination between groups of United States citizens on the basis of ancestry. This is in contravention of principles that have always been regarded as immutable and sacred to our way of life.

Although there is no indication that Murphy at this point had made a detailed study of the subject, he sensed, as the scholarship on the subject demonstrates beyond cavil, that the "facts" put forward by the military, and accepted by the Court, to justify the special treatment of Americans of Japanese descent were really nothing more than opinions based on prejudice and error. Evidence for the disloyalty of Japanese-Americans was singularly lacking, he observed, and "power," therefore, "should not be generated by the general considerations, applicable to many racial and cultural groups," that the Chief Justice had noted in his opinion.

Judging by a marginal comment he made on his copy of one of

[20] Murphy to Frankfurter, June 5 or 6, 1943, *ibid.*

the drafts of Stone's opinion, Murphy wondered if Stone's remarks about Japanese-language schools were equally applicable to "Catholic and other church schools."[21] "We have had foreign language schools in this country for generations, without considering their existence as grounds for racial discrimination," he observed in his dissent. The Michigan Justice found particularly galling the apparent belief of the military that "persons of Japanese descent had not been and could not be assimilated and that by and large they gave primary allegiance to the Empire of Japan." This view, Murphy thought, struck at the very basis of American democracy and was belied by his own experience in associating with a variety of ethnic groups. "To say that any racial or cultural group cannot be assimilated," he wrote, "is to admit that the great American experiment has failed, that our way of life has failed when confronted with the normal attachment of certain groups of people to other lands. . . . If people of Japanese extraction have shown an inclination to associate together and have not been fully assimilated, it is due largely to restrictions that have been placed upon them by law and social custom." No effort had been made "to isolate disloyal elements," but rather seventy thousand citizens had been set apart "because of the accident of ancestry." The nation had "not seen fit to do this to citizens of German and Italian ancestry. We have not done it to any other group of citizens."

Whereas Stone's opinion did not dispute the military judgment that the action taken against the Japanese-Americans was necessary to win the war, Murphy feared that what had been done might have baleful consequences both at home and abroad. The discrimination practiced was at variance with the principles for which the nation was fighting and might have "unfortunate repercussions among peoples in Asia and other parts of the East whose friendship and good-will we seek." He was appalled that what the United States had done bore "a melancholy resemblance to the treatment accorded to members of the Jewish race in Germany and other parts of Europe." It was, of course, necessary for the United States "to wage war to win, and to do it with all our might," but it would "avail us little to win the war on the battlefield and lose it at home. We do not win the war, on the contrary we lose it, if in the process of achieving military victory we destroy the constitutional safeguards

[21] Comment in Murphy's handwriting on draft of Stone opinion, *ibid.*

and the best traditions of our country. What we want to do is win it on the field and also win it at home."

When it became evident to his brethren early in June that Murphy intended to dissent in the Hirabayashi case, one of the Justices, anxious that the Court present a united front in this troublesome case, sought to appeal to his colleague to withdraw his dissent. "Please Frank—," he wrote, "with your eagerness for the austere functions of the Court and your desire to do all that is humanly possible to maintain and enhance the *corporate* reputation of the Court, why don't you take the initiative with the Chief in getting him to take out everything that either offends you or that you would want to express more irenically."[22] Murphy does not appear to have followed this advice, but judging from the futile effort of Justice William O. Douglas to persuade Stone to omit from his opinion "any suggestion of racial discrimination" and to indicate that the "individual member of the group" should be given the "opportunity at some stage . . . to show that he has been improperly classified," it is unlikely that Murphy, whose objections to the Chief Justice's opinion were even stronger than those of Douglas, could have persuaded Stone to alter his draft in any substantial way. At one point, Stone, as a matter of fact, had added a reservation to his opinion in order to satisfy Douglas, but, in the end, to hold other members of the Court, he had decided "to stand by the substance of my opinion."[23]

When Murphy seemed intent on going ahead with his dissent, the Justice who had previously appealed to him now advised his colleague that he would no longer attempt to dissuade him from writing, "not because I do not think it highly unwise but because I think you are immovable." But obviously stung by the language of Murphy's

[22] Note to Murphy [June 5, 1943], *ibid.*

[23] Mason, *Stone*, pp. 673-675. In the end, both Douglas and Wiley B. Rutledge wrote concurring opinions. In his concurrence, Douglas asserted that the Court was dealing with the problem of loyalty, not the problem of assimilation, that loyalty was not a matter of race, and that guilt was personal. The Court had decided only that the appellant could not test the validity of the orders applied to him by defying them. But, Douglas concluded, "Obedience to the military orders is one thing. Whether an individual member of a group must be afforded at some stage an opportunity to show that, being loyal, he should be reclassified is a wholly different question." Rutledge wished to take exception to Stone's suggestion, if it were so intended, that courts might not review the action a military officer chose to take regarding civilians in a military zone once he had found that an emergency had created the conditions justifying creation of the zone. The officer was entitled to wide discretion, "but it does not follow there may not be bounds beyond which he cannot go and, if he oversteps them, that the courts may not have power to protect the civilian citizen." *Hirabayashi* v. *U.S.*, 320 U.S. 81, pp. 105-109, 114.

proposed dissent, the Justice urged Murphy at least to soften the language of his opinion.

> Do you [he asked] really think it is conducive to the things you care about, including the great reputation of this Court, to suggest that everybody is out of step except Johnny, and more particularly that the Chief Justice and seven other Justices of this Court are behaving like the enemy and thereby playing into the hands of the enemy? Compassion is, I believe, a virtue enjoined by Christ. Well, tolerance is a long, long way from compassion—and can't you write your own views with such expressed tolerance that you won't make people think that when eight others disagree with you, you think their view means that they want to destroy the liberties of the United States and "lose the war" at home?[24]

These biting words seem to have had their effect on Murphy. He may also have been swayed by a question previously raised about his proposed dissent by Justice Reed, who had found Murphy's opinion "appealing" but who had nevertheless remained "unconvinced." "If there were substantial evidence that citizens of Japanese ancestry were generally disloyal," Murphy had written, "or had generally so conducted themselves in this area as to constitute a special menace to defense installations and to measures being taken or planned for the defense of the area, or had otherwise by their behavior furnished reasonable ground for dealing with them in this manner as a distinct group, the curfew order and other restrictions imposed on them might be defended and upheld against legal attack in the light of the conditions and the military situation which then prevailed." Commenting on this statement, Reed had remarked, "If you admit this you give your case away. Military protection only needs reasonable grounds, which this record has. You cannot wait for an invasion to see if loyalty triumphs."[25] At first Murphy advised his law clerk to omit the language that had prompted Reed's observation, since it possibly conceded "too much,"[26] but shortly thereafter, apparently swayed by the views of his colleagues, Murphy abandoned altogether the idea of writing a dissent. "I congratulate you," the most critical of his brethren soon was able to write to him, "on the wisdom of having been able to reach a concurrence."[27]

But the concurring opinion that Murphy wrote bore a striking

[24] Letter to Murphy, June 10, 1943, Box 132, Murphy Papers.
[25] Comment on draft of Murphy dissent, *ibid.*
[26] Murphy to Pickering, June 8, 1943, *ibid.*
[27] Undated note to Murphy, *ibid.*

resemblance to the dissenting opinion he had intended to issue. The published concurrence contained the same language as the unpublished dissent with regard to the applicability of the Bill of Rights in wartime, Murphy's abhorrence of distinctions based on ancestry, his faith in the American melting pot and his lack of concern about ethnic differences in American society, and the relationship of the due process clause to discrimination based on racial heritage. Now, however, Murphy found that the curfew order went only to "the very brink of constitutional power," whereas previously he had held that it went over the brink, and he eliminated the stinging sentence, "Until today our laws have been just and equal." In his proposed dissent he had asserted that the point at which discrimination became a denial of due process of law was reached "when we have one law for the majority of our citizens and another for those of a particular racial heritage," but in his concurrence he maintained that the point had simply been "dangerously approached."

In his concurring opinion, Murphy omitted the judgment contained in the dissent that the action of the military had been based on the belief that persons of Japanese descent were unassimilable and gave their primary loyalty to Japan, and he eliminated his criticism of the treatment of the problem of loyalty on a group rather than an individual basis. In view of the critical situation on the Pacific Coast in the spring of 1942, military authorities, he now reluctantly conceded, could "not be required to conform to standards of regulatory action appropriate to normal times. . . . Modern war does not always wait for the observance of procedural requirements that are considered essential and appropriate under normal conditions." The military had made "an allowable judgment" at the time the restriction had been imposed, which was not to say that such a restriction would still be valid.

In the concluding paragraphs of his concurring opinion, Murphy revealed how limited was the degree of his assent with the Chief Justice's opinion and how difficult it had been for him to join with the Court in upholding the curfew order.

> In voting for affirmance of the judgment [he wrote] I do not wish to be understood as intimating that the military authorities in time of war are subject to no restraints whatsoever, or that they are free to impose any restrictions they may choose on the rights and liberties of individual citizens or groups of citizens in those places which may be

designated as "military areas." While this Court sits, it has the inescapable duty of seeing that the mandates of the Constitution are obeyed. That duty exists in time of war as well as in time of peace, and in its performance we must not forget that few indeed have been the invasions upon essential liberties which have not been accompanied by pleas of urgent necessity advanced in good faith by responsible men. . . .

Nor do I mean to intimate that citizens of a particular racial group whose freedom may be curtailed within an area threatened with attack should be generally prevented from leaving the area and going at large in other areas that are not in danger of attack and where special precautions are not needed. Their status as citizens, though subject to requirements of national security and military necessity, should at all times be accorded the fullest consideration and respect. When the danger is past, the restrictions imposed on them should be promptly removed and their freedom of action fully restored.[28]

In the Hirabayashi case the Supreme Court had carefully avoided passing on the constitutionality of the exclusion of persons of Japanese ancestry from military areas on the west coast. When the matter was frontally presented to the Court in the Korematsu case, Justice Black, speaking for the majority and ignoring the guarded language of Stone's *Hirabayashi* opinion, cited the "principles we announced in the *Hirabayashi* case" as the basis for upholding the exclusion policy.[29] "Read this and perish!" Justice Murphy wrote to his law clerk. "The court has blown up on the Jap case—just [as] I expected it would."[30] If an effort was made by any of his brethren to persuade Justice Murphy to concur in the Korematsu case, the Murphy Papers provide no evidence of the fact. Such an attempt, we can be certain, would have been futile in any event. Murphy had concurred in *Hirabayashi* only with the greatest reluctance and, to a degree, against his better judgment.

One senses from his *Korematsu* dissent that he welcomed the opportunity to write again on the Japanese-American question and to state, as he had wished to state with regard to the curfew order, that the exclusion policy "goes over 'the very brink of constitutional power' and falls into the ugly abyss of racism." In a series of footnotes to his opinion, he now cited a number of "independent studies made

[28] Cf. *Hirabayashi* v. *U.S.*, 320 U.S. 81, pp. 109-114, and drafts of Murphy's proposed dissent in Box 132, Murphy Papers.

[29] *Korematsu* v. *U.S.*, 323 U.S. 214 (1944), 218. Jackson correctly noted in his dissent, "The Court is now saying that in *Hirabayashi* we did decide the very things we there said we were not deciding." *Ibid.*, p. 247.

[30] Murphy to Gene [Eugene Gressman], undated, Box 133, Murphy Papers.

by experts" to refute the racial and sociological "facts" that the Court had first presented in the Hirabayashi case to support the judgment of the military. The reasons advanced to justify the exclusion policy—the same reasons previously offered as the basis for the curfew order—were, Murphy correctly observed, "largely an accumulation of much of the misrepresentation, half-truths and insinuations that for years have been directed against Japanese-Americans by people with racial and economic prejudices," among whom was General DeWitt.[81] In *Korematsu*, Murphy was able to make partial amends for his troubled concurrence in *Hirabayashi* and to follow the line of reasoning and to reach the same conclusions as were embodied in his unpublished *Hirabayashi* dissent.

[81] *Korematsu* v. *U.S.*, 323 U.S. 214, pp. 233-242.

Errata

Amended by the author

Page 75 (original article page 195), second paragraph, first sentence should read: "The legal troubles of Gordon Hiragayashi, a native-born American of Japanese descent . . . ."

Page 88 (original article page 208), last paragraph, first sentence should read: "One senses from his *Korematsu* dissent that Murphy welcomed the opportunity . . . ."

II

GENESIS, EXODUS, AND LEVITICUS
GENEALOGY, EVACUATION, AND LAW[1]

HARROP A. FREEMAN

Neither interest in family trees, nor childhood reading of the innumerable "begats" in Genesis, nor the famous words of Justice Holmes, "Three generations of imbeciles are enough,"[2] accounts for the present attention to genealogy. Strange as it may seem, it is the association of the same three ideas identified with the first three books of the *Bible*—Genealogy, Evacuation, and Law—which brings us back to inquire into the bearing of ancestry upon a person's rights and the protection which he shall be afforded in society.

The presence of Jewish blood in one's veins, traced not merely to the third and fourth but to the *n*th generation has been the occasion in Germany for "days of broken glass," pogroms, mass evacuations, and slaughter. Many dismiss the insistent question, "Can these things *be* under law" by denying that Germany operates under a system of law. But the question cannot thus be downed.

On January 2, 1942, the President of the United States issued a statement: "Remember the Nazi technique: 'Pit race against race, religion against religion, prejudice against prejudice. Divide and conquer.' We must not let that happen

[1]If any who read this article look upon persons of Japanese ancestry with suspicion and hatred and accept the phrase, "A Jap's a Jap," then, if they would protect the basic liberties by which they themselves live, they must be the *more* cautious to see that these rights and liberties are protected to those whom they deem most objectionable. This has been well expressed by Macauley in his simile on liberty: "Liberty resembles the Fairy of Ariasto who, by some mysterious law of her nature, was condemned to appear at certain seasons in the form of a foul and poisonous snake. Those who injured her during the period of her disguise, were forever excluded from participation in the blessings which she bestowed. But to those who, in spite of her loathsome aspect, pitied and protected her, she afterwards revealed herself in the beautiful and celestial form which was natural to her, accompanied their steps, granted all their wishes, filled their houses with wealth, made them happy in love. . . . Such is the spirit of Liberty. At times she takes the form of a hateful reptile. She grovels, she hisses, she stings. But woe to those who in disgust shall venture to crush her! And happy are those who, having dared to receive her in her degraded and frightful shape, shall at length be rewarded by her in the time of her beauty and her glory."

As Judge Pound of New York said, "Although the defendant may be the worst of men, the rights of the best of men are secure only as the rights of the vilest and most abhorrent are protected." Quoted by Zechariah Chaffee, Jr., *The Bill of Rights Belongs to the People* (1942) 2 BILL OF RIGHTS REV. 92, 93.

The review by the United States Supreme Court of the saboteur trials is typical of the tenacity with which our legal system holds to this position. *Ex parte* Quirin, 317 U. S. 1, 63 Sup. Ct. 1 (1942). See Cushman, Ex parte *Quirin* et al.—*The Nazi Saboteur Case* (1942) 28 CORNELL L. Q. 54.

[2]Buck v. Bell, 274 U. S. 200, 47 Sup. Ct. 584 (1927).

here."[3] Yet within ninety days after those words were spoken we had evacuated 112,000 persons of Japanese ancestry, of whom 79,000 were American citizens, from five states on the west coast.

On May 10, 1942, three cases involving the constitutionality of this evacuation were argued before the United States Supreme Court.[4] These cases present the crucial issue of this war on the "home front."

It is to be hoped that the Supreme Court will squarely face the constitutionality of the action taken.[5]

I. The Statutes, Orders, and Proclamations

By the Act of July 6, 1798, authority was given the President to control "all natives, citizens, denizens, or subjects" of the enemy within the United States.[6] This was one of a series of acts known as the Alien and Sedition Laws. To certain parts of these acts, particularly those permitting the apprehension of persons "suspected to be dangerous to the peace and safety of the United States," and those interpreted as applying to alien friends as well as alien enemies, the opposition was immediate and vocal,[7] and resulted in substantial non-enforcement of these objectionable provisions.

---

[3]*Hearings before Select Committee Investigating National Defense Migration on Problems of Evacuation of Enemy Aliens and Others from Prohibited Military Zones Pursuant to H. R. Res. 113,* 77th Cong., 2d Sess. (1942) [hereafter referred to as *Tolan Committee Hearings*] 11,042; yet in government briefs we see this pitting of race against race, religion against religion, prejudice against prejudice. Brief for Appellee, 12-32, Hirabayashi v. United States, — U. S. —, 63 Sup. Ct. 1375 (1943).

[4]Hirabayashi v. United States; Korematsu v. United States; Yasui v. United States. All three cases were certified to the Supreme Court from the United States Circuit Court of Appeals for the Ninth Circuit.

[5]*Compare Ex parte* Vallandingham, 1 Wall. 243 (U. S. 1864), wherein certiorari was denied during the war, *with Ex parte* Milligan, 4 Wall. 2 (U. S. 1866). Should the Supreme Court uphold the constitutionality of the evacuation program, I would not, as many a more illustrious colleague has, assert that "the Constitution is gone." See collection of statements: Fairlie, *Has the Constitution Gone?* (1935) 33 Mich. L. Rev. 1037. One might, however, see the fulfillment of the words of Chief Justice Hughes uttered in 1920: "We may well wonder in view of the precedents now established, whether constitutional government as heretofore maintained in this republic could survive another great war, even victoriously waged" and might raise the question, if this is a "survival war," as President Roosevelt has said, for *what* are we surviving but those very rights asserted by the appellants in the three cases?

→ Since the preparation of this article, on June 21 the Supreme Court upheld the convictions in these cases based upon violation of the curfew regulations but expressly refused to decide the constitutionality of the evacuation program. Hirabayashi v. United States, 318 U. S. —, 63 Sup. Ct. 1375 (1943); Yasui v. United States, 318 U. S. —, 63 Sup. Ct. 1392 (1943). The question may again be presented in a case involving only the evacuation.

[6]40 Stat. 531 (1918), 50 U. S. C. § 21 (1940).

[7]Mr. Livingston, speaking in the Senate: 4 Elliot's Debates (1876) 440. The Virginia Resolution declaring the acts unconstitutional, passed December 24, 1798. *Id.* at 528. The Kentucky Resolutions of 1789 and 1799 were drafted largely by Jefferson.

In 1812 when our eastern coast was ravaged by Britain, proclamations were made prohibiting male British subjects over eighteen years of age from dwelling within forty miles of the eastern tidewater. This was enforced by voluntary removal to places farther inland, and by compulsion by the United States marshals upon refusal.[8]

In 1917, in furtherance of the Act of 1798, proclamations were issued under which alien enemies were controlled.[9] In broad outline these may be said to have liberalized the treatment of alien enemies, assured them that they would not be molested in their ordinary pursuits so long as they obeyed the law and acted loyally, and subjected them "to restraint, or to give security, or to remove and depart from the United States" if they failed so to act. The amendments of the Act of 1798 on April 16, 1918,[10] and in 1940 were slight, chiefly to broaden the application to females.[11]

It was under these laws that President Roosevelt made his proclamations of December 7 and 8, 1941,[12] restraining German, Italian, or Japanese aliens from violating the laws of the United States, giving aid or information to its enemies, or interfering in any way with its defense and making such aliens "liable to restraint, or to give security, or to remove and depart from the United States." Possession of various articles including firearms and cameras

---

Id. at 540. When these resolutions were circulated to the various states, those states which argued for the constitutionality of the laws did so on the basis that aliens did not have the rights of citizens, and conceded that the acts would be improper as applied to citizens, *e.g.*, the reply of Massachusetts. *Id.* at 534-535. This distinction was recognized in Madison's report to the Virginia House of Delegates in 1800. *Id.* at 554. The position taken by the states of Virginia and Kentucky resulted in a change of administration; the Alien and Sedition Laws were very laxly enforced; and the whole issue of detention of aliens or citizens disappeared from significance. In *Case of Fries,* 9 Fed. Cas. 826, No. 5,126 (C. C. D. Pa. 1799) the act was upheld on the theory of detention to prevent the commission of a crime. See 2 BEVERIDGE, THE LIFE OF JOHN MARSHALL (1919) 381 ff.; BOWERS, JEFFERSON AND HAMILTON: THE STRUGGLE FOR DEMOCRACY IN AMERICA, (1937) 375-80.

[8]Such action was upheld in *Lockington v. Smith,* 15 Fed. Cas. 758, No. 8, 448 (C. C. D. Pa. 1817); *Lockington's Case,* 1 Brightly 269 (Pa. Sup. Ct. 1813) and *Brown v. United States,* 8 Cranch 110, 121 (U. S. 1814).

[9]40 STAT. 1650, 1651 (1917); 40 STAT. 1716 (1917); 40 STAT. 1729 (1917). Although during the first World War indiscriminate and wholesale internment of aliens took place in Great Britain, Germany, and France, the Attorney General in the United States was governed by a determination of whether the particular alien was dangerous to the public safety. REP. ATT'Y GEN. (1918) 26.

[10]40 STAT. 531 (1918), 50 U. S. C. §§ 21-24 (1940).

[11]This act was upheld as to restraints upon and removal of alien enemies against the argument of lack of due process in *Minotto v. Bradley,* 252 Fed. 600 (N. D. Ill. 1918) and *DeLacey v. United States,* 249 Fed. 625 (C. C. A. 9th, 1918).

[12]Proclamation No. 2525, December 7, 1941, 6 FED. REG. 6321 (1941); H. R. REP. No. 2124, 77th Cong., 2d Sess. (1942) 294-97. Proclamation No. 2526, December 8, 1941, 6 FED. REG. 6323 (1941); H. R. REP. No. 2124, 297-298. Proclamation No. 2527, December 8, 1941, 6 FED. REG. 6324 (1941); H. R. REP. No. 2124, 299-300. Each preamble reads: "Whereas, section 21 of title 50 of the United States Code provides as follows:"

was forbidden; and travel, membership in organizations and similar activities were restricted. Execution of the regulations was vested in the Attorney General as to continental United States, the Virgin Islands, and Puerto Rico, and in the Secretary of War as to the Philippine and Hawaiian Islands, the Canal Zone, and after December 29, Alaska.[13] The Attorney General and Department of Justice immediately began to apprehend and detain alien enemies who were considered dangerous.[14] By Presidential Proclamation No. 2537, issued January 14, 1942,[15] all alien enemies were required to obtain identification certificates from the Attorney General under such regulations as he should deem necessary. The Attorney General issued orders and regulations clarifying and carrying out the presidential proclamations; through all

---

[13]Alaska was withdrawn from the jurisdiction of the Attorney General and transferred to the Secretary of War by Proclamation No. 2533, 7 FED. REG. 55 (1942); H. R. REP. No. 2124, 77th Cong., 2d Sess. (1942) 300.

[14]Attorney General Biddle stated his policy: "All alien enemies could be incarcerated. . . . But it was determined to be more in accordance with our American tradition, wiser, more humane, to hold only those who were dangerous to our safety, or who might become so." Collier's Magazine, March 21, 1942, also statement in H. R. REP. No. 2124, 77th Cong., 2d Sess. (1942) 28. This was in accordance with our tradition of 125 years, [see Koessler, *Enemy Alien Internment,* (1942) 57 POL. SCI. Q. 98] in recognition of the fact that "wholesale internment, without hearing and irrespective of the merits of individual cases, is the long and costly way round, as the British discovered by painful experience," and in the light of the justification of the policy of interning only dangerous aliens during the last war. REP. ATT'Y GEN. (1918) 27; Statement of Attorney General Biddle, H. R. REP. No. 2124, 27. Almost immediately after the outbreak of war, 5,000 aliens were interned. ANNUAL REP. F. B. I. (1941); New Republic, March 16, 1942, 355. By April 15, 1942, 8,010 aliens had been arrested. N. Y. Times, April 16, 1942, p. 7, col. 3. By June 30, 1942, 9,405 had been apprehended. ANNUAL REP. F. B. I. (1942) 6. By June, 1942, the F. B. I. had investigated 10,100 cases of reported sabotage, obtained 218 convictions and found not a "single foreign directed act of sabotage, except the sabotaging of Italian ships by their crews." *Id.* at 5. Hearing boards were established in all federal judicial districts [see (1942) 10 U. S. L. WEEK 2456] which by May, 1942, had released one-half of the 2,500 interned aliens who had appeared before them. Interpreter Release No. 198. Large amounts of contraband were seized from enemy aliens including "3,008 guns, 209,767 rounds of ammunition, 2,016 short wave radio receiving sets and 13 short wave radio transmitting sets, 522 other signalling devices" together with cameras, photographs, maps, and confidential information. ANNUAL REP. F. B. I. (1942) 6. Action was also taken by the Attorney General to reach naturalized former Axis nationals believed to be dangerous. N. Y. Times, March 26, 1942, p. 25, col. 2; N. Y. Times, July 8, 1942, p. 1, col. 3; (1942) 10 U. S. L. WEEK 2456. The F. B. I. during the year June 1941 to June 1942 smashed two major spy rings and held in custody awaiting trial members of other rings; it convicted 56 persons of espionage; and its efficiency in apprehending seven of the Nazi saboteurs landed on our shores in June, 1942, is one of the high points in the history of crime detection. ANNUAL REP. F. B. I. (1942) 5. There has never, to my knowledge, been any serious suggestion that the efforts of the Attorney General have been inadequate in controlling subversive activities of enemy aliens of whom there are upwards of one and one-half million, or American citizens of enemy stock of whom there are at least eleven million. In fact we may conclude that the Attorney General is more thorough than is General DeWitt for the General, in June, 1942, permitted German and Italian aliens to return to vital defense areas on the west coast from which the Attorney General removed them. N. Y. Times, June 29, 1942, p. 4, col. 1.

[15]7 FED. REG. 329 (1942); H. R. REP. No. 2124, 77th Cong., 2d Sess. (1942) 301.

of which runs the test of "source of danger."[16] Beginning with January 29, 1942, Attorney General Biddle designated limited areas on the west coast from which all alien enemies were to be excluded,[17] and on February 4 proclaimed the entire coastline of California from the Oregon border to Los Angeles and from 30 to 150 miles inland a "restricted area" from which aliens were not excluded but in which curfew was maintained and other restrictions were imposed as to all aliens.[18] These regulations necessitated the removal of approximately 10,000 German, Italian, and Japanese aliens and their re-establishment in jobs inland, not in concentration or relocation camps.[19]

On February 19, 1942, the President issued Executive Order No. 9066 transferring the prescription of military areas "from which any or all persons may be excluded" from the Department of Justice to the Secretary of War and military commanders whom he might designate.[20] On the same day, a bill, S. 2293, was introduced in the Senate to authorize "the taking into custody . . . of any or all Japanese"; this bill failed of passage.[21] The Department of Justice gave its opinion that American citizens of Japanese ancestry could not be removed under the President's executive orders.[22] The reason for transferring this control was stated to be that since the Attorney General was merely promulgating regulations for the areas recommended by the army, the division of authority and responsibility seemed undesirable and that "as a legal matter" and for the "acceptance [by] . . . the people generally" it would be better if regulation of citizens were "an exercise of the war power."[23]

---

[16](1942) 10 U. S. L. WEEK 2405; (1942) 10 U. S. L. WEEK 2425; (1942) 10 U. S. L. WEEK 2441; 7 FED. REG. 844 (1942); Interpreter Releases No. 60 *et seq.*, summarized in H. R. REP. No. 2124, 77th Cong., 2d Sess. (1942) 159-160.

[17]Two areas in San Francisco and Los Angeles on January 29th; sixty-nine areas around mouths of rivers, lighthouses, power plants, manufacturing locations on January 31st; fifteen areas in California around bridges, harbors, military reservations on February 2nd; forty-two additional strategic areas on February 4th; eighteen prohibited areas in Arizona on February 7th. H. R. REP. No. 2124, 77th Cong., 2d Sess. (1942) 302-314.

[18]*Id.* at 310-12. Both the prohibited and "restricted" areas were described and recommended by the War Department and that department made no recommendation for other areas.

[19]This removal was conducted by the aliens themselves. *Tolan Committee Hearings,* 11024 ff.

[20]7 FED. REG. 1407 (1942); H. R. REP. No. 2124, 77th Cong., 2d Sess. (1942) 314. This order does not recite the Alien Enemy Act of 1798 (as amended) as authorization, but depends on the President's powers as Commander-in-Chief to prevent espionage and sabotage.

[21]88 CONG. REC., February 19, 1942; *Report of the Committee on Immigration of the Senate,* S. REP. No. 1496, Calendar No. 1541 (1942).

[22]*Ibid.*

[23]H. R. REP. No. 2124, 77th Cong., 2d Sess. (1942) 166. Edward J. Ennis, Director of the Alien Enemy Control Unit, Department of Justice, testifying before the Senate Committee on Immigration on March 23 and 24, 1942. In the Tolan Committee Hearings, neither General DeWitt, the President, nor the Attorney General introduced any other

Between March 2 and March 21 by Public Proclamations No. 1 and No. 2, Lieutenant General DeWitt prescribed approximately the *same military areas* as had previously been established by the Attorney General and gave notice that "Such persons or classes of persons as the situation may require will by subsequent proclamation be excluded."[24]

The "instructions, rules, and regulations prescribed" by the Attorney General[25] "with respect to such prohibited and restricted areas," were "adopted and continued in full force and effect" and the responsibility of the F.B.I. to investigate alleged acts of espionage and sabotage was "not altered."[26] On March 15, when General DeWitt announced the formation of the wartime Civil Control Administration, he still referred only to "areas from which Axis enemies are to be removed."[27]

On March 21, at the request of the Secretary of War, Public Law No. 503 was adopted, which made any person guilty of a misdemeanor who knowingly did any act in a military area or zone contrary to the order of the Secretary of War or military commander.[28]

In the request for the bill and in the House and Senate discussion, the emphasis was at all times on the removal of *certain individuals from limited areas*.[29] And even as thus restricted serious question was raised as to the constitutionality of the law.[30]

---

statement as to the reasons for transfer of authority. The West Coast Congressional Delegation on February 13 had merely asked the President to determine the policy on "loyalty" rather than "citizenship." H. R. REP. No. 1911, 77th Cong., 2d Sess. (1942) 3.

24H. R. REP. No. 2124, 77th Cong., 2d Sess. (1942) 317-329.

25See notes 17, 18, 19 *supra*.

267 FED. REG. 2320 (1942); H. R. REP. No. 2124, 77th Cong., 2d Sess. (1942) 317-320. The N. Y. Times, March 4, 1942, p. 1, col. 3, reported General DeWitt as intending to evacuate gradually from an area 100 miles wide along the coast "all Japanese." This was coupled with a statement that he might require the complete evacuation of *all enemy aliens*. The proclamations required that any alien, "or any persons of Japanese ancestry" obtain and execute a change of residence notice, in addition to the travel permits required of all *aliens* by the Attorney General.

27H. R. REP. No. 1911, 77th Cong., 2d Sess. (1942) 10.

28Act of March 21, 1942, Pub. L. No. 503, 77th Cong., 2d Sess., c. 191 (1942); 56 STAT. 173, 18 U. S. C. A. § 97a (Supp. 1942).

2988 CONG. REC., part 2, pp. 27722-27725 (1942); H. R. REP. No. 1906, 77th Cong., 2d Sess. (1942) 2-3. Representative Costello for the House Military Affairs Committee said: " . . . to remove certain aliens as well as citizens from *areas in which war production is located and where military activities are being conducted.*" *Id.* at 2. Senator Reynolds, Chairman of the Senate Military Affairs Committee said: "we are asked to provide the department with authority to keep certain individuals from entering or leaving military zones. . . . "

30Senator Robert Taft said: "Mr. President, I think this is probably the 'sloppiest' criminal law I have ever read or seen anywhere. . . . I have no doubt that in peace time no man could ever be convicted under it because the Court would find that it was so indefinite and so uncertain that it could not be enforced under the Constitution." 88 CONG. REC., March 19, 1942, 2807; H. R. REP. No. 2124, 77th Cong., 2d Sess. (1942) 169.

It was only on March 24, 1942, after Public Law No. 503 had been passed, that the first discrimination between the German-Italian group and the Japanese was effected by General DeWitt's Public Proclamations No. 3 and 4. Proclamation No. 3 prescribed a curfew for "all alien Japanese, all alien Germans, all alien Italians, and all persons of Japanese ancestry" in Military Area No. 1 and directed that "no person of Japanese ancestry shall have in his possession or use" certain items including cameras and firearms.[31] Public Proclamation No. 4 prohibited "all alien Japanese and persons of Japanese ancestry" in Military Area No. 1 "from leaving that area for any purpose. . . ."[32]

On March 24 the first exclusion order[33] required all persons of Japanese ancestry to evacuate Bainbridge Island by March 30; the instructions which accompanied this order made it clear that persons could remove to areas of their own choosing or accept the government's provisions for "temporary residence."[34] On March 30, 1942, in announcing certain classes of aliens who might acquire exemption from exclusion orders, the General for the first time pointed out that evacuation was in prospect for practically all Japanese.[35] One hundred and five exclusion orders were issued, beginning March 30, the first ninety-nine covering specified limited sections of Military Area No. 1, and Public Proclamation No. 7 covering all the remainder of that Area.[36] The instructions accompanying these orders did not permit persons to remove to areas of their own choosing, although the movement was still referred to as "temporary residence."[37]

A War Relocation Authority was established by the President by Executive Order No. 9106[38] to arrange for the care and employment of the evacuees. Although about 7,000 evacuees are now on leave from the W.R.A. camps to work at various points,[39] small numbers have been permanently reestablished in inland homes,[40] and the War Relocation Authority has recently recognized

[31]*Id.* at 330-331.
[32]*Id.* at 331. This was said to be necessary "to insure the orderly evacuation and resettlement of Japanese *voluntarily* migrating from Military area No. 1." *Id.* at 165.
[33]*Id.* at 332-333.
[34]*Id.* at 333-334.
[35]Press Release, Wartime Civil Control Administration, March 30, 1942; H. R. Rep. No. 2124, 77th Cong., 2d Sess. (1942) 165. Application of the same exemption to Japanese would have left few to evacuate.
[36]N. Y. Times, June 8, 1942. By a queer coincidence Exclusion Order No. 57, under which Gordon Hirabayashi was required to move from Seattle, Washington, was issued on May 10, 1942, exactly one year to the day before his case attacking the constitutionality of the order was argued before the United States Supreme Court.
[37]H. R. Rep. No. 2124, 77th Cong., 2d Sess. (1942) 344.
[38]*Id.* at 315-317.
[39]Pacific Citizen, May 20, 1943, p. 6, col. 4.
[40]*Id.* at 1, col. 3.

that the establishment of relocation centers was a mistake and the evacuees should be freed,[41] the great bulk of the evacuees still reside, surrounded by barbed wire and armed guards, in concentration camps, euphemistically referred to as "relocation centers."

Although German and Italian aliens were, under the original orders of the Attorney General, excluded from certain areas on the west coast, they have, since June, 1942, been permitted to return to these areas.[42]

II. The Constitutional Problem

Two broad constitutional questions are presented, each of which involves determination of matters of degree or reasonableness. First, was the evacuation program under the conditions existing on the west coast in February to May of 1942 a constitutional exercise of the general power of the government, often referred to as the "police power"? Second, if the evacuation program was thus justified, was the method employed within constitutional authorization?[43] This approach accepts the oft repeated statement that the Constitution is an instrument drawn for government in war as well as in peace[44] and goes forward to essay an interpretation in keeping with the underlying concepts of the Constitution and our framework of government.

We, therefore, need to restate the elementary and basic premises of our system of jurisprudence before we attempt to formulate the rules of law to be applied or to examine the available facts to guide our determination of "degree" or "reasonableness."

Dean Roscoe Pound, in a recent address, lists "five characteristics of our Anglo-American law . . . (1) . . . supremacy of law . . . (2) subordination of the military to the civil power . . . (3) . . . emergencies do not suspend the constitution, (4) . . . there are fundamental individual rights, guaranteed and protected by the constitutions . . . (5) . . . the constitutions set up and the courts maintain a separation of powers."[45] I may be pardoned for briefly

---

41*Id.* at 1, col. 1-2.

42N. Y. Times, June 29, 1942, p. 4, col. 1.

43More attention will be given to the first question since the second has been fairly exhaustively argued in the Supreme Court whereas the first seems to me ultimately determinative.

44United States v. MacIntosh, 283 U. S. 605, 622, 51 Sup. Ct. 570, 574 (1931); Ashwander v. Tennessee Valley Authority, 297 U. S. 288, 326, 56 Sup. Ct. 466, 473 (1936); Schenck v. United States, 249 U. S. 47, 39 Sup. Ct. 247 (1919); Selective Draft Law Cases, 245 U. S. 366, 38 Sup. Ct. 159 (1918); Moyer v. Peabody, 212 U. S. 79, 85, 29 Sup. Ct. 235, 237 (1908); Hamilton v. Kentucky Distilleries Co., 251 U. S. 146, 161, 40 Sup. Ct. 106, 110 (1919); and Home Building and Loan Assn. v. Blaisdell, 290 U. S. 398, 54 Sup. Ct. 231 (1933); see also Hughes, *War Powers under the Constitution* (1917) 42 A. B. A. Rep. 232.

45Pound, *War and the Law* (1943) 14 Pa. Bar Ass'n Q. 110, 112-113.

documenting some portions of this most felicitous expression particularly pertinent to the consideration at hand. Any law student knows the expressions of the first characteristic; we may select the concise statement of Justice Miller in *United States v. Lee.*[46]

Of less familiarity, if we can judge from the attitude of the military, other executives, and even law writers,[47] is the second characteristic—the subordination of the military to the civil power. We need not go back to English precedent, though persuasive authority may also there be found. Our Declaration of Independence stated the eleventh charge against the King as justification for overthrowing a "government . . . destructive of these ends" for which "governments are instituted among men" that "he has affected to render the military independent of, and superior to the civil power." It was "Lord Dunmore's proclamation declaring his intention to execute martial law in that province" [Virginia] which caused the first representative government to be created in this country.[48] The change of wording in the proposed Constitution from "*make* war" to "*declare* war" was expressly for the purpose of "clogging rather than facilitating war; but for facilitating peace."[49] In the Constitutional Convention on August 20, 1787, proposals to be incorporated in the draft constitution were submitted to the Committee of Five; included was a provision: "The military shall always be subordinate to the civil power, and no grants of money shall be made by the legislature for supporting military land

---

[46]106 U. S. 196, 1 Sup. Ct. 240 (1882). "No man in this country is so high that he is above the law. No officer of the law may set that law at defiance, with impunity. All the officers of the Government, from the highest to the lowest, are creatures of the law and are bound to obey it. It is the only supreme power in our system of government, and every man, who by accepting office, participates in its functions, is only the more strongly bound to submit to the supremacy, and to observe the liabilities which it imposes upon the exercise of the authority which it gives." General Myron C. Cramer, Judge Advocate General, recently vivified this concept when he said: "Democracy and totalitarianism are gripped in a mighty battle and totalitarianism must be destroyed. Constitutional government must find within itself the powers necessary to its own preservation. In this total war, the *rule of law* rather than the rule of men must be preserved. This contrast in philosophy of government and in the rights of men is the world issue today." Cramer, *Trial of the Eight Saboteurs* (1942) 17 WASH. L. REV. 247.

[47]See Hatcher, *Martial Law and Habeas Corpus* (1940) 46 W. VA. L. Q. 187; *Tolan Committee Summary,* H. R. REP. No. 2124, 77th Cong., 2d Sess. (1942) 147; Edward J. Ennis, *id.* at 166; Hughes, *War Powers under the Constitution* (1917) 42 A. B. A. REP. 232, 238; Brief for government, Hirabayashi v. United States, 47. See also statements: Colonel Bendetsen, H. R. REP. No. 2124, 9; Tom C. Clark, representing General DeWitt, *Tolan Committee Hearing,* 11158; President Roosevelt in the preamble to his Executive Order No. 9066 [H. R. REP. No. 2124, 314] and in his Labor Day address to Congress [both of which are criticized by Edward S. Corwin in *American Government in War-Time* (1943) 37 AM. POL. SCI. REV. 18 ff.]. Similar expressions were frequent during the Civil War. RANDALL, CONSTITUTIONAL PROBLEMS UNDER LINCOLN (1926) 30.

[48]1 ELLIOT'S DEBATES (1876) 52.

[49]3 DOCUMENTARY HISTORY OF THE CONSTITUTION OF THE UNITED STATES OF AMERICA (1894-1905) 553, 554; 1 ELLIOT'S DEBATES (1876) 246.

forces for more than one year at a time."[50] As the Constitution wended its way through ratification in the states, either in debate or by proposed amendments, the various states showed their insistence upon the supremacy of civil over military power.[51]

Though extreme advocates of the war power have always asserted that the Constitution is not operative in wartime,[52] the Supreme Court has so often repeated the rule that war does not abolish constitutional protections and that

---

[50] 1 ELLIOT'S DEBATES (1876) 249. Apparently, from their report, the committee deemed the first portion sufficiently covered by the proposed constitution and inserted the second part in Art. I, § 8 of the Constitution. WARREN, THE MAKING OF THE CONSTITUTION (1928) 504-505.

[51] *Rhode Island* in its ratification declared: "at all times the military should be under strict subordination to the civil power." 1 WARREN, *op. cit. supra* note 50, at 336. *New York* did "declare and make known" its interpretation of the Constitution and necessary amendments to be adopted by its ratification: "At all times the military should be under strict subordination to the civil power." *Id.* at vol. 1, p. 237. The whole history of the ratification of the Constitution is one of insistence upon the inclusion of a bill of rights, the fear that central government would be too strong, the release of the *militia* from martial law except in time of war. Any opposition to these protections was not based on their undesirability but on the assertion that under our system of *delegated* powers no supremacy of the military, no infringement of basic rights, no usurpation of power could occur. ELLIOT'S DEBATES, vol. 1, pp. 325, 327, 334 ff.; vol. 2, pp. 32, 80, 123, 220, 251, 269, 316, 359, 398, 429, 435, 449, 455, 545 ff.; vol. 3, pp. 52, 317, 445, 449, 502, 649, 651, 660, are typical examples. See also Madison's remarks, 1st Cong., 1st Sess., June 8, 1789; also WARREN, *op. cit. supra* note 50, at 508, 769.

It should not be forgotten that the men whose views we are citing had just gone through seven years of war, the outstanding feature of which was the hindrances to our military effort by the weak Confederation government; yet with all this the one point on which more people agreed than on any other was the necessity for restricting the central power from encroaching upon their basic liberties by military authority or otherwise.

As Chafee, *Free Speech in the United States* (1941) 30, points out: "The first ten amendments were drafted by men who had just been through a war. The Third and Fifth Amendments expressly apply in war."

James A. Randall in *Constitutional Problems under Lincoln* (1926) 25-26 has summarized very concisely the American position of the military power:

> "Under the old German system, it was within the competence of the Kaiser to proclaim a 'state of war' throughout Germany, and thus to inaugurate a sweeping military regime. . . .
>
> "In contrast to this extension of executive action, during war, the Anglo-Saxon tendency has been always to emphasize the 'rule of law', and to regard the military power as subordinate to the civil. . . .
>
> "The law of military necessity, however, is not the typical American principle. To say that military force is not to be restrained by the superior power of law, is to quote the militaristic view as against that which has always prevailed here.
>
> "Even during war the person and property rights of the citizen, according to the Anglo-Saxon viewpoint, must be preserved."

Even many of the state constitutions contain provisions assuring the supremacy of the civil power; *e.g.* West Virginia.

Nor can it be gainsaid that the rationale of the majority and minority opinions in Ex parte *Milligan,* 4 Wall. 2 (U. S. 1866) is that the civil power is supreme over the military, a position which has not, so far as I am aware, been denied in any comment on the case.

[52] RANDALL, *op. cit. supra* note 47, at 30; H. R. REP. No. 2124, 77th Cong., 2d Sess. (1942) 147.

emergency cannot create power,[53] that the question should by now be foreclosed from debate.

It may be difficult to define or even to list all the fundamental rights to which individuals are entitled under our system of government. Certain it is that the principle that men are endowed with certain "inalienable rights" is well grounded.[54] A partial list of some of these rights possibly involved in our problem appears in Footnote 55 and may prove helpful.

A. Extent of the Federal Police Power in Wartime

The above enumeration of the basic characteristics of our system of law

---

[53]*Ex parte* Quirin, 317 U. S. 1, 63 Sup. Ct. 1 (1942); Schechter v. United States, 295 U. S. 495, 528, 55 Sup. Ct. 837, 842 (1934); Home Building and Loan Association v. Blaisdell, 290 U. S. 398, 54 Sup. Ct. 231 (1933); Highland v. Russell Car Co., 279 U. S. 253, 261, 49 Sup. Ct. 314, 316 (1928); United States v. Cohen Grocery Co., 255 U. S. 81, 88, 41 Sup. Ct. 298, 299 (1920); Hamilton v. Kentucky Distilleries Co., 251 U. S. 146, 155, 40 Sup. Ct. 106, 108 (1919); and *Ex parte* Milligan, 4 Wall. 2 (U. S. 1866).

[54]In Magna Charta in 1215; Bill of Rights in 1688; Declaration of Independence in 1776; Articles of Confederation, and state and Federal Constitutions.

[55]*Right of free movement.* Edwards v. California, 314 U. S. 160, 62 Sup. Ct. 164 (1941). "The conclusion that the *right of free movement* is a right of national citizenship stands on firm historical ground." See also Crandall v. Nevada, 6 Wall. 35, 48 (U. S. 1867); Williams v. Fears, 179 U. S. 270, 21 Sup. Ct. 128 (1900); United States v. Wheeler, 254 U. S. 281, 41 Sup. Ct. 133 (1920); and Note (1940) 40 COL. L. REV. 1032.

Right to occupation, home, family, knowledge, worship, etc. Meyer v. Nebraska, 262 U. S. 390, 399, 43 Sup. Ct. 625, 626 (1922): "While this court has not attempted to define with exactness the liberty thus guaranteed, the term has received much consideration, and some of the included things have been definitely stated. Without doubt, it denotes not merely freedom from bodily restraint, but also the right of the individual to contract, to engage in any of the common occupations of life, to acquire useful knowledge, to marry, establish a home and bring up children, to worship God according to the dictates of his own conscience, and, generally, to enjoy those privileges long recognized at common law as essential to the orderly pursuit of happiness by free men." See also Truax v. Raich, 239 U. S. 33, 41, 36 Sup. Ct. 7, 10 (1915).

Right to live and work where one wills. Allgeyer v. Louisiana, 165 U. S. 578, 589, 17 Sup. Ct. 427, 431 (1897); and Colgate v. Harvey, 269 U. S. 404, 56 Sup. Ct. 252 (1925).

Right to protection by government in the blessings of life, liberty, property and occupation. Corfield v. Coryell, 6 Fed. Cas. 546, No. 3,230 (C. C. E. D. Pa. 1823); Murray v. Schooner Charming Betsy, 2 Cranch 64 (U. S. 1804); and Butchers' Union S. H. & L. S. L. Co. v. Crescent City, 111 U. S. 746, 4 Sup. Ct. 652 (1883).

Right to peaceable assembly and petition for redress of grievances. United States v. Cruickshank, 92 U. S. 542 (1875).

Right to discuss national issues. Hague v. C. I. O., 307 U. S. 496, 59 Sup. Ct. 954 (1938).

Right to immunity from denial of voting or similar privileges because of race. United States v. Reese, 92 U. S. 214 (1875); and *Ex parte* Yarbrough, 110 U. S. 651, 4 Sup. Ct. 152 (1884).

These rights have been said to be "fundamental rights which belong to every citizen as a member of society" (United States v. Cruikshank, 92 U. S. 542 (1875), "vital to the maintenance of democratic institutions" (Schneider v. State of New Jersey, 308 U. S. 147, 161, 60 Sup. Ct. 146, 151 (1939), and "immutable principles of justice which inhere in the very idea of free government" (Holden v. Hardy, 169 U. S. 366, 389, 18 Sup. Ct. 383, 387 (1897).

and the rights of individuals suggests that the evacuation of persons of Japanese ancestry from the west coast has cut more deeply across those basic rights than has any government regulation thus far attempted. It is a commonplace that all private rights: property, personal, even life itself, are held subject to the public right of regulation in the common interest and that the right of private dominion is always in greater or less collision with the right of public control.[56] It would be simple to deny the government all right of control and make personal rights absolute, or to grant the government unfettered control and make government totalitarian. Our task, however, is more difficult, but more fruitful. As Aristotle remarks in the *Nicomachean Ethics*, Book V: "The easiest thing for the judge—or legislator—to do is to be strict and forbid. It is much more difficult to attempt to reconcile the seemingly irreconcilable, conserving the just in the claims of each side, and rendering a full measure of distributive justice to each." Our problem is to find a position sufficiently protective of the interests of government, which will nevertheless preserve to the individual citizen the greatest measure of those rights for the preservation of which government is itself formed.

We are indebted to government counsel in the *Hirabayashi* case, and to the recognized authority on martial law, Charles Fairman, for dispelling the idea which has pervaded much of the thinking on federal power in time of war, that declarations of "martial law" or "military necessity" or operations under "war powers" are determinative or even helpful in defining the extent of the power.[57] The plain and simple fact remains that we are exploring the boundaries of the "police power" under specific exigencies.

1. *Judicial Approach and Definition of Federal Police Power in Wartime.*—Thus having defined the problem and our basic legal concepts, we need to explore first the rules which have been developed to determine the right of the state, both in "peace" and "war," to regulate or prevent the use of one or

[56]West Coast Hotel Co. v. Parrish, 300 U. S. 379, 57 Sup. Ct. 578 (1936); Nebbia v. New York, 291 U. S. 502, 523-5, 54 Sup. Ct. 505, 510-11 (1933); Noble State Bank v. Haskell, 219 U. S. 104, 111, 31 Sup. Ct. 186, 188 (1910); Jacobson v. Massachusetts, 197 U. S. 11, 25 Sup. Ct. 358 (1904); and Commonwealth v. Alger, 7 Cush. 53 (Mass. 1851). See also 1 THAYER, CASES ON CONSTITUTIONAL LAW (1895) 693, and Walter Wheeler Cook, *What Is the Police Power* (1907) 7 COL. L. REV. 322.

[57]Charles Fairman, author of *The Law of Martial Rule and the National Emergency* (1930) 55 HARV. L. REV. 1253, is quoted as saying, concerning the evacuation: "The question in every case of military control would still be, (even with a martial law declaration) can the action complained of be justified as apparently reasonable and appropriate, under the circumstances, to the defense of the nation and the prosecution of the war?" San Francisco Chronicle, March 4, 1942, 14. See also the same position taken in his excellent article: Fairman, *The Law of Martial Rule and the National Emergency* (1942) 55 HARV. L. REV. 1253; Brief for Government, p. 42 ff., Hirabayashi v. United States. 2 WILLOUGHBY, THE CONSTITUTIONAL LAW OF THE UNITED STATES (2d ed. 1929) 1591.

more of an individual's basic rights, and, second, the attitude or general philosophy by which the courts have and should approach this inevitable collision of private and public rights.

Throughout the evacuation episode runs the attitude of unwillingness to trust any person of Japanese ancestry. Typical of this is the expression of Mayor Riley of Portland: "I wouldn't take a chance with one."[58] Is this the approach to the necessity for evacuation? Or is the approach that expressed in *Kwock Jan Fat v. White*,[59] "It is better that many Chinese immigrants should be improperly admitted than that one natural born citizen of the United States should be permanently excluded from his country."[60]

On April 13, 1943, General DeWitt, whose orders are the subject of our inquiry, was reported by the press as saying: "A Jap is a Jap . . . it makes no difference whether the Japanese is theoretically a citizen. . . . The west coast is too vulnerable . . . to take any chances."[61] On May 17, the Supreme Court in *John T. Regan v. King*,[62] recognized that there must continue to be a distinction between Japanese aliens and American citizens of Japanese ancestry and refused to reconsider the rule in the *Wong Kim Ark* case.[63] In Ex parte *Kawato*,[64] Justice Black speaking for the Court said:

> "Nothing in this record indicates, and we cannot assume, that he [a Japanese *alien*] came to America for any purpose different from that which prompted millions of others to seek our shores—a chance to make his home and work in a free country, governed by just laws, which promise equal protection to all who abide by them."

Clearly the Supreme Court requires a more trustful and fair attitude than has been employed by those who advocated and directed the evacuation.

Wherever individual rights and public rights are in opposition, there are two ends to the problem—the "protection of individual rights" end and the "risk to the public" end.[65] It would seem that the courts of a country which starts its history with the wording of the Declaration of Independence, carries the same philosophy into the Constitution, surrounds its civil and criminal procedures with protections of the individual, and is currently waging a war

---

[58]H. R. REP. No. 2124, 77th Cong., 2d Sess. (1942) 146.

[59]253 U. S. 454, 49 Sup. Ct. 566 (1919).

[60]This is in keeping with the maxim that "it is better that 99 guilty escape than that one innocent man be punished."

[61]See the Washington Post, April 15, 1943; San Francisco News, April 13, 1943, 1.

[62]— U. S. —, 63 Sup. Ct. 1168 (1943).

[63]United States v. Wong Kim Ark, 169 U. S. 649, 18 Sup. Ct. 456 (1897) holding that a person born in the United States of Japanese parents is an American citizen with all the attendant rights of citizenship, on a par with all other citizens.

[64]317 U. S. 69, 63 Sup. Ct. 115 (1942).

[65]Chaffee, *The Bill of Rights Belongs to the People* (1942) 2 BILL OF RIGHTS REV. 92.

for "four freedoms," which mean nothing except in terms of the individual, must always think first of the individual rights, rather than the risk. In my opinion the evacuation has been almost entirely the product of "risk fixation," to employ a psychologist's terminology.

It appears that the view which the Supreme Court has developed in dealing with personal or civil rights cases has properly laid stress on private protection, rather than public risk. The general presumptions of constitutionality of legislation and of the reasonableness of classification, and the refusal to inquire into the policy of legislation[66] are recognized as the essence of judicial restraint and separation of powers. But this concession does not explore the issue deeply enough to be determinative. It may be that the tendency has been to refuse to draw tight the circle of inviolability about property.[67] The increased vigilance in protecting personal rights is the more manifest.[68] Gradually the distinction has been worked out. A liberal construction of the first ten amendments in favor of the individual, was first recognized.[69] The power to abridge personal liberties became considered "the exception rather than the rule."[70] Astuteness in examining legislation charged to abridge basic freedoms was manifested in the same period.[71] A query was stated whether "there may be narrower scope for operation of the presumption of constitutionality when legislation appears on its face . . . to be within the first ten amendments."[72] In *Schneider v. Irvington*[73] the following year, Justice Roberts went on to answer the question:

"In every case, therefore, where legislative abridgement of the rights is asserted, the courts should be astute to examine the effect of the chal-

[66]United States v. Carolene Products Co., 304 U. S. 144, 153, 58 Sup. Ct. 778, 784 (1937); Borden's Farm Products Co. v. Baldwin, 293 U. S. 194, 209, 55 Sup. Ct. 187, 191-192 (1934); Corporation Commission v. Lowe, 281 U. S. 431, 50 Sup. Ct. 397 (1929); Green v. Frazier, 253 U. S. 233, 40 Sup. Ct. 499 (1919); and Dartmouth College v. Woodward, 4 Wheat. 518, 625 (U. S. 1820).

[67]United States v. Carolene Products, 304 U. S. 144, 58 Sup. Ct. 778 (1937); Home Building and Loan Association v. Blaisdell, 290 U. S. 398, 54 Sup. Ct. 231 (1933); Block v. Hirsh, 256 U. S. 135, 41 Sup. Ct. 458 (1920); and Hamilton v. Kentucky Distilleries, 251 U. S. 146, 40 Sup. Ct. 106 (1919).

[68]Thornhill v. Alabama, 310 U. S. 88, 60 Sup. Ct. 736 (1939); Schneider v. State of New Jersey, 308 U. S. 147, 60 Sup. Ct. 146 (1939); Hague v. C. I. O., 307 U. S. 496, 59 Sup. Ct. 954 (1939); and Johnson v. Zerbst, 304 U. S. 458, 58 Sup. Ct. 1021 (1937).

[69]Sgro v. United States, 287 U. S. 206, 53 Sup. Ct. 138 (1932); Grau v. United States, 287 U. S. 124, 53 Sup. Ct. 38 (1932); United States v. Lefkowitz, 285 U. S. 452, 52 Sup. Ct. 420 (1931); and Go-Bart Importing Co. v. United States, 282 U. S. 344, 51 Sup. Ct. 153 (1930).

[70]Herndon v. Lowry, 301 U. S. 242, 258, 57 Sup. Ct. 732, 739 (1936).

[71]DeJonge v. Oregon, 299 U. S. 353, 57 Sup. Ct. 255 (1936); and Lovell v. Griffin, 303 U. S. 444, 58 Sup. Ct. 666 (1937) and cases cited therein.

[72]United States v. Carolene Products Co., 304 U. S. 144, 152, 58 Sup. Ct. 778, 783 (1937).

[73]308 U. S. 147, 161, 60 Sup. Ct. 146, 150 (1939).

lenged legislation. Mere legislative preferences or beliefs respecting matters of public convenience may well support regulation directed at other personal activities, but be insufficient to justify such as diminishes the exercise of rights so vital to the maintenance of democratic institutions."

This decision has been generally accepted as giving "authoritative substance to the theory that there may be no room for the presumption of constitutionality, usually accorded state or municipal legislation, where the statute or ordinance interferes with a civil liberty as distinguished from legislative impairment of an economic privilege."[74] This approach has since been followed by the Court.[75] It is submitted that there is a real distinction authorizing this difference of approach. In the ordinary case involving constitutionality, government action either is being tested against the rule that this is a system of "express powers" or is being subjected to negative limitations. It is proper then that government should not be crippled, should be "allowed a little play in its joints."[76] And it is sufficient that only action which is arbitrary and wholly unreasonable be prevented. But individual rights are as specifically provided for by the Constitution as are governmental powers; in fact government was created for their protection. When these rights are involved, the *limitation* on governmental *power* is not as important as is the limitation on the individual's *rights*. Here the Court should itself determine, as umpire between government and the people, the extent to which the public need should be curtailed and some public risk be accepted and the degree to which cardinal freedoms should be overridden.

The Court in the *Hague* and *Schneider* decisions, carried this attitude protective of individual rights even further in a manner particularly applicable to the evacuation program. The evacuation has been justified as necessary to "protect" persons of Japanese ancestry against violence and vigilantism. In the *Hague* case the same argument was made, but the Court held that the city must meet threatened disorder by police protection instead of the more

---

[74](1940) 40 Col. L. Rev. 531, 532.

[75]Cantwell v. Connecticut, 310 U. S. 296, 60 Sup. Ct. 900 (1939); Carlson v. California, 310 U. S. 106, 60 Sup. Ct. 746 (1939); Thornhill v. Alabama, 310 U. S. 88, 60 Sup. Ct. 736 (1939); and Hague v. C. I. O., 307 U. S. 496, 59 Sup. Ct. 954 (1939). See also *Opelika, Fort Smith, Casa Grande* and other *Jehovah's Witnesses* cases of 1943. Note that this approach stems back to earlier dissenting opinions. Burns Baking Co. *et al.* v. Bryan, 264 U. S. 504, 517, 520, 44 Sup. Ct. 412, 415, 416 (1924); and Adams v. Tanner, 244 U. S. 590, 597, 600, 37 Sup. Ct. 662, 665, 666 (1916). Although in *Minersville School District v. Gobitis*, 310 U. S. 586, 60 Sup. Ct. 1010 (1939) the Court seemed less solicitous of these rights. The Court engaged in no presumption and now entertains the same question for reconsideration (following the unexpected suppression of Jehovah's Witness activity after the *Gobitis* case *supra*). As this article goes to press, the Supreme Court has on June 14, 1943, reversed the *Gobitis* case.

[76]Justice Holmes in Bain Peanut Co. v. Pinson, 282 U. S. 499, 51 Sup. Ct. 228 (1930).

efficient method of refusal of permits which interfered more seriously with the right of assembly. The threat of lawless conduct was no justification for depriving the innocent of their rights.[77] In the *Schneider* case the argument was advanced that the most efficient method of preventing fraudulent appeals, littering of the streets, and trespassing was to prohibit dissemination of information and pamphlets without permit. Again the Court required the less efficient method of ascertainment of guilt and punishment, since this would interfere less with basic liberties.[78] Against this background, what standing has the argument in favor of *evacuation that it constituted "protective custody"*[79] or that the difficulty of sorting out the disloyal by the method of individual hearing required the internment of the whole population of Japanese ancestry?

The two arguments most frequently used in support of the evacuation program and similar war measures: (1) that "the power to wage war is the power to wage war successfully,"[80] and (2) President Lincoln's homely words "by general law, life and limb must be protected, yet often a limb must be amputated to save a life; but a life is never wisely given to save a limb,"[81]

---

[77]This same argument of "prevention of conflict" was presented to the Supreme Court in *Buchanan v. Warley,* 245 U. S. 69, 81, 38 Sup. Ct. 16, 20 (1917), and was similarly rejected: "It is urged that this proposed segregation will promote the public peace by preventing race conflicts. Desirable as this is, and important as is the preservation of public peace, this aim cannot be accomplished by laws or ordinances which deny rights created or protected by the federal Constitution."

[78]"Any burden imposed upon the city authorities in cleaning and caring for the streets as an indirect consequence of such distribution results from the constitutional protection of the freedom of speech and press . . . , the public convenience in respect of cleanliness of the streets does not justify an exertion of the police power which invades the free communication of information and opinion secured by the Constitution." Schneider v. New Jersey, 308 U. S. 147, 162-3, 60 Sup. Ct. 146, 151 (1939).

"Frauds may be denounced as offenses and punished by law. Trespasses may similarly be forbidden. *If it is said that these means are less efficient and convenient than bestowal of power on police authorities to decide* what information may be disseminated from house to house, and who may impart the information, the answer is that considerations of this sort do not empower a municipality to abridge freedom of speech and press." *Id.* at 164, 60 Sup. Ct. at 152 (1939). This was but a more specific application of the rule stated in *Mountain Timber Co. v. Washington,* 243 U. S. 219, 37 Sup. Ct. 260 (1916); *Schlesinger v. Wisconsin,* 270 U. S. 230, 46 Sup. Ct. 260 (1925); and *Weaver v. Palmer Brothers,* 270 U. S. 402, 46 Sup. Ct. 320 (1925).

[79]There have been true cases of "protective custody" by the military. An example was the placing of Round Valley in California under martial law as a protection to the Indians, against whom the whites were committing outrages—but the Indians were left in the enjoyment of their property and liberties and the action was taken against the whites. See WAR OF THE REBELLION: OFFICIAL RECORDS OF THE UNION AND CONFEDERATE ARMIES, Series I, Vol. 50, pt. 2, pp. 218, 219, 310.

[80]Hughes, *War Powers under the Constitution* (1917) 42 A. B. A. REP. 237, 238; Brief for Government, p. 47 ff., Hirabayashi v. United States.

[81]Fairman, *The Law of Martial Rule and the National Emergency* (1942) 55 HARV. L. REV. 1253, 1278.

seem to carry argument by metaphor beyond what sound constitutional theory will justify. Amputation of a limb is a "last ditch" remedy and some risk to life is always attempted before this drastic step is taken. The "wage war successfully" argument is used to brand the war power as "plenary" and to suggest that the determination of the military commander must be accepted. This approach is wholly at variance with the attitude already discussed and its application to the exercise of war powers. From the earliest times, both in England[82] and in the United States[83] the courts refused to accept the decision of the military and insisted on making a determination *de novo* of the necessity for military action infringing individual liberties.[84] Viewed from every angle it appears that the Supreme Court has recognized, both in war and in peace, that private rights must be protected even at some risk to the public generally and that the Court shall decide when and to what extent such risk must be accepted.

Developed concurrently with the rule restricting the presumption of constitutionality, has been the rule that in order to justify the invasion of individual rights, the danger to society must be "imminent and impending" or "clear and present."[85] This test has been devised to meet not only the stress of peacetime emergency but also the exigencies of war. To most lawyers the history of the adoption of this rule during the last war is known: how exponents of civil liberties had previously insisted that no act was punishable unless

---

[82]The case of the Bristol Rights (S. T. U. S. III, 2-56), 1932, cited in 3 WILLOUGHBY, THE CONSTITUTIONAL LAW OF THE UNITED STATES (2d ed. 1929) 1591.

[83]Mitchell v. Harmony, 13 How. 115 (U. S. 1851): "It is not enough to show that he exercised an honest judgment, and took the property to promote the public service, he must also prove what the nature of the emergency was, or what he had reasonable grounds to believe it to be; and it will then be for the court and jury to say whether it was so pressing as to justify an invasion of private right."

[84]Although the unguarded dictum in *Luther v. Borden,* 7 How. 1 (U. S. 1849) and the severely criticized opinion in *Moyer v. Peabody,* 212 U. S. 78, 29 Sup. Ct. 235 (1909) relaxed this rule, the Court in *Sterling v. Constantin,* 287 U. S. 378, 53 Sup. Ct. 190 (1932) returned to its former position, relying on the *Mitchell* case. It may be that under the *Sterling* case there is a "range of honest judgment permitted," but the military judgment is not conclusive. See also Jones v. Securities Commission, 298 U. S. 1, 23, 56 Sup. Ct. 654, 660-661 (1935); and *Ex parte* Orzoco, 201 Fed. 106 (W. D. Tex. 1912).

[85]To the student of jurisprudence it will appear that this constitutes a replacement of the rule prevalent at the formation of this country and expressed particularly by Jefferson (see PADOVER, DEMOCRACY BY THOMAS JEFFERSON) which permitted of no group control by the "utilitarian" rule permitting governmental regulation of these rights in the common good, a position associated with Bentham and Mill, and more recently with Holmes and Brandeis. *See*: BENTHAM, THE BOOK OF FALLACIES; MILL, UTILITARIANISM; Holmes, J., in United States v. Schwimmer, 279 U. S. 644, 655, 49 Sup. Ct. 448, 451 (1928); Brandeis, J., in Whitney v. California, 274 U. S. 357, 375-77, 47 Sup. Ct. 641, 648-649 (1926). A more critical examination of the cases will prove that the replacement has not been complete and that the tendency is toward the preservation of the individual rights to the greatest extent possible, while permitting regulation only to the extent necessary.

it actually resulted in inciting criminal acts;[86] then in the five cases argued at the October term in 1918 the "clear and present danger" test was stated and such danger was found to exist in language that hindered the recruiting of an army;[87] in *Gitlow v. New York* it appeared the Court might accept suppression of "threatened danger in its incipiency" as the criterion;[88] *but the* Court finally reestablished the "clear and present danger" test to meet the exigencies of the depression years.[89] In the later cases the test has been applied to disapprove of governmental action[90] and sociological balancing of the preservation of individual rights against the risk to society has been employed, even when the danger was "present," to determine how "clear" was the danger.[91]

What is perhaps not so well understood is the relation of this test to the general power of government in time of war. If we go back to the earlier

---

[86]CHAFFEE, FREE SPEECH IN THE UNITED STATES (1941); Wechsler, *Symposium on Civil Liberties* (1941) 9 AM. L. SCHOOL REV. 881, 882; Murrish, *Protection of Free Speech under the Federal Constitution* (1940) 28 CALIF. L. REV. 733. In one of the first cases decided during the first World War, *Masses Publishing Co. v. Patten,* 244 Fed. 535 (S. D. N. Y. 1917), District Judge Learned Hand accepted this test.

[87]See Abrams v. United States, 250 U. S. 616, 40 Sup. Ct. 17 (1919); Schenck v. United States, 249 U. S. 47, 39 Sup. Ct. 247 (1919); Baer v. United States, 249 U. S. 47, 39 Sup. Ct. 247 (1919); Sugarman v. United States, 249 U. S. 182, 39 Sup. Ct. 191 (1918); Frohwerk v. United States, 249 U. S. 204, 39 Sup. Ct. 249 (1918); Debs v. United States, 249 U. S. 211, 39 Sup. Ct. 252 (1918).

[88]268 U. S. 652, 45 Sup. Ct. 625 (1924).

[89]Herndon v. Lowry, 301 U. S. 242, 57 Sup. Ct. 732 (1936). There are those who find a clear return to this rule as early as *Fiske v. Kansas,* 274 U. S. 380, 47 Sup. Ct. 655 (1927); *Whitney v. California,* 274 U. S. 357, 47 Sup. Ct. 641 (1927); and *Near v. Minnesota,* 283 U. S. 697, 51 Sup. Ct. 625 (1931).

[90]See cases cited note 89 *supra,* and see note 91 *infra.*

[91]In *Thornhill v. Alabama,* 310 U. S. 88, 102, 60 Sup. Ct. 736, 744 (1939), although the Court recognized a present danger, it balanced the value of labor organization against internal order and employer's rights and found certain risks worth taking "in the circumstances of our times . . . [to preserve] the area of free discussion that is guaranteed by the Constitution." And in *Lovell v. City of Griffin,* 303 U. S. 444, 58 Sup. Ct. 666 (1937); *Schneider v. New Jersey* (and companion cases), 308 U. S. 147, 60 Sup. Ct. 146 (1939); *De Jonge v. Oregon,* 299 U. S. 353, 57 Sup. Ct. 255 (1936) *Hague v. C. I. O.,* 307 U. S. 496, 59 Sup. Ct. 954 (1939); *Cantwell v. Connecticut,* 310 U. S. 296, 60 Sup. Ct. 900 (1939) though it was clear that some "danger" or risk was present, the Court permitted free discussion to prevail. Even in *Minersville School District v. Gobitis,* 310 U. S. 586, 60 Sup. Ct. 1010 (1939), where the alleged danger was to national unity, "an interest inferior to none in the hierarchy of legal values" (310 U. S. 586 at 595, 60 Sup. Ct. 1010 at 1013) the Court balanced values. The chief criticism of the case is that it overemphasizes the salute as a path to unity. Now the Court is reconsidering the case, a recognition that even in this area perhaps sufficient attention was not given to the need for protecting individual rights. On June 14, concurrently with this article's going to press, this case has been reversed on a rebalancing of considerations. The Court in *Bridges v. State of California,* 314 U. S. 252, 62 Sup. Ct. 190 (1941) and *Times-Mirror Co. v. Superior Court,* 314 U. S. 252, 62 Sup. Ct. 190 (1941) again followed the clear and present danger test and found no such danger as justified citation for contempt of court. In addition to considering the nearness of the danger, it considered the interest threatened and the likely effect of their decision.

precedents defining the limits and conditions for the exercise of military power, we shall find that here also is the requirement of proximity and degree, a "clear and present" or "immediate, imminent, and impending" danger. Thus Charles Fairman has traced the beginnings of "martial law" or military control to the common law principle of "measures necessary to preserve the realm and resist the enemy,"[92] which was analogous to protection of one's person or property by force.[93] Each of these principles recognized liability (lack of justification) if the danger was not real or imminent, or the means employed exceeded the occasion.[94]

If we look to the few cases in this country which have stated the rule to be applied in testing the action of the government through the military branch, as it affects individual rights, we shall find the same rule of "imminent danger" and the same tendency to balance the conflicting interests in favor of the individual and against government, which we have seen develop in other civil liberties fields.[95] Two cases involving the taking of property in time of war, on the claim that it was necessary in order to prevent its falling into the hands of the enemy, serve as illustrations. In *United States v. Russell,*[96] the rule was stated as "the public danger must be immediate, imminent, and impending, and the emergency in the public service must be extreme and imperative, and such as will not admit of delay or a resort to any other source of supply." In *Mitchell v. Harmony,*[97] since the evidence showed that the seizure was for the purpose of *using the property in a forthcoming campaign,* rather than to prevent the property coming into the enemy's possession, the plaintiff was allowed to recover against the military officer. The Court's language is very clear:

> "But in every such case the danger must be present or impending, and the necessity such as does not admit of delay or the intervention of the civil authority to provide the requisite means. It is impossible to define the particular circumstances in which the power may be lawfully exercised. Every case must depend on its own circumstances. It is the emergency

---

[92]Fairman, *The Law of Martial Rule and the National Emergency* (1942) 55 HARV. L. REV. 1253, 1256-1264.

[93]*Id.* at 1260.

[94]*Id.* at 1261-1264; 3 WILLOUGHBY, THE CONSTITUTIONAL LAW OF THE UNITED STATES (2d ed. 1929) 1591. A further recognition of this are the "Indemnity Acts" which have been passed in England and the United States, after military control which may have exceeded the necessity. 2 MAY, CONSTITUTIONAL HISTORY OF ENGLAND, 256-8; Mitchell v. Clark, 110 U. S. 633, 647, 4 Sup. Ct. 170 (1883); WAR OF THE REBELLION: O. R., *op. cit. supra* note 79, Series II, vol. 2, pp. 21-30.

[95]See notes 89, 90, and 91 *supra.*

[96]13 Wall. 623, 627-8 (U. S. 1871).

[97]13 How. 115 (U. S. 1851); see also Raymond v. Thomas, 91 U. S. 712, 716 (1875). The test of "imminent danger" is that fixed by the Constitution as justification for the use of a state's war powers. U. S. CONST. Art. I, § 10.

that gives the right, and the emergency must be shown before the taking can be justified. In deciding upon this necessity, the state of facts as they appeared at the time will govern the decision, because the officer in command must act upon the information of others as well as his own observation. And if, with such information as he can obtain, there is reasonable ground for believing that the peril is immediate or the necessity urgent, he may do what the occasion seems to require, and the discovery that he was mistaken will not make him a wrongdoer. It is not enough to show that he exercised an honest judgment, and took the property to promote the public service, he must also prove what the nature of the emergency was, or what he had reasonable grounds to believe it to be; and it will then be for the court and jury to say whether it was so pressing as to justify an invasion of private right."

More recently the Court made clear that a danger more proximate than has, I believe, been shown to exist on the west coast must be shown before military control will be justified. Said Chief Justice Hughes:[98]

"The absence of necessity for military order of the Governor . . . is established by showing that there was no actual uprising or showing of violence or anything more than threats of violence, breaches of the peace against oil producers, and that there was no closure of the courts or failure of civil authorities."

2. *The History of Governmental Action in Time of War.*—The other guide in determining rules to be applied now is found in action previously taken by the government in time of war, whether or not sustained by the courts, weighed in the light of the similarities or differences of the times. We can first dispose of a considerable number of cases, often cited but having little bearing on the problem in hand. Thus methods of raising an army and providing munitions fall within specific constitutional authorization.[99] War merely demonstrates the public nature of the use in taking private property with compensation in other cases.[100] Or national conditions created by war bring areas which were formerly subject to state police power under federal jurisdiction.[101] The control of enemy aliens and their property is recognized and

---

[98]Sterling v. Constantin, 287 U. S. 378, 53 Sup. Ct. 190 (1932).

[99]Ashwander v. Tennessee Valley Authority, 297 U. S. 288, 56 Sup. Ct. 466 (1935); Hamilton v. Regents of University of California, 293 U. S. 245, 55 Sup. Ct. 197 (1934); and Selective Draft Law Cases, 245 U. S. 366, 38 Sup. Ct. 159 (1917).

[100]Highland v. Russell Car and Snowplow Co., 279 U. S. 253, 49 Sup. Ct. 314 (1929); Dakota Central Tel. Co. v. South Dakota, 250 U. S. 163, 39 Sup. Ct. 507 (1919); and Northern Pacific Ry. Co. v. North Dakota, 250 U. S. 135, 39 Sup. Ct. 502 (1919).

[101]Block v. Hirsh, 256 U. S. 135, 41 Sup. Ct. 458 (1920) (rents); Hamilton v. Kentucky Distilleries Co., 251 U. S. 146, 40 Sup. Ct. 106 (1919) (sale of liquor); McKinley v. United States, 249 U. S. 397, 39 Sup. Ct. 324 (1918) (prostitution near army camps); and Stewart v. Kahn, 11 Wall. 493 (U. S. 1870).

plenary.[102] The freedom of speech cases have been treated elsewhere.[103]

On October 6, 1775, the first Congress of Delegates recommended to the colonial assemblies that they confine any person who might endanger the liberties of America.[104] We know that some Tories assisting the British cause were in fact apprehended. The Continental Congress was upheld in compelling the removal of articles which might fall into the hands of the enemy.[105] The history of the Alien and Sedition Laws and the practical restriction of these to alien enemies has already been noted.[106] Although Washington called out the army during the Whiskey Rebellion, arrests were made only on warrant issued by a judge for specific crimes shown,[107] and in the Burr Conspiracy of 1805 civil agencies alone were employed.[108] In 1812, our coastline was invaded by the British and our Capitol was burned; under these circumstances removal of alien enemies above tidewater was upheld.[109]

In the same period, the military attempted to control citizens, at least when they were suspected of being spies, but these were directed by the President to be released,[110] and General Jackson's imposition of military control at New Orleans over civilians, and his refusal to heed a writ of habeas corpus resulted in his having to pay a $1,000 fine.[111] The Mexican War saw no great interference with private rights, but the Supreme Court was astute to uphold an individual's right to recover for private property taken by a military officer without justification.[112]

It was the Civil War which produced a considerable amount of executive control over individual rights. James G. Randall has summarized some of the powers which President Lincoln claimed to exercise:

"As interpreted by President Lincoln, the war power specifically in-

---

[102]Central Trust Co. v. Garvan, 254 U. S. 554, 41 Sup. Ct. 214 (1920); The Prize Cases, 2 Black 635 (U. S. 1862); and Miller v. United States, 11 Wall. 268 (U. S. 1870).

[103]See notes 68-75, 91 *supra*. After *Schenck* and related cases, the other cases upholding convictions under the Espionage Act, United States v. Abrams, 250 U. S. 616, 40 Sup. Ct. 17 (1919), Schaeffer v. United States, 251 U. S. 466, 40 Sup. Ct. 259 (1919), Pierce v. United States, 252 U. S. 239, 40 Sup. Ct. 205 (1919), were by divided courts, with Holmes and Brandeis, whose views shaped the present rule of law, taking the position that the writings there denouncing our intervention in Russia and criticising our war strength should not be suppressed.

[104]1 ELLIOT'S DEBATES 50.

[105]Respublica v. Sparhawk, 1 Dall. 357 (U. S. 1788).

[106]See note 7 *supra,* and also 2 BEVERIDGE, THE LIFE OF JOHN MARSHALL (1919) 381 ff.; BOWERS, JEFFERSON AND HAMILTON: THE STRUGGLE FOR DEMOCRACY IN AMERICA (1937) 375 ff.

[107]2 MCMASTER, HISTORY OF THE PEOPLE OF THE UNITED STATES (1927) 190.

[108]2 WANDELL AND MINNEGERODE, AARON BURR (1927) 118-145.

[109]See notes 6 and 7 *supra.*

[110]*Ex parte* Quirin *et al.*, 317 U. S. 1, 63 Sup. Ct. 1 (1942).

[111]J. S. BASSETT, LIFE OF JACKSON (1911) 224-230.

[112]Mitchell v. Harmony, 13 How. 115 (U. S. 1851).

cluded the right to determine the existence of 'rebellion' and call forth the militia to suppress it; the right to increase the regular army by calling for volunteers beyond the authorized total; the right to suspend the habeas corpus privilege; the right to proclaim martial law; the right to place persons under arrest without warrant and without judicially showing the cause of detention; the right to seize citizens' property if such seizure should become indispensable to the successful prosecution of the war; the right to spend money from the treasury of the United States without congressional appropriation; the right to suppress newspapers; and the right to do unusual things by proclamation, especially to proclaim freedom to the slaves of those in arms against the Government."[113]

Lincoln's apology, in his letter of June 12, 1863, for the action taken is one of his great writings and goes far to prove the enormity of the emergency, rebellion in the North itself, and conditions in the courts bordering on anarchy.[114] Five famous cases growing out of the exercise by Lincoln and his generals of the war power are instructive: *Miller v. United States,*[115] Ex parte *Merryman,*[116] Ex parte *Vallandigham,*[117] Ex parte *Milligan,*[118] and the *Prize Cases.*[119]

The *Miller* case and the *Prize Cases* faced the problem that persons residing in the South were still technically citizens, but either they or their states were carrying on a rebellion against the central government. It was determined that a state of war existed between the southern states and the central government and that citizens of the southern states were "enemies" so that trading with them might be forbidden or their goods seized. Although this theory of "war" has been severely criticised, it is probably as good a description of a situation clearly "extra" the Constitution as any other phrase.

Merryman was a lieutenant drilling a secessionist company in Baltimore, a city in 1861 actively opposed to the war effort and on the only direct railroad line from the North to Washington. Troops moving through Baltimore to defend Washington had been set upon by a mob and the city and state had refused to take action;[120] General Cadwalader took Merryman into custody and, pursuant to presidential authorization, suspended the writ of habeas corpus. Chief Justice Taney wrote his famous opinion asserting that only

---

[113] RANDALL, CONSTITUTIONAL PROBLEMS UNDER LINCOLN (1926) 36.

[114] 8 NICOLAY AND HAY, WORKS (1894) 298.

[115] 11 Wall. 268 (U. S. 1870).

[116] 17 Fed. Cas. 144, No. 9,487 (1861).

[117] 1 Wall. 243 (U. S. 1863).

[118] 4 Wall. 2 (U. S. 1866).

[119] 2 Black 635 (U. S. 1862).

[120] See FAIRMAN, MR. JUSTICE MILLER AND THE SUPREME COURT (1939) ch. 4; THE WAR OF THE REBELLION: OFFICIAL RECORDS OF THE UNION AND CONFEDERATE ARMIES, Series 2, vol. 5, pp. 226, 564; and RANDALL, *op. cit. supra* note 47, at 161.

Congress could suspend the writ and leaving it to the President to cause "the civil process of the United States to be enforced." The sequel to that case must be noted. Soon after this opinion Merryman was transferred to a civil court, charged with treason (for destruction of railroad bridges), and released on bail. The case was delayed until after the war, when the charges were dropped.[121]

Vallandigham was an outstanding Copperhead politician of Ohio. Morgan and his raiders were preparing to, and later did, invade Ohio in the hopes of an uprising of Copperheads. General Burnside placed *Vallandigham* on trial before a military tribunal for the sentiments he had uttered. The district judge denied the writ of habeas corpus. A vain attempt to get the case to the Supreme Court for review was made by requesting certiorari of the military tribunal record, which of course was refused as not within the appellate jurisdiction of the Court. President Lincoln disapproved of Burnside's action, and commuted Vallandigham's sentence to removal within the Confederate lines, which was effected.[122]

Milligan held much the same position in Indiana which Vallandigham did in Ohio. He was arrested in 1864, charged and found guilty by a military tribunal of conspiracy to overthrow the government. Morgan had invaded Indiana in 1863; the Southern cause seemed at its height; the pro-Southern movement in Indiana was substantial. When the case came before the Supreme Court all the justices agreed that Congress had not authorized trial by the military commission. The majority, by Justice Davis, went on to deny that Congress had the power to establish "martial law" except in case of actual, present invasion. The minority, by Chief Justice Chase, believed that Congress had the power in cases of "imminent public danger," "where ordinary law no longer adequately secures public safety and private rights."[123] The *Milligan* case, though the dictum of the majority has been severely criticized, has been accepted as stating the applicable rules of law in other particulars as late as 1942.[124] The government now admits that both the majority and minority in

---

[121]SWISHER, ROGER B. TANEY (1935) 557; see also note 120 *supra*.

[122]For this history see: Fairman, *The Law of Martial Rule and the National Emergency*, (1942) 55 HARV. L. REV. 1253, 1283; RANDALL, *op. cit. supra* note 47, at 179; NICOLAY AND HAY, *op. cit. supra* note 114, at 336, 345, 360; and Lincoln's Letter to Burnside: "All the Cabinet regretted the necessity of arresting, for instance, Vallandigham—some perhaps doubting that there was a real necessity for it, but being done all are for seeing you through with it." THE WAR OF THE REBELLION: OFFICIAL RECORDS OF THE UNION AND CONFEDERATE ARMIES, Series 2, vol. 5, pp. 717, 657, 705.

[123]4 Wall. 2, 140, 142 (U. S. 1866). For the general background see Fairman, *The Law of Martial Rule and the National Emergency* (1942) 55 HARV. L. REV. 1253, 1284.

[124]*Ex parte* Quirin, 317 U. S. 1, 63 Sup. Ct. 1 (1942).

the *Milligan* case were right in holding that there was no "military necessity" for the action taken against Milligan.[125]

The Habeas Corpus Act of March 3, 1863, had the effect of recognizing the supremacy of the civil power and of freeing all persons held by the military if grand juries did not find indictments for crimes committed.[126] There were cases not as famous as the above five, and incidents other than cases, which show that in spite of the untoward danger to the Union, the basic civil rights were protected against military action.[127]

In both the Revolutionary War and the Civil War we have noted that persons who assisted those opposed to the government were interned by the military without trial. It appears to me that in the recent case of Ex parte *Quirin* the Court has recognized a distinction between this type of detention and the present detention of persons of Japanese ancestry. In the *Quirin* case the Court took great pains to point out that the military had a right to control Haupt, the only American citizen among the saboteurs, on the basis that "citizens who associate themselves with the military arm of the enemy government . . . are enemy belligerents," to be treated as such under the laws of war. Where, as in the Revolution, when there was no "citizenship," or in the Civil War when rebellion made "citizenship" only technical, it might be proper to treat those who assisted the "enemy" as "enemy belligerents." That would justify the military detention of persons of Japanese ancestry who could be proven to be aiding the enemy; but such determination of the military would be subject to judicial review just as at present judicial review may be had to

[125]Brief for Government, Hirabayashi v. United States.

[126]RANDALL, *op. cit. supra* note 47, at 163-164, 189.

[127]Abrams, *The Jeffersonian, Copperhead Newspaper* (1942) 2 BILL OF RIGHTS REV. 284 describes the recovery by the owner of the Jeffersonian due to its suppression. Roscoe Pound, *War and the Law* (1943) 14 PA. BAR ASS'N Q. 242, 248 depicts both the emergency background and official reaction thus: "In the summer of 1863, when Lee was moving on Pennsylvania, Morgan was preparing to invade Ohio, Rosecrans was stalled in Middle Tennessee, and Johnston was collecting an army in Grant's rear behind Vicksburg, there was an emergency if our country ever encountered one. But General Burnside's order suspending the Chicago Times for 'repeated expressions of disloyal and incendiary sentiments' was at once revoked by President Lincoln. In the summer of 1864, after Cold Harbor, after the operations about Petersburg seemed to have reached a standstill, when Sherman seemed to be making little headway toward Atlanta, and Early was in the Shenandoah Valley, a great political party was allowed to hold a convention which in its platform pronounced the war a failure. Joel Parker at the Harvard Law School was allowed to attack the legality of important items of the administration's policy. There was a protracted newspaper controversy between the partisans of Meade and of Sickles as to the conduct of the battle of Gettysburg. After Shiloh, during the long struggle to get a foothold back of Vicksburg, and after Cold Harbor, Grant was persistently attacked in the press. But the attacks were without effect on his imperturbable pursuit of his duty, and neither helped the South nor hindered the military operations of the North."

determine whether a person held as an alien is in fact such.[128]

Internment of aliens or citizens during World Wars I and II presents different questions in England and in the United States. In England during the First World War, the Defense of the Realm Act[129] provided for exclusion from defense areas, and internment on suspicion, of persons thought to be aiding the enemy. The power was lodged in *civilian* agencies. This action was upheld in *Bonnfeldt v. Phillips*[130] and *Rex v. Halliday.*[131] In 1941 the House of Lords upheld, in the *Liversidge* and *Greene* cases,[132] under the Emergency Powers Act of 1939[133] and Regulation 18B, internment on the order of the Home Secretary "with a view to preventing [the person held] acting in a manner prejudicial to the public safety or the defense of the realm." The decision turned on the issue of statutory construction: whether Parliament had lodged power in the Home Secretary to make the determination of prejudice or hostility objectively (of which courts should judge) or subjectively (of which he would be the sole judge). The court adopted the latter interpretation, but not without a strong dissent by Lord Atkin, who quoted the words of Pollock: "in a case in which the liberty of the subject is concerned we cannot go beyond the natural construction of the statute." The case was severely criticized by Carleton Kemp Allen in April, 1942.[134] In Canada, the War Measures Act[135] also lodges powers of exclusion in a *civilian,* the Minister of Justice, who has issued regulations excluding persons of the Japanese race from certain limited military areas, but not confining them in camps and allowing them to return to these areas under permit issued by the Mounted Police.[136] These cases must not be considered precedents for American action.[137] Although England has a tradition of protecting civil liberties not unlike our own, its constitutional theory is entirely different. Its Constitution is the Common Law as modified by Parliament; Parliament *is* the Constitution. As F. A. Brewing, in speaking of these statutes and decisions, says:

---

[128]*Ex parte* Quirin, 317 U. S. 1, 63 Sup. Ct. 1 (1942); *Ex parte* Gilroy, 257 Fed. 110 (S. D. N. Y. 1919); *Ex parte* Risse, 257 Fed. 102 (S. D. N. Y. 1919); and *Ex parte* Fronklin, 253 Fed. 984 (N. D. Miss. 1918). REP. ATT'Y GEN. (1918) 35.

[129]4 & 5 GEO. V, c. 29; 5 & 6 GEO. V, c. 8.

[130]35 T. L. R. 46 (K. B. 1916).

[131](1917) A. C. 260.

[132]Liversidge v. Anderson, [1942] A. C. 206, 3 All Eng. 338 (1941); Greene v. Secretary of State for Home Affairs, [1942] A. C. 284. Carr, *A Regulated Liberty* (1942) 42 COL. L. REV. 339.

[133]2 & 3 GEO. VI, c. 62.

[134]Allen, *Regulation 18b and Reasonable Cause* (1942) 58 L. Q. REV. 232.

[135]REV. STAT. CANADA 1927, c. 206.

[136]Canada Gazette, Extra No. 96, August 31, 1942.

[137]Note the action of the highest court of Australia setting aside, in a case involving Jehovah's Witnesses, virtually all subversive activities regulations. Associated Press, June 14, 1943.

"The Imperial Parliament is absolute and can, of course, abrogate any civil liberty it sees fit."[138] Even where power is thus absolute it should be noted that Parliament has lodged the discretion in *civilian* agencies, and has itself kept constant control by requiring that all regulations must be submitted to and passed by Parliament before they are effective.[139]

In the United States, World War I was marked by a liberalization of the treatment of enemy aliens so that only disloyal or dangerous aliens were interned;[140] even then the Attorney General pointed out in his 1918 report that such internment was "anomalous under the American judicial system, in that it provides for the summary exercise of executive authority. . . . "[141] A statute was proposed in Congress which would have divided this country into military districts subject to regulations to be adopted by the military commanders. President Wilson wrote to Senator Overman:

> "I am wholly and unalterably opposed to such legislation. . . . I think it is not only unconstitutional, but that in character it would put us nearly upon the level of the very people we are fighting. . . . It would be altogether inconsistent with the spirit and practice of America. . . . I think it is unnecessary and uncalled for. . . . "[142]

In *Stoutenburgh v. Frazier,*[143] the Court of Appeals of the District of Columbia gave the only consideration, of which I know, to an act of Congress permitting the confinement of a person as subject to suspicion. Although this was one of the wartime acts and could have been construed to authorize "preventive custody" only after a hearing, the court held the act unconstitutional. On December 26, 1941, Congress authorized the Commissioners of the District of Columbia to evacuate (but not to intern) persons from the District when "public interest or the safety of such persons creates the necessity therefor."[144]

In the fall of 1941 Attorney General Biddle spoke of the detention of aliens without trial:

> "The Hobbs Bill, introduced this year, authorizes, when and only as long as deportation is impossible, detention of the alien in federal institutions. . . . The bill is in a sense revolutionary, because it permits deten-

---

[138]Brewing, *Civil Liberties in Canada during Wartime* (1941) 1 BILL OF RIGHTS REV. 112.

[139]Note that only about 2,000 have been interned and over half of these have been released. Laski, *Civil Liberties in Great Britain in Wartime* (1942) 2 BILL OF RIGHTS REV. 243; and see note 138 *supra*.

[140]See note 9 and 10 *supra*.

[141]REP. ATT'Y GEN. (1918) 35.

[142]8 BAKER, WOODROW WILSON, LIFE AND LETTERS (1927-1939) 100.

[143]16 App. D. C. 229 (1900).

[144]Pub. L. No. 373, 77th Cong., 1st Sess. (Dec. 26, 1941), 55 STAT. 858 (1941).

tion without trial by jury. But where aliens cannot be deported some control is desirable. It is a very serious curb on civil rights made essential by the circumstances of war."[145]

The only act of Congress authorizing evacuation of internment relates solely to aliens.[146]

It has been reported in the newspapers that several American citizens of German ancestry have been directed by the Army to leave designated areas on the east coast, have refused to do so, and have not been prosecuted.[147] Can it be that the government realizes the inability to sustain such evacuation? Such action is even more open to criticism than "temporary mass evacuation," since it is only "preventive" in a technical sense. It amounts to a determination of individual guilt of ordinary civilian citizens without a hearing.

B. *Factual Background in Early 1942*

The necessity for determining reasonableness requires the examination of the situation—military, political, social—as it existed in early 1942, against which background the evacuation prescribed by General DeWitt must be tested.

No exercise of the federal police or war power has ever infringed on basic individual rights of citizens as directly as this evacuation, without being condemned as in excess of constitutional authorization. Unless, therefore, there were compelling reasons in the facts existing in early 1942 for the adoption of a new rule, we would have to conclude that evacuation of 79,000 American citizens from the five west coast states was an unconstitutional exercise of the federal police power, even in time of war.

The author recognizes that the whole military situation, perhaps even the part familiar to General DeWitt, cannot be known to the public until the end of the war. Yet the Supreme Court must, of necessity, render its judgment on some conception of military necessity. The government, without suggesting that it will reveal to the Court so much of the secret information as may be necessary to justify the evacuation, has attempted to uphold the program on the basis of "facts known to the public."[148] The author disclaims any knowledge of military tactics, but he is impelled to the belief that if "military necessity" is going to be decided on the basis of facts known to the public, then when *all* the facts known to the public are considered, far from compelling

[145]Attorney General Biddle, *Civil Rights in Times of Stress* (1941) 2 BILL OF RIGHTS REV. 13.

[146]40 STAT. 531 (1918), 50 U. S. C. § 21 (1940).

[147]Case of Mrs. Schuller, N. Y. Times, May 8, 1943. See also case of Wilcox on the west coast, CIVIL LIBERTIES QUARTERLY, A. C. L. U., June 1943.

[148]See Brief for Government, Hirabayashi v. United States.

the evacuation, they do not even support an "honest" (to use the Supreme Court's terminology) conclusion that evacuation of these citizens was required.

1. *The Government's Claim of Military Necessity.*—The government's justification[148] for the west coast evacuation is based upon the Japanese successes in the Pacific and the alleged nonassimilation of the west coast Japanese by reason of continuous anti-Japanese agitation, and the policy of exclusion of Japanese aliens from immigration, from the right to become naturalized, from intermarriage, from owning land. The government points to these, to the belief in Shintoism, the continuance of Japanese language schools in the United States, and the possible retention of dual citizenship, and concludes that "it is entirely possible that an unknown number of the Japanese may lack to some extent a feeling of loyalty toward the United States as a result of their treatment."[149] In the footnote it is admitted, however, that "most of the evacuees are loyal to this country."

2. *Factual Background Showing Lack of Military Necessity.*—

(a) The Military Background: At the same time that we were girding ourselves for battle in the Pacific, Germany and Italy also declared war on the United States. Both of these countries had experienced phenomenal military successes as even the briefest chronological history will recall. Italy had overrun Ethiopia and Albania;[150] had invaded British African colonies,[151] Egypt,[152] and Greece;[153] with the help of Germany had taken Crete[154] and other Mediterranean islands and completed the conquest of Greece,[155] thus challenging Britain's lifeline in the Mediterranean. Germany, rising from complete impotence in military matters in 1933, had occupied the Rhineland, annexed Austria,[156] forced upon England and France the Munich appeasement,[157] overrun Czechoslovakia,[158] and devastated Poland in twenty-six days.[159] She had occupied Denmark and invaded Norway,[160] crushed the Netherlands in four days[161] and Belgium in seventeen days;[162] had annihilated

[149]Brief for Government, p. 21, Hirabayashi v. United States.
[150]April 7, 1939. Note: Details for all these footnotes may be found in the N. Y. Times for the day given, or the day following.
[151]August 6, 1940.
[152]September 14, 1940.
[153]October 28, 1940.
[154]June 1, 1941.
[155]April 27, 1941.
[156]March 13, 1938.
[157]September 30, 1938.
[158]March 14, 1939.
[159]September 1-27, 1939.
[160]April 9, 1940.
[161]May 14, 1940.
[162]May 28, 1940.

the highly ranked French army and conquered France in ten days;[163] had launched and maintained mass air raids on England consistently for nearly six months[164] and had achieved a recognized superiority in the air over England. The Nazis had taken Rumania with her rich oil wells[165] and Bulgaria[166] and had moved to swift victories in Greece[167] and Crete.[168] Turning from England she had invaded Soviet Russia on a 2,000 mile front[169] and had piled victory upon victory—Kiev, Odessa, Kharkov, the outskirts of Moscow, Rostov. All these were completed immediately prior to March, 1942;[170] and she went on immediately to take Sevastopol, a large part of the Caucasus, and Stalingrad.[171] Surely in early 1942 the Germans seemed invincible.

Nor had the German efforts been limited to Europe. German armed ships and submarines had terrorized our Atlantic shipping and waters within 100 miles of our east coast. We recall a few of the better known examples: the raider *Deutschland* seized the *S.S. City of Flint;*[172] the *Graf Spee*[173] and the *Columbus*[174] were scuttled off the American coast; the *Robin Adair* was sunk by Nazi submarines in the South Atlantic;[175] the United States Destroyer *Kearny* was torpedoed off Iceland,[176] and the destroyer *Reuben James* was torpedoed nearer our coast.[177] The number and proximity to our coast of the sinkings became alarming by January, 1942.[178] So complete was Germany's freedom of movement in our waters that a submarine shelled the oil refineries on Amba, in the Caribbean, on February 16, 1942, and submarines were able to land saboteurs and quantities of explosives at two points on our coast on June 13 and 17.[179]

In Africa, Rommel launched his successful drive in January, 1942, which carried him by June to almost complete control of North Africa.[180] The Germans had large concentrations of planes, battleships, and submarines at

---

163June 3-13, 1940.
164Launched July 29, 1940; continued as heavy bombing until after January, 1941.
165November 23, 1940.
166March 1, 1941.
167April 6-27, 1941.
168June 1, 1941.
169June 22, 1941.
170September 20, October 14, October 17, October 25, November 22, 1941.
171May 11, 24, 26; July 2, 27; September 17, 1942.
172October 23, 1939.
173December 13, 1939.
174December 19, 1939.
175May 21, 1941.
176October 17, 1941.
177October 31, 1941.
178January 14, 1942.
179June 25, 1942.
180January 21; June 21, 25, 1942

bases nearer to New York than are Hawaii, Kiska, or Midway to the west coast. The roundup of two German spy and saboteur rings on the east coast was just being completed;[181] the influence and size of the German-American Bund were being revealed and quantities of munitions in their possession were being seized. By comparison, our successes in the Pacific were considerably greater than those in the Atlantic; and Japan, unable to conquer China in seven years, was a less formidable foe than Germany with her record of conquest after conquest of the most difficult antagonists. Certainly there was no clearer "military necessity" in the Pacific than there was in the Atlantic.[182]

(b) Census Statistics: One of the reasons given for the evacuation was the concentration of Japanese aliens and American citizens of Japanese ancestry in and around the major cities of the west coast. An examination of *Table A,* prepared from the census figures of 1940, shows how untenable on this basis evacuation of the Japanese without evacuation of the Germans and Italians really was.[183]

The government also placed some stress on the Japanese group being one of "late arrival" in this country with attendant exposure to Japanese indoctrination. *Table B* demonstrates no problem of "recent arrival" applicable to the Japanese which is not equally and more applicable to the Germans.

We have already noted that the government has emphasized that the preponderance of Japanese aliens in the upper age groups made them especially dangerous in indoctrination of the younger persons of Japanese ancestry.[184] The unsubstantial character of this "evidence" is seen by an examination of *Table C* which is a combination of figures on pages 91, 95, 236, 242, 243 of *H. R. Rep. No. 2124.* In fact General DeWitt exempted the *older* German and Italian aliens from his orders of evacuation.[185] We should recognize that since 5,000 males of Japanese ancestry between 21 and 35 were in the army,[186] 55,000 were females,[187] the alien males averaged 59 years of age,[188] and 31,000 evacuees were under the age of 15, the Japanese evacuation was largely of old men, women, and children.

---

[181] February 4, 1942, p. 12, col. 5; March 3, p. 13, col. 2; March 7, p. 1, col. 1; January 3, p. 1, col. 2.

[182] Brief for Government, p. 12 ff., Hirabayashi v. United States.

[183] Compare the ratio of German and Italian foreign born to Japanese aliens or citizens of German and Italian ancestry with those of Japanese ancestry. Compare also the east coast which is as important in war material production as the west. STATE DISTRIBUTION OF WAR SUPPLIES AND FACILITY CONTRACTS, June, 1940 through December, 1941, (issued January 18, 1942, by the O.P.M.); and the SUPPLEMENT, cumulative through February, 1943 (issued April 3, 1943).

[184] Note 16 *supra.*

[185] H. R. REP. No. 2124, 77th Cong., 2d Sess. (1942) 332.

[186] Letter of the President to the Secretary of War, February 1, 1942.

[187] H. R. REP. No. 2124, 77th Cong., 2d Sess. (1942) 94.

[188] *Id.* at 95.

TABLE A

CONCENTRATION OF JAPANESE, ITALIANS, AND GERMANS BY CITIES, 1940

WEST COAST

City	Los Angeles	Seattle	San Francisco	Oakland, Calif.	Portland, Oregon	Sacramento
Total population	1,191,182	368,302	634,536	302.163	305,394	105,958
Total "white" foreign born	215,248	59,612	130,271	42,661	38,608	12,144
German foreign born	17,528		14,977	3,682	3,947	931
Italian foreign born	13,256		24,036	5,707	2,685	1,962
Japanese foreign born	8,726	2,876	2,276	655	725	974
German alien	5,570	1,085	4,632	800	952	381
Italian alien	5,892	1,469	12,183	2,899	1,409	1,332
Japanese alien	8,726	2,876	2,276	655	725	974
Citizens of German ancestry‡	126,150	35,760	78,160	25,590	23,160	7,280
Citizens of Italian ancestry‡	73,180	20,264	44,302	14,500	13,120	4,120
Citizens of Japanese ancestry	14,595	4,099	3,044	1,135	955	1,905

EAST COAST

City	New York	Philadelphia	Baltimore	Pittsburgh	Chicago	Detroit
Total population	7,454,995	1,931,334	859,100	671,659	3,396,808	1,623,452
Total "white" foreign born	2,080,020	290,325	60,969	84,606	672,705	320,664
German foreign born	224,749	27,286	9,744	9,805	83,424	23,785
Italian foreign born	409,489	59,079	8,063	16,241	66,472	26,277
Japanese foreign born	1,456	*	*	*	*	*
German alien	91,240†	5,670	2,630	2,050	17,100	4,920
Italian alien	173,620†	21,250	6,430	5,840	23,460	11,400
Japanese alien	1,456	*	*	*	*	*
Citizens of German ancestry‡	1,248,000	174,190	36,580	50,760	404,820	192,400
Citizens of Italian ancestry‡	707,200	98,710	20,730	28,760	228,720	109,020
Citizens of Japanese ancestry	631	*	*	*	*	*

This chart is prepared from figures given in House Report No. 2124, pp. 100, 231, 233, 234, as supplemented by 16th Census of the United States, Population, Second Series Summary, pp. 108, 156.

*Negligible numbers—less than 500 (House Report No. 2124, p. 100).

†Computed figures based on percentage relation of aliens to foreign born for each group in the particular state (House Report No. 2124, p. 229).

‡Computed figures determined by multiplying the number of "white foreign born" in each city by the percentage (60% for the German, 34% for the Italians) showing the relationship of persons of German or Italian ancestry to the total "white foreign born" population of the United States (House Report No. 2124, pp. 229, 241).

TABLE B

ALIENS RESIDENT IN THE THREE WEST COAST STATES IN 1940 BY DATE OF LAST ARRIVAL

	1900 or prior	1910 or prior	1918 or prior	1935 or prior	Number who arrived after 1935
Italians	7.4%	41.3%	69.5%	94.5%	2,891
Japanese	3.0%	21.0%	46.7%	77.0%	6,671
Germans	11.0%	25.0%	35.0%	67.0%	7,710

This chart is prepared from House Report No. 2124, pp. 96, 235, 242.

Note: Japanese could not enter the United States for the *first* time after 1926 and therefore substantially all Japanese have had American contacts for at least 17 years. Among the Germans and Italians, the *last* arrival is most frequently also the *first*.

Note also: This chart does not show *foreign born;* it omits Germans and Italians who have become naturalized, whereas no Japanese may become naturalized.

TABLE C

AGE GROUP CONCENTRATION (THREE WEST COAST STATES)

	German	Japanese	Italian
Foreign born females over 45 yrs.	56.0%	50%	65.00%
Foreign born males over 45 yrs.	60.5%	75%	74.84%
Foreign born males over 55 yrs.	36.29%	45%	44.62%

This chart is prepared from House Report No. 2124, pp. 91, 95, 236, 242, 243.

TABLE D

RELATION OF ALIENS TO NATIVE BORN CITIZENS OF JAPANESE ANCESTRY IN THE UNITED STATES

	Total Japanese	Percentage native born citizens	Native born	Foreign born
1910	72,157	6.2	4,502	67,655
1920	111,010	26.7	29,672	81,338
1930	138,834	49.2	68,357	70,477
1940	126,947	62.7	79,642	47,305

This chart is a combination of the figures given in House Report No. 2124, pp. 60, 78, 94.

Nor can it be asserted that the Japanese group is not becoming integrated in our society. In spite of the fact that no Japanese alien may be naturalized and, therefore, their number remains constant, except as altered by death, *Table D* shows that the percentage of native born citizens of Japanese ancestry has steadily increased from 6.2 per cent of the total Japanese population in 1910 to 62.7 per cent in 1940.

The studies available suggest that the integration of Italians and Germans into our culture is even slower than that of the Japanese.[189]

The Japanese on the west coast were 45 per cent agricultural; the German and Italian aliens 3.43 per cent and 7.27 per cent agricultural and 52 per cent unemployed; the average value of a Japanese farm in California was $12,810.[190] Are these the Japanese people from whom we should anticipate the attempted overthrow of this government which forms the only protection for the value of their investments?

(c) Expectation of Sabotage or Fifth Column Activity and Effectiveness of Methods Other Than Evacuation to Prevent It: On December 7, 1941, the Territory of Hawaii was attacked and immediately martial law was declared. Yet at no time, though the Japanese in Hawaii constitute 38 per cent of the total population,[191] has it been deemed necessary to conduct any evacuation, and the evidence on record in the Tolan Committee Hearings at the time of General DeWitt's orders[192] was all to the effect that there was no sabotage

---

[189]On June 12, 1942, the census bureau reported that there were 3,360,740 native-born American citizens of German ancestry for whom German was the *mother tongue*. N. Y. Times, June 13, 1942. A careful study of Italians in San Francisco and the four surrounding counties, in 1935, reported 193,454 individuals of Italian ancestry "more or less directly in contact with Italian mores and habits of mind." RADIN, THE ITALIANS OF SAN FRANCISCO (1935) SERA Project 2-F2-98, pp. 38, 64. The studies showing the high degree of assimilation of the Japanese in the United States are of unimpeachable merit. See the monumental work, *The Second-Generation Japanese Problem* (1934) and companion volumes, by Professor E. K. Strong, who concludes at p. 6: "young Japanese are more readily assimilated than people of several European races." See also JAPAN REVIEW, April 1921, 93 ff.; SMITH, AMERICANS IN PROCESS: A STUDY OF OUR CITIZENS OF ORIENTAL ANCESTRY (1937); BEACH, ORIENTAL CRIME IN CALIFORNIA (1932); MEARS, RESIDENT ORIENTALS ON THE AMERICAN PACIFIC COAST (1928).

[190]These figures are taken from H. R. REP. No. 2124, 77th Cong., 2d Sess. (1942) pp. 88, 104, 117, 131, 135, 237, 244.

[191]*Id.* at 91.

[192]Stafford Austin, Chairman of all rural districts on Oahu under the Office of Civilian Defense said: " . . . at no time prior to or subsequent to December 7, 1941, have any signs of sabotage or subversive activities or 'blinker' signals been reported to affiant." *Id.* at 51. The Attorney General informed the Tolan Committee that "Mr. John Edgar Hoover, Director of the Federal Bureau of Investigation, has advised me there was no sabotage committed there [Hawaii] prior to December 7, on December 7, or subsequent to that time." *Id.* at 49. Henry L. Stimson, Secretary of War said: " . . . the War Department has received no information of sabotage committed by Japanese during the attack on Pearl Harbor." *Id.* at 48. John Burns of Honolulu, in charge of the espionage

or fifth column activity before, during, or after Pearl Harbor and that persons of Japanese ancestry behaved with remarkable loyalty.

The evidence of loyalty in the United States is equally clear. Persons who should know are direct in their testimony as to the loyalty of American citizens of Japanese ancestry in the United States.[193] Government counsel conceded, in the circuit court of appeals in the *Hirabayashi* case, that there had been no acts of sabotage or disloyalty by American citizens of Japanese ancestry.[194]

Another attempt to find a justification for the evacuation of American citizens of Japanese ancestry is the government's argument that these American citizens retain a "dual citizenship" in Japan. There are five considerations which completely dispose of any justification based upon this alleged lack of loyalty. First, the government's position is directly contrary to and at variance with the express statutory law of the United States and decisions of the Supreme Court.[195] Second, every authority on international law recognizes the possibility of "dual citizenship" of a person born outside the country of his parents' nationality as a result of the two systems of law for determining citizenship: place of birth (*jus soli*) and ancestry or descent (*jus sanguinis*), since nearly all countries employ *both* systems.[196] Third, it needs to be said that Japan has gone farther in surrendering its claim to allegiance from persons of Japanese ancestry born elsewhere than in Japan than has almost any other country, certainly much farther than Germany and Italy.[197] Fourth,

bureau, working with the office of military intelligence and F. B. I., summarized his investigations: " . . . there has not been either on or since December 7 any authenticated case of sabotage or any authenticated case of group activity on behalf of Japan on the part of persons of Japanese ancestry, citizen or alien, residing on the Island of Oahu." *Id.* at 58.

[193]The New Year's 1942 edition of the Pacific Citizen is made up almost entirely of statements by political and other leaders on the west coast attesting to this loyalty. Milton Eisenhower, then Director of the War Relocation Authority, had no difficulty in recognizing 85 per cent of the American born as loyal. (Testimony before House Subcommittee on Appropriations, June 15, 1942.)

[194]Brief for Hirabayashi, p. 6, Hirabayashi v. United States.

[195]Rev. Stat. § 1999 (1875), 8 U. S. C. § 800 (1940); 54 Stat. 1168, 8 U. S. C. § 801 (1940); and United States v. Wong Kim Ark, 169 U. S. 649, 18 Sup. Ct. 456 (1897).

[196]Flourney (assistant solicitor in the Department of State), *Dual Nationality and Election* (1921) 30 Yale L. J. 545; 3 John Bassett Moore, A Digest of International Law (1906) 518-751; Stowell, International Law (1931) 190-91; Hyde, International Law Chiefly As Interpreted and Applied by the United States (1922) 658, 666-669; Borchard, The Diplomatic Protection of Citizens Abroad (1927) 18-19, 550-580. Moore, The Doctrine of Expatriation (1905) 234-236; Borchard, The Diplomatic Protection of Citizens Abroad (1927) 19. Our chief struggles as to dual allegiance have been with European nations, some of which absolutely refuse expatriation, and some of which require permission for expatriation and fulfillment of military service requirements. Borchard, *op. cit. supra*, at 546; Moore, American Diplomacy (1905) 190-91.

[197]Prior to 1885 Japan forbade all immigration. In 1899 Japan enacted the first nationality law, which adopted the European rule of *jus sanguinis*. Expatriation under

against great odds and at considerable cost to themselves, American citizens of Japanese ancestry have, to a degree not shown by any other "foreign" group, surrendered any claim to Japanese citizenship.[198] Fifth, even if dual citizenship did exist in theory, it is recognized by international law that in fact only the nation of *domicile* is effective in claiming the person's allegiance.[199]

Considerable point has been made of the attendance of children of Japanese parents at Japanese language schools, implying that these are something unique in this country and hotbeds of Japanese indoctrination.[200] One who advances this argument must be unmindful of constitutional history. *Meyer v. Nebraska*[201] and *Bartels v. Iowa*[202] certainly recognized the right to teach German in a school. It is only necessary to read quickly what studies have been made of the Japanese language schools to be clear in several conclusions, all of which definitely disprove any subversive influence or disloyalty of these schools. Similar schools are conducted by Chinese, Germans, and others.[203] The Japanese schools were originally formed either to teach Christianity or to expedite the Americanization of the Japanese.[204] They form a necessary part of fitting the young people for vocations, since anti-Japanese agitation

this law was possible but difficult. The Japanese associations in the United States from 1914 to 1923 petitioned the Japanese government to facilitate expatriation. Finally, in 1924, Imperial Ordinance No. 262 conceded American citizenship to any person born in the United States of Japanese parents and *denied to him Japanese citizenship* unless within 14 days of birth he is registered by his parents. All impartial observers recognize that far from Japan having "made every effort" (through the doctrine of dual citizenship) to retain the allegiance of the American Japanese, as the government asserts at p. 12 of the *Yasui* brief (Brief for Government, p. 12, Yasui v. United States) it has in fact been the most liberal of governments in surrendering that allegiance. See for a recognition of this and the laws here summarized: STRONG, JAPANESE IN CALIFORNIA (1933) 45-47; KIYO SUE INUI, THE UNSOLVED PROBLEM OF THE PACIFIC (1927) 300-320; WILLIAM C. SMITH, AMERICANS IN PROCESS: A STUDY OF OUR CITIZENS OF ORIENTAL ANCESTRY (1937) 132-137; FLOURNOY AND HUDSON, A COLLECTION OF NATIONALITY LAWS OF VARIOUS COUNTRIES (1929) 385-387; KIYOSHI K. KAWAKAMI, THE REAL JAPANESE QUESTION (1921) 174-187. YAMAMOTO ICHIHASHI, THE JAPANESE IN THE UNITED STATES (1932) 323; MEARS, RESIDENT ORIENTALS ON THE AMERICAN PACIFIC COAST (1928) 107-11; 2 FOREIGN RELATIONS OF THE UNITED STATES (1924) 412.

[198]Even before 1924 when the Japanese law required the registration of every child with the Japanese consul and the children were automatically citizens, Raymond Leslie Buell found that only two-thirds were being registered. Buell, *Some Legal Aspects of the Japanese Question* (1923) 17 AM. J. INT. L. 29, 34.

[199]BORCHARD, *op. cit. supra* note 196, at 589; 3 MOORE, A DIGEST OF INTERNATIONAL LAW (1906) 518; STOWELL, INTERNATIONAL LAW (1931) 187; R. W. Flournoy, *Revision of Nationality Laws of the United States* (1940) 34 AM. J. INT. L. 43-44.

[200]Brief for Government, pp. 28-31, Hirabayashi v. United States.

[201]262 U. S. 390, 43 Sup. Ct. 625 (1922).

[202]262 U. S. 404, 43 Sup. Ct. 628 (1922).

[203]PALMER, ORIENTALS IN AMERICAN LIFE (1934) 51.

[204]Colket, *Suppressing Japanese Schools in Hawaii,* The Nation, November 22, 1922; W. C. SMITH, AMERICANS IN THE MAKING (1939) 302; BELL, PUBLIC SCHOOL EDUCATION OF SECOND-GENERATION JAPANESE (1935) 20; and REUTER, RACE AND CULTURE CONTRACTS (1934).

requires them to work for Japanese.[205] The splendid record of the Japanese for honesty, good citizenship, lack of disorganization and demoralization is directly traced to these schools.[206] Impartial research categorically denies that they are intended to perpetuate the traditions or policies of Japan and verifies the cooperation of these schools in selecting textbooks and courses of study "which correspond to the spirit of Americanism."[207]

Shintoism as "loyalty to the emperor" seems to have been overemphasized. The brief, *amicus curiae*, of the Japanese American Citizens League shows that pictures of Washington or Lincoln were sometimes hung in a Shinto temple and revered as forefathers who had contributed greatly to human advancement, and that the majority of Japanese American citizens were in fact Christian.[208]

There is respectable evidence to the effect that there would have been no substantial danger of civil disorder had the Japanese been left on the west coast;[209] and that the evacuation has been detrimental to productive war capacity,[210] thus not only failing to be a "military necessity" but proving to be a military "blunder."

When the final history of the Japanese evacuation is written, it will almost certainly appear that decisions were made on misinformation, assumptions, prejudices, half truths, when excellent, scientifically accurate material was available.[211]

I make no apology for the detailed examination of facts, for constitutional issues are now largely problems of social justice and degree. On the basis of the *known* facts the author cannot conclude that the military situation or conditions on the west coast in early 1942 could honestly be deemed to require

205Marian Svensrud, *Attitudes of the Japanese Toward Their Language Schools* (1933) 17 SOCIOLOGY AND SOCIAL RESEARCH 259-264; STRONG, THE SECOND-GENERATION JAPANESE PROBLEM (1934) 6; MEARS, RESIDENT ORIENTALS ON THE AMERICAN PACIFIC COAST (1928) 358; and Stearns, *History of the Japanese People in California*, Thesis Series 4, University of Oregon (1937).

206SMITH, AMERICANS IN THE MAKING (1939) 302; BELL, *op. cit. supra* note 204, at 106.

207BELL, *op. cit. supra* note 204, at 20-24; SVENSRUD, *op. cit. supra* note 205; MILLIS, THE JAPANESE PROBLEM IN THE UNITED STATES (1915) 265; and YAMATO ICHIHASHI, *op. cit. supra* note 197.

208Brief for Japanese American Citizens League, *Amicus Curiae*, pp. 49-53, Hirabayashi v. United States.

209*Id.* at 53-60.

210*Id.* at 115-125.

211Typical of the misinformation are the following: Senator Johnson said: ". . . we have 130,000 Japanese, 30,000 of whom are citizens." 88 CONG. REC., FEBRUARY 17, 1942, 1371. Representative Ford said: "Every child of Japanese parentage, born in the States . . . is still a citizen of Japan." 88 CONG. REC., February 18, 1942, 1458. There is some evidence that the demand for evacuation came from certain professional anti-orientals and selfish economic groups. Frank Taylor, *The People Nobody Wants*, Sat. Eve. Post, May 9, 1943, 24; *Tolan Committee Hearings*, 11, 242, 11432. This article refers to only a small portion of the accurate material available; see the Congressional Library catalogue in Washington for considerable additional material.

evacuation of 79,000 American citizens of Japanese ancestry, as a matter of "military necessity" or otherwise. The very foundation of upholding the evacuation and internment as an exercise of the federal police power, therefore, must fail.

C. If Evacuation Was Justified, Was the Method Employed Constitutional?

Although our conclusion has been that the situation in early 1942 did not justify the federal government in authorizing the evacuation program under any of its powers, we now approach the next issue as though the program could have been sustained if the proper authorization and procedure were followed.

1. *Preliminary Issues.*—There are several arguments debated by counsel for the government and counsel for the evacuees which need to be disposed of before facing the major issues. The first involves the question whether either as Commander in Chief or under his power to enforce the laws, the President could carry out the evacuation without congressional authorization; and if not, did Congress properly authorize the President and the military commanders to carry forward the program. That the President has asserted "war powers" other than those derived from congressional empowerment or the direct provisions of the Constitution cannot be doubted.[212] Lincoln took a somewhat similar position, but the Supreme Court has never sustained this claim. In the *Prize Cases* Congress had ratified the earlier action taken by Lincoln, and the Court upheld the blockade on the basis of congressional action.[213] The position of the government is not confirmed by any of the authoritative text writers on presidential power[214] or by the decisions of the Supreme Court, which expressly held in *Brown v. United States*[215] that the President did not have the right to seize alien enemy *property* when Congress had only empowered him to detain alien enemies. Nor does it appear that the executive action can be sustained without congressional authorization on the basis of the President's power under Article II, Section 3, of the Constitution to see that

[212]See the President's Labor Day address to Congress asking the repeal of the Emergency Price Control Act: "In the event that the Congress should fail to act, and act adequately, I shall accept the responsibility, and I will act . . . When the war is won, the powers under which I act automatically revert to the people—to whom they belong." And see preamble to Executive Order No. 9066, Feb. 21, 1942, 7 FED. REG. 1407 (1942).

[213]2 Black 635, 671 (U. S. 1862).

[214]RANDALL, CONSTITUTIONAL PROBLEMS UNDER LINCOLN (1926); CORWIN, THE PRESIDENT: OFFICE AND POWERS (1940); and BERDAHL, WAR POWERS OF THE EXECUTIVE IN THE UNITED STATES (1921).

[215]8 Cranch 110 (U. S. 1814). See also *Ex parte* Milligan, 4 Wall. 2 (U. S. 1866); and the dissent in *Luther v. Borden*, 7 How. 1 (U. S. 1849), considered a classic on "War Powers."

the laws are faithfully executed.[216] As to the evacuation of "citizens," there was no law which the President could "execute."[217]

Congressional authorization or ratification is extremely difficult to spell out in this case.[218] None of the actions taken prior to March 21, when Public Law No. 503 was adopted by Congress, had suggested that evacuation of any "loyal citizens" was to take place. Not only was Congress advised that executive evacuation under the existing congressional laws could not apply to citizens, but it refused to pass an express authorization for evacuation of all Japanese. Public Law No. 503 should be interpreted as punishing only those who refused to obey a constitutional order. The attempt to find ratification in the appropriation of money to care for the evacuees in the Congressional Appropriation Act of July 25, 1942, is a "grasping at straw" argument since obviously whether properly or improperly evacuated the government had to care for them. That Congress could have authorized or ratified executive action is well recognized.[219]

2. *Delegation of Legislative Power*.—If it be assumed that Public Law No. 503 constituted congressional authorization, does that law stand the test of constitutionality? At its head are leveled two criticisms: (1) the act unconstitutionally delegates legislative authority; and (2) for a criminal law the statute is too indefinite. Although the limits placed upon delegation of legislative authority in *Panama Refining Co. v. Ryan*[220] and the *Schechter* case[221] are recognized, it is the author's belief that the delegation here involved would be upheld. It could be argued from the circumstances surrounding its enactment that the law clearly shows the general policy (protection of military areas) and a primary standard (the prevention of any act which might unreasonably hamper the war effort). All cases recognize that Congress need only legislate "as far as was reasonably practicable."[222] It is certainly true that great discretion may be lodged in the executive in the fields of foreign

216The cases relied upon are In re *Debs,* 158 U. S. 564, 15 Sup. Ct. 900 (1894); In re *Neagle,* 135 U. S. 1, 10 Sup. Ct. 658 (1890); and Ex parte *Siebold,* 100 U. S. 371 (1879).

217See concession of the government, Brief for Government, p. 51, Yasui v. United States.

218See notes 14, 13-29 *supra* for proof of following statements.

219Mitchell v. Clarke, 110 U. S. 633, 4 Sup. Ct. 170 (1884); Graham v. Goodcell, 282 U. S. 409, 51 Sup. Ct. 186 (1930); Swayne & Hoyt, Ltd. v. United States, 300 U. S. 297, 301, 57 Sup. Ct. 478, 479 (1936); and particularly Tiaco v. Forbes, 228 U. S. 549, 556, 33 Sup. Ct. 585, 586 (1912) (upholding legislative ratification of a prior executive deportation).

220293 U. S. 388, 55 Sup. Ct. 241 (1934).

221A. L. A. Schechter Poultry Corp. v. United States, 295 U. S. 495, 55 Sup. Ct. 837 (1934).

222United States v. Chemical Foundation, 272 U. S. 1, 47 Sup. Ct. 1 (1926); and Red "C" Oil Mfg. Co. v. Board of Agriculture, 222 U. S. 380, 394, 32 Sup. Ct. 152, 156 (1911).

relations and war where decisions have to be made on "confidential information."[223] Senator Taft was right in characterizing Public Law No. 503 as a "sloppy" criminal law, but appellants' argument of indefiniteness fails to take into account that if Congress had fixed the general standard and properly authorized administrative agencies to fill in the details, the indefiniteness of the definition of the acts made criminal must inhere in the regulations, as well as the law.[224] The cases in which statutes were clearly indefinite and were not made definite by regulations are not applicable.[225] Although the words "any act" in Public Law No. 503, and "any or all persons" or "military areas" in Executive Order No. 9066 seem indefinite, a construction which would interpret these in the light of the preamble of Executive Order No. 9066 as reasonably related to "espionage or sabotage" would seem sound. *United States v. Grimaud*[226] recognized the right of Congress to make the violation of the Secretary of Agriculture's regulations, *not yet issued,* a misdemeanor.

3. *Due Process.*—There are, however, other more serious objections to the method employed in the evacuation regardless of whether congressional authorization was necessary or given. Did the program lack "due process" either by reason of improper discrimination or lack of hearing? It neither advances our discussion nor does it sound convincing in the mouth of the government to assert that the fifth, unlike the fourteenth amendment, contains no "equal protection" clause, even though the Supreme Court has used this language on countless occasions.[227] Equal protection before the law is the

[223]United States v. Curtiss-Wright Export Corp., 299 U. S. 304, 57 Sup. Ct. 216 (1936); United States v. Chemical Foundation, 272 U. S. 1, 47 Sup. Ct. 1 (1926); Northern Pacific Ry. Co. v. North Dakota, 250 U. S. 135, 39 Sup. Ct. 502 (1918); Union Bridge Co. v. United States, 204 U. S. 364, 27 Sup. Ct. 367 (1906); Field & Co. v. Clark, 143 U. S. 649, 12 Sup. Ct. 495 (1891); and The Thomas Gibbons, 8 Cranch 420 (U. S. 1814).

[224]United States v. Eaton, 144 U. S. 677, 12 Sup. Ct. 764 (1891) (recognizing the right of Congress to make the violation of an administrative regulation a crime); United States v. Curtiss-Wright Export Corp., 299 U. S. 304, 57 Sup. Ct. 216 (1936) (a criminal law imposing a penalty for selling arms); McKinley v. United States, 249 U. S. 397, 39 Sup. Ct. 324 (1918) (a criminal law whereby an administrative officer prescribed the area around army camps from which prostitutes were excluded); Gorin v. United States, 312 U. S. 19, 61 Sup. Ct. 429 (1940); Kay v. United States, 303 U. S. 1, 58 Sup. Ct. 468 (1937); and Kentucky Whip & Collar Co. v. Illinois Central R. Co., 299 U. S. 534, 57 Sup. Ct. 277 (1936).

[225]Lanzetta v. New Jersey, 306 U. S. 451, 59 Sup. Ct. 618 (1936); Connally v. General Construction Co., 269 U. S. 385, 46 Sup. Ct. 126 (1925); and United States v. Cohen Grocery Co., 255 U. S. 81, 41 Sup. Ct. 298 (1920).

[226]220 U. S. 506, 31 Sup. Ct. 480 (1910).

[227]Detroit Bank v. United States, 317 U. S. 329, 63 Sup. Ct. 297 (1943); Sunshine Anthracite Coal Co. v. Adkins, 310 U. S. 381, 401, 60 Sup. Ct. 907, 916 (1939); Currin v. Wallace, 306 U. S. 1, 14, 16, 59 Sup. Ct. 379, 386, 387 (1938); United States v. Carolene Products Co., 304 U. S. 144, 151, 58 Sup. Ct. 778, 783 (1937); Stewart Machine Co. v. Davis, 301 U. S. 548, 584, 57 Sup. Ct. 883 (1930); and LaBell Iron Works v. United States, 256 U. S. 377, 392, 41 Sup. Ct. 528, 532 (1920).

very warp and woof of our Federal System;[228] the President has declared racial discrimination abhorrent to our institutions[229] and has said that "we are fighting, as our fathers have fought, to uphold the doctrine that all men are created equal in the sight of God";[230] congressional acts of long standing have expressly forbidden discrimination based on race or color;[231] and even in the cases where a distinction between the fifth and fourteenth amendments has been stated, I do not find the court approving federal action which would not have been upheld under the "equal protection" clause of the fourteenth amendment as state action.[232]

The rules of permissible classification should perhaps be restated. A reasonable and causal relationship not based on conjecture must be shown between the basis of classification and a proper object to be accomplished.[233] The class may be limited in its scope and not cover every possible source of danger; the limits may be determined by degree of evil or danger, but the classification should operate equally upon all in like circumstances.[234] A practical classification may be sustained though lacking in scientific exactness or uniformity.[235]

---

[228] DECLARATION OF INDEPENDENCE, par. 2; Preamble to U. S. CONST.; U. S. CONST. Art. I, § 2, cl. 1 and 2, § 3, cl. 1 and 3, § 9, cl. 3, 4, and 8; Art. II, § 1, cl. 3 and 5; Art. IV, § 2, cl. 1, § 4; U. S. CONST. Amend. I to X, XIII, XIV, XV, XIX. Truax v. Corrigan, 257 U. S. 312, 332, 42 Sup. Ct. 124, 129 (1921).

[229] Exec. Order No. 8802, June 25, 1941, 6 FED. REG. 3109 (1941).

[230] *War and Peace Aims*, Special Supplement No. 1, UNITED NATIONS REVIEW, January 30, 1943, 6.

[231] REV. STAT. §§ 1977-79 (1875), 8 U. S. C. §§ 41, 42, 43 (1940); 35 STAT. 1092 (1909), 18 U. S. C. §§ 51, 52 (1940).

[232] See note 227 *supra, e.g., Steward Machine Co. v. Davis*, 301 U. S. 548, 57 Sup. Ct. 883 (1930) upheld the Federal Social Security System; similar state action was also upheld. W. H. Chamberlin v. Andrews, Ind. Comm., 271 N. Y. 1, 2 N. E. (2d) 22 (1936), *aff'd*, 299 U. S. 515, 57 Sup. Ct. 122 (1936). The Court has also said that the same protection is afforded under "due process" as under "equal protection." Heiner v. Donnan, 285 U. S. 312, 326, 52 Sup. Ct. 358, 361 (1931); Coolidge v. Long, 282 U. S. 582, 51 Sup. Ct. 306 (1930); and Sims v. Rives, 84 F. (2d) 871 (App. D. C., 1936), *cert. den.* 298 U. S. 682, 56 Sup. Ct. 960 (1936).

[233] Hartford Steam Boiler Insp. & Ins. Co. v. Harrison, 301 U. S. 459, 57 Sup. Ct. 838 (1936); West Coast Hotel v. Parrish, 300 U. S. 379, 57 Sup. Ct. 578 (1936); Mayflower Farms v. Ten Eyck, 297 U. S. 266, 56 Sup. Ct. 457 (1935); Sunday Lake Iron Co. v. Wakefield, 247 U. S. 350, 38 Sup. Ct. 495 (1917); Chicago Dock and Canal Co. v. Fraley, 228 U. S. 680, 33 Sup. Ct. 715 (1912); and Atchison, Topeka and Santa Fe R. Co. v. Matthews, 174 U. S. 96, 19 Sup. Ct. 609 (1898). *Compare Grosjean v. American Press Co.*, 297 U. S. 233, 56 Sup. Ct. 444 (1935) where the number of outlets had no proper relation to the result to be accomplished *with State Board of Tax Commissioners v. Jackson*, 283 U. S. 527, 51 Sup. Ct. 540 (1930) where it did.

[234] State Board of Tax Commissioners v. Jackson, 283 U. S. 527, 51 Sup. Ct. 540 (1930); Silver v. Silver, 280 U. S. 117, 50 Sup. Ct. 57 (1929); Farmers & M. Bank v. Federal Reserve, 262 U. S. 649, 43 Sup. Ct. 651 (1922); Hall v. Geiger-Jones Co., 242 U. S. 539, 37 Sup. Ct. 217 (1916); Miller v. Wilson, 236 U. S. 373, 35 Sup. Ct. 342 (1914); Patsone v. Pennsylvania, 232 U. S. 138, 34 Sup. Ct. 281 (1913); and Barbier v. Connelly, 113 U. S. 27, 5 Sup. Ct. 357 (1884).

[235] Continental Baking Co. v. Woodring, 286 U. S. 352, 52 Sup. Ct. 595 (1931); Chi-

A margin of error is allowed, and the mere fact that some incidental innocent articles or transactions may be found in the class will not upset the plan.[236] Although aliens, like citizens, are entitled in time of peace to the equal protection of the laws, alienage may have a reasonable relationship to a proper object so as to form a proper basis of classification.[237] There is a presumption of reasonableness of classification.[238]

In a strict sense the classification here is not on a basis of race, for all Chinese, Japanese, Pacific islanders, and Indians are of the same race,[239] but is a segregation on nationality of forebears. Yet, in a real sense the criterion is race; for the basic assumption is made that the portion of our citizenry descending from Mongolians who are our enemies may be disloyal, but those descending from "white" enemies are not. I would not agree that race may *never* be made the subject of valid classification;[240] but it can be asserted with some finality that to the problem of "defense" race has no reasonable relationship. The concurring opinion of Justices Douglas and Jackson in the *Edwards* case[241] has stated: "we should say now, and in no uncertain terms, that a man's mere property status, without more, cannot be used by a state to test, qualify or limit his right as a citizen of the United States . . . the mere state of being without funds is a neutral fact—constitutionally an irrelevance, *like race,* creed, or color." This is in accord with a long line of cases which hold that thus far we have never found a situation in which race is a proper basis of classification.[242] Just as race has no bearing on a person's tendency to

---

cago Dock and Canal Co. v. Fraley, 228 U. S. 680, 33 Sup. Ct. 715 (1912); Mutual Loan Co. v. Martell, 222 U. S. 225, 32 Sup. Ct. 74 (1911): and Orient Insurance Co. v. Daggs, 172 U. S. 557, 19 Sup. Ct. 281 (1898).

[236]Euclid v. Ambler Realty Corp., 272 U. S. 365, 47 Sup. Ct. 114 (1926); Jacob Ruppert v. Caffey, 251 U. S. 264, 40 Sup. Ct. 141 (1919); and Purity Extract & Tonic Co. v. Lynch, 226 U. S. 192, 33 Sup. Ct. 44 (1912).

[237]Clarke v. Deckebach, 274 U. S. 392, 47 Sup. Ct. 630 (1926) (operation of pool room); Terrace v. Thompson, 263 U. S. 197, 44 Sup. Ct. 15 (1923) (ownership of land); Crane v. New York, 239 U. S. 195, 36 Sup. Ct. 83 (1915) (employing aliens on public work); and Patsone v. Pennsylvania, 232 U. S. 138, 34 Sup. Ct. 281 (1913) (shooting wild life).

[238]United States v. Carolene Products Co., 304 U. S. 144, 58 Sup. Ct. 788 (1937); Borden's Farm Products Co. v. Baldwin, 293 U. S. 194, 55 Sup. Ct. 187 (1934); Silver v. Silver, 280 U. S. 117, 50 Sup. Ct. 57 (1929); and Clarke v. Deckebach, 274 U. S. 392, 47 Sup. Ct. 630 (1926).

[239]Kroeber, Kinship in the Philippines (1921) 47; Kroeber, Anthropology (1923) 44; Linton, The Study of Man (1936) 43; Stibbe, An Introduction to Physical Anthropology (1930) 160; and Ales Hrdlicka, *No Certain Way to Tell Japanese from Chinese* (1941) 40 Science News Letter 394.

[240]An example could be imagined wherein one race had such a higher percentage of leprosy cases that members of that race could be compelled to submit to examination or other means of eradication of the disease. If some characteristic of the *race* made it more susceptible, the race would form a basis of classification.

[241]Edwards v. California, 314 U. S. 160, 62 Sup. Ct. 164 (1941).

[242]Missouri *ex rel.* Gaines v. Canada, 305 U. S. 337, 59 Sup. Ct. 232 (1938); Yu Cong

sabotage or disloyalty, so neither nationality of forebears nor even former nationality of the person is pertinent.[243]

An equally critical omission of the evacuation program was the failure to afford hearings to determine the necessity for moving or confining any individual. If the Supreme Court ever intended, in *Whitney v. California*,[244] to restrict a person's rights on the theory of "guilt by association" it has certainly now reversed its position and required a determination of individual guilt or innocence.[245] It has also always been a cardinal principle that before any individual can be confined, punished, or even deprived of valued rights, a fair hearing must be accorded and personal guilt or obligation ascertained.[246] Although an army intelligence officer has made a report finding 75 per cent of the Japanese on the west coast loyal and advocating individual examinations;[247] although the advisability of individual hearings was recognized in the Tolan reports;[248] although individual hearings were accorded aliens other than Japanese[249] and the vast group of Italian aliens and Italian citizens was cleared of disloyalty by October, 1942,[250] and German aliens were permitted to return to the west coast by July;[251] although the English experience of 73,353 hearings in six months resulting in 569 persons interned, 6,782 re-

---

Eng v. Trinidad, 271 U. S. 500, 528, 46 Sup. Ct. 619, 627 (1925); Truax v. Corrigan, 257 U. S. 312, 42 Sup. Ct. 124 (1921); Buchanan v. Warley, 245 U. S. 60, 38 Sup. Ct. 16 (1917); and Yick Wo v. Hopkins, 118 U. S. 356, 6 Sup. Ct. 1064 (1886). Cases like *Gong Lum v. Rice,* 275 U. S. 78, 48 Sup. Ct. 91 (1927) and *Plessy v. Ferguson,* 163 U. S. 537, 16 Sup. Ct. 1138 (1895) which permits segregation if the colored race is given equal accommodations are decided on the basis that where there is equality there is no discrimination.

243In *United States v. Wong Kim Ark,* 169 U. S. 649, 18 Sup. Ct. 456 (1897), it was declared that Congress had no authority "to restrict the effective birth, declared by the Constitution to constitute sufficient and complete right to ciizenship on the basis of nationality of the parents." And in *Skinner v. Oklahoma,* 316 U. S. 535, 541, 62 Sup. Ct. 1110, 1113 (1941) the Court said: " . . . when the law lays an unequal hand on those who have committed intrinsically the same quality of offense and sterilizes one and not the other, it has made as invidious a discrimination as if it had selected a particular race, or *nationality* for oppressive treatment." In Ex parte *Kawato,* 317 U. S. 69, 63 Sup. Ct. 115 (1943) the Court was unwilling to assume disloyalty, even of an *alien* Japanese.

244274 U. S. 357, 47 Sup. Ct. 641 (1926).

245*Ex parte* Quirin, 317 U. S. 1, 63 Sup. Ct. 1 (1942); Herndon v. Lowry, 301 U. S. 242, 57 Sup. Ct. 732 (1936); and DeJonge v. Oregon, 299 U. S. 353, 57 Sup. Ct. 255 (1936).

246CHAFFEE, FREE SPEECH IN THE UNITED STATES (1941) 470-485. Skinner v. Oklahoma, 316 U. S. 535, 62 Sup. Ct. 1110 (1941); Morgan v. United States, 304 U. S. 1, 14, 58 Sup. Ct. 773, 776 (1937); Buck v. Bell, 274 U. S. 200, 47 Sup. Ct. 584 (1926); Truax v. Corrigan, 257 U. S. 312, 42 Sup. Ct. 124 (1921); and Hurtado v. California, 110 U. S. 516, 4 Sup. Ct. 111 (1884).

247Harper's Magazine, October, 1942, 489.

248H. R. REPS. No. 1911, 2124, 77th Cong., 2d Sess. (1942).

249See note 14 *supra.*

250Statement of Attorney General, October 12, Order October 14, 1942.

251See note 14 *supra.*

stricted, and 64,254 freed completely was known,[252] and although the army was able to test loyalty individually when it inducted evacuees into the army,[253] it was unable to stop long enough in March, 1942, to permit of individual hearings to test the loyalty of American citizens. That greater consideration should be accorded aliens than citizens is a bitter pill for *any* citizen to swallow. Even if congressional action and the orders of the President were legal, the action of General DeWitt would constitute administration "by public authority with an evil eye and an unequal hand" so as to deprive the whole evacuation program of constitutional validity.[254]

D. The Problem of Detention

Although strictly beyond the scope of this article, brief reference should be made to the continued detention of the evacuees.[255] Detention stands on less firm ground from the beginning than does evacuation for neither congressional act nor executive order authorize detention;[256] and it rests, therefore, solely on General DeWitt's orders. Further, even if evacuation were legal, it is so only on the theory that it was "preventive," not "punitive,"[257] and was employed to gain time necessary to determine whether there were any specific persons of Japanese ancestry whose presence on the west coast would be dangerous. It was, as Professor Cushman has remarked,[258] a vast quarantine program falling alike on those who have or may have been exposed to a disease, justifiable only for such time as would permit of individual examinations. But the facts demonstrate that evacuation has been used otherwise than as "preventive." Although citizens of Japanese ancestry have proved most cooperative in camps,[259] and detention centers have been recognized as mistakes,[260] only a few individual hearings, though practicable for the entire group, have been held. Regulations (requiring a job, sponsor, distance from war facilities, *etc.*) at present make it virtually impossible for very many to be resettled in ordinary civilian life; General DeWitt and some west coast congressmen have been making it clear that the evacuees will not return to

[252] H. R. Rep. No. 2124, 77th Cong., 2d Sess. (1942) 280-281.
[253] See note 193 *supra*.
[254] Yick Wo v. Hopkins, 118 U. S. 356, 6 Sup. Ct. 1064 (1886).
[255] A word of appreciation should be spoken concerning the efforts of the War Relocation Authority, which seems to have done an excellent job against the most trying opposition.
[256] See Exec. Order No. 9066, Feb. 21, 1942, 7 Fed. Reg. 1407 (1942).
[257] Fairman, *The Law of Martial Rule and the National Emergency* (1942) 55 Harv. L. Rev. 1253, at 1302.
[258] Cushman, *Our Interned Citizens: The Problem of Japanese Evacuation* (1943) 23 B. U. L. Rev. —.
[259] See statements of Milton S. Eisenhower and Edward J. Ennis. Brief for the Japanese American Citizens League, *amicus curiae*, p. 8, Hirabayashi v. United States.
[260] See note 41 *supra*.

the west coast alive.[261] David Reisman, in an excellent article,[262] has emphasized that the protection of civil liberties now depends upon an aggressive willingness of individuals to use the processes of government and their own abilities to prevent those inimical to civil liberties from using governmental processes to suppress liberties. Let us be fair and admit that our hasty action has left us with a difficult problem on our hands and that even if evacuation could be supported, detention cannot. We, therefore, should be prompt in making good to these citizens their loss from improper detention. Make it good in dollars and cents, yes; and in willingness to arrange for their discharge from camps and also to help them found new homes.

E. Miscellaneous Objections to Evacuation

Certain other objections to the evacuation program should be noted though space will not permit an analysis of them.

1. *Construction Treason.*—A tenable case can be made against the evacuation program as being the equivalent of constructive treason. The argument would run something like this: evacuation and detention were in the nature of punishment; espionage or sabotage, or disloyalty to the government constitutes treason. The Constitution specifies how treason shall be proved and that guilt shall not work corruption of blood to the person's descendants; the United States in the trial of Aaron Burr and in Ex parte *Bollman,*[263] rejected the doctrine of "constructive treason" by which a person could be punished as though for treason without a trial. The confinement of citizens of Japanese ancestry either amounts to assuming treason of the parents (aliens) and working a corruption of blood, or to "constructive treason."

2. *Bill of Attainder.*—Of similar type is the argument that evacuation operated as a bill of attainder in violation of Article I, section 9, clause 3. In the *Test Oath Cases,*[264] (civil war) the court held that to deprive a person of any basic rights as a citizen (such as right to office) on the basis of his former sympathy with the enemy, was unconstitutional as a bill of attainder. To deprive a citizen of his rights because of his ancestry, which is a "past fact" over which he has no control, is of the same nature.

3. *Slavery and Involuntary Servitude.*—The government claims that the evacuation is not punishment for crime. There is considerable evidence to show that the evacuees have been made to work at substantially less than the

[261]Reply brief for Hirabayashi, appendix, Hirabayashi v. United States.
[262]Reisman, *Civil Liberties in a Period of Transition* (1942) 3 PUBLIC POLICY 33-96.
[263]4 Cranch 75 (U. S. 1807).
[264]4 Wall. 277 (U. S. 1866).

regular pay available in private employment[265] and that they have been permitted to work only where assigned by the government. The only compulsory service for the government which has ever been considered legal has been service in the armed forces, on jury, in repair of roads, and in similar service.

III. The Position of the Alien Japanese

Because I cannot agree with the position expressed in nearly all articles to date that "so far as enemy aliens are affected, *no* constitutional question is presented,"[266] I shall make brief reference to the alien problem. I recognize that the war power *within constitutional limits* is as to the alien almost plenary. That disposes of any objection under sections IIA and IIB of this article. It does *not* answer the issue raised in IIC. The right to classify on the basis of alienage demands that alienage shall have a proper relation to the purpose to be accomplished. The purpose here is to prevent disloyal activities. The government has conceded that "alien Japanese individually conceived to be dangerous to the safety of the country *have been* apprehended and interned."[267] Internment of other Japanese aliens could have no reasonable relation to safety. This, together with the discriminatory treatment accorded alien Japanese, as compared to Germans and Italians, may yet require the holding that the evacuation failed to accord alien Japanese the rights guaranteed them, as recently as 1942, by the Supreme Court in Ex parte *Kawato,*[268] and admirably summarized with extensive citation by Reuben Oppenheimer.[269]

IV. Conclusion

Some may wish to shave off all nonessential protrusions from the streamlined war machine so that nothing may interfere with the speed of its progress. They forget that some of the things they term "gadgets" will be helpful, if not indispensable, when the machine gets its destination—in fact, the trip may well have been futile without them. The history of America has been one of jealously guarding the civil liberties of its people, even at the price of reduction in speed of its governmental or war machine.

If we would, therefore, preserve those principles which this country has designated its "war aims," then it would appear that the evacuation cannot be upheld.

---

[265]H. R. Rep. No. 2124, 77th Cong., 2d Sess. (1942).

[266]Fairman, *The Law of Martial Rule and the National Emergency* (1942) 55 Harv. L. Rev. 1253, 1301; Comment, *Alien Enemies and Japanese Americans* (1942) 51 Yale L. J. 1316, 1322; and Cushman, *Our Interned Citizens: The Problem of Japanese Evacuation* (1943) 23 B. U. L. Rev. —.

[267]Brief for Government, p. 51, Yasui v. United States. The reports of the F. B. I. are to the same effect. Annual Rep. F. B. I. (1941); Annual Rep. F. B. I. (1942).

[268]317 U. S. 69, 63 Sup. Ct. 115 (1943).

[269]Oppenheimer, *The Constitutional Rights of Aliens* (1941) 1 Bill of Rights Review 100 ff.

BOOK REVIEW

"OTHER NON-WHITES" IN AMERICAN LEGAL HISTORY: A REVIEW OF *JUSTICE AT WAR*

JUSTICE AT WAR by Peter Irons. New York: Oxford University Press, 1983. Pp. xiii, 407 pp. $18.95.

*Reviewed by Neil Gotanda**

In January of 1983, Fred Korematsu, Gordon Hirabayashi, and Minor Yasui filed writs of error coram nobis in federal district courts in Washington, Oregon, and California to set aside their World War II convictions for criminal violations of curfew and evacuation orders.[1] One of their lawyers is Peter Irons, a professor of political science at the University of California at San Diego. Irons' representation of Korematsu comes at a time when Japanese-Americans have begun to pursue aggressively a campaign not only to publicize a largely forgotten injustice but also to seek monetary compensation in redress.

In *Justice at War*, Irons demonstrates that the Justice Department and the Department of War were well aware that the Japanese-Americans on the West Coast did not constitute a serious security threat. He shows that not only was this information not presented to the Supreme Court, but discredited allegations of sabotage and spying by Japanese-Americans in Hawaii and California were submitted to the Court despite prosecutors' knowledge of their falsity (pp. 202–06, 279–84).

Irons focuses on the lawyers involved in these cases. The most compelling sections of *Justice at War* are the conversations of the various attorneys and judges, which Irons reconstructs using newly uncovered documents. Three subjects at the center of his account prove especially interesting: the differences between the Justice Department lawyers and those at the Department of War, the maneuverings of Roger Baldwin on behalf of the ACLU, and the conference discussions of the Supreme Court. While the book's detail occasionally makes reading tedious, this is a minor inconvenience as the inherent drama of the events draws the reader past such description toward the book's conclusion.

Irons' understanding of the Japanese-American concentration camps is straightforward.[2] He condemns the camps, and seeks to show

---

* Associate Professor of Law, California Western School of Law. B.S. 1967, Stanford University; J.D. 1972, University of California, Berkeley; LL.M. 1980, Harvard University.

1. Their convictions were upheld in Korematsu v. U.S., 323 U.S. 214 (1944); Hirabayashi v. United States, 320 U.S. 81 (1943); Yasui v. United States, 320 U.S. 115 (1943).

2. I follow current practice in using "concentration camp" or "camps" in prefer-

that as a result of prosecutorial misconduct in the cases that came before the Supreme Court, the Justice Department and the Department of War were involved in "a legal scandal without precedent in the history of American law" (p. viii). This condemnation of governmental and judicial conduct toward the Japanese has been shared by almost every legal writer examining the camps since 1944.[3]

Irons' account, however, is ultimately unsatisfying. His essential problem is a failure to reexamine the events clearly and critically. First, the theme of ethical conflict among lawyers and the theme of the conflict between politics and law are not presented clearly, and prove inconsistent and confusing. The lawyers at times are reduced to morality play actors, either good or evil depending on whether they opposed or supported the camps. At other times, they are portrayed as facing legal ethical dilemmas which either remain unclarified or seem to be incorrectly stated (pp. 284–92, 347–48). Similarly, when Irons refers to the "dominance of politics over law" (p. 269), he simplistically uses law as an external source of ethical good that is contrasted with less wholesome electoral politics and bureaucratic infighting (pp. 45, 55, 269).

Second and more importantly, he fails to show how these events can be viewed in a broader scope. He presents history only as a product of individuals' actions, describing these events exclusively in terms of the actions of lawyers and judges. This approach to history extends particularly to his understanding of race. For Irons, racism is situated in the personal prejudices of the lawyers and officials involved. Viewing questions of race in American legal history in such a limited fashion is inadequate.

While Irons has added substantially to the body of material available to examine these events, his simple repetition of historical and legal conclusions is insufficient. To bring forth new documents which show that there were some individuals who were racist, others who did not care, and others who did care, is important. But those facts were already known, albeit not in as much detail. When considered against a search for some context that will prove meaningful for those affected by the events, Irons' interpretive efforts to portray ethical conflict among

---

ence to other descriptive terms such as "relocation" or "evacuation." The phrase has created its own minor controversies. In the *Korematsu* opinion, for example, Justice Black objects to Justice Robert's use of the term "concentration camp." 323 U.S. at 214, 223, 226, 230.

3. See, e.g., R. Daniels, Concentration Camps, North America (1981) (originally published as Concentration Camps, USA); M. Grodzins, Americans Betrayed (1949); J. TenBroek, Prejudice, War and the Constitution (1954); M. Weglyn, Years of Infamy (1976); Commission on Wartime Relocation and Internment of Civilians, Personal Justice Denied, Final Report (Washington D.C. 1982); Dembitz, Racial Discrimination and the Military Judgment: The Supreme Court's *Korematsu* and *Endo* Decisions, 45 Colum. L. Rev. 175 (1945); Freeman, Genesis, Exodus and Leviticus: Genealogy, Evacuation and Law, 28 Cornell L.Q. 414 (1943); Rostow, The Japanese American Cases—A Disaster, 54 Yale L.J. 489 (1945).

lawyers and to make history solely the product of individual actions fall disappointingly short.

A broader historical perspective is possible. Such a perspective would view the status of Japanese-Americans in American legal history as one component in the larger question of race in American law. While in legal history the discussion of race has usually meant analysis of the legal condition of Blacks, I suggest an examination of non-Black racial minorities—"Other non-Whites"—as a distinct mode of analysis. One of the critical features of legal treatment of Other non-Whites has been the inclusion of a notion of "foreignness" in considering their racial identity and legal status. This previously unexamined dimension of the relationship between race and law helps shed light not only on the episode Irons studies, but on contemporary debate as well.

In the last two decades, the treatment of Other non-Whites has grown from a regional issue to a divisive national issue. Japanese- and Chinese-Americans have played a significant role in the legal clarification of the status of Other non-Whites. It is within this dynamic—the evolution of the treatment of Other non-Whites—that the concentration camp cases are best understood. My thesis is that these cases were crucial steps in the development of the complex links of the social and legal categories of race and alienage. Most important in this development has been the persistence of the view that even American-born non-Whites were somehow "foreign." This undeserved stigma became, and may remain, an unarticulated basis for the legal treatment of these groups, leading to unfair and often shocking consequences.

Other non-Whites, with the exception of Native Americans, did not enter American life until the mid-nineteenth century, when small but increasing numbers of Mexicans, Chinese, and Japanese became, by annexation and immigration, part of the American nation. Early in the 1850s, while the country grappled with slavery and the inability to reconcile an increasingly divergent North and South, Chinese immigrants began to arrive. Initially, the Chinese were welcomed. Within a short time, however, their presence gave rise throughout the West to widespread, sustained, violent campaigns to expel them. Anti-Chinese sentiments surfaced first around 1852 in the mining areas, but quickly spread throughout the West.[4]

The racial status of the Chinese, sufficiently different from the status of Blacks and Indians to require an opinion of the California Supreme Court, was soon clarified. In 1854, in *People v. Hall*[5], a criminal defendant challenged the admissibility of the testimony of Chinese

4. See, e.g., S. Miller, The Unwelcome Immigrant: The American Image of the Chinese 1784–1882 (1969); E. Sandmeyer, The Anti-Chinese Movement in California (1939); A. Saxton, The Indispensible Enemy: Labor and the Anti-Chinese Movement in California (1971); McClain, The Chinese Struggle for Civil Rights in Nineteenth Century America: The First Phase, 1850–1870, 72 Calif. L. Rev. 529 (1984).

5. 4 Cal. 399 (1854).

witnesses used to convict a "free white citizen" of murder. At issue was the applicability of an 1850 statute, passed before California's formal admittance into the Union, providing that "no Black, or Mulatto person, or Indian shall be allowed to give evidence in favor of, or against a White man."[6] In a short but expansive opinion, the California Supreme Court reversed the conviction and held the Chinese testimony inadmissible under the statute.[7] To do so, the court expanded "Indian" to include Chinese and any other Asiatic, and as an alternative basis for its decision, made Black a generic term, meaning any non-White, thus including Chinese.[8]

Yet there remained ambiguity as to the appropriate racial classification for the Chinese. The 1850 statutes had not mentioned Chinese at all, and the court in *People v. Hall* had to go out of its way to achieve its desired result. The opinion did make clear, however, that the Chinese were an "alien" group, distinguishable from white Americans in a way different from Blacks:

> The anomalous spectacle of a distinct people, living in our community, recognizing no laws of this State except through necessity, bringing with them their prejudices and national feuds, in which they indulge in open violation of law; whose mendacity is proverbial; a race of people whom nature has marked as inferior, and who are incapable of progress or intellectual development beyond a certain point, as their history has shown; differing in language, opinions, color, and physical conformation; between whom and ourselves nature has placed an impassable difference, is now presented, and for them is claimed, not only the right to swear away the life of a citizen, but the further privilege of participating with us in administering the affairs of our Government.[9]

In 1898, the United States Supreme Court addressed the question of whether Chinese children born in the United States were American citizens in *United States v. Wong Kim Ark*.[10] In a seven to two decision, Justice Gray wrote that a child born to two Chinese nationals legally present on American soil was an American citizen under the fourteenth amendment. The Court thus resolved that Chinese were not to become a semipermanent colony of foreign "guest workers" squatting upon American soil, but could evolve as Americans.

Yet there was also a vigorous dissent by Chief Justice Fuller in which Justice Harlan joined.[11] Fuller quoted with approval from an

---

6. Id. at 399.
7. Id. at 405.
8. Id. at 400–02.
9. Id. at 404–05.
10. 169 U.S. 649 (1898).
11. Harlan's agreement with Fuller reflected the sharp opposition to the assimilation of Chinese into American society that Harlan had voiced the year before in *Plessy v. Ferguson* in which, contrasting the status of Chinese and Blacks, he had stated:

earlier decision concerning the Chinese, which had stated that the presence

> of large numbers of Chinese laborers, of a distinct race and religion, remaining strangers in the land, residing apart by themselves, tenaciously adhering to the customs and usages of their own country, unfamiliar with our institutions, and apparently incapable of assimilating with our people, might endanger good order, and be injurious to the public interests . . . .[12]

In the majority and dissent in *United States v. Wong Kim Ark* can be found the paradoxical aspects of the emergent Chinese-American racial identity. Chinese-Americans *are* American citizens, yet as Fuller and Harlan emphasize, they retain a dimension of "foreignness"—they are unassimilable strangers. This inclusion of foreignness into racial identity, begun with the Chinese, became a part of the American understanding of Japanese immigrants as well,[13] and also part of the explicitly racial classification "Oriental."

The concentration camp cases were a continuation of this racial association of Other non-Whites with foreignness. The idea that many Japanese would "actually" sympathize with Japan as a matter of "human nature" was a simple extension of that racial tradition and was shared by many Americans. Yet Justice Black, writing for the majority, denied that the issue in *Korematsu v. United States* was racial. "Our task

---

> There is a race so different from our own that we do not permit those belonging to it to become citizens of the United States. Persons belonging to it are, with few exceptions, absolutely excluded from our country. I allude to the Chinese race. But by the [segregation] statute in question, a Chinaman can ride in the same passenger coach with white citizens of the United States, while citizens of the black race in Louisiana . . . who are entitled, by law, to participate in the political control of the State and nation . . . and who have all the legal rights that belong to white citizens, are yet declared to be criminals, liable to imprisonment, if they ride in a public coach occupied by citizens of the white race.

Plessy v. Ferguson, 163 U.S. 537, 561 (1896).

12. United States v. Wong Kim Ark, 169 U.S. 649, 731 (1898) (quoting Fong Yue Ting v. United States, 149 U.S. 698, 717 (1893)).

13. Anti-Japanese agitation was marked by charges of "foreignness" although the charges differed in some ways from those leveled against the Chinese. While charging that the Japanese were alien and foreign, anti-Japanese agitators also expressed fears that the Japanese were too anxious to stay and take up residence. This raised again the unpleasant prospect of a resident "unassimilable" racial minority. Ownership of property by Japanese was thus a focal point of attention throughout the pacific coast. Efforts to restrict Japanese land ownership in California resulted in passage of Alien Land Laws in 1913 and by initiative in 1920. See R. Daniels, Politics of Prejudice 46–91 (1962); F. Chuman, The Bamboo People 38–51, 72–89 (1976). Ownership or control of real property was prohibited to "aliens ineligible for citizenship," a legal circumlocution applicable primarily to Asian immigrants. The Supreme Court in 1923 upheld California's Alien Land Law. See Terrace v. Thompson, 263 U.S. 197 (1923); Porterfield v. Webb, 263 U.S. 225 (1923). The Court thus reinforced foreignness within the racial identity of Chinese-Americans and Japanese-Americans. Foreign bloodlines remained a permissible basis for detrimental governmental distinctions.

would be simple," he wrote, "were this a case involving the imprisonment of a loyal citizen in a concentration camp because of racial prejudice . . ."[14] But Black then went on to ascribe a sufficient likelihood of disloyalty to justify special treatment for "all citizens of Japanese ancestry." Unstated is the assumption that the greater likelihood of disloyalty is "natural" given the foreign nature of the non-White Japanese.

In sum, the separability of the juridical categories of "citizen" and "alien" is clear, as is the parallel social distinction between "American" and "foreign." But when the individuals concerned are Other non-Whites, the racial considerations render the "natural" coincidence of citizen and American much less certain. A Japanese-American citizen in 1942 was easily considered "foreign," thus making possible the judgment that likelihood of disloyalty was high enough to justify wholesale internment.

The linkage between Other non-Whites and foreignness has eroded considerably in Supreme Court decisions since the end of World War II. Within a few years of the sharp criticism by legal commentators of the camp cases,[15] the Court reconsidered the status of aliens and began to restrict the use by states of various discriminatory provisions.[16] Congress, too, reexamined the racial barriers to naturalization and in 1952, as part of the Immigration and Nationality Act, removed the "whites only" condition for naturalization.[17] Moreover, when the Supreme Court addressed Black and White race relations in *Brown v. Board of Education*[18] and *Bolling v. Sharpe*,[19] the Court drew upon dicta in *Hirabayashi* and *Korematsu* to argue that racial classifications were inherently suspect and required careful judicial scrutiny. In so doing, the Court brought Japanese-Americans and Other non-Whites into the emerging mainstream of racial jurisprudence and minimized the constitutional distance between them and Blacks. In recent years, the Court has continued this evolution by examining the status of aliens, at first moving toward making alienage a suspect category,[20] and then having second thoughts about such a tie between race and

14. 323 U.S. 214, 223 (1944).

15. See, e.g., Dembitz, supra note 3; Freeman, supra note 3; Rostow, supra note 3.

16. In Oyama v. California, 332 U.S. 633 (1948), the Supreme Court struck down certain presumptions used to enforce California's Alien Land Law and suggested that the underlying alienage classification might be improper. Takahashi v. Fish & Game Comm'n, 334 U.S. 410 (1948), similarly struck down California's prohibition on issuance of commercial fishing licenses to persons ineligible for citizenship.

17. Immigration and Nationality Act, ch. 477, 66 Stat. 634 (1952). While generally considered to be restrictive legislation, the Act did contain this liberalizing provision.

18. 347 U.S. 483 (1954).

19. 347 U.S. 497 (1954).

20. See Graham v. Richardson, 403 U.S. 365 (1971); Sugarman v. Dougall, 413 U.S. 634 (1973); In re Griffiths, 413 U.S. 717 (1973); Nyquist v. Mauclet, 432 U.S. 1 (1977).

alienage.[21]

The camp cases stand at the end of the "old" line of Supreme Court cases beginning with *United States v. Wong Kim Ark*. *Hirabayashi* and *Korematsu* threw into sharp relief the stigma of foreignness that is associated with Japanese-Americans. Yet *Korematsu* has subsequently become the standard precedential reference for strict judicial scrutiny of suspect categories. The camp cases thus also begin the newer approach in which Other non-Whites are considered as American as Blacks, sharing racial minority status with minimal associations of foreignness. Yet, this "newer" approach is by no means clearly accepted. Foreignness as a dimension of the racial identity of Other non-Whites remains a serious social question.[22]

Thus, interpretation of the camp cases should involve an inquiry into the question of race in American legal history. *Justice at War* provides useful new information about the decisionmaking process which led to the camps, and such new information should bring with it deeper understandings. Unfortunately, Irons fails to provide such perspectives because his simplistic moral condemnation lacks the interpretive power to move us beyond previous commentaries. His historical vision similarly fails to move beyond individuals to institutional dimensions.

In 1944, immediately after its publication, the *Korematsu* opinion was seen as a serious threat to our civil liberties. Without suggesting that the threat to individual liberty is any less an issue today, we should continue to expand our inquiry into our governmental decisionmaking process. The work that remains is to better understand the racial component of that process. Perhaps when we genuinely confront how racially conditioned governmental actions take place, we may seriously ask whether or not we can avoid a repetition of the events of 1942.

---

21. See Ambash v. Norwick, 427 U.S. 347 (1976); Foley v. Connelie, 435 U.S. 291 (1978); Cabell v. Chavez-Salido, 454 U.S. 432 (1982); Toll v. Moreno, 458 U.S. 1 (1982).

22. During the 1984 congressional debates over proposed immigration reform, Hispanics and others objected to penalties on employers who hire undocumented aliens, arguing that the penalties would create an increased potential for racial discrimination against Hispanics. The underlying fear is that many employers would assume that other non-White job applicants were foreigners, or be forced to make such an assumption to avoid criminal prosecution. The result would be codification of the presumption of foreignness for other non-Whites. For accounts of the debates, see N.Y. Times, Sept. 27, 1984, at B10, col. 1; N.Y. Times, Sept. 20, 1984, at A23, col. 6.

FANCY DANCING IN THE MARBLE PALACE

*Peter Irons**

On October 12, 1944, Solicitor General Charles Fahy took the podium to urge that the United States Supreme Court affirm the criminal conviction of Fred Korematsu. The crime involved was the defendant's refusal in 1942 to obey one of the military "exclusion orders" that forced the mass evacuation from the west coast and subsequent internment of more than 110,000 Americans of Japanese ancestry. Before he left the podium that afternoon, Fahy had knowingly and deliberately misled the Court on two crucial issues raised in this historic case. By an initial margin of one vote and a final six-to-three division, the Court affirmed the use of military power that Fahy had passionately defended.[1] The Court thus sanctioned the expulsion from their homes and detention of American citizens on the basis of ancestry alone. Fahy's role in the *Korematsu* case casts a shadow over an otherwise distinguished career as government counsel and federal judge.[2]

The transcript of Fahy's argument in *Korematsu* has an intrinsic historical importance that justifies its publication in this journal. Its recent discovery, which I recount briefly below, adds to the record of a seminal case in constitutional law, one that balanced the factors of race and ancestry against the weight of military control over civilians. But the discovery of this transcript in March 1985 has contemporary significance as well. It came too late to influence

* Associate Professor of Political Science, University of California, San Diego. I should note here that I have served as counsel to both Fred Korematsu and Gordon Hirabayashi in their *coram nobis* efforts.

1. Korematsu v. United States, 323 U.S. 214 (1944). Justice William O. Douglas voted at conference to reverse, and prepared a dissenting opinion, but he later switched his vote and withdrew his opinion. For an account of the Supreme Court deliberations in this case, see P. IRONS, JUSTICE AT WAR 319-42 (1983).

2. Lest anyone conclude that I harbor a personal animus toward Fahy, see my admiring account of his service (1935-1940) as general counsel of the National Labor Relations Board in P. IRONS, THE NEW DEAL LAWYERS 226-89 (1982). In that book, I characterized Fahy as a "legal craftsman" and wrote: "His job was to enforce a statute through the presentation of carefully selected cases in the courts, with meticulous attention to detail and the formulation of narrowly drawn issues the keys to success." *Id.* at 235. For tributes to Fahy after his death by his colleague, Judge David L. Bazelon, and former clerk, Prof. Sherman L. Cohn, see 68 GEO. L. J. i-vii (1979).

the outcome of Korematsu's recent challenge to his wartime conviction, a challenge that persuaded a federal judge in October 1983 to vacate the conviction and grant a *coram nobis* petition based on charges of governmental misconduct.[3] Although this transcript had been sought for inclusion in Korematsu's petition, other documents from government files (discussed below) formed a sufficient basis for Judge Patel to grant the petition, since the government did not contest the misconduct charges on their merits.[4]

The transcript has since been placed in evidence in a related case, a similar *coram nobis* petition filed by Gordon Hirabayashi to challenge his 1942 convictions for violation of military curfew and exclusion orders. Affirming the convictions in 1943, the Supreme Court ruled only on the curfew issue and evaded until the next year the exclusion issue raised in *Korematsu*.[5] At a two-week hearing in Seattle in June 1985, Judge Donald Voorhees allowed the transcript in evidence since Fahy had made specific reference in his *Korematsu* argument to the "military necessity" claims made at length in the *Hirabayashi* brief. Over the vehement objection of government lawyers that the Fahy transcript lacked relevance to the *Hirabayashi* case, Judge Voorhees permitted Hirabayashi's counsel to question the veracity and ethical propriety of Fahy's argument.

This brief commentary on the Fahy transcript will first discuss the circumstances of its discovery, a question raised at the recent hearing in Seattle. I will then try to place the transcript into the context of the heated arguments within the Justice Department over the issues of "military necessity" and detention, disputes that began during preparation of the government's Supreme Court brief in *Hirabayashi* in early 1943 and continued through argument of *Korematsu* in late 1944. Finally, I will discuss the relevance of this transcript to the recent *coram nobis* cases, two of which are pending at this writing, and conclude with my assessment of Fahy's behavior in these historic wartime cases.

I

I think a brief word on the discovery and provenance of this transcript is in order, although I will shorten this tale of legal archaeology. I began working in 1981 on a book about the Japanese

3. Korematsu v. United States, 584 F. Supp. 1406 (N.D. Cal. 1984) (opinion of Judge Marilyn Hall Patel).

4. The government's two-page response to the Korematsu petition called the evacuation and internment an "unfortunate episode" in American history and included a general denial of the misconduct allegations. Judge Patel considered the government's response "tantamount to a confession of error:" 584 F. Supp. at 1413.

5. Hirabayashi v. United States, 320 U.S. 81 (1943).

internment cases. This book, published in 1983 as *Justice At War*, recounts in considerable detail the events discussed in this commentary, and provides the primary source citations to documents I cite below to the book, for ease in reference.[6] Early in my research, I requested from the Justice Department access to the original litigation files in these cases. Sad to say, I was first told, the files had been "lost" many years ago and no record existed of their disposition.

But with an unaccustomed—at least to me—diligence, Justice staffers continued their search and informed me in October 1981 that the missing litigation files had been located in a most unlikely place, the files of the Immigration and Naturalization Service, an agency with no involvement in these cases.[7] When the internment case files were finally opened for my inspection, they had not been screened and rested in three dusty cardboard boxes. Virtually the first document I examined in the *Korematsu* file was a memorandum from a Justice Department lawyer that warned Solicitor General Fahy not to pass on War Department "lies" to the Supreme Court. Another document in the *Hirabayashi* file cautioned Fahy that he risked the "suppression of evidence" in withholding crucial military intelligence reports from the Supreme Court.[8]

As I dug through these files, I eagerly looked for transcripts of the Supreme Court arguments that Fahy made in both *Hirabayashi* and *Korematsu*. My appetite was whetted by finding a partial transcript of his argument in the *Endo* case, which was argued the same day as *Korematsu*.[9] The *Korematsu* file included records showing that Fahy had ordered a transcript of his argument and that two copies had been delivered to his office. But I could find neither copy in the Justice Department files provided to me, and I had no reason to believe the file was incomplete.

Needless to say, the documents I found in the files became an

---

6. P. IRONS, *supra* note 1.

7. Although the internment cases were prosecuted at trial by local U.S. attorneys and appeals would normally be handled by the Criminal Division of the Justice Department, they were actually handled by the Alien Enemy Control Unit, headed by Edward Ennis; this unit was established in December 1941 and abolished in 1946. Perhaps the title of this unit led someone to assume that its records should be sent to INS.

8. *See* P. IRONS, *supra* note 1, at 204, 288.

9. Ex parte Endo, 323 U.S. 283 (1944). This case involved a habeas corpus challenge by a young woman who filed a petition after reporting for detention. The Supreme Court unanimously concluded that Congress had not authorized the indefinite detention of concededly loyal citizens, although the constitutional challenge to detention was not decided.

I should add that before 1955, when the Court began recording and transcribing all oral arguments (transcripts and tape recordings are available from the National Archives), counsel were responsible for hiring a private court-reporting firm if they wanted transcripts. Fahy only had his own argument transcribed, and those of his opponents in *Korematsu* are not available.

important part of my book. After I interviewed the defendants in the criminal cases, their reaction to these records led to my involvement as an attorney in efforts to vacate their convictions on the ground of governmental misconduct. In order to complete the *coram nobis* petitions, I continued my search for the missing *Korematsu* transcript; no record existed of a similar *Hirabayashi* transcript. Over the next three years, I searched through Fahy's files in the Roosevelt Library in Hyde Park, New York, and in the Library of Congress. I asked the clerk and librarian of the Supreme Court to look for the transcript. I enlisted the help of Fahy's family in searching their attics and basements. Fahy's opponent in the case, Charles Horsky, looked in his files. I tracked down in Florida the retired head of the court-reporting firm that made the transcript. Fahy's secretary in 1944 and the person who ordered the transcripts, Mildred Fanebust, still worked in 1983 as secretary to Solicitor General Rex Lee. Ms. Fanebust searched her files, but neither she nor anyone else could locate the elusive document.

Almost convinced that my quest was quixotic, I gave up for two years. But, during a quick visit to the National Archives in March 1985, I asked an archivist if I could look again at the card-index file on the *Korematsu* case, from which I obtained the case-file number in 1981. The cards had since been closed to researchers, she told me, but she volunteered to look through them for me. Had I ever looked, she asked, for an "enclosures" file in the case? When I professed ignorance, she said that bulky or superfluous documents were often stored in such files. Five minutes later, a staffer at the Federal Records Center just outside Washington called back to report that the missing transcript was indeed in the "enclosures" file, the existence of which had never been disclosed in response to my longstanding Freedom of Information Act and judicial discovery requests.[10]

10. I recount this otherwise unexciting tale of the unanticipated reward for persistence because of its possible legal significance. The Justice Department lawyer now defending the government against Hirabayashi's misconduct charges has argued vehemently—and thus far with no success—that the *coram nobis* petition is barred by laches; the relevant evidence has been available for almost forty years, he claims, and Hirabayashi has waited too long to reopen his case. But at the Seattle hearing before Judge Voorhees in June 1985, Edward Ennis testified about the substance of the Fahy transcript, and I testified about my continuing search for it in countering the government's laches defense. Although Judge Voorhees has not ruled on the petition at this writing, he indicated in comments to the government's disputatious lawyer that he found no merit in the laches argument. At the conclusion of the hearing, Judge Voorhees asked both sides for posttrial briefs in place of closing arguments and set October 4, 1985, as the deadline for briefs. His ruling on the petition will come at some time after this date.

II

Let me turn now to the substance of Fahy's argument in *Korematsu*, and discuss what the transcript adds to our knowledge and understanding of the issues in the case. Two issues in particular require some background and context. The year before this argument, Fahy had defended the conviction of Gordon Hirabayashi on the ground that "military necessity" had justified the curfew and exclusion orders as a means of preventing espionage and sabotage by Japanese Americans. The government's *Hirabayashi* brief had outlined this argument at length, and the Supreme Court had unanimously echoed Fahy's claims that Army "findings of danger from espionage and sabotage," combined with findings of "disloyalty" among the Japanese Americans, made the curfew order an "appropriate measure" to deal with these dangers.[11]

When Fahy returned to the Court to defend the exclusion order that Korematsu had challenged, he concealed from the Justices all evidence of an internal insurrection that he had quashed just days before. At issue was the veracity of the *Final Report* of General John L. DeWitt, who had both urged and implemented the mass evacuation and internment of Japanese Americans. DeWitt had flatly charged in this official report to the Secretary of War that evidence of espionage by Japanese Americans had prompted his evacuation recommendation. DeWitt's report cited "hundreds of reports nightly of signal lights visible from the coast, and of intercepts of unidentified radio transmissions" from the mainland to offshore Japanese submarines as justification for evacuation.[12]

These espionage charges aroused the suspicions of two Justice Department lawyers, Edward Ennis and John Burling, who were responsible for the government's *Korematsu* brief. The two men prevailed on Attorney General Francis Biddle to ask the FBI and the Federal Communications Commission to investigate DeWitt's charges. Both agencies reported to Biddle that exhaustive investigations had completely refuted the espionage claims. FBI Director Hoover found no evidence that Japanese Americans "have been associated with any espionage activity ashore or that there has been any illicit shore-to-ship signaling, either by radio or lights."[13] FCC Chairman James L. Fly assured Biddle that his agency had investigated hundreds of reports of suspicious radio transmissions and that DeWitt's charges "indicating the existence of illicit radio signaling

11. 320 U.S. at 103-04.

12. P. IRONS, *supra* note 1, at 279.

13. *Id.* at 281.

cannot be regarded as well-founded."[14]

The reports from Hoover and Biddle prompted Ennis and Burling to draft a footnote for the *Korematsu* brief, designed to wave a red flag before the Justices. DeWitt's claim of "military necessity" and his supporting evidence of "the use of illegal radio transmitters and . . . shore-to-ship signaling by persons of Japanese ancestry," Ennis and Burling wrote, were "in conflict with information in the possession of the Department of Justice. In view of the contrariety of the reports on this matter we do not ask the Court to take judicial notice of the recital of those facts contained in the Report."[15] It is hard to imagine a more pointed refutation of a document on which the government proposed to rest its defense of DeWitt's exclusion orders.

What happened after Ennis and Burling drafted their confessional footnote can only be summarized here. Before the printed brief reached the Supreme Court, Assistant Secretary of War John J. McCloy read the drafted footnote, called Fahy to complain, and prevailed on the Solicitor General to stop the printing presses and remove the offending language. After some indecision, Fahy complied with McCloy's request and finally approved substitute language drafted by Herbert Wechsler, who headed the War Division of the Justice Department. Assuring McCloy's deputy that he revised the footnote to refer to DeWitt's report "in the gentlest conceivable way," Wechsler settled on language that "we rely upon the Final Report only to the extent that it relates" to details of the evacuation process.[16] Fahy endorsed Wechsler's revision and sent the brief back to the printers.

The transcript of Fahy's argument shows the extent to which DeWitt's report dominated the proceeding. Charles Horsky and Wayne Collins, who shared the podium for Korematsu, both had denounced the report the previous day. Collins had argued that the public record provided no evidence of espionage or sabotage by Japanese Americans and that DeWitt's orders thus lacked any "rational basis" in fact; Horsky had also intimated that Fahy doubted the veracity of DeWitt's espionage charges. Pointing the Court to the watered-down footnote in the government's brief, Horsky suggested that it could be read as an admission that no evidence existed to support the charges.[17]

Fahy's response to these pointed remarks went far beyond the

14. *Id.* at 283.
15. *Id.* at 284-92.
16. *Id.* at 291.
17. For an account of the oral arguments of Horsky and Collins, see *id.* at 311-15. This account is based on the handwritten notes made by an Army observer and an interview with

limits of partisan argument. Answering the claims that DeWitt had "no rational basis on which to make a military judgment," Fahy denied that "a single line, a single word, or a single syllable" in the report could cast any doubt on the "military necessity" basis for DeWitt's orders. Later in his argument, Fahy assured Chief Justice Stone that "the report proves the basis for the exclusion orders." This flat endorsement of a document the FBI and FCC had torn to shreds, and that Ennis had urged Fahy to repudiate, turned on its head the Solicitor General's dismissal of his opponents' arguments as "a neat little piece of fancy dancing." Fahy himself danced around the ambiguous footnote in denying that "the Government has repudiated the military necessity of the evacuation." His claim that "no person in any responsible position has ever taken a contrary position" was a flat out lie. As Director of the Alien Enemy Control Unit, Ennis had consistently opposed the evacuation, and Attorney General Biddle had made a futile objection to Roosevelt that the War Department had no military basis for the forced exodus of civilians.[18]

III

Fahy also misled the Court in his argument on a second major issue in the *Korematsu* case, that of the link between evacuation and detention. As the transcript demonstrates, this was a complex issue that confounded the Justices, spilled over into the related *Endo* case, and rested on evidence absent from the skimpy trial court record in the case. The issue of detention as a concomitant of evacuation was nonetheless crucial; Fahy's concession in *Endo* that Congress had not authorized the indefinite detention of loyal citizens forced him in *Korematsu* to minimize the connection. Fahy was fortunate that Wayne Collins, who defended Korematsu at trial, had neglected this question and had failed to build a record of documents and testimony for the Supreme Court.[19]

What Fahy wanted to avoid in his *Korematsu* argument was any admission that Fred Korematsu, had he reported as ordered to the assembly center at the required time on May 9, 1942, would have been held in detention for a period that had no predetermined limit. John Burling, who began his work on the *Korematsu* brief with the belief that Japanese Americans had been offered the chance

Horsky. My account of Fahy's argument, at 315-17, is also based on the Army notes and is now superseded by the full transcript.

18. On Biddle's objections, see *id.* at 61-63. For Biddle's later reflections about his failure to press his objections more strongly, see F. BIDDLE, IN BRIEF AUTHORITY 213-26 (1962).

19. P. IRONS, *supra* note 1, at 295.

for "voluntary migration" from the west coast, changed his mind after a careful review of the record. "Contrary to the assumptions upon which we in this office have been going for some time," he wrote Fahy in April 1944, "the original detention was not ordered as a mere temporary expedient, to be in effect for a few days while the persons were removed from California, nor was it in any sense hypothetical or speculative." The offer of "voluntary migration" had in fact expired before Korematsu was compelled to report for evacuation.[20]

Burling pleaded with Fahy before the argument not to adopt the War Department's claim, expressed in the DeWitt report, that the "voluntary migration" program was abandoned *after* exclusion orders were issued, and only *after* the governors of interior states refused to allow Japanese Americans to settle in their states. Burling knew, and told Fahy, that detention in assembly centers, and subsequent indefinite detention in the relocation camps, had been decided as a policy well *before* the hostile governors met with Army officials on April 7, 1942. Burling's point was that Korematsu, even before he became subject to the exclusion order, could not have left his home town, and that on May 9 it became a crime to remain there. His only lawful alternative to this dilemma was indefinite detention.

When he appeared before the Supreme Court, Fahy ignored Burling's protests and claimed that mass evacuation was adopted only because voluntary migration "did not work" and only after "a great wall of hostility" was erected in the Rocky Mountain states. Fahy's bathetic evocation of "old men, old women, babies in arms, and children" being forced into the deserts and mountains "at the point of a bayonet" was, in truth, not the unfortunate alternative to voluntary migration but the calculated policy of the government.

Fahy additionally misled the Court in denying that Korematsu, had he reported for evacuation, would necessarily "have found himself in a relocation center." That is exactly where Korematsu would have found himself. Fahy's vague reference to "security reasons" as a basis for the release of some Japanese Americans from the assembly centers evaded two issues. First, release was entirely discretionary with Army officials; and second, those released were required to remain outside the west coast area designated as a "military zone." Nothing in the record countered Burling's claim that Korematsu faced an indefinite period of detention. Fahy knew this as he assured the Court that even after issuance of the exclusion order Korematsu "could go where he would." Korematsu could go

20. *Id.* at 293. The full account of this dispute between Burling and Fahy is at 292-302.

only one place on May 9, 1942, and that was a barbed-wire compound.

IV

What impact might this transcript have on the pending *coram nobis* cases that seek vacation of criminal convictions in 1942? The *Korematsu* petition has reached final decision, on a record that did not include this petition. But the available record, including many of the documents cited above, provided Judge Patel with adequate grounds for granting the petition. The full record of the case, she wrote, demonstrates the falsity of the espionage charges in the DeWitt report and "is replete with protestations of various Justice Department officials that the government had the obligation to advise the courts of the contrary facts and opinions." She also found "substantial support in the record that the government deliberately omitted relevant information and provided misleading information in papers before the court." Without direct reference to Fahy, but with his role in the case obvious from the context, Judge Patel concluded that the "judicial process is seriously impaired when the government's law enforcement officers violate their ethical obligations to the court."[21]

In the two pending cases, *Yasui* and *Hirabayashi*, the Fahy transcript might well affect the outcomes, at least in adding to the weight of the evidence already before the courts. The attention given to the transcript at the recent *Hirabayashi* hearing in Seattle, and the interest that Judge Voorhees displayed in the testimony of Edward Ennis about his wartime conflicts with Charles Fahy, suggest that the transcript might well figure in the final opinion on the petition. The glaring disparity between the FBI and FCC refutations of DeWitt's espionage charges and Fahy's unbending endorsement of the report certainly adds weight to the misconduct charges raised in the *Hirabayashi* petition.

Finally, what does this argument transcript tell us about Fahy? He emerges from the printed page as a masterful advocate. Soft-spoken and short in stature, and nicknamed "Whispering Charlie"

---

21. 584 F. Supp. at 1418, 1420. I should note that the Supreme Court decided a third criminal case, one that challenged only the curfew and that was decided along with *Hirabayashi.* Yasui v. United States, 320 U.S. 115 (1943). The trial judge had held both that the curfew regulation was unconstitutional as applied to citizens and that Yasui had renounced his citizenship by employment in the Japanese consulate before the war. The Supreme Court sustained the conviction but reversed the judge on both issues. In February 1984, Judge Robert Belloni of the U.S. District Court in Portland, Oregon, vacated Yasui's conviction on the government's motion and dismissed his petition. Yasui has appealed this latter ruling and the case is currently pending before the Ninth Circuit Court of Appeals.

by the press, Fahy was adept in fielding (and sometimes deflecting) questions from the bench. He projected an image of openness and brought a reputation for integrity to the Supreme Court podium. What went wrong in the *Korematsu* case? Did he go over the line of ethical behavior, as Judge Patel's opinion concludes? And was Fahy the only culpable official?

The first of these questions is difficult to answer in this brief commentary. The conclusion that Fahy became trapped between the wartime pressures that gripped the entire government and his professional and ethical obligations is both true and simplistic. Such a formula could excuse virtually any act, as was attempted in far more extreme cases after the war by the Nuremberg defendants. Fahy was, however, a man of deep respect for military authority; he was inclined to defer to arguments of "military necessity" and to men like John McCloy. One need not borrow from psychobiography to understand Fahy as a man lacking in subtlety, who felt strongly that Japanese Americans should share the "hardships of millions of our people in the fight of the nation for its life."

The complete record leaves no real doubt that Fahy misled the Supreme Court in the *Korematsu* case. Edward Ennis, who ultimately swallowed his protests and signed the government's brief, told Judge Voorhees in 1985 that the crucial footnote represented "the narrowest path available to avoid censurable misconduct." Harried by his opponents and questioned sharply by the Court, Fahy strayed from that narrow path in vouching for every line, word, and syllable of DeWitt's tainted report.

One might argue that Fahy had no obligation to disclose the FBI and FCC reports to the Supreme Court, as Ennis and Burling urged. One might also note that Collins and Horsky, in defending Korematsu, cast doubt on DeWitt's espionage charges, doubt that was echoed in the scathing dissent of Justice Frank Murphy.[22] Full disclosure of every piece of potentially exculpatory evidence was not then a prosecutorial duty, and Ennis accepted Fahy's power to overrule him on this issue. However culpable in misleading the Court, Fahy had acceded to McCloy's protest over the original red-flag footnote. McCloy, who supported mass evacuation in 1942 and defended it four decades later as "retribution" for Pearl Harbor,

22. 323 U.S. 214, 233 (1944) (Murphy, J., dissenting). In addition to blasting the majority opinion as a "legalization of racism," Murphy tore the DeWitt report to shreds and noted that "there was no adequate proof that the Federal Bureau of Investigation and the military and naval intelligence services did not have the espionage and sabotage situation well in hand" at the time of evacuation. *Id.* at 241-42. This point, which echoed the arguments of Collins and Horsky, hardly answers the claim that Fahy had a positive duty to inform the Court of the existence and contents of the FBI and FCC reports.

certainly bears more blame and shame than Fahy in the final calculus of ethical failure.

But the Solicitor General speaks for the American people when he appears before the Supreme Court in its marble palace. If he does not address the Court fully and honestly, he betrays the people's trust and undermines the Court's integrity. In this case, Charles Fahy bears the responsibility for persuading the Court to violate the rights of Fred Korematsu, a man of quiet dignity who waited forty years for his ultimate vindication.

JUSTICE, WAR, AND THE JAPANESE-AMERICAN EVACUATION AND INTERNMENT

Book Review of—
JUSTICE AT WAR: THE STORY OF THE JAPANESE AMERICAN INTERNMENT CASES

By Peter Irons. New York: Oxford University Press, 1983. Pp. xiii, 407. $18.95.

Reviewed by Arval A. Morris*

With all the advantages of hindsight, the shameful episode that saw the militarily ordered exclusion and internment of over 112,000 Japanese Americans during World War II without declaration of martial law looms as one of the greatest mass deprivations of civil liberties by the American government since slavery.[1] This harsh, vast, and discriminatory program of uprooting and imprisoning concededly loyal Americans was initiated by an Army general and avidly ordered and supervised by civilian heads of the War Department, defended by the Justice Department, authorized by President Roosevelt, endorsed by Congress, approved by the Supreme Court, and, sadly, supported by the people. The additional sad fact is that except for General DeWitt, every person important to producing this shameful affair was a lawyer. Today, the wartime episode of the Japanese Americans remains an evil blotch upon our national legal conscience and our law, just as it was yesterday and surely will continue to be until the

* Professor of Law, University of Washington.

1. This shameful episode embodied "one of the most sweeping and complete deprivations of constitutional rights in the history of this nation." Korematsu v. United States, 323 U.S. 214, 235 (1944) (Murphy, J., dissenting). Thirty-six years later, Congress, by the Commission On Wartime Relocation and Internment Act, Pub. L. No. 96-317, § 2, 94 Stat. 964 (1980), created a commission composed of former members of Congress, the Supreme Court, the Cabinet, and distinguished private citizens. The Commission held hearings in cities across the United States, reviewed documents and made findings and conclusions that were unanimous. In general, the Commission found that "[t]he broad historical causes which shaped these decisions [regarding exclusion and detention] were race prejudice, war hysteria, and a failure of political leadership; that there was substantial credible evidence, known at the time, from governmental agencies indicating that the claimed ground of military necessity was defective and did not warrant the exclusion and detention of ethnic Japanese," and that as a result, "[a] grave injustice was done to American citizens and resident aliens of Japanese ancestry who, *without individual review or any probative evidence against them,* were excluded, removed and detained by the United States during World War II." PERSONAL JUSTICE DENIED: REPORT OF THE COMMISSION ON WARTIME RELOCATION AND INTERNMENT OF CIVILIANS 18 (1982) (emphasis added).

Japanese-American cases[2] are repudiated, overruled, and completely excised from our law.

The entire Japanese-American program of World War II violated and degraded our nation's proper legal concerns for individual citizens, qua individuals, and subverted the basic values that inform and sustain a democracy. It disparaged fundamental principles—that guilt is individual; that citizens are to be treated with equal care and concern; that all Americans have the constitutional right to move about freely, to live and work where they choose, and to establish and maintain homes and families—all of which are not to be impaired except on an individual basis, and only

2. Korematsu v. United States, 323 U.S. 214 (1944), Yasui v. United States, 320 U.S. 115 (1943), Hirabayashi v. United States, 320 U.S. 81 (1943). A fourth case and the last one decided at the end of World War II, *Ex parte* Endo, 323 U.S. 283 (1944), resulted in a hollow victory. Mitsuye Endo's challenge was to the internment itself, and was based on habeas corpus. "We are of the view that Mitsuye Endo should be given her liberty" wrote Justice Douglas for a unanimous court, *Ex parte Endo* at 297, but, in "reaching that conclusion we do not come to the underlying constitutional issues" because "we conclude that, whatever power the War Relocation Authority may have to detain other classes of citizens, it has no authority to subject citizens who are concededly loyal to its leave procedures." *Id.* "A citizen who is concededly loyal presents no problem of espionage or sabotage" because "[h]e who is loyal is by definition not a spy or a saboteur" and "detention which has no relationship to [those] objective[s] is unauthorized." *Id.* at 302. Thus, Mitsuye Endo, a concededly loyal American, obtained her freedom after more than two years of detention because her detention was beyond the scope of all agency authority and not because her constitutional rights had been violated: the War Relocation Authority's power "to detain a citizen or grant him a conditional release as protection against espionage or sabotage is exhausted *at least when his loyalty is conceded;*" if the Court held otherwise, it "would transform an espionage or sabotage measure into something else" and "[t]hat was not done by Executive Order No. 9066 or by the Act of March 21, 1942, which ratified it." *Id.* at 302 (emphasis added).

The Court announced its *Endo* and *Korematsu* decisions on Monday, December 18, 1944. One day earlier, in an unusual and amazingly timed press release by General Henry Pratt, then head of the Western Defense Command, the War Department announced that "[t]hose persons of Japanese ancestry whose records have stood the test of Army scrutiny during the past two years," (i.e., they concededly were loyal) would be released from internment after January 2, 1945, and would be "permitted the same freedom of movement throughout the United States as other loyal citizens and law-abiding aliens." Thus, by revoking the mass exclusion orders issued in 1942, General Pratt preempted the Supreme Court by one day, freeing Mitsuye Endo and many additional thousands of Japanese Americans, after two years of unjustified incarceration.

Was the timing of General Pratt's announcement fortuitous? Peter Irons thinks not, basing his remarks on an interview with Robert Daniels, an historian who relies on an unnamed source working in the office of John J. McCloy, Assistant Secretary of War at the time, and claims that "McCloy was told by Frankfurter of the date the Supreme Court would issue the *Endo* opinion." Despite the hearsay quality of their evidence, Irons reasonably concludes, "it seems clear that the War Department was privy to the Court's timetable in the *Korematsu* and *Endo* cases." P. IRONS, JUSTICE AT WAR: THE STORY OF THE JAPANESE AMERICAN INTERNMENT CASES, 345 (1983) [hereinafter cited as IRONS]. Endo's "victory" was hollow, and all Americans lost too.

For scathing condemnations of these decisions, see Dembitz, *Racial Discrimination and the Military Judgment*, 45 COLUM. L. REV. 175 (1945); Rostow, *The Japanese-American Cases—A Disaster*, 54 YALE L.J. 489 (1945); *see also* M. GRODZINS, AMERICANS BETRAYED 354 (1949). Peter Irons states that in the years since the publication of the Dembitz and Rostow articles "not a single legal scholar or writer has attempted a substantive defense of the Supreme Court opinions." IRONS, *supra* at 371.

after adequate notice, hearing, fair trial, and due process of law. The Japanese-American program destroyed basic and precious constitutional rights of equal participation in a democracy—the right to peaceably assemble; the rights of free speech and press; the rights freely to read, learn, and hear; the right to vote, seek and hold public office; and the right and responsibility to defend, with one's life if need be, one's beloved native land. German Americans or Italian Americans were not subjected to this kind of wartime treatment. Dangerously, it loosened judicial control of the military. In summary, from a legal point of view, the shameful "Japanese-American episode culminated in a constitutional sanctification of these deprivations by the highest court in the land—a court dedicated to justice, defense of the Constitution, determination of the powers and limitations of government, and protection of the rights of men."[3]

In his highly recommended book, *Justice at War*,[4] Peter Irons shows in absorbing detail how it all happened. In the course of tracing the legal histories of the four most important cases,[5] he also shows how the exclusion and detention program tainted and tarnished almost everyone involved—including a long list of certified liberal heroes. Moreover, he presents disturbing evidence that the Supreme Court was factually misled by the Justice Department.

On an inadequate factual basis, President Roosevelt and a Congress dominated by New Deal Democrats gave carte blanche power to a bona fide racist, Lieutenant General John L. DeWitt, to subject Japanese Americans to military control and detention without first declaring martial law. Earl Warren, then Governor of California, firmly supported the program. A distinguished group of New Deal liberals serving on the national American Civil Liberties Union (ACLU) governing bodies—people like Morris Ernst, Whitney North Seymour, Corliss Lamont, John Dos Passos, Elmer Rice, and Max Lerner—subordinated civil liberties to wartime considerations and to a strong political loyalty to President Roosevelt. By a two-to-one vote, the national ACLU supported the government's right during wartime "to establish military zones and to remove persons, either citizens or aliens, from such zones when their presence may endanger national security, even in the absence of a declaration of martial law."[6]

---

3. J. TENBROEK, E. BARNHART & F. MATSON, PREJUDICE, WAR AND THE CONSTITUTION 325-26 (1968). This altogether excellent book is required reading for any serious understanding of this period.

4. IRONS, *supra* note 2. Peter Irons is a graduate of Harvard Law School and holds a Ph.D. from Boston University. He is the author of THE NEW DEAL LAWYERS (1982), and is a professor of political science at the University of California, San Diego.

5. *See supra* note 2.

6. IRONS, *supra*, note 2, at 129. (quoting from the Frank-Greene ACLU resolution). There was a qualifying condition attached, but, nevertheless, the basic point remains. The qualifier stated that the

To their credit, ACLU people at the local level, such as Mary Farquharson and Art Barnett in Seattle and Ernest Besig in San Francisco, strongly resisted and refused to withdraw,[7] thereby maintaining an ACLU presence in the cases. Their position was that without a formal declaration of martial law, the government could not authorize the military to remove any citizen from his home. The national ACLU board disagreed, and would only allow its lawyers to argue the narrower claim that the military orders were invalid because they were applied in a racially discriminatory manner. The national board ultimately prevailed over the local ACLU's in the cases.

Edward Ennis, who was then in the Department of Justice's Alien Enemy Control Unit and later served with distinction as general counsel of the ACLU, repeatedly closeted his conscience and suppressed his best judgment, carrying out the dictates of his superiors. The response of the National Lawyers Guild was even more disappointing. It "actually backed the wartime internment program."[8] "The Guild's wartime policy was that of 'complete support' for Roosevelt and the American-Soviet alliance . . . ."[9]

Justice Hugo Black, with the approval of William O. Douglas, wrote the opinion for the Supreme Court in *Korematsu*,[10] in which he denied that the Japanese-American internment program was based in any part on racial antagonism. He argued that it had been based on reasonable congressional and military judgments, relying on General DeWitt's *Final*

national ACLU would support evacuation "only if directly necessary to the prosecution of the war or defense of national security" and only if evacuation was "based upon a classification having a reasonable relationship to the danger intended to be met," presumably on the ground of some "reasonable basis." *Id.* at 129. Within a week, on June 22, 1942, the national ACLU board promptly moved to implement its new policy instructing its Director, Roger Baldwin, to inform the ACLU's local West Coast branches that "local committees are not free to sponsor cases in which the position is taken that the government has no constitutional right to remove citizens from military areas" and to "advise the defendants in the test cases already brought to arrange, if they desire, for counsel who will be free to raise other constitutional issues." *Id.* at 130.

7. Mary Farquharson and Art Barnett acted as volunteers in the *Hirabayashi* case. Farquharson wrote Baldwin from Seattle saying she was "shocked . . . to learn of the action" of the national ACLU Board. IRONS, *supra* note 2, at 131. Ernest Besig, Director, Northern California ACLU, refused to back down, writing Baldwin from San Francisco that "we feel compelled to proceed as before, because we cannot in good conscience withdraw from the [*Korematsu*] case at this late date." He pointedly reminded the national ACLU director that "at the time of our intervention in this case we were acting in complete accord with the Board's position." *Id.* On the other side of the local ledger, Karl Bendetsen of Aberdeen, Washington, a very important figure in the evacuation program as an advisor to General DeWitt, displayed a "single-minded and efficient dedication" to the "internment campaign." *Id.* at 31; *see also id.* at 38 *passim*, 58, 64 *passim*

8. *Id.* at 181.

9. *Id.* "Weighed against the wartime sacrifices of the Soviet people, the plight of the Japanese Americans failed to evoke the Guild's sympathy for the test cases." *Id.*

10. Korematsu v. United States, 323 U.S. 214 (1944).

Report on the evacuation stating that there were disloyal Japanese Americans; that their identity could not reliably and quickly be determined; that the West Coast was vulnerable to enemy attack, which disloyal Japanese Americans surely would aid; and that, therefore, an emergency justified the internment measures, which lasted more than two years.

Justices Black, Douglas, and others apparently ignored the racism stated in General DeWitt's *Final Report*. He had branded *all* people of Japanese ancestry as members of "an enemy race," and apparently the Justices ignored General DeWitt's reasoning when, after admitting he had no evidence of Japanese-American complicity in actions against the United States, he incredibly concluded that "[t]he very fact that no sabotage has taken place to date is a disturbing and confirming indication that such action will be taken"![11] Justices Roberts, Murphy, and Jackson dissented, with Justice Murphy declaring: "I dissent, therefore, from this legalization of racism."[12] How did it all happen?

CREATING THE PROGRAM

In a surprise attack Japan bombed Pearl Harbor on December 7, 1941. Thereafter, no internal group appeared to be a more likely target for retribution than the 117,000 Japanese-American citizens and Japanese aliens who lived on the West Coast. The initial response, however, was surprisingly sane;[13] but "some six weeks after Pearl Harbor . . . the tide of public opinion abruptly shifted."[14] On January 16, 1942, Congressman Leland Ford, in identical letters to Secretary of the Navy Frank Knox, and Federal Bureau of Investigation (FBI) Director J. Edgar Hoover, urged that "all Japanese, whether citizens or not, be placed in inland concentration camps."[15] The manager of the Grower-Shipper Vegetable Association, a powerful force in some congressional districts, replied to the charge that they wanted "to get rid of the Japs for selfish reasons" by saying that "[w]e might as well be honest. We do. It's a question of whether the white man lives on the Pacific Coast or the brown man."[16]

11. IRONS, *supra* note 2, at 59, *quoting* DeWitt's Final Report, U.S. Dept. of War, *Final Report, Japanese Evacuation From the West Coast, 1942* 33–34 (Wash. D.C., 1943).

12. *Korematsu*, 323 U.S. at 242 (Murphy, J. dissenting).

13. For example, from December 8, 1941, until January 30, 1942, the *Los Angeles Times* published editorials urging that "there be no precipitation, no riots, no mob law"; that "thousands of Japanese here and in other coast cities" were "good Americans, born and educated as such"; and that the Japanese American Citizens League had promised its "fullest cooperation and its facilities" to the government. The *Times* cautioned "Let's Not Get Rattled." IRONS, *supra* note 2, at 6.

14. IRONS, *supra* note 2.

15. *Id.*

16. *Id.* at 39–40.

President Roosevelt's commission to investigate the attack on Pearl Harbor, headed by Supreme Court Justice Owen J. Roberts, released its report, which was headlined in the Sunday papers on January 25, 1942. The Roberts Commission placed blame for lack of preparedness squarely on the Army and Navy commanders in Hawaii; but it also stated that Japanese consular officials in Hawaii were part of an espionage ring that included "persons having no open relations with the Japanese foreign service," and that this ring had sent "information about U.S. military installations to the Japanese Empire."[17] This flatly asserted imputation of disloyal acts to Japanese Hawaiians has never been documented, but the racial die was cast. It was all downhill after that. On January 28, 1942, the *Los Angeles Times* editorialized that "the rigors of war demand proper detention of Japanese and their immediate removal from the most acute danger spots" on the West Coast.[18] Much earlier, on January 4, in a private conference, General DeWitt, referring to Japanese Americans and displaying his racism confessed "no confidence in their loyalty whatsoever"; he clarified his remarks saying, "I am speaking now of the native born Japanese . . . 42,000 in California alone."[19] Thereafter, he became adamant about excluding and relocating them.[20] It had become a "military necessity."

Despite his basic humanitarian impulses, President Roosevelt's record reflects "a limited awareness of and attention to the plight of racial minorities."[21] Segregation was the law of the land. Politically, he depended on Southerners who had racist values and dominated Congress, and he himself was insensitive to the plight of Japanese Americans. In early February, 1942, on the urgings of General DeWitt, Secretary of War Henry L. Stimson decided to put a question to Roosevelt: "Is the President willing to authorize us to move Japanese citizens as well as aliens from restricted areas?"[22] Without answering this question directly, Roosevelt told Stimson "to go ahead on the line that [he] thought the best." Imme-

17. *Id.*

18. *Id.* at 7.

19. *Id.* at 37.

20. Indeed, General DeWitt sought to influence the drafting of the congressional statute that ratified and implemented Roosevelt's Executive Order 9066 evacuating all persons of Japanese descent from the West Coast: "the recommendation of General DeWitt [was] that prison terms be mandatory" for Japanese Americans found in any military zone where their presence had been prohibited by military orders. Calling attention to the fact that "you have greater liberty to enforce a felony than you have to enforce a misdemeanor," Karl Bendetson, General DeWitt's aid, "quoted DeWitt as saying that 'you can shoot a man to enforce a felony.' " John J. McCloy "shrank from [DeWitt's] trigger-happy proposal, and changed the statute that Bendetson had drafted to reduce the crime of violating a military order to a misdemeanor . . . ." *Id.* at 65.

21. *Id.* at 57.

22. *Id.*

diately thereafter, Stimson's assistant, John J. McCloy, telephoned Karl Bendetsen at the Presidio in California, reporting that "we have carte blanche to do what we want to as far as the President's concerned."[23] Mass racial evacuation was on the move.

On February 19, 1942, seventy-four days after Pearl Harbor, President Roosevelt, as President and as Commander-in-Chief, signed Executive Order 9066. This order conferred authority on the War Department to designate military zones "from which any or all persons may be excluded" and to place further restrictions on "the right of any person to enter, remain in, or leave" such areas at the discretion of military authorities.[24] Thus a racially based, military system of curfew, exclusion and internment of concededly loyal Americans was authorized without any declaration of martial law. The rights of more than 112,000 Japanese Americans became subject to military edict when, on March 2, 1942, General DeWitt declared the entire West Coast a "designated military area." His aide, Karl Bendetsen, prepared the initial draft of a ratifying statute for Congressional approval. Congress hurriedly ratified the president's action when it passed Public Law 503—with only a ten minute "debate" in the House and scarcely more in the Senate—incorporating the standards of Executive Order 9066, criminalizing their violation and also criminalizing violations of military orders, such as the curfew and evacuation proclaimed thereunder.[25]

HIRABAYASHI AND RACIAL CURFEW

A meager handful of cases were generated, and space permits only two to be considered.[26] Gordon Hirabayashi, now in his mid-sixties and a Professor of Sociology in Canada, was a student at the University of Washington in March, 1942, living with a dozen other students in a dormitory at the campus YMCA. He was and is a pacifist of principled com-

23. *Id.* at 58.

24. *Id.* at 63.

25. *Id.* at 66–67. Conservative Senator Robert Taft (Ohio) raised the sole objection in the Senate saying,

> I think this is probably the 'sloppiest' criminal law I have ever read or seen anywhere. It does not say who shall prescribe the restrictions. It does not say how anyone shall know that the restrictions are applicable to that particular zone. It does not appear that there is any authority given to anyone to prescribe any restriction . . . . I have no doubt that in peacetime no man could ever be convicted under it, because the Court would find that it was so indefinite and so uncertain that it could not be enforced under the Constitution.

Id. at 68. On March 18, 1942, President Roosevelt signed Executive Order 9102, which created the War Relocation Authority as the civilian agency jointly responsible with the War Department for the evacuation program.

26. I deal only with the *Hirabayashi* and *Korematsu* cases. For example, I do not deal with Yasui v. United States, 320 U.S. 115 (1943).

mitment and deep religious conviction.[27] Like all Japanese Americans, Hirabayashi was subject to General DeWitt's curfew order, requiring him to be at home each night from 8 p.m. until 6 a.m. One night when it was about five minutes of eight and he was running to his campus YMCA room he stopped and said to himself: "Why the hell am I running back? Am I an American or not? Why am I running back and nobody else is?"[28] He kept a diary of his activities, and he first broke curfew on May 4, 1942, when his friend, Helen Blom, had come to the YMCA to return his raincoat, and "though it was 8 p.m.," he walked her home.[29]

Hirabayashi came to the firm conclusion that he would use himself to test the constitutionality, not of General DeWitt's curfew order, but of General DeWitt's more important orders excluding Japanese Americans from the West Coast—an order under which Hirabayashi was subject to evacuation on May 16—and the order interning them in detention camps. Twenty-four years old, Hirabayashi arrived at the FBI office in Seattle on May 16, 1942, accompanied by his lawyer, Arthur Barnett. Hirabayashi told a surprised Special Agent Francis Manion that he must live by his "Christian principles," that he felt it his "duty to maintain the democratic standards for which this nation lives," and that "[t]herefore, I must refuse this order for evacuation,"[30] thereby raising a most fundamental constitutional question. Agent Manion confiscated Hirabayashi's briefcase containing the diary in which he admitted to three curfew violations between May 4 and May 10.

---

27. Gordon Hirabayashi grew up in an unconventional family. Born in 1918 . . . Hirabayashi absorbed the pacifism of his parents, Japanese natives who belonged to a sect called "Friends of the World." Similar in organizational form and belief to the Quakers, . . .

> When he entered the University of Washington in Seattle in 1937, Hirabayashi became active in the student religious and social-action groups . . . . He assumed a leadership role in the student YMCA and the Japanese Student's Club, and also joined the Seattle chapter of the Japanese American Citizens League. . . .
>
> As a YMCA officer, Hirabayashi traveled to New York City in the summer of 1940 to attend the "President's School" held jointly at Columbia University and the affiliated Union Theological Seminary. Here he was exposed in seminars and discussions to the social activism and pacifism preached by A.J. Muste, Evan Thomas (brother of Socialist Party leader Norman Thomas), and John Nevin Sayre, all members of the pacifist Fellowship of Reconciliation (FOR). This experience proved to be a turning point in the young student's life. "I guess I was ready for that philosophy," Hirabayashi later said. "It all made sense, and the fact that my parents belonged to this non-conformist, non-church Christian group from Japan made these things at home to me." Hirabayashi joined the FOR, which advocated nonviolent but militant resistance to war and military service, during his summer in New York. On his return to Seattle, he registered with his local draft board as a conscientious objector and was granted that status without objection.

IRONS, *supra* note 2, at 89–90.

28. *Id.* at 90.

29. *Id.* at 91.

30. *Id.* at 88.

Gordon Hirabayashi was convicted on two counts—of intentionally violating both General DeWitt's evacuation order and his curfew order. On appeal, the Ninth Circuit Court of Appeals did not decide his case but certified questions of law about each conviction to the Supreme Court of the United States, making possible a definitive, early decision on the constitutional foundations of the Japanese-American evacuation and internment program.[31] In an opinion on June 21, 1943, by Chief Justice Stone, the Supreme Court ducked the hard issues of evacuation and internment. It upheld Hirabayashi's conviction for violating curfew but refused to review the more profound question of his conviction for refusing to submit to evacuation, on the weak ground that Hirabayashi's "sentences on the two counts are to run concurrently and conviction on the second [curfew] is sufficient to sustain the sentence."[32] Thus, the Court put off the more difficult evacuation-internment issue, and it took an additional year before returning in Korematsu's case.

The most fundamental issue decided by the Court in *Hirabayashi* was whether General DeWitt's curfew order could be selectively applied on a racial basis solely to all Japanese Americans and Japanese aliens whether loyal or not. Hirabayashi admitted that a military curfew applying generally to all persons in a "designated military zone" was an obvious device protecting against sabotage, most readily committed during the hours of darkness. But General DeWitt's curfew was selective, applying only to persons of Japanese descent. As such, it was both over- and under-inclusive because thousands upon thousands of Japanese Americans were concededly loyal Americans and needlessly subjected to the curfew, while disloyal persons not of Japanese descent were not subject to the curfew at all. Thus, General DeWitt's curfew order was squarely grounded on race. Nevertheless, the Supreme Court upheld it.

Five years earlier in the famous Footnote Four to the *Carolene Products* case,[33] Justice Stone stated that there

> may be a narrower scope for operation of the presumption of constitutionality when legislation appears . . . directed at . . . racial minorities: whether prejudice against discrete and insular minorities may be a special condition, which tends seriously to curtail the operation of those political processes ordinarily to be relied upon to protect minorities, and which may call for a correspondingly more searching judicial inquiry.[34]

Ignoring this doctrine requiring heightened judicial scrutiny of racial

31. Hirabayashi v. United States, 320 U.S. 81, 83–85 (1943).
32. *Id.* at 105.
33. United States v. Carolene Products Co., 304 U.S. 144 (1938).
34. *Id.* at 152 n.4 (citations omitted).

classifications, Chief Justice Stone ruled in *Hirabayashi* that governmental action need only have a minimal rational basis.[35]

Irons' research into Supreme Court archives and draft opinions reveals that "Stone swallowed whole the 'racial characteristics' arguments made in the briefs submitted by the Justice Department and the three West Coast states and repeated by Solicitor General Fahy in his oral argument before the Supreme Court."[36] Citing undocumented allegations that Japanese Americans posed a fifth-column danger, their location in designated military areas, their supposed adherence to Shintoism (Hirabayashi had become a Quaker), and the system of dual citizenship maintained by the law of Japan, Stone concluded that "the nature and extent of the racial attachments of our Japanese inhabitants to the Japanese enemy were consequently matters of grave concern" to General DeWitt.[37] Stone recognized that hostile governmental action based on race is odious, but "measures necessary for the successful prosecution of the war" could override constitutional concerns "and justify military orders 'which place citizens of one ancestry in a different category from others.' "[38] The Court referred four times to "facts and circumstances" that General DeWitt possibly may have relied upon to " 'show that one racial group more than another' constituted 'a greater source of danger' to the Army's wartime efforts."[39]

Irons observes that Stone wrote his *Hirabayashi* opinion without knowledge of the Office of Naval Intelligence's *Ringle Report* (known to General DeWitt), which estimated the highest number of Japanese Americans "who would act as sabateurs or agents" of Japan as "less than three percent of the total, or about 3500 in the entire United States." Moreover, according to the report, "the most dangerous [were] already in custodial detention" or were "well known to the Naval Intelligence service or the F.B.I." Ringle summarized: "In short, the entire 'Japanese Problem' has been magnified out of its true proportion, largely because of the physical characteristics of the people [and] should be handled on the basis of the *individual*, regardless of citizenship, and *not* on a racial basis." Ringle proposed immediate individualized hearings for residents of internment camps for deciding whether grounds existed to retain the

35. "Our investigation here does not go beyond the inquiry whether, in the light of all the relevant circumstances preceding and attending their promulgation, the challenged orders and statute afforded a reasonable basis for the action taken in imposing the curfew." Hirabayashi v. United States, 320 U.S. 81, 101 (1943).

36. IRONS, *supra* note 2, at 235.

37. *Id.* at 235–36.

38. *Id.* at 236.

39. *Id.*. "Gordon Hirabayashi became the only defendant forced back into prison by the Supreme Court." IRONS, *supra* note 2, at 250–51.

individual as "potentially dangerous."[40] Thus, as General DeWitt knew but the Court did not, the Navy believed that 90% of the Army's evacuation of Japanese Americans was unnecessary and that a maximum of about 10,000 persons necessary for evacuation could readily be identified, most by name.

Edward Ennis discovered the *Ringle Report* and made it known to Solicitor General Fahy a few days before the government filed its brief in the *Hirabayashi* case, saying that "one of the crucial points" in *Hirabayashi* forced the government to argue "that individual selective evacuation would have been impractical and insufficient," but now "we have positive knowledge that the only Intelligence agency responsible for advising General DeWitt gave him advice directly to the contrary."[41] Ennis advised Fahy that the Justice Department had "a duty to advise the Court of the existence of the Ringle memorandum . . . [and] that any other course of conduct might approximate the suppression of evidence."[42] But the government's brief did not bring the *Ringle Report* to any court's attention. Instead, it argued that the rationale behind General DeWitt's orders was "not the loyalty or disloyalty of individuals but the danger from the residence of the class as such within a vital military area" and that "the identities of the potentially disloyal were not readily discoverable" and "virtually impossible" to determine on the basis of individualized hearings.[43]

In addition to the *Ringle Report*, Stone "did [not] know that General DeWitt's own intelligence staff had disclaimed any involvement by Japanese Americans in espionage or sabotage in the period after Pearl Harbor"[44] because responsible authorities failed to bring that information to the Court's attention. Because the Court practically abdicated review of military and Congressional action in *Hirabayashi*—"it is not for any court to sit in review of the wisdom of their action or substitute its judgment for theirs"[45]—Irons laconically observes: "Whether knowledge of this material would have affected the outcome of the case is a matter of speculation."[46] Korematsu's case was left. Perhaps the Court would treat

40. Quotations from the *Ringle Report* come from IRONS, *supra* note 2, at 203.

41. *Id.* at 204.

42. *Id.*

43. *Id.* at 205.

44. *Id.* at 249.

45. Hirabayashi v. United States, 320 U.S. 81, 93 (1943).

46. IRONS, *supra* note 2, at 249. The *Los Angeles Times* hailed *Hirabayashi* as "heartening news for the Pacific Slope, where opinion has held with similar unanimity that the presence of any Japs here is dangerous in wartime" and that "[a]gitation for the return of Japs to the Pacific Coast—which has gained recruits in high circles in Washington—gets its devastating answer from the clear analysis of the situation by Justice Stone . . . ." *Id.* at 251–52. Representative J. Leroy Johnson of California proposed to Congress a postwar deportation of all Japanese aliens and "disloyal" citizens as "a way

the government's power to exclude an American from his home differently from the government's power to keep him in it.

KOREMATSU AND RACIAL EVACUATION

Three weeks after his evacuation deadline, confident he would not be recognized as a Japanese American, Fred Toyosaburo Korematsu, a 23-year-old welder from Oakland, was picked up on May 30, 1943, by the San Leandro police as he was walking down the street with his Caucasian fiancée, Ida Boitano. Being in love, they had decided that Korematsu should have "plastic surgery on his eyes and nose" during March 1942, and that thereafter, "it might be possible for them to go to Arizona and get married without anyone suspecting [he] was Japanese." Otherwise, "he feared violence should anyone discover that he, a Japanese, was married to an American girl."[47] Korematsu told the police that he was of "Spanish-Hawaiian" origin, born in Las Vegas, and that his parents and family had been killed in a fire that burned their house to the ground. His story quickly fell apart when he was confronted with the facts that he spoke no Spanish and that his draft card had been clumsily altered with an ink eradicator. Thereafter, Korematsu confessed all. As the Supreme Court described it, "[n]o question was raised as to [Korematsu's] loyalty to the United States."[48] He "was convicted . . . for remaining in San Leandro, California, a 'Military Area,' contrary to Civilian Exclusion Order No. 34 of the Commanding General [DeWitt] of the Western Command, U.S. Army, which directed that after May 9, 1942, all persons of Japanese ancestry should be excluded from that area."[49] Thus, racially based exclusion and internment finally came before the Court.

Upholding Korematsu's conviction in December 1944, Justice Black's Opinion for the Court recognized that the "Hirabayashi conviction and this one thus rest on the . . . same basic executive and military orders, all of which orders were aimed at the twin dangers of espionage and sabotage."[50] The Court further recognized that exclusion from one's home "is a far greater deprivation" than being confined to it under a curfew

---

to get rid of a group that may make future trouble." Governor Earl Warren said that "nothing more destructive to our defense could happen than to release the potential fifth columnists" and that the return of Japanese Americans to the West Coast would result in sabotage and "a second Pearl Harbor in California." *Id.* at 250–51.

47. IRONS, *supra* note 2, at 95. The FBI report made at the time of his arrest describes Korematsu as a member of the "yellow race," observing: "Scars or marks—Cut scar on the forehead, lump between eyebrows on nose." *Id.* at 94.

48. Korematsu v. United States, 323 U.S. 214, 216 (1944).

49. *Id* at 215–16.

50. *Id.* at 217.

order.[51] Nevertheless, Korematsu's case was controlled by *Hirabayashi*. In light of the decision in *Hirabayashi*, Black wrote, "we are unable to conclude that it was beyond the war power of Congress and the Executive to exclude those of Japanese ancestry from the West Coast area at the time they did"[52] because, judged under *Hirabayashi's* minimal rationality test, "exclusion from a threatened area, no less than curfew, has a definite and close relationship to the prevention of espionage and sabotage"[53] and "because we could not reject the finding of the military authorities that it was impossible to bring about an immediate segregation of the disloyal from the loyal."[54]

Refusing to go behind General DeWitt's decision and evaluate its reasonableness, the Court strained to separate the inseparable in order to duck the issue whether concededly loyal Americans, like Korematsu, could be interned, which was the inevitable next step after evacuation for almost every Japanese American. The Court restricted its *Korematsu* holding strictly to the question of evacuation, ruling that because Korematsu had "not been convicted of failing to report or to remain in an assembly or relocation center, we cannot in this case determine the validity of those separate provisions of [DeWitt's] order."[55]

Like Stone in *Hirabayashi*, Justice Black wrote his *Korematsu* opinion without knowledge of the *Ringle Report*, which Edward Ennis had denounced as "suppression of evidence,"[56] and without knowledge that General DeWitt's own intelligence staff had disclaimed any involvement by Japanese Americans in espionage or sabotage in the period after Pearl Harbor. A more disturbing concern was the shameful way in which General DeWitt's *Final Report* on the evacuation was suppressed in *Hirabayashi* but ultimately brought to the attention of the Supreme Court in *Korematsu*. There were two versions of this report, and Irons charges that in "one significant respect, the differences between the two versions resulted in misleading the Supreme Court in all three of the criminal test cases that challenged General DeWitt's military orders."[57]

General DeWitt's *Final Report* was ten months in preparation by a team headed by Colonel Karl Bendetsen, a lawyer from Aberdeen, Washington. Assistant Secretary of War John J. McCloy, a lawyer, received two printed and bound copies of it on its publication date of April 19,

51. *Id.* at 218.
52. *Id.* at 217–18.
53. *Id.* at 218.
54. *Id.* at 219.
55. *Id.* at 222.
56. *See supra* note 40 and accompanying text.
57. IRONS, *supra* note 2, at 207. The cases are *Hirabayashi*, *Korematsu*, and Yasui v. United States, 320 U.S. 115 (1943).

1943, well before briefs were due in the Supreme Court in the *Hirabayashi* case, and a covering letter from Bendetsen saying that he understood "there is an urgent need of the material contained therein for use in the preparation of the Federal Government's briefs in the cases now pending before the Supreme Court of the United States challenging the constitutionality of the entire program."[58]

McCloy's hasty review of the report caused him concern for two reasons, both damaging to the government's case. First, in the Foreward, General DeWitt flatly stated his opposition to the return of Japanese Americans to the West Coast at any time during the war irrespective of their loyalty.[59] Second, in the important chapter entitled the "Need for Military Control and for Evacuation," General DeWitt asserted, contrary to the Navy's *Ringle Report*, that it had been "impossible to establish the identity of the loyal and the disloyal with any degree of safety." But, his next sentence claimed: "It was not that there was insufficient time in which to make such a determination; it was simply a matter of facing the realities that a positive determination could not be made, that an exact separation of the 'sheep from the goats' was unfeasible."[60] Thus, General DeWitt appears to have been clearly motivated by racial considerations, that is, he believed Japanese Americans were such a "tightly-knit racial group" that separation of the disloyal from the loyal simply could not have been accomplished by any means. Yet earlier, in his January 27, 1943 memo to the War Department, General DeWitt flatly stated that "there was time to determine loyalty" once the evacuation group had been collected together into assembly centers. Equally damaging was General DeWitt's admission that lack of time to conduct individualized loyalty hearings had not been a factor in his decision ordering evacuation.[61] Thus, the "military necessity" argument of the government became vulnerable to the counter-argument that the Japanese Americans had been deprived of individualized hearings and due process of law.

McCloy telephoned Bendetsen, who told McCloy that only ten copies of the *Final Report* had been printed, published, and transmitted. Under pressure from McCloy and Bendetsen, General DeWitt accepted McCloy's revision of the offending two sentences, now to read: "To complicate the situaton no ready means existed for determining the loyal

58. IRONS, *supra* note 2, at 207, quoting Ernest Besig, director of the Northern California ACLU.

59. *Id.*

60. *Id.* at 208.

61. *Id.*

and the disloyal with any degree of safety. It was necessary to face the realities—a positive determination could not have been made.''[62]

McCloy's editing eliminated General DeWitt's admission that lack of time to conduct loyalty hearings had not been a factor in General DeWitt's evacuation decision. He disingenuously had substituted for General DeWitt's racism claim (that in any event it was impossible to determine loyalty) the explanation that the Army had no ''ready means'' by which to perform the task. The other copies of General DeWitt's published *Final Report* were tracked down and removed from War Department files. ''McCloy then instructed General DeWitt to submit a second transmittal letter along with a revised and reprinted version of the *Final Report*, as if the first had never existed.''[63] Knowing DeWitt was in the process of preparing his report, the Justice Department earlier filed a pending request with the War Department for all ''published material'' on the evacuation. The Justice Department wanted to use General DeWitt's report to prepare its brief in *Hirabayashi*. Although available, McCloy took no chances and failed to provide either version of the report to the Justice Department. ''Protected by a 'Confidential' security label, the *Final Report* was not in fact released by the War Department until January 1944, more than seven months after the Supreme Court decided'' *Hirabayashi*,[64] but in time for the Court's consideration of *Korematsu*.

Ennis was outraged when General DeWitt's *Final Report* was made public because, *inter alia*, the War Department had deliberately withheld it during preparation of the *Hirabayashi* brief. Moreover, he discovered that there were two versions of the report and that his pending request for ''published materials'' had not been honored. To avoid possible embarrassment with the Supreme Court, ''Ennis asked Attorney General Biddle late in January 1944 to put the FBI to work on a dissection of the DeWitt report.''[65] General DeWitt's report claimed that ''substantially every ship leaving a West Coast port was attacked by an enemy submarine in the weeks after Pearl Harbor,'' but J. Edgar Hoover reported that

> there is no information in the possession of this Bureau . . . which would indicate that the attacks made on ships or shores in the area immediately after Pearl Harbor have been associated with any espionage activity ashore or that there has been any illicit shore-to-ship signaling, either by radio or lights.[66]

The last item was important to Ennis because General DeWitt's *Final*

62. *Id.* at 210.
63. *Id.* at 211.
64. *Id.* at 212.
65. *Id.* at 280.
66. *Id.* at 280–81.

Report cited "the interception of illicit radio transmissions" as military justification for his evacuation order, and also charged the Justice Department had "impeded" his search for clandestine transmitters. Ennis wrote to the Federal Communications Commission (FCC) chairman, asking for information.

George Sterling, head of the FCC's Radio Intelligence Division, earlier had submitted a memorandum dated January 9, 1942, a month before General DeWitt formally recommended mass evacuation, and recounted a meeting he had with General DeWitt. "Gen'l. DeWitt . . . seemed to believe that the woods were full of Japs with transmitters," so "I proceeded to tell him and his staff" about FCC radio-monitoring operations.[67] Referring to General DeWitt's operation, Sterling said, "Frankly, I have never seen an organization that was so hopeless to cope with radio intelligence requirements."[68]

After his meeting with Sterling, General DeWitt requested that the War Department establish "a joint Radio Intelligence Center for the purpose of coordinating radio interecept . . . information now being collected by the Army, Navy and FCC." This was done, and the FCC operated sixteen monitoring stations on the West Coast capable of pinpointing "the exact house and room in which a transmitter was located if necessary." Moreover, early in 1942, at General DeWitt's request, the FCC set up fully equipped roving coastal patrols, and by July 1, 1942, they had investigated 760 reports of suspicious radio signals—641 turned out not to be radio signals at all and the remaining 119 were traced to Army and Navy transmitters. "No cases involved signals which could not be identified." Sterling was emphatic, "there wasn't a single illicit station and General DeWitt knew it." On April 4, 1944, the FCC chairman, James L. Fly, after reviewing all his sources of information wrote to Attorney General Biddle, summarizing that "[t]here were no radio signals reported to the Commission which could not be identified, or which were unlawful"; moreover, "the Commission knows of no evidence of any illicit radio signaling in this area during the period in question."[69]

The FBI and FCC bombs went off in a memo about the *Korematsu*

67. *Id.* at 283.

68. *Id.*

> The personnel is unskilled and untrained. Most are privates who can read only ten words a minute. They know nothing about signal identification, wave propagation and other technical subjects, so essential to radio intelligence procedure. They take bearings with loop equipment on Japanese stations in Tokio . . . and report to their commanding officers that they have fixes on Jap agents operating transmitters on the West Coast. These officers, knowing no different, pass it on to the General and he takes their word for it. It's pathetic to say the least.

Id.

69. *Id.* at 284.

brief from John Burling to Solicitor General Fahy. "The unmistakable inference of the *Final Report*, Burling wrote, was that among the most important factors making evacuation necessary were reports of frequent signaling by unlawful radio transmitters from Japanese Americans on shore to submarines at sea."[70] But the FBI had concluded that "no one was ever seen signaling from a house or elsewhere to Japanese ships off shore," and the FCC had found that "there was no unlawful radio signaling going on." Thus, Burling concluded,

> we are now . . . in possession of substantially incontrovertible evidence that the most important statements of fact advanced by General DeWitt to justify the evacuation and detention were incorrect, and furthermore that General DeWitt had cause to know, and in all probablility did know, that they were incorrect at the time he embodied them in his final report to General Marshall.[71]

Burling, believing the Department had an ethical obligation to the Court to refrain from citing and relying on the *Final Report*, sent the final draft of the Justice Department's *Korematsu* brief to the printer containing a footnote clearly designed to alert the Supreme Court to the Department's disavowal and repudiation of much of General DeWitt's *Final Report*.[72] But the printing presses were stopped when John J. McCloy of the War Department discovered the footnote and asked Solicitor General Fahy to remove it. Burling and Ennis balked. Assistant Attorney General Herbert Wechsler, a Columbia Law Professor and a supervisor of the *Korematsu* brief, played the role of negotiator and drafted a new, compromise footnote acceptable to the War Department and to Solicitor General Fahy. It eliminated any reference to the espionage allegations in General DeWitt's report: "We have specifically recited in this brief the facts relating to the justification for the evacuation, of which we ask the Court to take judicial notice; and we rely upon the *Final Report* only to the extent it relates to such facts."[73] Thereafter it would take a very clever and discriminating reading between the lines to discern any substantive

70. *Id.* at 285.

71. *Id.*

72. The footnote read:

> The Final Report of General DeWitt is relied on in this brief for statistics and other details concerning the actual evacuation and the events that took place subsequent thereto. The recital of the circumstances justifying the evacuation as a matter of military necessity, however, is in several respects, particularly with reference to the use of illegal radio transmitters and to shore-to-ship signaling by persons of Japanese ancestry, in conflict with information in the possession of the Department of Justice. In view of the contrariety of the reports on this matter we do not ask the Court to take judicial notice of the recital of those facts contained in the Report.

Id. at 286.

73. *Id.* at 290–91.

criticisms of General DeWitt's *Final Report*. On October 5, 1944, the *Korematsu* brief was delivered to the Supreme Court, and misleading semantic ambiguity triumphed over outright repudiation of General DeWitt's *Final Report*.[74]

The Court in *Korematsu* based its opinion on General DeWitt's unfounded "finding" that Japanese Americans posed a danger of espionage on the West Coast. The Court never went behind General DeWitt's "finding" to evaluate its underlying facts independently. Perhaps Burling's footnote would have moved the Court to do so, but perhaps not. Given the Supreme Court's obvious willingness to uphold the Japanese-American evacuation, would its decision in *Korematsu* have been any different if Burling's disputed footnote had been included?[75]

Except for the roles played by a few lawyers, the World War II internment of Japanese Americans does not bring glory to the legal profession. Some forty lawyers took part, over a period of three years, in the legal battles that preceded and followed Roosevelt's signing of Executive Order 9066, and twenty judges ruled on the cases that challenged the internment orders. The outcome of the cases reflected the failure of the legal system that should have stood as a bulwark resisting racism, war hysteria, and the failure of leadership at high levels of government. Instead, it was lawyers and judges who produced what Justice Jackson aptly labeled in his *Korematsu* dissent, "a loaded weapon ready for the hand of any authority that can bring forward a plausible claim of an urgent need." Although *Korematsu* may currently be honored in the breach, its fundamental claim of "pressing public necessity"—short of a declaration of martial law—was upheld by the Court in 1944. It is capable of revival and

74. Although the Court ducked the issue, the Justice Department's brief also addressed the question of detention, seeking to defend it. The brief relied almost completely on General DeWitt's *Final Report's* "preventive detention" rationale, the facts for which Burling, in a memorandum to Fahy, identified as "erroneous" and stated that the Justice Department "can support detention only if it informs the Court, on the strength of its own reputation for veracity," that General DeWitt's claim was factually true. But, "we know DeWitt has made false statements in his evacuation report and we therefore should not take the position in court" that the *Final Report* constituted an adequate defense of detention. *Id.* at 294.

75. Despite the government's argument that evacuation was a military necessity, a number of authors have shown that it was not and that the military knew or should have known the falsity of the claim at the time of the evacuation. *See, e.g.*, F. BIDDLE, IN BRIEF AUTHORITY (1962); M. GRODZINS, AMERICANS BETRAYED (1949). By going behind General DeWitt's "finding," the Court could have ordered that this evidence be produced. But, the Court chose not to. The recently established congressional commission to investigate the internment concluded that no military necessity existed and attrributed the evacuation and internment primarily to political pressures. *See* PERSONAL JUSTICE DENIED, *supra* note 1. In 1983, bills were introduced and referred to the judiciary committees of both houses of Congress to provide payments to Japanese Americans displaced in World War II. *See* S. 1520, 98th Cong., 1st Sess. (1983); S. 2116, 98th Cong., 1st Sess. (1983); H.R. 3387, 98th Cong., 1st Sess. (1983); H.R. 4110, 98th Cong., 1st Sess. (1983).

repetition. The Japanese-American exclusion cases should be reversed, repudiated, and excised from our law.

THE ANCIENT WRIT OF *CORAM NOBIS*

Res judicata is the principal doctrine standing in the way of reopening the *Hirabayashi* and *Korematsu* cases. In such circumstances, a writ of *coram nobis*[76] is an appropriate remedy by which a court can correct errors in criminal convictions where other remedies are precluded. The writ is also designed to maintain public confidence in the integrity of the administration of justice. It is appropriate where the procedure by which guilt was ascertained is under attack, and is reserved for extraordinary circumstances. Federal Rule of Civil Procedure 60(b) abolishes common law writs, including the writ of *coram nobis* in civil cases, but the writ is still available to reopen criminal proceedings where other relief is not available. The petition for a writ of *coram nobis* is properly filed in the federal district court in which the conviction was earlier obtained, even though its earlier judgment has been appealed and affirmed by the Supreme Court.

On the basis of the evidence he uncovered in researching this book, Peter Irons and others have prepared petitions of *coram nobis* in the Japanese-American cases. Korematsu's case in the United States District Court for the Northern District of California came to decision on April 19, 1984.[77] At oral argument, although the government refused to confess error, it ''acknowledged the exceptional circumstances involved and the injustice suffered by [Korematsu] and other Japanese Americans.''[78] The court considered the 1982 findings of the Commission on Wartime Relocation and Internment of Civilians that at the time of Roosevelt's

76. The source of a federal court's power to grant *coram nobis* relief lies in the All Writs Act, 28 U.S.C. § 1651(a) (1948). For discussion, see United States v. Morgan, 346 U.S. 502 (1954); James v. United States, 459 U.S. 1044 (1982) (Brennan, J., dissenting from a denial to grant certiorari); Chresfield v. United States, 381 F. Supp. 301, 302 (E.D. Pa. 1974). In *James*, Justice Brennan stated that in ''criminal cases, the writ of *coram nobis* itself remains available whenever resort to a more usual remedy would be inappropriate.'' 459 U.S. at 1047. He observed that ''28 U.S.C. § 2255, which has taken over most of the function of the writ of *coram nobis* in federal criminal procedure, only applies to collateral attacks on underlying sentences, and could not be employed to vacate and reenter an order denying a motion under Rule 35.'' 459 U.S. at 1047, n.5. By a parity of reasoning, *coram nobis* is not defeated in cases like Korematsu's by the habeas corpus provisions of 28 U.S.C. § 2255 because habeas corpus is not an adequate remedy where the sentence has been served and the ''in custody'' requirements of 28 U.S.C. § 2255 cannot be met. In these unusual circumstances, the extraordinary writ of *coram nobis* is appropriate to correct fundamental errors and prevent injustice. United States v. Correa De-Jesus, 708 F.2d 1283 (7th Cir. 1983).

77. Korematsu v. United States, No. Cr-27635 W, (N.C. Cal., April 19, 1984) (memorandum opinion, available on Lexis).

78. *Id.* at 26.

executive order and General DeWitt's military orders "there was substantial credible evidence from a number of federal civilian and military agencies contradicting the report of General DeWitt that military necessity justified exclusion and internment of all persons of Japanese ancestry without regard to individual identification of those who may have been potentially disloyal."[79] In granting the writ of *coram nobis*, the court also looked to the original record and to the new evidence uncovered by Irons, observing:

> there is substantial support in the record that the government deliberately omitted relevant information and provided misleading information in papers before the court. The information was critical to the court's determination, although it cannot now be said what result would have obtained had the information been disclosed. Because the information was of the kind peculiarly within the government's knowledge, the court was dependent upon the government to provide a full and accurate account. Failure to do so presents [a] "compelling circumstance" . . . . The judicial process is seriously impaired when the government's law enforcement officers violate their ethical obligations to the court.[80]

This decision did not reach any errors of law that had been argued by Korematsu because, the court ruled, the writ "was not used to correct legal errors and this court has no power, nor does it attempt, to correct any such errors."[81] Thus, the Supreme Court's decision in *Korematsu* still stands as the law of the case. But Fred Korematsu now has his writ of *coram nobis*, and he knows it can be issued "only under circumstances compelling such action to achieve justice" and to correct "errors of the most fundamental character."[82] Yet he and all Americans must also know that the most fundamental error will not be corrected until the Supreme Court repudiates and reverses its Japanese-American decisions, replacing them with Mr. Justice Murphy's dissent in *Korematsu*:[83]

> Racial discrimination in any form and in any degree has no justifiable part whatever in our democratic way of life. It is unattractive in any setting but it is utterly revolting among a free people who have embraced the principles set forth in the Constitution of the United States. All residents of this nation are kin in some way by blood or culture to a foreign land. Yet they are primarily and necessarily a part of the new and distinct civilization of the United States. They must accordingly be treated at all times as the heirs of the American experiment and as entitled to all the rights and freedoms guaranteed by the Constitution.

79. *Id.* at 19.
80. *Id.* at 26–27.
81. *Id.* at 27.
82. United States v. Morgan, 346 U.S. 502, 511–12 (1954).
83. Korematsu v. United States, 323 U.S. 214, 242 (1944) (Murphy, J., dissenting).

Moving for Redress

And Justice for All: An Oral History of the Japanese American Detention Camps. By John Tateishi.* *New York: Random House*, 1984. Pp. 260.

Philip Tajitsu Nash†

Over 120,000 Americans of Japanese ancestry were forcibly removed from their west coast homes in 1941 and early 1942, suffering extensive property, income, and psychological damage as a result. Despite reports to the contrary by staff members of the F.B.I. and Office of Naval Intelligence,[1] military leaders argued that persons of Japanese ancestry posed special dangers of espionage and sabotage which justified first curfews and travel restrictions, and eventually removal and incarceration[2] of all Japanese Americans. President Roosevelt issued Executive Order 9066 on February 19, 1942, giving military commanders the power to remove any and all persons from so-called military areas and then provide for the persons so excluded.[3] Men, women, and children, citizens and aliens alike, were then imprisoned without indictment or inquiry as to loyalty, simply because the military felt that time pressures precluded individual loyalty in-

* National Redress Director, Japanese American Citizens' League, Los Angeles, California.

† Charles H. Revson Urban Legal Fellow, Max E. and Filomen M. Greenberg Center for Legal Education and Urban Policy, City College of New York, and Staff Attorney, Asian American Legal Defense and Education Fund, New York.

1. *See* P. IRONS, JUSTICE AT WAR 280–81 (1983) (discussing FBI report); *id.* at 202–03 (discussing Office of Naval Intelligence report).

2. Euphemisms used by some wartime officials obscured what was done to Japanese Americans. "Incarceration," a term that includes "forced removal" and subsequent "detention," provides a more accurate description than the neutral, even beneficent, "relocation."

Similarly, it is time to stop calling the barbed-wire enclosures "assembly centers" or "relocation centers." They were concentration camps. The term "concentration camps" is defined simply as camps where "prisoners of war, enemy aliens, and political prisoners are confined." AMERICAN HERITAGE DICTIONARY 304 (1982). President Roosevelt himself used the words "concentration camps," Press Conference No. 853 (Oct. 20, 1942), *reprinted in* 20 FDR COMPLETE PRESS CONFERENCES 155–57 (1972), as did numerous high-ranking government officials as early as October of 1940. *See, e.g.*, Memorandum from Frank Knox, Secretary of the Navy, to President Franklin D. Roosevelt (Oct. 9, 1940); Letter from Representative Leland M. Ford to Henry L. Stimson, Secretary of War (Jan. 16, 1942); Letter from E. P. Carville, Governor of Nevada, to Lieutenant General J. L. DeWitt (Feb. 21, 1942) (copies of letters on file with author). These and other revealing documents were discovered through the arduous labors of several independent researchers over the years, including Michi Weglyn, Aiko Herzig-Yoshinaga, Jack Herzig, Roger Daniels, and Peter Irons.

To avoid confusion with the atrocities committed by Nazi Germany, we should stop using euphemisms in that context also, and call Hitler's camps what they were: "death camps."

3. P. IRONS, *supra* note 1, at 63. This power was used against the entire Japanese community of the west coast. Criminal sanctions were imposed for violations of Exec. Order 9066. Pub. L. No. 77-503, 56 Stat. 173 (1942). *See* P. IRONS, *supra* note 1, at 64–68.

vestigations. No comparable time pressures were asserted by the military on the east coast. Italian and German Americans there were given individual loyalty hearings, not incarcerated en masse.[4] The Supreme Court accepted these fallacious arguments of "military necessity" in *Hirabayashi v. United States*,[5] *Yasui v. United States*,[6] and *Korematsu v. United States*,[7] decisions that effectively allowed the ordeal of incarceration to continue even beyond the end of World War II. Victims were kept behind barbed wire for an average of thirty months.

Scores of books and articles have been written to document the wartime incarceration.[8] For the first time, however, John Tateishi's *And Justice for All: An Oral History of the Japanese American Detention Camps*[9] brings the ordeal to life in the words of a representative cross-section of the victims. Tateishi was himself incarcerated in a concentration camp and has been active most recently as the National Redress Director of the Japanese American Citizens' League (JACL). After a brief introduction, he records the words of twelve women and eighteen men, who hailed from cities all over the western United States and ranged in age from twelve to thirty-eight when taken to the camps in 1942.

Even for those familiar with the overall story, each page brings forth new insight: the massive pre-war F.B.I. surveillance; the heartbreak of leaving the only home some had ever known; the exploitation by neighbors when household items had to be sold on less than a week's notice; the barbed wire, guard towers, and armed sentries found in every camp; the harsh living conditions; the intrafamily disputes caused by the recruitment of U.S. army combat troops from within the camps; the heroic European Theatre exploits of the Japanese American soldiers in the 442nd Regimental Combat Team and the 100th Battalion,[10] who served despite the irony of fighting for a freedom which their families did not enjoy; and the

---

4. F. CHUMAN, THE BAMBOO PEOPLE: THE LAW AND JAPANESE-AMERICANS 159-60 (1976).

5. 320 U.S. 81 (1943).

6. 320 U.S. 115 (1943).

7. 323 U.S. 214 (1944).

8. *See, e.g.*, F. CHUMAN, *supra* note 4; R. DANIELS, CONCENTRATION CAMPS U.S.A.: JAPANESE AMERICANS AND WORLD WAR II (1970); P. IRONS, *supra* note 1; J. TEN BROEK, E. BARNHART & F. MATSON, PREJUDICE, WAR AND THE CONSTITUTION (1968); M. WEGLYN, YEARS OF INFAMY: THE UNTOLD STORY OF AMERICA'S CONCENTRATION CAMPS (1976); Dembitz, *Racial Discrimination and the Military Judgement: The Supreme Court's* Korematsu *and* Endo *Decisions*, 45 COLUM. L. REV. 175 (1945).

9. J. TATEISHI, AND JUSTICE FOR ALL: AN ORAL HISTORY OF THE JAPANESE AMERICAN DETENTION CAMPS (1984) [hereinafter cited by page number only].

10. In seven major European campaigns, the 442nd Regimental Combat Team, which included men who volunteered from behind barbed wire, suffered 9486 casualties, or more than 300 percent of its original infantry strength. Other Japanese Americans served the United States in the Office of Strategic Services, the Office of War Information, and the Women's Army Corps. COMMISSION ON WARTIME RELOCATION AND INTERNMENT OF CIVILIANS, PERSONAL JUSTICE DENIED 257-59 (1982) [hereinafter cited as REPORT].

tragedy of families forced to rebuild their lives, often from scratch, after leaving the camps.

Consider the irony of this imprisonment for the two-thirds who were American citizens by birth:

> I was born in San Diego. I went to Calexico Elementary School, Calexico Junior High School, and graduated valedictorian from Calexico High School; and then I went to UCLA. I went out to UCLA, but I couldn't get in any of the dorms. They didn't allow Orientals in any of the dorms at UCLA, and so I ended up staying at the YWCA in Boyle Heights and commuted back and forth all those years. I got married in April 1939, graduated from UCLA in June, and then I looked for a job.
>
> First I worked in a market as a cashier and then I took a civil service exam for the city. I passed it and got a job with the city in the Fingerprint and Identification Bureau. I had been working there for a short time when Pearl Harbor was attacked and they said it wasn't convenient to have Japanese working in that division, so I got transferred to the Jefferson Branch Library for six weeks and then I was terminated. They didn't give me a reason for it but I knew why. It was because I was Japanese. . . .
>
> It was really sort of unbelievable. You know, when you go to college you have very high ideals of democracy, and when you have your rights taken away, it is really a shock. I kept saying all along, we're American citizens and the government couldn't possibly put us into camps. I really didn't believe it would happen until it did.[11]

Listen to a pregnant mother, forced to live in a horse stall:

> Before the war I lived in Fresno, California. When evacuation came I was married and had two children, seven and five years old. I was also pregnant three months. The day of the order was just about two weeks after my operation for a tumor. The doctor tried to have my internment delayed for a while and so asked the provost marshal, but it was not allowed. So I closed the bookshop, closed the home with my two children, carried whatever we were allowed to carry, and left for the camp, which was Fresno Assembly Center. . . .
>
> And you know what the summer is like in Fresno—110 degrees—and we were living under a low, tar-paper roof. The floors were built right on top of the racetrack. And there was the manure, and there were cracks in the floor, so that every bit of summer heat, every minute of the day when you're in the barracks, pushed the smell up. . . .
>
> Then there was the food poisoning. I was the first to succumb. We

11. Pp. 108-09 (statement of Mabel Ota).

usually had to wait something like an hour or an hour and a half to be served any food. The line was so long, and things were so disorganized that they didn't have a system that let a sick woman or an older person or pregnant woman go first. They never did that. So if you wanted to eat something, you stayed in line in that hot sun for an hour, sometimes two hours, and by the time you got your food and took it back to your barracks to eat, the food was spoiled. The first time I got food poisoning I thought I would lose the baby, and I was confined to the hospital, which had no roof, no windows, hardly anything. I was there for about a week and barely sustained the baby's life. When I was released to the barracks, I was under the doctor's care for about a month. But all during that pregnancy, I hemorrhaged, and so the doctor thought that if the baby was to be born, it might not be normal.[12]

Even as the incarceration of Japanese Americans was taking place, however, it was recognized as both a direct violation of the victims' constitutional rights and a serious threat to the freedom of every American. For example, in 1945, Eugene V. Rostow, professor and later Dean of the Yale Law School, wrote a seminal article, both insightful and prescient, asserting that "[a]ll in all, the internment of the West Coast Japanese is the worst blow our liberties have sustained in many years."[13] Rostow identified five propositions necessarily underlying the program of Japanese incarceration: (1) protective custody is a permissible form of imprisonment; (2) political opinions may justify such imprisonment; (3) members of a given ethnic group, citizens and resident aliens alike, may be presumed to possess ideas sufficiently dangerous to justify their imprisonment; (4) in times of war or national emergency, the military may decide, independently, what political ideas justify imprisonment and which ethnic groups can be presumed to possess them; and (5) such a determination by the military can be executed without regard to the protections of the Bill of Rights, including the rights to indictment, trial by jury, confrontation of witnesses, defense counsel, and the privilege against self-incrimination.[14] Acceptance of these propositions posed the "utmost potential menace"[15] to the concept that punishment should be meted out only for individual behavior—a concept embodied in American, and indeed all, constitutional law.

Rostow concluded his condemnation of the incarceration by advocating three forms of reparation, to ensure both compensation of actual victims and prevention of similar violations in future years. To protect the civil

12. Pp. 124-26 (statement of Violet de Cristoforo).
13. Rostow, *The Japanese American Cases—A Disaster*, 54 YALE L.J. 489, 490 (1945).
14. *Id.* at 532.
15. *Id.*

rights of Japanese Americans upon leaving the camps, Rostow argued, the federal government was obliged to prosecute vigilantes and other bigots under the authority of the Civil Rights Act of 1870.[16] Second, Rostow maintained that the government should provide financial indemnity for the enormous property losses suffered by victims as a result of their forced removal.[17] And finally, Rostow urged the Supreme Court to reverse *Hirabayashi*, *Yasui*, and *Korematsu*, as it had reversed earlier benighted decisions, so as to reaffirm the constitutional rights abandoned in those wartime opinions.[18]

Professor Rostow and others joined with Japanese Americans to procure financial redress, and in 1948, Congress enacted a sincere but inadequate program.[19] The full measure of Rostow's prophetic arguments, however, has come to life only in the past few years, in the form of a comprehensive movement on three fronts to gain more complete redress for the victims. Both legislative activity[20] and a class action lawsuit[21] seek a more equitable, if still token, compensation for losses suffered. On the third front, three related but distinct *coram nobis*[22] lawsuits have already resulted in two victories—federal district court vacations of the wartime criminal convictions of exclusion-order resister Fred Korematsu and curfew-order resister Minoru Yasui.

It is in the context of this redress movement that *And Justice for All* makes its most tangible contribution. Social justice movements, like good lawsuits, require not only coherent theories and adequate documentation, but compelling fact patterns as well. Precedents and documents sing through the voices of victims. In bringing to life the personal dimension of this major American tragedy, *And Justice for All* serves more than its intended historical purpose. With empathy comes greater vaccination of a new generation of Americans against the dangers of a recurrence. Moreover, a presentation of the personal losses suffered improves the case for individualized compensation. Whether that compensation comes through

---

16. *Id.* at 533.
17. *Id.*
18. *Id.*
19. The Japanese American Evacuation Claims Act, Pub. L. No. 80-886, 62 Stat. 1231 (1948) (codified as amended at 50 U.S.C. app. §§ 1981-1987 (1982)), gave persons of Japanese ancestry the right to claim from the government any real and personal property losses—but not loss of income or losses due to pain and suffering—which occurred as a consequence of governmental action. The Act required an elaborate proof of loss, however, which was virtually impossible for claimants to provide, given the frantic "evacuation sales" held in the few days between notice of the move and the move itself. The Act thus resulted in awards of only $37 million. This amount is far below what would have been full and fair compensation for actual economic losses. COMMISSION ON WARTIME RELOCATION AND INTERNMENT OF CIVILIANS, PERSONAL JUSTICE DENIED, PART 2: RECOMMENDATIONS 7 (1983) [hereinafter cited as RECOMMENDATIONS].
20. *See infra* pp. 748-50.
21. *See infra* pp. 750-51.
22. *See infra* pp. 751-52.

legislation for direct compensation or a judicial opinion awarding compensatory damages, the oral histories will play an important role.

Discussion of the redress concept, public since the late 1960's, resulted in several piecemeal successes in the 1970's: a successful nationwide campaign to repeal Title II of the Internal Security Act of 1950,[23] which had allowed the establishment of camps to detain individuals purely on the basis of suspicion of disloyalty; a 1972 amendment to the Social Security Act which deemed Japanese Americans to have earned and contributed to the Social Security system while so detained;[24] and a 1978 amendment to the federal civil service retirement provisions, allowing Japanese Americans credit for time spent in detention after the age of eighteen.[25] Autonomous state and local campaigns to procure token amounts of compensation for former government employees who were dismissed or forced to resign during World War II because of their Japanese ethnicity succeeded in California, Washington State, Los Angeles County, and San Francisco.[26]

The national legislative campaign for direct financial redress reached Congress in the summer of 1979. With the backing of Japanese American and other organizations, prominent members of Congress, including House Majority Leader James Wright (D.-Tex.) and Senator Daniel Inouye (D.-Hawaii), introduced H.R. 5499[27] and S. 1647.[28] These bills proposed a commission to determine whether any wrong was committed against those American citizens and permanent resident aliens affected by President Franklin D. Roosevelt's Executive Order 9066.[29]

Some members of the Japanese American community who felt the government's wartime wrongs to be self-evident saw the two Commission bills as unnecessary and dilatory. Instead they proposed a direct compensation bill, H.R. 5977,[30] which was introduced on their behalf by Representative Michael Lowry (D.-Wash.) in November of 1979.

After Congressional hearings, the Commission bills were passed on

23. Pub. L. No. 81-831, 64 Stat. 1019 (1950), *repealed by* Act of Sept. 25, 1971, Pub. L. No. 92-128, § 2(a), 85 Stat. 348. *See* M. Weglyn, *supra* note 8, at 325 n.13.

24. Social Security Amendments of 1972, Pub. L. No. 92-603, § 142, 86 Stat. 1367 (codified at 42 U.S.C. § 431 (1982)). *See* Recommendations, *supra* note 19, at 7.

25. Act of Sept. 22, 1978, Pub. L. No. 95-382, 92 Stat. 727 (codified at 5 U.S.C. § 8332(k) (1982)). *See* Recommendations, *supra* note 19, at 7.

26. *See* Recommendations, supra note 19, at 7.

27. H.R. 5499, 96th Cong., 1st Sess., 125 Cong. Rec. 26,884 (1979).

28. S. 1647, 96th Cong., 1st Sess., 125 Cong. Rec. 22,333 (1979).

29. This determination was to be made by a federal commission, which would then make recommendations to Congress as to an appropriate remedy. *See Carter Sent Commission Redress Bill*, Pacific Citizen, Aug. 1-8, 1980, at 1, col. 1.

30. H.R. 5977, 96th Cong., 1st Sess., 125 Cong. Rec. 33,966 (1979). Titled the "World War II Japanese American Human Rights Violation Redress Act," this bill directed the Justice Department to locate all internees, determine how long they had been interned, and then pay each $15,000 plus $15 per day of internment.

July 31, 1980.[31] The new law established a commission that ultimately included nine distinguished public figures, including former Supreme Court Justice Arthur Goldberg and several former congresspersons. After hearing the testimony of over 750 persons in twenty days of nationwide hearings, the Commission issued its report, *Personal Justice Denied*,[32] on February 24, 1983. The report's independent assessment of economic losses and its recommendations for direct monetary redress payments to surviving Japanese Americans and Aleut[33] victims, as well as other remedies, were published on June 16, 1983.[34]

The Commission's conclusions were similar to those of Professor Rostow:

> The promulgation of Executive Order 9066 was not justified by military necessity, and the decisions which followed from it—detention, ending detention and ending exclusion—were not driven by analysis of military conditions. The broad historical causes which shaped these decisions were race prejudice, war hysteria and a failure of political leadership.[35]

Flowing from these conclusions, the following recommendations for redress were sent to Congress: first, passage of a joint congressional resolution recognizing the grave injustice of the incarceration and offering a national apology to its victims; second, presidential pardon of persons convicted of either curfew violations or refusal to accept discriminatory treatment; third, liberal review by executive agencies of applications by Japanese Americans for restitution of positions, status, or entitlements lost as a consequence of events between December 1941 and 1945; fourth, establishment of a special foundation to sponsor research and educational activities designed to illuminate the causes of, specifically, the incarceration of Japanese Americans and, more generally, similar abuses of civil liberties by governmental action in times of national stress; and fifth, appropriation of $1.5 billion to provide per capita compensatory payments of $20,000 to each surviving victim, as well as to fund the research and educational activities just described.[36]

Soon after the Commission released its findings and recommendations

---

31. Commission on Wartime Relocation and Internment of Civilians Act, Pub. L. No. 96-317, 94 Stat. 964 (1980).

32. Report, *supra* note 10.

33. The Aleuts were removed from their island homes to camps with "deplorable" conditions. Of 876 persons removed, an estimated 10% died from epidemics and inadequate health care and sanitation facilities. *Id.* at 18–23.

34. Recommendations, *supra* note 19.

35. Report, *supra* note 10, at 18.

36. Recommendations, *supra* note 19, at 8–10.

in 1983, five bills were introduced to implement some or all of these recommendations.[37] The two that most closely conformed to the Commission's recommendations[38] were eventually co-sponsored in the 98th Congress by 106 representatives and twenty senators.[39] Although Congress failed to act on this legislation in 1984, the sponsors plan to resubmit these bills when the 99th Congress convenes in early 1985.[40] Whether the bills not directly conforming to the Commission's recommendations[41] will be reintroduced in the 99th Congress remains to be seen.

In the second major arena of action for redress, in March of 1983 nineteen named Japanese American plaintiffs filed suit in *Hohri v. United States*[42] on behalf of a potential class of 120,000 victims. For each class member, the complaint demanded $10,000 for each of twenty-two causes of action. Without addressing the issue of class certification, Judge Louis Oberdorfer dismissed the suit on May 17, 1984,[43] holding that the case was barred by the six-year statute of limitations for actions against the United States.[44] Diligent plaintiffs could have brought suit in the 1950's, Judge Oberdorfer reasoned, because circumstantial evidence that the internment was not justfied by governmental claims of "military necessity" was available as early as 1949. Plaintiffs need not have waited for the documents and information revealed in the 1983 Commission Report, *Personal Justice Denied*.[45]

Criticizing Judge Oberdorfer's decision, named plaintiff William Hohri observed:

> We had just come out from behind barbed wire and had been intimidated by losing four major cases[46] in the Supreme Court based on

37. H.R. 4110, 98th Cong., 1st Sess., 129 CONG. REC. H8193 (daily ed. Oct. 6, 1983), and S. 2116, 98th Cong., 1st Sess., 129 CONG. REC. S16606 (daily ed. Nov. 17, 1983), conformed exactly to the Commission's recommendations. H.R. 4322, 98th Cong., 1st Sess., 129 CONG. REC. H9256 (daily ed. Nov. 4, 1983), would have provided redress only to the Aleutian and Pribilof Island native peoples who were forcibly removed from their homes to camps in southeast Alaska. H.R. 3387, 98th Cong., 1st Sess., 129 CONG. REC. H4329 (daily ed. June 22, 1983), advocated compensation for both Japanese Americans and Aleuts, but differed from the Commission's recommendations by advocating compensation to both surviving victims and heirs of deceased victims. S. 1520, 98th Cong., 1st Sess., 129 CONG. REC. S8915 (daily ed. June 22, 1983), was introduced as a discussion bill only, and, while favoring compensation, did not specify a particular formula.

38. H.R. 4110, *supra* note 37; S. 2116, *supra* note 37.

39. *Redress Bills Backed by 106 Reps, 20 Senators*, Pacific Citizen, Oct. 12, 1984, at 1, col. 1.

40. *Id.* It is noteworthy that, forty years later, there are four Japanese-American members of Congress who are helping to support the redress legislation. They are Representatives Robert Matsui (D.-Cal.) and Norman Mineta (D.-Cal.), and Senators Daniel Inouye (D.-Hawaii) and Spark Matsunaga (D.-Hawaii).

41. H.R. 4322, *supra* note 37; H.R. 3387, *supra* note 37; S. 1520, *supra* note 37.

42. 586 F. Supp. 769 (D.D.C. 1984), *appeal docketed*, No. 83-0750 (D.C. Cir. Aug. 15, 1984).

43. *Id.*

44. 28 U.S.C. § 2401(a) (1982).

45. *Hohri v. United States*, 586 F. Supp. at 786-90.

46. *Ex parte* Endo, 323 U.S. 283 (1944) (a victory of sorts, holding that no concededly loyal

> what we now recognize as fraudulent evidence of "military necessity." Given those problems, the political climate of the early 1950's, and the need to start rebuilding our broken lives, how could we have been expected to mount a major lawsuit of this kind?[47]

The National Council for Japanese American Redress, the organizational sponsor of the lawsuit, subsequently voted to file an appeal on which oral argument is expected in the early spring of 1985.[48]

The third major development in the redress movement involves the use of *coram nobis,* the common law writ of error, to reopen the *Korematsu,*[49] *Yasui,*[50] and *Hirabayashi*[51] convictions. This writ, which federal courts can grant under 28 U.S.C. § 1651(a),[52] allows one who has served time for a criminal conviction to petition the court, no matter how long after the conviction, for a vacation of that conviction. Such vacation will be granted if there is evidence of prosecutorial impropriety, or if there are "exceptional circumstances" or errors that "result in a complete miscarriage of justice."[53]

On January 19, 1983, after extensive research into recently declassified government documents showing such impropriety, Fred Korematsu filed a *coram nobis* petition. Minoru Yasui and Gordon Hirabayashi followed his action shortly thereafter in separate but related cases. After hearing arguments, Judge Marilyn Hall Patel of the Northern District of California vacated Korematsu's conviction on grounds of prosecutorial misconduct, including "deliberately omitt[ing] relevant information and provid[ing] misleading information in papers before the [Supreme C]ourt."[54] The information was "critical to the [C]ourt's determination"[55] because it dealt with the military necessity justification for the forced removal and incarceration. Because the common law writ can be used only to correct errors of fact, however, Judge Patel did not reach any errors of law. Meanwhile, Judge Robert C. Belloni of the District of Oregon vacated the wartime

---

American could be detained, although plaintiff's habeas corpus petition, filed in July of 1942, was not granted until December of 1944); *Korematsu*, 323 U.S. 214 (1944); *Yasui*, 320 U.S. 115 (1943); *Hirabayashi*, 320 U.S. 81 (1943).

47. Nash, *Japanese American Redress Movement Wins and Loses*, 114 N.J.L.J., Aug. 2, 1984, at 117, col. 1.

48. 6 National Council for Japanese American Redress Newsletter, Aug. 1984, at 3.

49. 323 U.S. 214 (1944).

50. 320 U.S. 115 (1943).

51. 320 U.S. 81 (1943).

52. Federal courts "may issue all writs necessary or appropriate in aid of their respective jurisdictions and agreeable to the usages and principles of law." 28 U.S.C. § 1651(a) (1982).

53. Korematsu v. United States, 584 F. Supp. 1406, 1419 (N.D. Cal. 1984). *See also* United States v. Morgan, 346 U.S. 502, 511-12 (1954) (*coram nobis* to be used "only under circumstances compelling such action to achieve justice" and to correct "errors of the most fundamental character").

54. *Korematsu*, 584 F. Supp. at 1420.

55. *Id.*

curfew violation of Minoru Yasui.[56] Nevertheless, because Judge Belloni refused to examine the merits of Yasui's petition and to address the issue of governmental misconduct, Yasui filed an appeal with the Ninth Circuit on October 31, 1984, arguing that the court has a duty to address the merits of the case.[57] His appeal will be heard in the spring of 1985. Finally, without addressing the merits of the *Hirabayashi* case, Judge Donald Voorhees in the Western District of Washington has ruled that an evidentiary hearing shall be held.[58] Thus the Supreme Court may yet, as Professor Rostow and others had hoped, have a chance formally to overturn its wartime camp decisions.

Although groups ranging from the American Bar Association[59] to the United Methodist Church,[60] and from the American Legion[61] to the American Friends Service Committee[62] have already taken positions supporting various facets of the current redress movement, much more public education will be needed, on the grassroots level and in the halls of Congress, until the redress movement's goals are reached. By sensitively and accurately recreating the human dimension of the Japanese American incarceration, *And Justice for All* fills a gap in the existing redress literature, and in so doing, helps to advance both of these important goals: individualized compensation for each victim and the sensitizing of a new generation to the dangers of a recurrence.

And Justice for All is not without its shortcomings, however. First, the stories are not placed in any thematic categories that would help the public, and especially a new generation of legislators and judges, to conceptualize the enormity of the losses suffered and to formulate strategies for monetary redress. For example, if the Bill of Rights had been used as an outline, personal stories would substantiate allegations of deprivations of due process, through the lack of personal loyalty hearings; denials of equal protection, through the fact that a similar en masse roundup of German- or Italian-Americans did not take place; the denial of just compensation for property losses directly attributable to the government's removal orders; the denial of the right of freedom of religion, through discriminatory constraints on the practice of Buddhism; the denial of freedom of speech and press, through restrictions on the right to meet and commu-

56. Yasui v. United States, No. 83-151 (D. Or. Jan. 26, 1984).
57. Yasui v. United States, *appeal docketed*, No. 84-3730 (9th Cir. Oct. 31, 1984).
58. *See* Nash, *supra* note 47.
59. *See American Bar Assn. Recognizes Wrong But Rejects Reparations*, Pacific Citizen, Aug. 17, 1984, at 1, col 1.
60. *See $7500 to NCJAR From Church Body*, The New York Nichibei, Mar. 5, 1981, at 1, col. 5.
61. *See Nisei Vets' Resolution Passed by Am. Legion*, Pacific Citizen, Sept. 14, 1984, at 1, col. 1.
62. *See American Friends Service Tells Hayakawa To Stop Using Its Name Against Redress*, The New York Nichibei, Nov. 1, 1984, at 1, col. 1.

nicate in the camps; and so forth. Alternatively, the stories could have been grouped by legal causes of action, including alleged violations of the right to privacy, due to restrictions on family and interspousal behavior imposed by the confines of life in the barracks; of the right to freedom from assault and battery, due to the physical abuse—and even death[63]—caused by camp guards; and of the right to protection from involuntary servitude, because work done by Japanese Americans in the camps was grossly undercompensated, given prevailing wartime wages.[64] While any one victim's experience certainly involves a combination of these violations, leaving them uncategorized overwhelms rather than activates the reader.

The second shortcoming, related to the literary genre, is more serious. Excerpted lives, by definition, lack depth. Such excerpts frequently sensationalize by emphasizing experiences outside the norm and give no hard data to reinforce the conclusions of victims, many of whom did not comprehend what was happening to them at the time. Because it lacks explanatory footnotes, adeqate visual aids,[65] reproductions of government documents, and a postscript on the current redress movement, the book does not provide the context necessary to a meaningful understanding of such a collection of excerpts.

What overview the introduction does provide is too short and too superficial, especially in the wake of the findings in *Personal Justice Denied*.[66] Tateishi places the blame for the government's incarceration policy squarely where it belongs: with President Roosevelt, who signed Executive Order 9066. Nevertheless, Tateishi misses the opportunity to provide scholarly evidence to support his tacit conclusions, instead simply echoing the Commission's statement that "race prejudice, war hysteria and a failure of political leadership"[67] inspired the promulgation of this order and other governmental actions.

The introduction also omits any mention of key events illustrating how

63. *See* REPORT, *supra* note 10, at 179.

64. *Id.* at 165-69. The *Hohri* complaint incorporates these and other innovative causes of action, Complaint at 35-39, Hohri v. United States, 586 F. Supp. 769 (D.D.C. 1984), appeal docketed, No. 83-0750 (D.C. Cir. Aug. 15, 1984). Many have been discussed and developed over the years by Asian American attorneys and legal groups, including the Asian Law Caucus in Oakland, California, the Asian American Legal Defense and Education Fund in New York, the Bay Area Attorneys for Redress in San Francisco, the Asian Law Association in Seattle, and the Asian/Pacific Bar Association of California. Major community organizations that have participated in these and other redress discussions include the Japanese American Citizens' League, based in San Francisco; the National Coalition on Redress/Reparations, based in Los Angeles; the National Council for Japanese American Redress, based in Chicago; and the Washington Coalition on Redress, based in Seattle. Space limitations prevent the recognition of each individual whose volunteer efforts made this movement possible.

65. There is only one map (showing locations of the camps), p. 2, and one contemporary photograph (of newspaper headlines announcing the impending incarceration), following p. 132.

66. REPORT, *supra* note 10.

67. *Id.* at 18.

the mass incarceration program profoundly violated American values of justice and individual liberty. For example, President Roosevelt had received reports in November of 1941 from a State Department investigator, Curtis B. Munson, confirming F.B.I. and Office of Naval Intelligence reports that, because of a "remarkable degree of loyalty to the United States among Japanese Americans,"[68] there was no Japanese "problem" on the west coast. Even assuming there had been a "military necessity" for the forced removal in 1942, top War Department officials privately acknowledged as early as May of 1943, nineteen months before the December 1944 announcement that the camps were to be closed,[69] that such necessity no longer existed.[70] Both the original incarceration and the continued imprisonment ran completely counter to our fundamental values of justice and individual liberty.

Another example of how these values were undermined is the plan to use Japanese Americans, including United States citizens, as "hostages" to be "bartered" to Japan in return for Caucasian P.O.W.'s, and the equally nefarious plan to uproot and transport certain Japanese nationals from Central and South American countries to the United States. These two plans, discussed at the highest levels of government, were first publicly described nine years ago in Michi Weglyn's pathbreaking book, *Years of Infamy*.[71] A discussion of the extent to which the treatment of Japanese Americans so completely ignored personal liberties and individual rights would have strengthened the impact of these individual stories immensely. The context of these histories is crucial to our understanding that they are unique, yet representative. Each is one individual's story, yet it illustrates treatment of the class; and although a group was punished, it is individuals who suffered.

* * *

Despite the book's shortcomings, the movement for Japanese American redress can only be aided by *And Justice for All* and the heroic stories of its thirty protagonists. Read in tandem with more factual and analytical works, it will go far towards vaccinating a new generation against a recurrence and laying the foundation for an individualized compensation program—whether judicial or legislative. The extraordinary power of the human voice, as captured by *And Justice for All*, makes real the threat

68. Curtis B. Munson, Japanese on the West Coast (report attached to memorandum from John Franklin Carter to Pres. Franklin D. Roosevelt (Nov. 7, 1941)), *reprinted in Hearings Before Joint Comm. on Investigation of the Pearl Harbor Attack*, 79th Cong., 1st Sess. 2680, 2682 (1946).

69. *See* REPORT, *supra* note 10, at 235 (discussing Proclamation No. 21, Dec. 17, 1944).

70. *Id.* at 223, 419 n.33 (quoting memorandum from Colonel Karl R. Bendetsen to Lieutenant General John L. DeWitt (May 3, 1943)).

71. M. WEGLYN, *supra* note 8, at 54-66.

posed by the Japanese American internment experience to the civil liberties and property rights of every American.

One can only hope that, with those voices ringing in our ears, we as a nation will move swiftly to correct the injustices recognized forty years ago and vindicate those rights through the classic legal remedy, money damages. And, with the individual hardships better understood, the necessity of an individualized rather than a class-based remedy will become even clearer. To provide otherwise would be to replicate the class-based nature of the original injustice.

THE YALE LAW JOURNAL

VOLUME 54 JUNE, 1945 NUMBER 3

THE JAPANESE AMERICAN CASES—A DISASTER

EUGENE V. ROSTOW†

"He [the King of Great Britain] has affected to render the Military independent of and superior to the Civil Power."

THE DECLARATION OF INDEPENDENCE

"War is too serious a business to be left to generals."

CLEMENCEAU

I

OUR war-time treatment of Japanese aliens and citizens of Japanese descent on the West Coast has been hasty, unnecessary and mistaken. The course of action which we undertook was in no way required or justified by the circumstances of the war. It was calculated to produce both individual injustice and deep-seated social maladjustments of a cumulative and sinister kind.[1]

---

† Professor of Law, Yale University.

The following short-form citations will be used: TOLAN COMMITTEE HEARINGS: *Hearings before House Select Committee Investigating National Defense Migration pursuant to H. Res. 113*, 77th Cong., 2d Sess. (1942); TOLAN COMMITTEE REPORTS (Preliminary) and (Fourth Interim): H. R. REP. No. 1911 (Preliminary Report and Recommendations) and H. R. REP. No. 2124 (Fourth Interim Report), 77th Cong., 2d Sess. (1942); DEWITT FINAL REPORT: U. S. ARMY, WESTERN DEFENSE COMMAND, FINAL REPORT, JAPANESE EVACUATION FROM THE WEST COAST, 1942 (1943, released 1944).

1. See Message from the President of the United States, Segregation of Loyal and Disloyal Japanese in Relocation Centers, Report on S. Res. 166, 78th Cong., 1st Sess., S. Doc. No. 69 (1943); TOLAN COMMITTEE REPORTS (Preliminary and Fourth Interim); MCWILLIAMS, PREJUDICE (1944); MCWILLIAMS, WHAT ABOUT OUR JAPANESE AMERICANS (1944); LEIGHTON, THE GOVERNING OF MEN (1945); An Intelligence Officer, *The Japanese in America: The Problem and the Solution* (1942) 185 HARPER'S 489; Miyamoto, *Immigrants and Citizens of Japanese Origin* (1942) 223 ANNALS 107; Fisher, *What Race Baiting Costs America* (1943) 60 CHRISTIAN CENTURY 1009; Heath, *What About Hugh Kiino?* (1943) 187 HARPER'S 450; "*Issei, Nisei, Kibei*" (April 1944) 29 FORTUNE 8; Bellquist, *Report on the Question of Transferring the Japanese from the Pacific Coast* (1942) 29 TOLAN COMMITTEE HEARINGS 11240; La Violette, *The American-Born Japanese and the World Crisis* (1941) 7 CAN. J. ECON. & POL. SCI. 517; Redfield, *The Japanese-Americans*, in OGBURN (ed.), AMERICAN SOCIETY IN WARTIME (1943) 143; Stonequist, *The Restricted Citizen* (1942) 223 ANNALS 149.

The War Relocation Authority has compiled an admirable bibliography on Japanese and Japanese Americans in the United States; Parts I and II were published November 7, 1942, and Part III August 14, 1943. *The Pacific Citizen*, a newspaper published in Salt Lake

All in all, the internment of the West Coast Japanese is the worst blow our liberties have sustained in many years. Over one hundred thousand men, women and children have been imprisoned, some seventy thousand of them citizens of the United States, without indictment or the proffer of charges, pending inquiry into their "loyalty." They were taken into custody as a military measure on the ground that espionage and sabotage were especially to be feared from persons of Japanese blood. They were removed from the West Coast area because the military thought it would take too long to conduct individual loyalty investigations on the ground. They were arrested in an area where the courts were open, and freely functioning. They were held under prison conditions in uncomfortable camps, far from their homes, and for lengthy periods—several years in many cases. If found "disloyal" in administrative proceedings they were confined indefinitely, although no statute makes "disloyalty" a crime; it would be difficult indeed for a statute to do so under a Constitution which has been interpreted to minimize imprisonment for political opinions, both by defining the crime of treason in extremely rigid and explicit terms, and by limiting convictions for sedition and like offenses.[2] In the course of relocation citizens have suffered severe property losses, despite some custodial assistance by the Government.[3] Perhaps 70,000 persons are still in camps, "loyal" and "disloyal" citizens and aliens alike, more than three years after the programs were instituted. Although the process of relocation has been recently accelerated, many will remain in the camps at least until January 2, 1946.[4]

By the time the question reached the Supreme Court, the crisis which was supposed to justify the action had passed. The Court faced two issues: should it automatically accept the judgment of the military as to the need for the relocation program, or should it require a judicial

---

City by the Japanese American Citizens League is an indispensable source of material on events and attitudes with respect to the process of evacuation, internment and relocation.

2. See Cramer v. United States, 65 Sup. Ct. 918 (U. S. 1945) (treason). For the evidence required to justify imprisonment for attacking the loyalty of the armed forces, see Hartzel v. United States, 322 U. S. 680 (1944). It is notable that persons—citizens or aliens—who actively propagandize in favor of the Axis cause cannot be convicted of sedition, nor placed into protective custody, although loyal citizens of Japanese descent can be arrested and held in preventive custody for periods of more than three years. See also Keegan v. United States, 65 Sup. Ct. 1203 (U. S. 1945), which reverses the conviction of active members of the German-American Bund, a Nazi organization, for conspiracy to obstruct the draft. Apparently the defendants included persons of German nationality as well as of German descent, *id.* at 1212. As for the difficulty of obtaining individual exclusion orders against persons—usually naturalized citizens—with strong German political affiliations, see cases cited *infra* note 13.

3. On the handling of evacuees' property see War Relocation Authority, A Statement on Handling of Evacuee Property (May 1943); DeWitt Final Report, c. xi; Tolan Committee Reports (Fourth Interim) 173–97.

4. See Myer, *The WRA Says "Thirty"* (1945) 112 New Republic 867.

investigation of the question? Was there factual support for the military judgment that the course of the war required the exclusion and confinement of the Japanese American population of the West Coast? Clearly, if such steps were not necessary to the prosecution of the war, they invaded rights protected by the Third Article of the Constitution, and the Fifth and Sixth Amendments.

If the Court had stepped forward in bold heart to vindicate the law and declare the entire program illegal, the episode would have been passed over as a national scandal, but a temporary one altogether capable of reparation. But the Court, after timid and evasive delays, has now upheld the main features of the program.[5] That step converts a piece of war-time folly into political doctrine, and a permanent part of the law. Moreover, it affects a peculiarly important and sensitive part of the law. The relationship of civil to military authority is not often litigated. It is nonetheless one of the two or three most essential elements in the legal structure of a democratic society. The Court's few declarations on the subject govern the handling of vast affairs. They determine the essential organization of the military establishment, state and federal, in time of emergency or of war, as well as of peace. What the Supreme Court has done in these cases, and especially in *Korematsu v. United States*, is to increase the strength of the military in relation to civil government. It has upheld an act of military power without a factual record in which the justification for the act was analyzed. Thus it has created doubt as to the standards of responsibility to which the military power will be held. For the first time in American legal history, the Court has seriously weakened the protection of our basic civil right, the writ of habeas corpus. It has established a precedent which may well be used to encourage attacks on the civil rights of citizens and aliens, and may make it possible for some of those attacks to succeed. It will give aid to reactionary political programs which use social division and racial prejudice as tools for conquering power. As Mr. Justice Jackson points out, the principle of these cases "lies about like a loaded weapon ready for the hand of any authority that can bring forward a plausible claim of an urgent need." [6]

---

5. Hirabayashi v. United States, 320 U. S. 81 (1943); Korematsu v. United States, 323 U. S. 214 (1944); *Ex parte* Mitsuye Endo, 323 U. S. 283 (1944). See FAIRMAN, THE LAW OF MARTIAL RULE (2d ed. 1943) 255–61; Dembitz, *Racial Discrimination and the Military Judgment* (1945) 45 COL. L. REV. 175; Fairman, *The Law of Martial Rule and the National Emergency* (1942) 55 HARV. L. REV. 1253; Freeman, *Genesis, Exodus and Leviticus: Genealogy, Evacuation and the Law* (1943) 28 CORN. L. Q. 414; Graham, *Martial Law in California* (1942) 31 CALIF. L. REV. 6; Lerner, *Freedom: Image and Reality* in SAFEGUARDING CIVIL LIBERTY TODAY (1945); Watson, *The Japanese Evacuation and Litigation Arising Therefrom* (1942) 22 ORE. L. REV. 46; Wolfson, *Legal Doctrine, War Power and Japanese Evacuation* (1944) 32 KY. L. J. 328; Comment (1942) 51 YALE L. J. 1316; Note (1943) 11 GEO. WASH. L. REV. 482.

6. Korematsu v. United States, 323 U. S. 214, 246 (1944).

The opinions of the Supreme Court in the Japanese American cases do not belong in the same political or intellectual universe with Ex parte *Milligan*,[7] *DeJonge v. Oregon*,[8] *Hague v. CIO*,[9] or Mr. Justice Brandeis' opinion in the *Whitney* case.[10] They threaten even more than the trial tradition of the common law and the status of individuals in relation to the state. By their acceptance of ethnic differences as a criterion for discrimination, these cases will make it more difficult to resolve one of the central problems in American life—the problem of minorities. They are a breach, potentially a major breach, in the principle of equality. Unless repudiated, they may encourage devastating and unforeseen social and political conflicts.

II

What General DeWitt did in the name of military precaution within his Western Defense Command was quite different from the security measures taken in Hawaii or on the East Coast—although both places were more active theatres of war in 1942 than the states of Washington, Oregon, California, and Arizona, which comprised the Western Defense Command.

On the East Coast, and in the United States generally, enemy aliens were controlled without mass arrests or evacuations, despite a considerable public agitation in favor of violent action. A registration of aliens had been accomplished under the Alien Registration Act of 1940, and the police authorities had compiled information about fascist sympathizers among the alien population, as well as about those who were citizens. "On the night of December 7, 1941," the Attorney General has reported, "the most dangerous of the persons in this group were taken into custody; in the following weeks a number of others were apprehended. Each arrest was made on the basis of information concerning the specific alien taken into custody. We have used no dragnet techniques and have conducted no indiscriminate, large-scale raids."[11] Immediately after Pearl Harbor restrictions were

7. 4 Wall. 2 (U.S. 1867).

8. 299 U.S. 353 (1937).

9. 307 U.S. 496 (1939).

10. Whitney v. California, 274 U. S. 357, 372–80 (1927). See Professor Riesman's thoughtful essay, *Civil Liberties in a Period of Transition*, in (1942) 3 PUBLIC POLICY 33; CHAFEE, FREE SPEECH IN THE UNITED STATES (1941) *passim*, especially pp. 440–90; Lusky, *Minority Rights and the Public Interest* (1942) 52 YALE L. J. 1.

11. ANNUAL REPORT OF THE ATTORNEY GENERAL FOR FISCAL YEAR ENDED JUNE 30, 1942 (1943) 14. In the first few weeks of war, 2,971 enemy aliens were taken into custody, 1,484 Japanese, 1,256 Germans and 231 Italians. See N. Y. Times, Jan. 4, 1942, § IV, p. 8, col. 3. The basic Presidential proclamations on the treatment of enemy aliens appear in 6 FED. REG. 6321, 6323, 6324 (1941). Regulations under them are issued from time to time by the Attorney General. See, *e.g.*, 7 FED. REG. 844 (1942). See TOLAN COMMITTEE REPORTS (Fourth Interim) 25; Biddle, *Taking No Chances*, Collier's, March 21, 1942, p. 21;

imposed upon the conduct of all enemy aliens over 14 years of age. They were forbidden the Canal Zone and certain restricted military areas thereafter to be specified. They were not to leave the country, travel in a plane, change their place of abode, or travel about outside their own communities, without special permission. They were forbidden to own or use firearms, cameras, short-wave radio sets, codes, ciphers or invisible ink. The District Attorneys were given broad discretion to allow aliens of enemy nationality to carry on their usual occupations, under scrutiny, but without other restriction. A new registration of aliens of enemy nationality was conducted. The basic object of the control plan was to keep security officers informed, but otherwise to allow the aliens almost their normal share in the work and life of the community.

Aliens under suspicion, and those who violated the regulations, were subject to summary arrest on Presidential warrant. "The law," the Attorney General said, "does not require any hearing before the internment of an enemy alien. I believed that nevertheless, we should give each enemy alien who had been taken into custody an opportunity for a hearing on the question whether he should be interned." [12] Those arrested were therefore promptly examined by voluntary Alien Enemy Hearing Boards, consisting of citizens appointed for the task by the Attorney General. These Boards could recommend that individuals be interned, paroled, or released unconditionally. This operation was smoothly conducted, with a minimal interference with the standards of justice in the community. Of the 1,100,000 enemy aliens in the United States, 9,080 had been examined by the end of the fiscal year 1943; 4,119 were then interned, 3,705 paroled, 1,256 released, and 9,341 were still in custody. On June 30, 1944, the number in custody had been reduced to 6,238. The number of those interned was then 2,525, those paroled, 4,840, and those released, 1,926.[13]

Lasker, *Friends or Enemies?* (1942) 31 SURVEY GRAPHIC 277; Rowe, *The Alien Enemy Program—So Far* (Summer 1942) 2 COMMON GROUND 19; Bentwich, *Alien Enemies in the United States* (1943) 163 CONTEMP. REV. 225; Comment (1942) 51 YALE L. J. 1316.

12. ANNUAL REPORT OF THE ATTORNEY GENERAL FOR FISCAL YEAR ENDED JUNE 30, 1942 (1943) 14.

13. The number in custody is greater than the number interned by reason of the inclusion of members of internees' families who request internment, as well as certain alien enemy seamen and alien enemies held for Central and South American countries. See ANNUAL REPORT OF THE ATTORNEY GENERAL FOR FISCAL YEAR ENDED JUNE 30, 1944 (1945) 8.

A small number of citizens and enemy aliens suspected of a propensity for espionage or sabotage by reason of their political opinions were ordered removed from designated security areas both on the East and West Coasts under the statute of March 21, 1942, cited *infra* note 27. This process met with notable judicial resistance. Schueller v. Drum, 51 F. Supp. 383 (E. D. Pa. 1943); Ebel v. Drum, 52 F. Supp. 189 (D. Mass. 1943); Scherzberg v. Maderia, 57 F. Supp. 42 (E. D. Pa. 1944). *Cf.* Labedz v. Kramer, 55 F. Supp. 25 (D. Ore. 1944); Ochikubo v. Bonesteel, 57 F. Supp. 513 (S. D. Calif. 1944). See also United States v. Meyer, 140 F. (2d) 652 (C. C. A. 2d, 1944); Alexander v. DeWitt, 141 F. (2d) 573 (C. C. A.

In Hawaii a somewhat different procedure was followed, but one less drastic than the evacuation program pursued on the West Coast. Immediately after Pearl Harbor martial law was declared in Hawaii, and the commanding general assumed the role of military governor. Courts were reopened for some purposes shortly after the bombing raid, but the return of civil law to Hawaii has been a slow, controversial process, not yet complete. During the period of three and a half years after Pearl Harbor, military power was installed in Hawaii, constitutionally or not, and the normal controls against arrest on suspicion were not available.[14] The population of Hawaii is 500,000, of whom some 160,000, or 32%, were of Japanese descent. Despite the confusions of the moment in Hawaii, only 700 to 800 Japanese aliens were arrested and sent to the mainland for internment. In addition, fewer than 1,100 persons of Japanese ancestry were transferred to the mainland to relocation centers. These Japanese were arrested on the basis of individual suspicion, resting on previous examination or observed behavior, or they were families of interned aliens, transferred voluntarily. Of those transferred from Hawaii to the mainland, 912 were citizens, the rest aliens.[15] Even under a regime of martial law, men were arrested as individuals, and safety was assured without mass arrests.

These procedures compare favorably in their essential character with the precautions taken in Britain and France. The British procedure was the model for our general practice in dealing with enemy aliens. The British Government began in 1939 by interning only those enemy aliens who were on a "security list." Others were subjected to minor police restrictions, pending their individual examination by especially established tribunals. One hundred and twelve such tribunals were set up, under citizens with legal experience, to examine all enemy aliens in Britain. There was an appeals advisory committee to advise the Home Secretary in disputed cases. Aliens were divided into three classes: those judged dangerous were interned; if judged doubtful

---

9th, 1944). The standards developed in these cases to justify the exclusion of persons from military areas as dangerous now closely correspond to those applied in sedition cases. Exclusion will be sustained, that is, only on a showing of "clear and present danger," of aid to the enemy, something more than opinions alone.

14. See FAIRMAN, THE LAW OF MARTIAL RULE (2d ed. 1943) 239–55; LIND, THE JAPANESE IN HAWAII UNDER WAR CONDITIONS (1942); Anthony, *Martial Law in Hawaii* (1942) 30 CALIF. L. REV. 371, (1943) 31 CALIF. L. REV. 477; Frank, *Ex parte Milligan v. The Five Companies: Martial Law in Hawaii* (1944) 44 COL. L. REV. 639; Coggins, *The Japanese-Americans in Hawaii* (1943) 187 HARPER'S 75; Fisher, *Our Two Japanese American Policies* (1943) 60 CHRISTIAN CENTURY 961; Henderson, *Japan in Hawaii* (1942) 31 SURVEY GRAPHIC 328; Horne, *Are the Japs Hopeless?*, Sat. Eve. Post, Sept. 9, 1944, p. 16; LIND, ECONOMIC SUCCESSION AND RACIAL INVASION IN HAWAII (1936); LIND, AN ISLAND COMMUNITY (1938); Smith, *Minority Groups in Hawaii* (1942) 223 ANNALS 36.

15. Communication from the Hon. Abe Fortas, Under Secretary of the Interior, June 28, 1945.

in their loyalty, they were subjected to certain continuing restrictions, especially as to travel, and the ownership of guns, cameras and radios; those deemed entirely loyal to the Allied cause were freed without further restraint. At first 2,000 enemy aliens on a black list were interned. But the entire group was then examined individually, and by March 1940 only 569 of approximately 75,000 aliens were ordered interned. During the panic period of 1940, a new screening was undertaken, to intern all those of doubtful loyalty, and other measures of mass internment were undertaken. Beginning as early as July 1940, however, the policy of wholesale internment was modified, and releases were granted, either generally, or on certain conditions—the proved politics of the internee, his joining the Auxiliary Pioneer Corps, his emigration, and so on.[16] The maximum number interned, during July 1940, was about 27,000 of a total enemy alien population (German, Austrian and Italian) of about 93,000. By September 1941, the number of internees dropped to about 8,500. At the same time, the British undertook to arrest certain British subjects on suspicion alone, under the Emergency Powers Act of 1939. A constitutional storm was aroused by this procedure, which was finally resolved in favor of the government.[17] The general pattern of British security practice was thus to treat enemy aliens on an individual basis, and to arrest British subjects of fascist tendencies in a limited number, and then only on strong personal suspicion.

In France all men enemy aliens between the ages of 17 and 65 were interned in 1939. After a good deal of confusion and complaint, and a vigorous parliamentary protest, many were screened out, either upon joining the Foreign Legion, or, for older men, upon examination, and sponsorship by French citizens. Further parliamentary criticism in December 1939 led to relief for the internees, but the crisis of May and June, 1940, produced mass internment. In France, though less effectively than in Britain, the principle of internment on an individual basis was the objective of policy, if not always its norm.[18]

But on the West Coast the security program was something else again. A policy emerged piecemeal, apparently without sponsors or

---

16. *Report, The Position of Aliens in Great Britain During the War* (1942) 31 TOLAN COMMITTEE HEARINGS 11861; Koessler, *Enemy Alien Internment: With Special Reference to Great Britain and France* (1942) 57 POL. SCI. Q. 98; Kempner, *The Enemy Alien Problem in the Present War* (1940) 34 AM. J. INT. L. 443; Kohn, *Legal Aspects of Internment* (1941) 4 MOD. L. REV. 200; Feist, *The Status of Refugees* (1941) 5 MOD. L. REV. 51.

17. Liversidge v. Anderson [1942] A. C. 206; Greene v. Secretary of State [1942] A. C. 284; Keeton, *Liversidge v. Anderson* (1942) 5 MOD. L. REV. 162; Allen, *Regulation 18B and Reasonable Cause* (1942) 58 L. Q. REV. 232; Goodhart, Notes (1942) 58 L. Q. REV. 3, 9, and *A Short Replication* (1942) 58 L. Q. REV. 243; Holdsworth, Note (1942) 58 L. Q. REV. 1; Carr, *A Regulated Liberty* (1942) 42 COL. L. REV. 339, and *Crisis Legislation in Britain* (1940) 40 COL. L. REV. 1309.

18. See Koessler, *supra* note 16, at 114 *et seq.*

forethought. By May 1, 1942, it had become a policy of evacuating all persons of Japanese ancestry from the West Coast, and confining them indefinitely in camps located away from the coastal area. After some hesitation, General DeWitt proposed evacuation. Quite clearly, a conflict took place between the military authorities on the West Coast and some of the representatives of the Department of Justice over the justification for such action.[19] But no one in the Government would take the responsibility for overruling General DeWitt and the War Department which backed him up.

The dominant factor in the development of this policy was not a military estimate of a military problem, but familiar West Coast attitudes of race prejudice. The program of excluding all persons of Japanese ancestry from the coastal area was conceived and put through by the organized minority whose business it has been for forty-five years to increase and exploit racial tensions on the West Coast. The Native Sons and Daughters of the Golden West, and their sympathizers, were lucky in their general, for General DeWitt amply proved himself to be one of them in opinion and values. As events happened, he became the chief policy maker in the situation, and he has caused more damage even than General Burnside in 1863, whose blunderings with Vallandigham, the Ohio Copperhead, were the previous high in American military officiousness.[20]

In the period immediately after Pearl Harbor there was no special security program on the West Coast for persons of Japanese extraction, and no general conviction that a special program was needed.[21] Known enemy sympathizers among the Japanese, like white traitors and enemy agents, were arrested. There was no sabotage on the part of persons of Japanese ancestry, either in Hawaii or on the West Coast. There was no reason to suppose that the 112,000 persons of Japanese descent on the West Coast, 1.2% of the population, constituted a greater menace to safety than such persons in Hawaii, 32% of the Territory's population. Their access to military installations was not substantially different in the two areas; their status in society was quite similar;

---

19. See DEWITT FINAL REPORT at 3, 7, 19. Mr. Tom Clark (now the Attorney General) stated that mass evacuation was not contemplated as necessary on Feb. 23, 1942. 29 TOLAN COMMITTEE HEARINGS 11164.

20. See 2 SANDBURG, ABRAHAM LINCOLN, THE WAR YEARS (1939) 160–5. President Lincoln wrote to General Burnside, "All the Cabinet regretted the necessity of arresting for instance Vallandigham—some perhaps doubting that there was a real necessity for it, but being done all were for seeing you through with it." Lincoln arranged to have Vallandigham passed through the Confederate lines and banished. RANDALL, CONSTITUTIONAL PROBLEMS UNDER LINCOLN (1926) 176–9. The text of Lincoln's remarks is given somewhat differently by Sandburg and Randall. See also KLAUS, THE MILLIGAN CASE (1929) 12–6.

21. See Rowell, *Clash of Two Worlds* (1942) 31 SURVEY GRAPHIC 9, 12; MCWILLIAMS, PREJUDICE (1944) 108–14; TOLAN COMMITTEE REPORTS (Fourth Interim) 154–6; An Intelligence Officer, *The Japanese in America: The Problem and the Solution* (1942) 185 HARPER'S 489.

their proved record of loyalty in the war has been the same. Although many white persons were arrested, and convicted, as Japanese agents, no resident Japanese American has so far been convicted of sabotage or espionage as an agent of Japan.[22]

After a month's silence, the professional anti-Oriental agitators of the West Coast began a comprehensive campaign. There had been no sabotage in the area, although there was evidence of radio signaling from unknown persons within the area to enemy ships at sea. The West Coast Congressional delegation, led by Senator Hiram Johnson, memorialized the Administration in favor of excluding all persons of Japanese lineage from the coastal area. Anti-Oriental spokesmen appeared as witnesses before the Tolan Committee,[23] and later the Dies Committee,[24] and they explained the situation as they conceived of it to General DeWitt.[25] Some of the coast newspapers, and particularly those owned by William Randolph Hearst, took up the cry. Politicians, fearful of an unknown public opinion, spoke out for white supremacy. Tension was intensified, and doubters, worried about the risks of another Pearl Harbor, remained silent, preferring too much caution to too little. An opinion crystallized in favor of evacuating the Japanese. Such action was at least action, promising greater relief from tension than the slow, patient work of military preparation for the defense and counter-attack. German and Italian aliens were too numerous to be arrested or severely confined, and they were closely connected with powerful blocs of voters. There were too many Japanese Americans in Hawaii to be moved. The 100,000 persons of Japanese descent on the West Coast thus became the chief available target for the release of frustration and aggression.

Despite the nature of the emergency, the military refused to act without fuller legal authority. Executive Order No. 9066 was issued on February 19, 1942, authorizing the Secretary of War, and military commanders he might designate, to prescribe "military areas" in their discretion, and either to exclude any or all persons from such areas, or to establish the conditions on which any or all such persons might enter, remain in or leave such areas.[26] Lieutenant General J. L. DeWitt, head of the Western Defense Command, was ordered on February 20, 1942, to carry out the policy of the Executive Order. During the first two weeks of March, more than three months after Pearl Harbor, General DeWitt issued orders in which he announced that he would subse-

22. See McWilliams, Prejudice (1944) 111.

23. 29 Tolan Committee Hearings 10973, 11061, 11068, 11087, 11111; 30 *id.* at 11303–6, 11314–21, 11325; 31 *id.* at 11642.

24. *Hearings before Special Committee on Un-American Activities on H. Res. 282*, 78th Cong., 1st Sess. (1943), vols. 15, 16.

25. 31 Tolan Committee Hearings 11643; *Hearings before Special Committee on Un-American Activities, supra* note 24, vol. 15, p. 9207.

26. 7 Fed. Reg. 1407 (1942).

quently exclude "such persons or classes of persons as the situation may require" from the area.

But the Army's lawyers wanted more authority than the Executive Order. With inevitable further delays, a statute was therefore obtained prescribing that

> ". . . whoever shall enter, remain in, leave, or commit any act in any military area or military zone prescribed, under the authority of an Executive order of the President, by the Secretary of War, or by any military commander designated by the Secretary of War, contrary to the restrictions applicable to any such area or zone or contrary to the order of the Secretary of War or any such military commander, shall, if it appears that he knew or should have known of the existence and extent of the restrictions or order and that his act was in violation thereof, be guilty of a misdemeanor and upon conviction shall be liable to a fine of not to exceed $5,000 or to imprisonment for not more than one year, or both, for each offense." [27]

The statute thus authorized the exclusion of people from the military areas. It said nothing about their subsequent confinement in camps. This omission was seized upon in Ex parte *Endo* as a crucial fact limiting the power of the Government to hold persons shifted under military orders to relocation centers.[28]

Starting on March 27, 1942, almost four months after Pearl Harbor, the first actual restrictions were imposed. A policy of encouraging the Japanese to move away on a voluntary and individual basis had shown signs of producing confusion and irritation.[29] It was decided to have a uniform and comprehensive program of governmentally controlled migration. At first Japanese aliens and citizens of Japanese ancestry were subjected to the same controls applied to German and Italian aliens. Citizens of German and Italian descent were left free. Early in April, the first of a series of civilian exclusion orders were issued. They applied only to Japanese aliens and citizens of Japanese descent, who were to be excluded altogether from West Coast areas, ordered to report to control stations, and then confined in camps conducted by the newly organized War Relocation Authority, which became an agency of the Department of Interior on February 16, 1944.[30]

---

27. 56 STAT. 173 (1942), 18 U. S. C. § 97a (Supp. 1943).

28. *Ex parte* Mitsuye Endo, 323 U. S. 283, 300–1 (1944).

29. See DEWITT FINAL REPORT, c. ix. But see Fisher, *Japanese Colony: Success Story* (1943) 32 SURVEY GRAPHIC 41.

30. Public Proclamations No. 1, 7 FED. REG. 2320 (1942), No. 2, 7 FED. REG. 2405 (1942), No. 3, 7 FED. REG. 2543 (1942), and other public proclamations established restrictions on travel, residence, and activities for enemy aliens and citizens of Japanese extraction. Civilian Exclusion Order No. 1, March 24, 1942, 7 FED. REG. 2581 (1942), and subsequent exclusion orders established the basis of evacuation. Civilian Exclusion Order No. 34, 7 FED. REG. 3967 (1942), was the basis of Korematsu's case. The War Relocation Authority was established by Executive Order 9102, 7 FED. REG. 2165 (1942).

The rules and policies of these camps were perhaps the most striking part of the entire program. Despite the humanitarian character of the WRA, which was from the beginning intrusted to high-minded and well-meaning men, a policy for discharging Japanese was developed which encouraged lawlessness and refused support to the simplest constitutional rights of citizens and aliens. It was originally thought that the camps would give temporary haven to some Japanese refugees from the West Coast who could not easily arrange new homes, jobs and lives for themselves. Then it was decided to make a stay in the camps compulsory, so as to facilitate the loyalty examinations which were supposed to have been too difficult and prolonged to conduct on the West Coast. Further, it was wisely decided that a loyalty "screening" would facilitate relocation and combat anti-Japanese agitation. The fact that all released evacuees had been approved, so far as loyalty was concerned, gave practical support to their position in new communities. Japanese aliens and citizens of Japanese origin found by this administrative process to be disloyal were confined indefinitely in a special camp. Persons of Japanese descent found to be loyal were to be released from the camps upon the satisfaction of certain conditions. As applied to citizens especially, those conditions upon the right to live and travel in the United States are so extraordinary as to require full statement:

> "In the case of each application for indefinite leave, the Director, upon receipt of such file from the Project Director, will secure from the Federal Bureau of Investigation such information as may be obtainable, and will take such steps as may be necessary to satisfy himself concerning the applicant's means of support, his willingness to make the reports required of him under the provisions of this part, the conditions and factors affecting the applicant's opportunity for employment and residence at the proposed destination, the probable effect of the issuance of the leave upon the war program and upon the public peace and security, and such other conditions and factors as may be relevant. The Director will thereupon send instructions to the Project Director to issue or deny such leave in each case, and will inform the Regional Director of the instructions so issued. The Project Director shall issue indefinite leaves pursuant to such instructions.
>
> "(f) A leave shall issue to an applicant in accordance with his application in each case, subject to the provisions of this Part and under the procedures herein provided, as a matter of right, where the applicant has made arrangements for employment or other means of support, where he agrees to make the reports required of him under the provisions of this Part and to comply with all other applicable provisions hereof, and where there is no reasonable cause to believe that applicant cannot successfully maintain employment and residence at the proposed destination, and no reasonable

ground to believe that the issuance of a leave in the particular case will interfere with the war program or otherwise endanger the public peace and security.

"(g) The Director, the Regional Director, and the Project Director may attach such special conditions to the leave to be issued in a particular case as may be necessary in the public interest." [31]

In other words, loyal citizens were required to have official approval of their homes, jobs and friends before they were allowed to move. They had to report subsequent changes of address, and remain under scrutiny almost amounting to parole. Officials were required to ascertain that community sentiment was not unfavorable to the presence of such citizens before they were permitted to enter the community. The briefs in behalf of the United States before the Supreme Court in the *Korematsu* and *Endo* cases explain the kind of evidence regarded as sufficient to uphold a finding of unfavorable community sentiment, and a suspension of the relocation process: the introduction of anti-Japanese bills in the local legislature, the occurrence of riots or other lawless episodes, and similar expressions of minority opinion.[32]

This policy played a part in encouraging the growth and violent expression of race antagonisms in American society. The forces of the national government were not devoted to protecting and vindicating what *Edwards v. California* had recently upheld as the privilege of a United States citizen, or indeed of any resident, to move freely from state to state, without interference.[33] Local lynch spirit was not con-

31. War Relocation Authority, *Issuance of Leave for Departure from a Relocation Area*, 7 FED. REG. 7656, 7657 (1942). These regulations were revised in detail from time to time, but their basic policy was not substantially altered. See War Relocation Authority, Administrative Notice No. 54 (Summary of Leave Clearance Procedures), March 28, 1944. The basic security data on an evacuee is provided by the FBI and other intelligence agencies, not by independent investigation. This data is supplemented by his answers to questionnaires, particularly as to his loyalty to the United States, and by field investigations in doubtful cases. These field investigations include interviews with the evacuee. An appeal is provided to a Board of Appeals for leave clearance, consisting of citizens not employed by the War Relocation Authority. This Board has the power to advise the Director. Actually, leave was granted *pending inquiry* in cases where the applicant did not have an adverse FBI record; had answered the loyalty questions affirmatively; was not a Shinto priest; and had not spent the larger part of his life in Japan. Thus in fact Japanese Americans were given permission to leave the camps and, after the decision in the *Endo* case, to return to their homes, on the basis of very little information, beyond their answers to questionnaires, which was not available on the West Coast in 1942. Administrative Notice No. 54, *supra*. See discussion of issues in the report of the House Special Committee on Un-American Activities, H. R. REP. No. 717, 78th Cong., 1st Sess. (1943) 13–6, 25.

32. Brief for United States, pp. 35–6, *Ex parte* Mitsuye Endo, 323 U. S. 283 (1944); Brief for United States, p. 15, Korematsu v. United States, 323 U. S. 214 (1944).

33. Edwards v. California, 314 U. S. 160 (1941). Justices Douglas, Black, Murphy and Jackson concurred specially on the ground that California's ban on indigent migrants

trolled and punished by the agencies of law enforcement. On the contrary, it was encouraged to manifest itself in words and unpunished deeds. The threat of lawlessness was allowed to frustrate the legal rights of colored minorities unpopular with small and articulate minorities of white citizens. In March 1943, a small number of Japanese returned to their homes in Arizona, which had been removed from the military zone, without substantial incident.[34] In the spring of 1945, however, the Ku Klux Klan spirit in California had been manifested in at least twenty major episodes of arson or intimidation.[35] The War Relocation Authority has been consistently and effectively on the side of facilitating resettlement and combatting race prejudice. Yet the terms of its leave regulations constituted an extraordinary invasion of citizens' rights, as the Supreme Court later held. They were a practical compromise, under the circumstances, but a compromise nonetheless, with social forces which might better have been opposed head-on.

Studies are beginning to appear about conditions within the camps.

---

from the South West was not only an unconstitutional interference with commerce, but a violation of privileges and immunities of national citizenship. See Myers, *Federal Privileges and Immunities: Application to Ingress and Egress* (1944) 29 CORN. L. Q. 489.

34. See ENCYCLOPAEDIA BRITANNICA BOOK OF THE YEAR: 1944 (1944) 47.

35. *Are Japs Wanted?*, Newsweek, May 28, 1945, p. 33. Including minor episodes, there were 59 such incidents by the end of April 1945. See N. Y. Times, May 6, 1945, § IV, p. 7, col. 4. Some of the episodes are terroristic shooting by night riders; others are arson, the desecration of cemeteries, posting of approbious handbills, etc.; still others are commercial boycotts, like the refusal of Portland, Ore., vegetable merchants (largely of Italian descent) to buy farm produce from a Japanese American farmer. See Pacific Citizen, May 5, 1945, p. 5, col. 4. See also N. Y. Times, Jan. 11, 1945, p. 4, col. 7; *id.*, Jan. 21, 1945, p. 4, col. 3; *id.*, Feb. 17, 1945, p. 2, col. 5; *id.*, Feb. 25, p. 26, col. 4; *id.*, March 18, 1945, p. 17, col. 1. Both West Coast judges and juries have so far tended to acquit persons charged with violence directed against the Japanese, often after confessions by defendants and inflammatory appeals by defense counsel. See Pacific Citizen, April 28, 1945, p. 1, col. 4; p. 4, col. 1 ("This is a white man's country"); (1945) 160 THE NATION 531, 598. Labor leaders, historically one of the strongest anti-Japanese groups in West Coast life, are in the forefront of resistance to the return of the Japanese to their homes. See, *e.g.*, the position of Dave Beck, reported in The Pacific Citizen, April 21, 1945, p. 4, col. 2; p. 5, col. 4.

Strong reactions of opinion and of citizens groups in favor of protecting the rights of Japanese Americans have been manifested, led by Secretary of War Stimson, Secretary of Interior Ickes, and the staff of the War Relocation Authority. See Pacific Citizen, April 7, 1945, p. 1, col. 1, quoting Secretary Ickes' forceful statement of April 4, 1945; Pacific Citizen, April 14, 1945, p. 2, col. 1 (Secretary Stimson's remarks at press conference of April 5). Many West Coast groups have been organized to oppose the Klan movement in the Far West. See Pacific Citizen, April 28, 1945, p. 7, col. 1; *id.*, April 21, 1945, p. 3, col. 1. See excellent speech of Attorney General Robert W. Kenny of California, delivered to a convention of California sheriffs, calling on law enforcement officers to protect the legal rights of returning Japanese Americans. N. Y. Times, March 18, 1945, p. 17, col. 1; Pacific Citizen, March 24, 1945, p. 1, col. 4; *id.*, March 31, 1945, p. 5, col. 1 (partial text of Mr. Kenny's speech); Beshoar, *When Good Will Is Organized* (Spring 1945) 5 COMMON GROUND 19; Pacific Citizen, March 3, 1945, p. 6, col. 1 (speech by Joe E. Brown before Commonwealth Club of San Francisco in behalf of fair play for Japanese Americans); Time, May 28, 1945, p. 13 (Quakers aid returned evacuees in Oregon).

They make it plain that the camps were in fact concentration camps, where the humiliation of evacuation was compounded by a regime which ignored citizens' rights, and the amenities which might have made the relocation process more palatable.[36]

Thus there developed a system for the indefinite confinement and detention of Japanese aliens and citizens of Japanese descent, without charges or trial, without term, and without visible promise of relief. By May 1942, it was compulsory and self-contained. On pain of punishment under the Act of March 21, 1942, all had to leave the West Coast through Assembly Centers and the Relocation Centers. Counsel in the *Hirabayashi* case called it slavery; Mr. Justice Jackson said it was attainder of blood.[37] The Japanese radio discussed it at length, finding in the system ample propaganda material for its thesis that American society was incapable of dealing justly with colored peoples.

III

Attempts were made at once to test the legality of the program. The district courts and the circuit courts of appeals had a good deal of difficulty with the issues. Although troubled, they generally upheld both the exclusion of Japanese aliens and citizens from the West Coast, and at least their temporary confinement in WRA camps.[38]

The question of how and on what grounds the Supreme Court should dispose of the cases was one of broad political policy. Would a repudiation of the Congress, the President and the military in one aspect of their conduct of the war affect the people's will to fight? Would it create a campaign issue for 1944? Would it affect the power, status and prestige of the Supreme Court as a political institution? How would a decision upholding the Government influence civil liberties and the condition of minorities? A bench of sedentary civilians was reluctant to overrule the military decision of those charged with carrying on the war. Conflicting loyalties, ambitions and conceptions of the Court's duty undoubtedly had their part in the positions the justices took.

The issue first came before the Supreme Court in May 1943, and the first cases, *Hirabayashi v. United States* and *Yasui v. United States*, were decided on June 21, 1943.[39] No Japanese submarines had been detected off the West Coast for many months. Midway was won; Libya, Tripolitania and Tunisia had been conquered. Guadalcanal and a good deal of New Guinea were in Allied hands. The posture of the war had changed profoundly in a year. We had suffered no defeats since the

36. See Leighton, *op. cit. supra* note 1.

37. Brief for Northern California Branch of the American Civil Liberties Union, p. 93; Korematsu v. United States, 323 U. S. 214, 243 (1944).

38. See, *e.g.*, United States v. Yasui, 48 F. Supp. 40 (D. Ore. 1942); Korematsu v. United States, 140 F. (2d) 289 (C. C. A. 9th, 1943).

39. 320 U. S. 81 and 115 (1943).

fall of Tobruk in July 1942, and we had won a long series of preliminary victories. Our forces were poised for the offensive. The phase of aggressive deployment was over.

The problem presented to the Supreme Court was thus completely different from that which confronted worried legislators and officials in the bleak winter and spring of 1942. Invalidation of the exclusion and confinement programs would do no possible harm to the prosecution of the war. The Court could afford to view the issues in full perspective. The war powers of the legislative and executive must of course be amply protected. But the special concerns of the Supreme Court for the development of constitutional law as a whole could be given proper weight, free of the pressure of the Pearl Harbor emergency.

It was only half the truth to say that the cases had to be decided as if the date of decision were February 1942. It was not in fact the date of decision, and could not be made so. The issue was not only whether the military should have excluded the Japanese in the spring of 1942, but whether the Court should now validate what had been done. As many episodes in the history of the United States eloquently attest, these are different issues. The problem of the Court in the *Hirabayashi* case was not that of General DeWitt in 1942, but an infinitely more complex one. Whether it faced the issues or tried to ignore them, whether it decided the cases frankly or obliquely, by decision or evasion, the Court could not escape the fact that it was the Supreme Court, arbiter of a vast system of rules, habits, customs and relationships. No matter how inarticulate, its decision could not be confined in its effect to the United States Reports. It would necessarily alter the balance of forces determining the condition of every social interest within range of the problems of the cases—the power of the military and the police; our developing law of emergencies, which is beginning to resemble the French and German law of the state of siege; the status of minorities and of groups which live by attacking minorities; the future decision of cases in police stations and lower courts, involving the writ of habeas corpus, the equal rights of citizens, the protection of aliens, the segregation of racial groups, and like questions.

In a bewildering and unimpressive series of opinions, relieved only by the dissents of Mr. Justice Roberts and of Mr. Justice Murphy in *Korematsu v. United States*,[40] the Court chose to assume that the main issue of the cases—the scope and method of judicial review of military decisions—did not exist. In the political process of American life, these decisions were a negative and reactionary act. The Court avoided the risks of overruling the Government on an issue of war policy. But it weakened society's control over military authority—one of the polarizing forces on which the organization of our society depends. And it

40. 323 U.S. 214, 225, 233 (1944).

solemnly accepted and gave the prestige of its support to dangerous racial myths about a minority group, in arguments which can be applied easily to any other minority in our society.

The cases are worth separate statement, for they are by no means alike. In *Hirabayashi v. United States* the Court considered a conviction based on the Act of March 21, 1942, for violating two orders issued by General DeWitt under authority of the Executive Order of February 19, 1942. Gordon Hirabayashi, a citizen of the United States and a senior in the University of Washington, was sentenced to three months in prison on each of two counts, the sentences running concurrently. The first count was that Hirabayashi failed to report to a control station on May 11 or May 12, 1942, for exclusion from the duly designated military area including Seattle, his home. The first count thus raised the legality of the compulsory transportation of an American citizen from one of the military areas to a WRA camp, and of his indefinite incarceration there. The second count was that on May 9, 1942, he had violated a curfew order, by failing to remain at home after 8 P.M., within a designated military area, in contravention of a regulation promulgated by the military authority. The Court considered the violation of the second count first, upheld the curfew order and the sentence imposed for violating it. Since the two sentences were concurrent, it said, there was no need to consider the conviction on the first count.

In fact, of course, the Court was entirely free to consider the first count if it wanted to. It would have been normal practice to do so. Its refusal to pass on the more serious controversy cannot be put down to wise and forbearing judicial statesmanship. This was not the occasion for prudent withdrawal on the part of the Supreme Court, but for affirmative leadership in causes peculiarly within its sphere of primary responsibility. The social problems created by the exclusion and confinement of the Japanese Americans of the West Coast states increased in seriousness with every day of their continued exclusion. The rabble-rousers of California now were demanding the permanent exclusion of all persons of Japanese ancestry from the West Coast area. They were living at peace, altogether free of the threat of Japanese invasion. Yet they were still successful in their efforts to keep the Japanese out. The business and professional capital of the Japanese was being profitably used by others. Intelligent and resourceful competitors had been removed from many markets. At the expense of the Japanese, vested interests were being created, entrenched, and endowed with political power. All these interests would resist the return of the Japanese by law if possible, if not, by terror. The refusal of the Supreme Court to face the problem was itself a positive decision on the merits. It gave strength to the anti-Oriental forces on the West Coast, and made a difficult social situation more and more tense. A full assertion of the ordinary rights of citizenship would have shamed and weakened the lynch spirit. It would have fortified the party of law and order. Instead, that party was confused and weakened by the vacillation of the Court.[41]

The reasoning of the Court itself contributed to the intensification of social pressure.

In the *Hirabayashi* case the Court held that its problem was the scope of the war power of the national government. The extent of Presidential discretion was not presented as a separate issue, because the statute of March 21, 1942, and appropriation acts under it, were passed with full knowledge of the action taken and proposed by General DeWitt, and thus fully authorized the curfew. Both Congress and the Executive were held to have approved the curfew as a war measure, required in their judgment because espionage and sabotage were especially to be feared from persons of Japanese origin or descent on the West Coast during the spring of 1942.

The premise from which the Court's argument proceeded was the incontestable proposition that the war power is the power to wage war successfully. The State must have every facility and the widest latitude in defending itself against destruction. The issue for the Court, the Chief Justice said, was whether at the time "there was any substantial basis for the conclusion" that the curfew as applied to a citizen of Japanese ancestry was "a protective measure necessary to meet the threat of sabotage and espionage which would substantially affect the war effort and which might reasonably be expected to aid a threatened enemy invasion." [42] The formulation of the test followed the lines of the Court's familiar doctrine in passing on the action of administrative bodies: was there "reasonable ground" for those charged with the responsibility of national defense to believe that the threat was real, and the remedy useful? The orders of the commander, the Court held, were based on findings of fact which supported action within the contemplation of the statute. The findings were based on an informed appraisal of the relevant facts in the light of the statutory standard, and published as proclamations. The circumstances, the Court said, afforded a sufficiently rational basis for the decision made.

The "facts" which were thus held to "afford a rational basis for decision" were that in time of war "residents having ethnic affiliations with an invading enemy may be a greater source of danger than those of different ancestry," and that in time of war such persons could not readily be isolated and dealt with individually.[43] This is the basic factual hypothesis on which all three cases rest.

The first part of this double-headed proposition of fact is contrary

41. See materials cited *supra* note 34.
42. 320 U. S. 81, 95 (1943).
43. *Id.* at 101–2.

to the experience of American society, in war and peace.[44] Imagine applying an ethnic presumption of disloyalty in the circumstances of the Revolution or the Civil War! In the World War and in the present war, soldiers who had ethnic affiliations with the enemy—German, Austrian, Hungarian, Finnish, Roumanian, Bulgarian, Japanese and Italian—fought uniformly as Americans in our armed forces, without any suggestion of group disloyalty. As a generalization about the consequences of inheritance, as compared with experience, in determining political opinions, the Supreme Court's doctrine of ethnic disloyalty belongs with folk proverbs—"blood is thicker than water"—and the pseudo-genetics of the Nazis. It is flatly contradicted by the evidence of the biological sciences, of cultural anthropology, sociology, and every other branch of systematic social study, both in general, and with specific reference to the position of Japanese groups on the West Coast. The most important driving urge of such minority groups is to conform, not to rebel. This is true even for the American minorities which are partially isolated from the rest of society by the bar of color.[45] The desire to conform is stronger than resentments and counter-reactions to prejudice and discrimination. Insecure and conscious of the environment as a threat, such minorities seek to establish their status by proving themselves to be good Americans. The younger generation rejects the language, customs and attitudes of the older. The exemplary combat records of the Japanese American regiments in Italy and in France is a normal symbol of their quest for security within the environment. It is an expected part of the process of social adjustment, repeated again and again in our experience with minorities within American society. By and large, men and women who grow up in the American cultural community are Americans in outlook, values and basic social attitudes. This is the conclusion of the scientific literature on the subject. It has been the first tenet of American law, the ideal if not always the practice of American life.

To support its contrary opinion, the Supreme Court undertook a

44. Compare the opinion of Mr. Justice Black, for a unanimous Court, in *Ex parte* Kumezo Kawato, 317 U. S. 69, 73 (1942).

45. See *infra*, pp. 520–3 and materials cited *supra* notes 1 and 14; Wirth, *The Problem of Minority Groups* in LINTON (ed.), THE SCIENCE OF MAN IN THE WORLD CRISIS (1945) 347; MYRDAL, AN AMERICAN DILEMMA (1944) cc. 3, 33–9, app. 10; SHERMAN, BASIC PROBLEMS OF BEHAVIOR (1941) 289–91; MEAD, AND KEEP YOUR POWDER DRY (1942) cc. 3, 46; WARNER AND SROLE, THE SOCIAL SYSTEMS OF AMERICAN ETHNIC GROUPS (1945) 283–4; BENEDICT, PATTERNS OF CULTURE (1934) especially cc. 1–3, 7, 8; BENEDICT, RACE: SCIENCE AND POLITICS (1940); LOCKE AND STERN (eds.), WHEN PEOPLES MEET (1942) cc. 7–12; MIYAMOTO, SOCIAL SOLIDARITY AMONG THE JAPANESE IN SEATTLE (1939); DOLLARD, CASTE AND CLASS IN A SOUTHERN TOWN (1937), cc. 12–6; THOMPSON (ed.), RACE RELATIONS AND THE RACE PROBLEM (1939); STONEQUIST, THE MARGINAL MAN, A STUDY IN PERSONALITY AND CULTURE CONFLICT (1937), cc. 3–4, particularly pp. 101–6; Cox, *Race and Caste: A Distinction* (1945) 50 AM. J. SOC. 360, 365–6; MACIVER (ed.), GROUP RELATIONS AND GROUP ANTAGONISMS (1944) pt. 1.

review of its own intuitions, without a judicial record before it, and without serious recourse to available scientific studies of the problem. Kiplingesque folklore about East and West is close to the heart of the opinions. The Japanese, the Court said, had been imperfectly assimilated; they constituted an isolated group in the community; their Japanese language schools might be sources of Japanese propaganda. Moreover, the discriminatory way in which the Japanese on the West Coast were treated may have been regarded as contributing to Japanese solidarity, preventing their assimilation, and increasing in many instances their attachments to Japan and its institutions.[46]

There was no testimony or other evidence in the record as to the facts which governed the judgment of the military in entering the orders in question. They were not required to support the action they had taken by producing evidence as to the need for it. Nor were they exposed to cross-examination. By way of judicial research and notice the Court wrote four short paragraphs to explain "some of the many considerations" which in its view might have been considered by the military in making their decision to institute a discriminatory curfew.[47]

The second part of the Court's basic premise of fact was that it was impossible to investigate the question of loyalty individually. As to the validity of this proposition there was neither evidence in the record nor even discussion by the Court to indicate a basis for the conclusion which might appeal to a reasonable man, or even to a choleric and harassed general, faced with the danger of invasion and the specter of his own court martial. The issue was dismissed in a sentence. "We cannot say that the war-making branches of the Government did not have ground for believing that in a critical hour such persons could not readily be isolated and separately dealt with, and constituted a menace to the national defense and safety, which demanded that prompt and adequate measures be taken to guard against it." [48] In view of the history of security measures during the war, it would not have been easy to establish strong grounds for such a belief. There were about 110,000 persons subject to the exclusion orders, 43% of them being over 50 or under 15.[49] At the time of the exclusion orders, they had lived in California without committing sabotage for five months after Pearl Harbor. The number of persons to be examined was not beyond the capacities of individual examination processes, in the light of experience with such security measures, both in the United States and abroad.[50] The fact was that the loyalty examinations finally under-

---

46. 320 U. S. 81, 98 (1943). See *infra*, pp. 520–3. Such fears arising from sentiments of guilt are of special interest to the student of social psychology.

47. *Id.* at 99.

48. *Ibid.*

49. DEWITT FINAL REPORT, at 403–4.

50. See *supra*, pp. 494–5.

taken in the Relocation Authority camps consisted in large part of filling out a questionnaire, and little more, except in cases of serious doubt as to loyalty. Most of those released from the camps were given their freedom on the basis of little information which was not available on the West Coast in 1942.[51]

Actually, the exclusion program was undertaken not because the Japanese were too numerous to be examined individually, but because they were a small enough group to be punished by confinement. It would have been physically impossible to confine the Japanese and Japanese Americans in Hawaii, and it would have been both physically and politically impossible to undertake comparable measures against the 690,000 Italians or the 314,000 Germans living in the United States. The Japanese were being attacked because for some they provided the only possible outlet and expression for sentiments of group hostility. Others were unable or unwilling to accept the burden of urging the repudiation of a general's judgment which he placed on grounds of military need.

The *Hirabayashi* case states a rule which permits some judicial control over action purporting to be taken under military authority. It proposes that such action be treated in the courts like that of administrative agencies generally, and upheld if supported by "facts" which afford "a rational basis" for the decision. For all practical purposes, it is true, the *Hirabayashi* case ignores the rule; but the Court did go to great lengths to assert the principle of protecting society against unwarranted and dictatorial military action. *Korematsu v. United States* seems sharply to relax even the formal requirement of judicial review over military conduct. Korematsu, an American citizen of Japanese descent, was convicted under the Act of March 21, 1942 for violating an order requiring his exclusion from the coastal area. The Court held the problem of exclusion to be identical with the issue of discriminatory curfew presented in the *Hirabayashi* case. There, it said, the Court had decided that it was not unreasonable for the military to impose a curfew in order to guard against the special dangers of sabotage and espionage anticipated from the Japanese group. The military had found, and the Court refused to reject the finding, that it was impossible to bring about an immediate segregation of the disloyal from the loyal. According to Mr. Justice Black, the exclusion orders merely apply these two findings—that the Japanese are a dangerous lot, and that there was no time to screen them individually. Actually, there was a new "finding" of fact in this case, going far beyond the situation considered in the *Hirabayashi* case. The military had "found" that the curfew provided inadequate protection against the danger of sabotage and espionage. Therefore the exclusion of all Japanese, citizens and aliens

51. See note 31 *supra*.

alike, was thought to be a reasonable way to protect the Coast against sabotage and espionage. Mr. Justice Black does not pretend to review even the possible foundations of such a judgment. There is no attempt in the *Korematsu* case to show a reasonable connection between the factual situation and the program adopted to deal with it.

The Court refused to regard the validity of the detention features of the relocation policy as raised by the case. Korematsu had not yet been taken to a camp, and the Court would not pass on the issues presented by such imprisonment. Those issues, the Court said, are "momentous questions not contained within the framework of the pleadings or the evidence in this case. It will be time enough to decide the serious constitutional issues which petitioner seeks to raise when an assembly or relocation order is applied or is certain to be applied to him, and we have its terms before us." [52] This is a good deal like saying in an ordinary criminal case that the appeal raises the validity of the trial and verdict, but not the sentence, since the defendant may be out on probation or bail. It is difficult to understand in any event why this consideration did not apply equally to the evidence before the Court on the issue which the Court conceded was raised by the pleadings, *i.e.*, the decision of the General to exclude all Japanese from the Defense Area. On this problem there was literally no trial record or other form of evidence in the case.

There were four other opinions in *Korematsu v. United States*. Mr. Justice Roberts and Mr. Justice Murphy dissented on the merits, in separate opinions. Mr. Justice Roberts said that while he might agree that a temporary or emergency exclusion of the Japanese was a legitimate exercise of military power, this case presented a plan for imprisoning the Japanese in concentration camps, solely because of their ancestry, and "without evidence or inquiry" as to their "loyalty and good disposition towards the United States."[53] Such action, he said, was clearly unconstitutional.

Mr. Justice Murphy's substantial opinion does not join issue with the opinion of the Court on the central problem of how to review military decisions, but it does contend that the military decisions involved in this case were unjustified in fact. The military power, he agreed, must have wide and appropriate discretion in carrying out military duties. But, "like other claims conflicting with the asserted constitutional rights of the individual, the military claim must subject itself to the judicial process of having its reasonableness determined and its conflicts with other interests reconciled. . . .

> "The judicial test of whether the Government, on a plea of military necessity, can validly deprive an individual of any of his con-

52. 323 U.S. 214, 222 (1944).
53. *Id.* at 226.

> stitutional rights is whether the deprivation is reasonably related to a public danger that is so 'immediate, imminent, and impending' as not to admit of delay and not to permit the intervention of ordinary constitutional processes to alleviate the danger. . . . Civilian Exclusion Order No. 34, banishing from a prescribed area of the Pacific Coast 'all persons of Japanese ancestry, both alien and non-alien,' clearly does not meet that test. Being an obvious racial discrimination, the order deprives all those within its scope of the equal protection of the laws as guaranteed by the Fifth Amendment. It further deprives these individuals of their constitutional rights to live and work where they will, to establish a home where they choose and to move about freely. In excommunicating them without benefit of hearings, this order also deprives them of all their constitutional rights to procedural due process. Yet no reasonable relation to an 'immediate, imminent, and impending' public danger is evident to support this racial restriction which is one of the most sweeping and complete deprivations of constitutional rights in the history of this nation in the absence of martial law." [54]

The action taken does not meet such a test, Justice Murphy argues, because there was no reasonable ground for supposing that all persons of Japanese blood have a tendency to commit sabotage or espionage, nor was there any ground for supposing that their loyalty could not have been tested individually where they lived. A review of statements made by General DeWitt before Congressional committees and in his Final Report to the Secretary of War clearly reveals that the basis of his action was "an accumulation of much of the misinformation, half-truths and insinuations that for years have been directed against Japanese Americans by people with racial and economic prejudices." [55] These are compared with the independent studies of experts, and shown to be nonsensical. The supposed basis for the exercise of military discretion disappears, and the case for the order falls.

Mr. Justice Jackson wrote a fascinating and fantastic essay in nihilism. Nothing in the record of the case, he said very properly, permits the Court to judge the military reasonableness of the order. But even if the orders were permissible and reasonable as military measures, he said, "I deny that it follows that they are constitutional." [56]

> "I should hold that a civil court cannot be made to enforce an order which violates constitutional limitations even if it is a reasonable exercise of military authority. The courts can exercise only the judicial power, can apply only law, and must abide by the Constitution, or they cease to be civil courts and become instruments of military policy.

54. *Id.* at 234–5.
55. *Id.* at 239. See discussion *infra*, pp. 520–3.
56. *Id.* at 245.

> "Of course the existence of a military power resting on force, so vagrant, so centralized, so necessarily heedless of the individual, is an inherent threat to liberty. But I would not lead people to rely on this Court for a review that seems to me wholly delusive. The military reasonableness of these orders can only be determined by military superiors. If the people ever let command of the war power fall into irresponsible and unscrupulous hands, the courts wield no power equal to its restraint. The chief restraint upon those who command the physical forces of the country, in the future as in the past, must be their responsibility to the political judgments of their contemporaries and to the moral judgments of history.
>
> "My duties as a justice as I see them do not require me to make a military judgment as to whether General DeWitt's evacuation and detention program was a reasonable military necessity. I do not suggest that the courts should have attempted to interfere with the Army in carrying out its task. But I do not think that they may be asked to execute a military expedient that has no place in law under the Constitution. I would reverse the judgment and discharge the prisoner." [57]

Thus the Justice proposes to refuse enforcement of the statute of March 21, 1942. Apparently, in this regard at least, the statute would be treated as unconstitutional. The prisoner would then be taken to the camp and kept there by the military, and all judicial relief would be denied him.

It is hard to imagine what courts are for if not to protect people against unconstitutional arrest. If the Supreme Court washed its hands of such problems, for what purposes would it sit? The idea that military officers whose only authority rests on that of the President and the Congress, both creatures of the Constitution, can be considered to be acting "unconstitutionally" when they carry out concededly legitimate military policies is Pickwickian, to say the least. For judges to pass by on the other side, when men are imprisoned without charge or trial, suggests a less appealing analogy. The action of Chief Justice Taney in Ex parte *Merryman* is in a more heroic tradition of the judge's responsibility.[58]

What Justice Jackson is saying seems to be this: Courts should refuse to decide hard cases, for in the hands of foolish judges they make bad law. The ark of the law must be protected against contamination. Therefore law should not be allowed to grow through its application to the serious and intensely difficult problems of modern life, such as the punishment of war criminals or the imprisonment of Japanese

---

57. *Id.* at 247–8.

58. *Ex parte* Merryman, 17 Fed. Cas. 144, No. 9487 (D. Md. 1861). See SWISHER, ROGER B. TANEY (1935) c. 26.

Americans. It should be kept in orderly seclusion, and confined to problems like the logical adumbration of the full faith and credit clause, and other lawyers' issues.[59] The problems which deeply concern us should be decided outside the courts, even when they arise as the principal and inescapable issues of law suits. Judges are thus to be relieved of the political responsibilities of their citizenship and their office. They will be allowed to pretend that the judicial function is to "interpret" the law, and that law itself is a technical and antiquarian hobby, not the central institution of a changing society.

Mr. Justice Frankfurter concurred specially, answering Mr. Justice Jackson's dissent. "To talk about a military order that expresses an allowable judgment of war needs by those entrusted with the duty of conducting war as 'an unconstitutional order' is to suffuse a part of the Constitution with an atmosphere of unconstitutionality," he said.[60] But one of the first issues of the case was whether or not the military order in question did express an "allowable judgment of war needs." That was the question which the Court was compelled to decide, and did decide, without benefit of the testimony of witnesses, or a factual record, and without substantial independent study on its own motion.

Ex parte *Endo* was the next stage in the judicial elucidation of the problem.[61] In Ex parte *Endo*, decided on December 18, 1944, an adjudication was finally obtained on about one half the question of the validity of confining Japanese aliens and citizens in camps. The case was a habeas corpus proceeding in which an American citizen of Japanese ancestry sought freedom from a War Relocation Center where she was detained, after having been found loyal, until the Authority could place her in an area of the country where local disorder would not be anticipated as a result of her arrival. The Court held that the statute, as rather strenuously construed, did not authorize the detention of persons in the petitioner's situation, although temporary detention for the purpose of investigating loyalty was assumed to be valid as an incident to the program of "orderly" evacuation approved in the *Korematsu* case.

The purpose of the statute under which exclusion and detention were accomplished, the Court said, was to help prevent sabotage and espionage. The act talks only of excluding persons from defense areas. It does not mention the possibility of their detention. While the Court assumes that an implied power of temporary detention may be ac-

59. See Jackson, *Full Faith and Credit—The Lawyer's Clause of The Constitution* (1945) 45 Col. L. Rev. 1. See also Northwestern Bands of Shoshone Indians v. United States, 65 Sup. Ct. 690, 700–2 (U. S. 1945); Jackson, *The Rule of Law Among Nations* (1945) 31 A. B. A. J. 290, 292–3. Compare his report to the President on trials for war criminals, N. Y. Times, June 8, 1945, p. 4.

60. 323 U. S. 214, 224–5 (1944).

61. 323 U. S. 283 (1944).

cepted, as an incident in the program of exclusion, for the purpose of facilitating loyalty examinations, such an implied power should be narrowly confined to the precise purpose of the statute, in order to minimize the impact of the statute on the liberties of the individual citizen. The authority to detain a citizen as a measure of protection against sabotage and espionage is exhausted when his loyalty is established. The persistence of community hostility to citizens of Japanese descent is not a ground for holding them in camp under the present statute. The disclosure of the full scope of the detention program to various committees of the Congress, including appropriation committees, was held not to support a ratification by the Congress of what was done. The basis of this conclusion was the extraordinarily technical proposition that the appropriation acts which might have been considered to ratify the entire program were lump-sum appropriations, and were not broken down by items to earmark a specific sum for the specific cost of detaining citizens found to be loyal pending their relocation in friendly communities. In this respect the reasoning of the Court is contrary to that in the *Hirabayashi* case, where Congressional ratification of the plans of the executive branch was established in a broad and common-sense way. Justices Roberts and Murphy concurred specially, urging that the decision be based on the constitutional grounds stated in their opinions in the *Korematsu* case, rather than on the statutory interpretation underlying Justice Douglas' opinion.

IV.

The many opinions of the three Japanese cases do not consider the primary constitutional issues which are raised by the West Coast anti-Japanese program as a whole. This was a program which included (a) a discriminatory curfew against Japanese persons; (b) their exclusion from the West Coast; (c) their confinement pending investigations of loyalty; and (d) the indefinite confinement of those persons found to be disloyal. These measures were proposed and accepted as military necessities. Their validity as military measures was an issue in litigation. By what standards are courts to pass on the justification for such military action? Were those standards satisfied here?

The conception of the war power under the American Constitution rests on the experience of the Revolution and the Civil War. It rests on basic political principles which men who had endured those times of trouble had fully discussed and carefully articulated. The chief architects of the conception were men of affairs who had participated in war, and had definite and sophisticated ideas about the role of the professional military mind in the conduct of war.

The first and dominating proposition about the war power under the Constitution is that the Commander-in-Chief of the armed forces is a civilian and must be a civilian, elected and not promoted to his office. The subordination of the military to the civil power is thus primarily assured. In every democracy the relationship between civil and military power is the crucial social and political issue on which its capacity to survive a crisis ultimately depends. Inadequate analysis of this problem, and inadequate measures to deal with it, led to the downfall of the Spanish Republic, and gravely weakened the Third French Republic. British experience, especially during the first World War, puts the problem in dramatic perspective.[62] In its own proper sphere of tactics, the professional military judgment is decisive. In waging war the larger decisions—the choice of generals, the organization of command, the allocation of forces, the political, economic and often strategic aspects of war—these have to be made by responsible civilian ministers.[63] Clemenceau's famous remark, quoted at the head of this article, is not a witticism, but the first principle of organizing democracy for war. It reflects a balanced view of the proper relation in policy-making between the expert and the practical man. It expresses a keen sense of the supremacy of civil power in a republic. The image of Napoleon is never far from the surface of French political consciousness. France's experience with Pétain has once more underscored the danger. In our own national life recurring waste and incompetence in the handling of war problems—in the Mexican War, the Civil War, and the Spanish-American War—led to important reforms in the organization of the War Department under Elihu Root, and further developments under later Secretaries of War.[64] The process of achieving adequate organization and control is by no means complete.

The second political principle governing the exercise of the war power in a democracy is that of responsibility. Like every other officer of government, soldiers must answer for their decisions to the system of law, and not to the Chief of Staff alone. Where, as in the Japanese exclusion program, military decisions lead to conflicts between individuals and authority, the courts must adjudicate them. Even if Mr. Justice Jackson's doctrine of the judicial function is accepted, the courts will adjudicate nonetheless, by refusing relief, and thus decide

62. See WAR MEMOIRS OF DAVID LLOYD GEORGE (1933–1937), vol. 6, c. 10 (*Some Reflections on the Functions of Governments and Soldiers Respectively in a War*); vol. 1, cc. 5, 6, 9, 10, 14, 15; vol. 2, cc. 8–10, 17–9; vol. 3, cc. 3–6, 9–11; vol. 4, cc. 9–11, 13; vol. 5, cc. 6, 8; CHURCHILL, THE WORLD CRISIS (1931) cc. 4, 19, 38, pp. 733–45; WILKINSON, WAR AND POLICY (1910) 259–300; WRIGHT, AT THE SUPREME WAR COUNCIL (1921); Rogers, *Civilian Control of Military Policy* (1940) 18 FOREIGN AFFAIRS 280.

63. See PALMER, WASHINGTON, LINCOLN, WILSON, THREE WAR STATESMEN (1930) 224–7, 282–3; PALMER, AMERICA IN ARMS (1941) 145–6; De Weerd, *Civilian and Military Elements in Modern War* in CLARKSON AND COCHRAN, WAR AS A SOCIAL INSTITUTION (1941) 95. See also MCKINLEY, DEMOCRACY AND MILITARY POWER (2d ed. 1941); VAGTS, A HISTORY OF MILITARISM (1937).

64. See 1 JESSUP, ELIHU ROOT (1938) 240–64; ROOT, THE MILITARY AND COLONIAL POLICY OF THE UNITED STATES (1916); Rogers, *op. cit. supra*, note 62, at 288–91.

cases in favor of the military power. The problem is the scope of the military power, and means for assuring its responsible exercise. It is not a problem which can be avoided by any verbal formula.

Most occasions for the exercise of authority in the name of military need will not present justiciable controversy. When a general attacks or retreats in the field, sends his troops to the right or to the left, he may have to justify his decision to a court martial, but not often to a court. On the other hand some steps deemed to be required in war do raise the kind of conflict over property or personal rights which can be presented to the courts. A factory or business may be taken into custody, prices and wages may be established, whole classes of activity, like horse-racing, temporarily forbidden. Without stopping for an over-nice definition of the terms, these are justiciable occasions—situations in which courts have customarily decided controversies, and determined the legality of official action when such problems were implicit in the conflicts presented to them.[65] It is essential to every democratic value in society that official action taken in the name of the war power be held to standards of responsibility under such circumstances. The courts have not in the past, and should not now, declare the whole category of problems to be political questions beyond the reach of judicial review. The present Supreme Court is dominated by the conviction that in the past judicial review has unduly limited the freedom of administrative action. But surely the permissible response to bad law is good law, not no law at all. The Court must review the exercise of military power in a way which permits ample freedom to the Executive, yet assures society as a whole that appropriate standards of responsibility have been met.

The issue for judicial decision in these cases is not lessened or changed by saying that the war power includes any steps required to win the war. The problem is still one of judgment as to what helps win a war. Who is to decide whether there was a sensible reason for doing what was done? Is it enough for the General to say that at the time he acted, he honestly thought it was a good idea to do what he did? Is this an example of "expertise," to which the courts must give blind deference?[66] Or must there be "objective" evidence, beyond the General's state of mind, to show "the reasonable ground for belief" which the *Hirabayashi*

---

65. See, *e.g.*, Block v. Hirsh, 256 U. S. 135 (1921); Bowles v. Willingham, 321 U. S. 503 (1944); Home Building & Loan Ass'n v. Blaisdell, 290 U. S. 398 (1934); Yakus v. United States, 321 U. S. 414 (1944); Montgomery Ward & Co. v. United States, C. C. A. 7th, June 8, 1945.

66. Railroad Commission of Texas v. Rowan & Nichols Oil Co., 310 U. S. 573 (1940), *mod.*, 311 U. S. 614 (1941); Railroad Commission v. Rowan & Nichols Oil Co., 311 U. S. 570 (1941). *Cf.* Thompson v. Consolidated Gas Corp., 300 U. S. 55 (1937); Note (1942) 51 YALE L. J. 680.

case says is necessary?[67] Should such evidence be available before the action is taken? Should the rule be a procedural one that the general has to consider evidence, and then come to a decision, or should it be only that at the subsequent trial suitable evidence is available to justify the result? As the Chief Justice remarked, the Constitution "does not demand the impossible or the impractical."[68] The inquiry should be addressed to the rationality of the general's exercise of his judgment as a general, not as a master in chancery. It should give full and sympathetic weight to the confusion and danger which are inevitable elements in any problem presented for military decision.

Unless the courts require a showing, in cases like these, of an intelligible relationship between means and ends, society has lost its basic protection against the abuse of military power. The general's good intentions must be irrelevant. There should be evidence in court that his military judgment had a suitable basis in fact. As Colonel Fairman, a strong proponent of widened military discretion, points out: "When the executive fails or is unable to satisfy the court of the evident necessity for the extraordinary measures it has taken, it can hardly expect the court to assume it on faith."[69]

The *Hirabayashi* case proposes one test for the validity of an exercise of military power. Even though that test is not applied in the *Hirabayashi* case, and is roughly handled in the *Korematsu* case, it is not hopelessly lost. As the Court said in *Sterling v. Constantin*, the necessity under all the circumstances for a use of martial power "is necessarily one for judicial inquiry in an appropriate proceeding directed against the individuals charged with the transgression."[70]

Perhaps the closest judicial precedent and analogy for the Japanese American cases is *Mitchell v. Harmony*, which arose out of the Doniphan raid during the Mexican war. The plaintiff was a trader, whose wagons, mules, and goods were seized by the defendant, a lieutenant colonel of

---

67. See note 17 *supra*. For recent treatments of administrative and executive findings by various justices of the Supreme Court in cognate, if not directly comparable situations, see Schneiderman v. United States, 320 U. S. 118 (1943); ICC v. Inland Waterways, 319 U. S. 671 (1943); FPC v. Hope Natural Gas Co., 320 U. S. 591 (1944); Connecticut Light & Power Co. v. FPC, 65 Sup. Ct. 749 (U. S. 1945); Bridges v. Wixon, 65 Sup. Ct. 1443 (U. S. 1945).

68. Hirabayashi v. United States, 320 U. S. 81, 104 (1943).

69. FAIRMAN, THE LAW OF MARTIAL RULE (2d ed. 1943) 217–8. See also *id.* at 47–9, 103–7; Fairman, *The Law of Martial Rule and the National Emergency* (1942) 55 HARV. L. REV. 1253, 1259–61, 1272. The test is put by WIENER, A PRACTICAL MANUAL OF MARTIAL LAW (1940) 26–7, for "the hapless Guardsman who commands the troops," as "What can you justify afterwards?". See Comment (1936) 45 YALE L. J. 879. The statute of March 21, 1942 should be interpreted to pose the same issue, despite its broad language.

70. 287 U. S. 378, 398 (1932). *Id.* at 401: "What are the allowable limits of military discretion, and whether or not they have been overstepped in a particular case, are judicial questions." Certain cases, though technically distinguishable, seem to proceed from different hypotheses. Martin v. Mott, 12 Wheat. 19 (U. S. 1827); The Prize Cases, 2 Black 635 (U. S. 1862); Moyer v. Peabody, 212 U. S. 78 (1909).

the United States Army, during the course of the expedition. The plaintiff, who wanted to leave the Army column and trade with the Mexicans, was forced to accompany the troops. All his property was lost on the march and in battle. The action was of trespass, for the value of the property taken, and for damages. The defenses were that the control of the trader and the destruction of his property were a military necessity, justified by the circumstances of the situation. After a full trial, featured by depositions of the commanding officers, the jury found for the plaintiff.

> "The defence has been placed . . . on rumors which reached the commanding officer and suspicions which he appears to have entertained of a secret design in the plaintiff to leave the American forces and carry on an illicit trade with the enemy, injurious to the interests of the United States. And if such a design had been shown, and that he was preparing to leave the American troops for that purpose, the seizure and detention of his property, to prevent its execution, would have been fully justified. But there is no evidence in the record tending to show that these rumors and suspicions had any foundation. And certainly mere suspicions of an illegal intention will not authorize a military officer to seize and detain the property of an American citizen. The fact that such an intention existed must be shown; and of that there is no evidence.
>
> "The 2d and 3d objections will be considered together, as they depend on the same principles. Upon these two grounds of defence the Circuit Court instructed the jury, that the defendant might lawfully take possession of the goods of the plaintiff, to prevent them from falling into the hands of the public enemy; but in order to justify the seizure the danger must be immediate and impending, and not remote or contingent. And that he might also take them for public use and impress them into the public service, in case of an immediate and pressing danger or urgent necessity existing at the time, but not otherwise.
>
> "In the argument of these two points, the circumstances under which the goods of the plaintiff were taken have been much discussed, and the evidence examined for the purpose of showing the nature and character of the danger which actually existed at the time or was apprehended by the commander of the American forces. But this question is not before us. It is a question of fact upon which the jury have passed, and their verdict has decided that a danger or necessity, such as the court described, did not exist when the property of the plaintiff was taken by the defendant. And the only subject for inquiry in this court is whether the law was correctly stated in the instruction of the court; and whether any thing short of an immediate and impending danger from the public enemy, or an urgent necessity for the public service, can justify the taking of private property by a military commander to prevent it from falling into the hands of the enemy or for the purpose of converting it to the use of the public.

"The instruction is objected to on the ground, that it restricts the power of the officer within narrower limits than the law will justify. And that when the troops are employed in an expedition into the enemy's country, where the dangers that meet them cannot always be foreseen, and where they are cut off from aid from their own government, the commanding officer must necessarily be intrusted with some discretionary power as to the measures he should adopt; and if he acts honestly, and to the best of his judgment, the law will protect him. But it must be remembered that the question here, is not as to the discretion he may exercise in his military operations or in relation to those who are under his command. His distance from home, and the duties in which he is engaged, cannot enlarge his power over the property of a citizen, nor give to him, in that respect, any authority which he would not, under similar circumstances, possess at home. And where the owner has done nothing to forfeit his rights, every public officer is bound to respect them, whether he finds the property in a foreign or hostile country, or in his own.

"There are, without doubt, occasions in which private property may lawfully be taken possession of or destroyed to prevent it from falling into the hands of the public enemy; and also where a military officer, charged with a particular duty, may impress private property into the public service or take it for public use. Unquestionably, in such cases, the government is bound to make full compensation to the owner; but the officer is not a trespasser.

"But we are clearly of opinion, that in all of these cases the danger must be immediate and impending; or the necessity urgent for the public service, such as will not admit of delay, and where the action of the civil authority would be too late in providing the means which the occasion calls for. It is impossible to define the particular circumstances of danger or necessitv in which this power may be lawfully exercised. Every case must depend on its own circumstances. It is the emergency that gives the right, and the emergency must be shown to exist before the taking can be justified.

"In deciding upon this necessity, however, the state of the facts, as they appeared to the officer at the time he acted, must govern the decision; for he must necessarily act upon the information of others as well as his own observation. And if, with such information as he had a right to rely upon, there is reasonable ground for believing that the peril is immediate and menacing, or the necessity urgent, he is justified in acting upon it; and the discovery afterwards that it was false or erroneous, will not make him a trespasser. But it is not sufficient to show that he exercised an honest judgment, and took the property to promote the public service; he must show by proof the nature and character of the emergency, such as he had reasonable grounds to believe it to be, and it is then for a jury to say, whether it was so pressing as not to admit of delay; and the occasion such, according to the information upon which he acted, that private rights must for the time give way to the common and public good.

"But it is not alleged that Colonel Doniphan was deceived by false intelligence as to the movements or strength of the enemy at the time the property was taken. His camp at San Elisario was not threatened. He was well informed upon the state of affairs in his rear, as well as of the dangers before him. And the property was seized, not to defend his position, nor to place his troops in a safer one, nor to anticipate the attack of an approaching enemy, but to insure the success of a distant and hazardous expedition, upon which he was about to march.

"The movement upon Chihuahua was undoubtedly undertaken from high and patriotic motives. It was boldly planned and gallantly executed, and contributed to the successful issue of the war. But it is not for the court to say what protection or indemnity is due from the public to an officer who, in his zeal for the honor and interest of his country, and in the excitement of military operations, has trespassed on private rights. That question belongs to the political department of the government. Our duty is to determine under what circumstances private property may be taken from the owner by a military officer in a time of war. And the question here is, whether the law permits it to be taken to insure the success of any enterprise against a public enemy which the *commanding* officer may deem it advisable to undertake. And we think it very clear that the law does not permit it." [71]

Applied to the circumstances of the Japanese exclusion cases, these precedents require that there be a showing to the trial court of the evidence upon which General DeWitt acted, or evidence which justifies his action under the statute and the constitution. Nor will it do to say that there need be only enough evidence to prove his good faith, or to provide a possible basis for the decision. This was the contention expressly overruled in *Mitchell v. Harmony*.[72] The varying formulae about presumptions, and the quantum of proof required in different classes of cases, merely conceal the court's problem. There must be evidence enough to satisfy the court as to the need for the grave and disagreeable action taken—arrest on vague suspicion, denial of trial, and permanent incarceration for opinions alone. The standard of reasonableness, here as elsewhere, is one requiring a full evaluation of all circumstances. But the law is not neutral. It has a positive preference for protecting civil rights where possible, and a long-standing suspicion of the military mind when acting outside its own sphere. In protecting important social values against frivolous or unnecessary interference by generals, the courts' obligations cannot be satisfied by a scintilla of evidence, or any other mechanical rule supposed to ex-

71. Mitchell v. Harmony, 13 How. 115, 133–5 (U. S. 1851).
72. *Id.* at 119–20.

plain the process of proof. There must be a convincing and substantial factual case, in Colonel Fairman's phrase, to satisfy the court of "the evident necessity" for the measures taken.

No matter how narrowly the rule of proof is formulated, it could not have been satisfied in either the *Hirabayashi* or the *Korematsu* cases. Not only was there insufficient evidence in those cases to satisfy a reasonably prudent judge or a reasonably prudent general: there was no evidence whatever by which a court might test the responsibility of General DeWitt's action, either under the statute of March 21, 1942, or on more general considerations. True, in the *Hirabayashi* case the Court carefully identified certain of General DeWitt's proclamations as "findings," which established the conformity of his actions to the standard of the statute—the protection of military resources against the risk of sabotage and espionage. But the military proclamations record conclusions, not evidence. And in both cases the record is bare of testimony on either side about the policy of the curfew or the exclusion orders. There was every reason to have regarded this omission as a fatal defect, and to have remanded in each case for a trial on the justification of the discriminatory curfew, and of the exclusion orders.

Such an inquiry would have been illuminating. General DeWitt's Final Report and his testimony before committees of the Congress clearly indicate that his motivation was ignorant race prejudice, not facts to support the hypothesis that there was a greater risk of sabotage among the Japanese than among residents of German, Italian, or any other ethnic affiliation. The most significant comment on the quality of the General's report is contained in the Government's brief in *Korematsu v. United States*. There the Solicitor General said that the report was relied upon "for statistics and other details concerning the actual evacuation and the events that took place subsequent thereto. We have specifically recited in this brief the facts relating to the justification for the evacuation, of which we ask the Court to take judicial notice, and we rely upon the Final Report only to the extent that it relates such facts." [73] Yet the Final Report embodies the basic decision under review, and states the reasons why it was actually undertaken. General DeWitt's Final Recommendation to the Secretary of War, dated February 14, 1942, included in the Final Report, is the closest approximation we have in these cases to an authoritative determination of fact. In that Recommendation, General DeWitt says:

> "In the war in which we are now engaged racial affinities are not severed by migration. The Japanese race is an enemy race and

73. Brief for United States, p. 11, n. 2, Korematsu v. United States, 323 U. S. 214 (1944). See Brief for United States, p. 23, *Ex parte* Mitsuye Endo, 322 U. S. 233 (1944). It was peculiarly inappropriate to decide these cases on the basis of judicial notice alone. Borden's Farm Products Co., Inc. v. Baldwin, 293 U. S. 194 (1934); United States v. Carolene Products Co., 304 U. S. 144 (1938); Polk Co. v. Gloser, 305 U. S. 5 (1938). See Comment (1936) 49 HARV. L. REV. 631.

while many second and third generation Japanese born on United States soil, possessed of United States citizenship, have become 'Americanized,' the racial strains are undiluted. To conclude otherwise is to expect that children born of white parents on Japanese soil sever all racial affinity and become loyal Japanese subjects, ready to fight and, if necessary, to die for Japan in a war against the nation of their parents. That Japan is allied with Germany and Italy in this struggle is no ground for assuming that any Japanese, barred from assimilation by convention as he is, though born and raised in the United States, will not turn against this nation when the final test of loyalty comes. It, therefore, follows that along the vital Pacific Coast over 112,000 potential enemies, of Japanese extraction, are at large today. There are indications that these are organized and ready for concerted action at a favorable opportunity. The very fact that no sabotage has taken place to date is a disturbing and confirming indication that such action will be taken." [74]

In his Final Report to the Secretary of War General DeWitt adduces somewhat more evidence than the absence of sabotage to prove its special danger. His report, and the briefs for the United States in *Hirabayashi v. United States* and *Korematsu v. United States* emphasize these points as well: The Japanese lived together, often concentrated around harbors and other strategic areas. They had been discriminated against, and it was suggested that their resentment at such treatment might give rise to disloyalty. Japanese clubs and religious institutions played an important part in their social life. Japanese language schools were maintained to preserve for the American-born children something of the cultural heritage of Japan. The Japanese Government, like that of Italy, France, and many other countries, asserted a doctrine of nationality which was thought to result in claims of dual citizenship, and thus to cast doubt on the loyalty of American citizens of Japanese descent. There were some 10,000 Kibei among the population of the West Coast, Japanese Americans who had returned to Japan for an important part of their education, and who were thought to be more strongly affiliated with Japan in their political outlook than the others.[75]

---

74. DeWitt Final Report at 34. See also *id.* at vii, 7–24. Some of the reasoning used to justify the discriminatory treatment of the Japanese Americans can only be described as astounding in its terms, and in its refusal to consider or to evaluate available sociological data. See, *e.g.*, Fairman, The Law of Martial Rule (2d ed. 1943) 260 ("Fundamental differences in mores have made them inscrutable to us"); Watson, *The Japanese Evacuation and Litigation Arising Therefrom* (1942) 22 Ore. L. Rev. 46, 47 ("Their mental and emotional responses are understood by but few of our people and in general the Japanese presents an inscrutable personality").

75. See Tolan Committee Reports (Preliminary) 16. Such persons were of course individually known, through travel records and otherwise.

Much of the suspicion inferentially based on these statements disappears when they are more closely examined. In many instances the concentration of Japanese homes around strategic areas had come about years before, and for entirely innocent reasons. Japanese fishing and cannery workers, for example, were compelled by the canneries to live on the waterfront, in order to be near the plants in which they worked. Japanese truck gardeners rented land in the industrial outskirts of large cities in order to be as close as possible to their markets. They rented land for agricultural purposes under high tension lines—regarded as a very suspicious circumstance—because the company could not use the land for other purposes. The initiative in starting the practice came from the utility companies, not from the Japanese.[76] Despite discrimination against the Japanese, many had done well in America. They were substantial property owners. Their children participated normally and actively in the schools and universities of the West Coast. Their unions and social organizations had passed resolutions of loyalty in great number, before and after the Pearl Harbor disaster.[77] It is difficult to find real evidence that either religious or social institutions among the Japanese had successfully fostered Japanese militarism, or other dangerous sentiments, among the Japanese American population. The Japanese language schools, which the Japanese Americans themselves had long sought to put under state control, seem to represent little more than the familiar desire of many immigrant groups to keep alive the language and tradition of the "old country;" in the case of Japanese Americans, knowledge of the Japanese language was of particular economic importance, since so much of their working life was spent with other Japanese on the West Coast.[78]

There were of course suspicious elements among the Japanese. They were known to the authorities, which had for several years been checking the security of the Japanese American population. Many had been individually arrested immediately after Pearl Harbor, and the others were under constant surveillance. We had many intelligence officers who knew both the language and the people well. So far as the police were concerned, there was no substance to the man-in-the-street's belief that all Orientals "look alike." [79] On the contrary, the Japanese

76. See McWilliams, Prejudice (1944) 119–21; 29 Tolan Committee Hearings 11225.

77. See Tolan Committee Reports (Preliminary) 15 ("We cannot doubt, and everyone is agreed, that the majority of Japanese citizens and aliens are loyal to this country"); An Intelligence Officer, *The Japanese in America: The Problem and the Solution* (1942) 185 Harper's 489.

78. See McWilliams, Prejudice (1944) 121–2.

79. See, *e.g.*, 31 Tolan Committee Hearings 11631; Denman, J., dissenting, Korematsu v. United States, 140 F. (2d) 289, 302–3 (C. C. A. 9th, 1943). As for the knowledge of the situation possessed by security officers, see 31 Tolan Committee Hearings 11697–702; An Intelligence Officer, *loc. cit. supra* note 77. A considerable percentage—perhaps

were a small and conspicuous minority on the West Coast, both individually and as a group. They would have been an unlikely source of sabotage agents for an intelligent enemy in any case.

Apart from the members of the group known to be under suspicion, there was no evidence beyond the vaguest fear to connect the Japanese on the West Coast with the unfavorable military events of 1941 and 1942. Both at Pearl Harbor and in sporadic attacks on the West Coast the enemy had shown that he had knowledge of our dispositions. There was some signaling to enemy ships at sea, both by radio and by lights, along the West Coast. It was said to be difficult to trace such signals because of limitations on the power of search without warrant. There had been several episodes of shelling the coast by submarine, although two of the three such episodes mentioned by General DeWitt as tending to create suspicion of the Japanese Americans had taken place after their removal from the Coast. These were the only such items in the Final Report which were not identified by date.[80] And it was positively known that no suspicions attached to the Japanese residents for sabotage at Pearl Harbor before, during or after the raid.[81] Those subsequently arrested as Japanese agents were all white men. "To focus attention on local residents of Japanese descent, actually diverted attention from those who were busily engaged in espionage activity." [82]

It is possible that the absence of a trial on the facts may permit the Court in the future to distinguish or to extinguish the Japanese American cases; for in these cases the defendants did not bring forth evidence, nor require the Government to produce evidence, on the factual justification of the military action. Whoever had the burden of going forward, or of proof, Government or defendant, the burden was not met.[83] Not even the *Korematsu* case would justify the exclusion of such

---

19%—of the evacuees gave negative answers to the loyalty questions in their questionnaires. Many of those answers were expressly referred to the treatment the Japanese had received in being uprooted and imprisoned. It is estimated that many more of the answers were directly or indirectly referable to the shock of evacuation and confinement. See *Hearings before Committee on Immigration and Naturalization on H. R. 2701, 3012, 3489, 3446, and 4103*, 78th Cong., 1st Sess. (1944) 36–43. Basically, of course, the issue is to a considerable extent irrelevant. Disloyalty is not a crime, even in the aggravated form of enthusiastic propaganda for the Axis cause. See note 2 *supra*. At most, it is a possible ground for interning enemy aliens, see N. Y. Times, June 27, 1945, p. 15, col. 7, but hardly a sufficient ground for excluding individuals from strategic areas. See note 13 *supra*.

80. DeWitt Final Report at 18; N. Y. Times, June 23, 1942, p. 1, col. 4; p. 9, col. 4; *id.*, Sept. 15, 1942, p. 1, col. 3; p. 10, col. 5.

81. See McWilliams, Prejudice (1944) 144.

82. *Id.* at 111.

83. In applying the doctrine of *Mitchell v. Harmony*, the burden of proof in fact falls on the Government, claiming the privileges of the emergency. Whatever is said about the presumption of constitutionality of statutes, or the interest of the court in not substituting its judgment on the facts for that of the qualified executive or legislative authority, where the

evidence, nor the denial of a defendant's request to call the General as a witness. A future case may therefore create a better record for establishing appropriate criteria of judicial control over military conduct, and for applying such criteria to better purpose.

A trial on the factual justification of the curfew and exclusion orders would require the Court to confront Ex parte *Milligan*,[84] which it sought to avoid in all three of the Japanese cases. Ex parte *Milligan* represents an application to a large and common class of semi-military situations of what Chief Justice Stone articulated in the *Hirabayashi* case as a "rule of reason" governing the scope of military power. The military power, the Chief Justice said, included any steps needed to wage war successfully. The justices in the majority in Ex parte *Milligan* declared in effect that it would be difficult, if not impossible, to convince them that there was or could be a military necessity for allowing the military to hold, try, or punish civilians while the civil courts were open and functioning. And it held further that it is for the judges, not the generals, to say when it is proper under the Constitution to shut the courts, or to deny access to them.

Ex parte *Milligan* is a monument in the democratic tradition, and should be the animating force of this branch of our law. At a time when national emergency, mobilization and war are more frequent occurrences than at any previous period of our history, it would be difficult to name a single decision of more fundamental importance to society. Yet there is a tendency to treat Ex parte *Milligan* as outmoded, as if new methods of "total" warfare made the case an anachronism.[85] Those who take this view have forgotten the circumstances of the Civil War. Fifth columns, propaganda, sabotage and espionage were more generally used than in any war since the siege of Troy, and certainly more widely used than in the second World War.

Ex parte *Milligan* illustrates the point. Milligan was convincingly charged with active participation in a fifth column plot worthy of Hitler or Alfred Hitchcock. A group of armed and determined men were to seize federal arsenals at Columbus, Indianapolis and at three points in Illinois, and then to release Confederate prisoners of war held in those states. Thus they would create a Confederate army behind the Union lines in Tennessee. Milligan and his alleged co-conspirators

---

justification for extraordinary behavior rests on a showing of extraordinary circumstances, it will finally be the Government's burden to bring in the evidence of emergency, or take the risk of not persuading the court. See, *e.g.*, cases cited *supra* notes 13, 72, and 73.

84. 4 Wall. 2 (U. S. 1867). See Frank, *Ex parte Milligan v. The Five Companies: Martial Law in Hawaii*, 44 COL. L. REV. 639 (1944); KLAUS, THE MILLIGAN CASE (1929); FAIRMAN, MR. JUSTICE MILLER AND THE SUPREME COURT (1939) c. 4.

85. Brief for Respondent, pp. 45–8, *Ex parte* Quirin, 317 U. S. 1 (1942); *Ex parte* Ventura, 44 F. Supp. 520, 522–3 (W. D. Wash. 1942). For a moderate view see Schueller v. Drum, 51 F. Supp. 383, 387 (E. D. Pa. 1943). *Cf.* Frank, *supra* note 84, at 639.

acted in Indiana, Missouri, Illinois, and in other border states. Their strategy had a political arm. The Union was to be split politically, and a Northwest Confederation was to be declared, friendly to the South, and embracing Illinois, Wisconsin, Iowa, Kansas, Indiana and Minnesota. This plan was not an idle dream. It was sponsored by a well-financed society, the Sons of Liberty, thought to have 300,000 members, many of them rich and respectable; the planned uprising would coincide with the Chicago convention of the Democratic Party, which was sympathetic to abandoning the war, and recognizing the Confederacy.[86]

The unanimous Court which freed Milligan for civil trial was a court of fire-eating Unionists. Mr. Justice Davis, who wrote for the majority, was one of President Lincoln's closest friends, supporters and admirers. The Chief Justice, who wrote the opinion for the concurring minority, was a valiant and resolute supporter of the war, whatever his shortcomings in other respects. The Court had no difficulty in freeing Milligan, and facing down the outcry of radical Republicans which was provoked by the decision. The issue dividing the Court in the *Milligan* case was parallel in some ways to the problem presented by the Japanese exclusion program under the statute of March 21, 1942. Congress had passed a statute in 1863 permitting the President to suspend the privilege of habeas corpus in a limited way whenever, in his judgment, the public safety required it, holding prisoners without trial for a short period. If the next sitting of the grand jury did not indict those held in its district, they were entitled to release under the statute.

The statute was in fact a dead letter, although the Court did not consider that aspect of the situation in deciding Milligan's case.[87] Milligan had been arrested by the military. The grand jury had not returned an indictment against him at its next sitting. He had nonetheless been tried by a military commission, and sentenced to death. The minority of the Court urged his release according to the terms of the statute, because no indictment had been presented against him. The Court, however, freed him for normal criminal trial on broader grounds. The controlling question of the case, the Court said, was whether the military commission had jurisdiction to try Milligan. This question was considered without express reference to the statute of 1863, as such, but on the evidence which might justify the exercise of martial law powers either under the statute or otherwise. The only constitutional reason, the Court said, for denying Milligan the trial provided for in the Third Article of the Constitution, and in the Fifth and Sixth Amendments, is that such a trial could not physically be conducted.

86. See KLAUS, THE MILLIGAN CASE (1929) 27–33.
87. See RANDALL, CONSTITUTIONAL PROBLEMS UNDER LINCOLN (1926) 167.

So long as the courts are open, persons accused of crime, and not subject to the laws of war as members of the armed forces or enemy belligerents, must be brought before the courts, or discharged. Ex parte *Milligan* therefore holds Milligan's trial before a military commission to be unconstitutional, despite the President's action under the first section of the Act of 1863. The factual situation was not such as to justify the exercise of martial law powers, even for temporary detention, and certainly not for trial. Ordinary civilians could be held for military trial only when the civil power was incapable of acting—during an invasion, for example, or during a period of severe riot or insurrection.

> "It is difficult to see how the *safety* of the country required martial law in Indiana. If any of her citizens were plotting treason, the power of arrest could secure them, until the government was prepared for their trial, when the courts were open and ready to try them. It was as easy to protect witnesses before a civil as a military tribunal; and as there could be no wish to convict, except on sufficient legal evidence, surely an ordained and established court was better able to judge of this than a military tribunal composed of gentlemen not trained to the profession of the law."
>
> "It is claimed that martial law covers with its broad mantle the proceedings of this military commission. The proposition is this: that in a time of war the commander of an armed force (if in his opinion the exigencies of the country demand it, and of which he is to judge) has the power, within the lines of his military district, to suspend all civil rights and their remedies, and subject citizens as well as soldiers to the rule of *his will;* and in the exercise of his lawful authority cannot be restrained, except by his superior officer or the President of the United States.
>
> "If this position is sound to the extent claimed, then when war exists, foreign or domestic, and the country is subdivided into military departments for mere convenience, the commander of one of them can, if he chooses, within his limits, on the plea of necessity, with the approval of the Executive, substitute military force for and to the exclusion of the laws, and punish all persons, as he thinks right and proper, without fixed or certain rules.
>
> "The statement of this proposition shows its importance; for, if true, republican government is a failure, and there is an end of liberty regulated by law. Martial law, established on such a basis, destroys every guarantee of the Constitution, and effectually renders the 'military independent of and superior to the civil power'—the attempt to do which by the King of Great Britain was deemed by our fathers such an offence, that they assigned it to the world as one of the causes which impelled them to declare their independence. Civil liberty and this kind of martial law cannot endure together; the antagonism is irreconcilable; and, in the conflict, one or the other must perish." [88]

---

88. 4 Wall. 2, 127, 124–5 (U. S. 1867).

The Court's dismissal of Ex parte *Milligan* in Ex parte *Endo* requires some analysis. The Court said, "It should be noted at the outset that we do not have here a question such as was presented in Ex parte *Milligan,* 4 Wall. 2, or in Ex parte *Quirin,* 317 U. S. 1, where the jurisdiction of military tribunals to try persons according to the law of war was challenged in *habeas corpus* proceedings. Mitsuye Endo is detained by a civilian agency, the War Relocation Authority, not by the military. Moreover, the evacuation program was not left exclusively to the military; the Authority was given a large measure of responsibility for its execution and Congress made its enforcement subject to civil penalties by the Act of March 21, 1942. Accordingly, no questions of military law are involved." [89]

The proposition is extraordinary. Under penalty of imprisonment, the orders before the Court in Ex parte *Endo* required that enemy aliens and citizens of Japanese blood be removed from their home and confined in camps. If found to be "disloyal," they were kept in the camps indefinitely. If found to be "loyal," they were kept in the camps as long as was necessary for the Authority to place them in friendly communities.

The problems of Ex parte *Milligan* are avoided by the simplest of expedients. In Ex parte *Milligan* the Court said that the military could not constitutionally arrest, nor could a military tribunal constitutionally try, civilians charged with treason and conspiracy to destroy the state by force, at a time when the civil courts were open and functioning. Under the plan considered in the Japanese American cases, people not charged with crime are imprisoned for several years without even a military trial, on the ground that they have the taint of Japanese blood. Why doesn't the *Milligan* case apply *a fortiori?* If it is illegal to arrest and confine people after an unwarranted military trial, it is surely even more illegal to arrest and confine them without any trial at all. The Supreme Court says that the issues of the *Milligan* case are not involved because the evacuees were committed to camps by military orders, not by military tribunals, and because their jailers did not wear uniforms. It is hard to see any sequence in the sentences. The Japanese Americans were ordered detained by a general, purporting to act on military grounds. The military order was enforceable, on pain of imprisonment. While a United States marshal, rather than a military policeman, assured obedience to the order, the ultimate sanction behind the marshal's writ is the same as that of the military police: the bayonets of United States troops. It is hardly a ground for distinction that the general's command was backed by the penalty of civil imprisonment, or that he obtained civilian aid in running the relocation camps. The starting point for the program was a military

89. 323 U. S. 283, 297–8 (1944).

order, which had to be obeyed. It required enemy aliens and citizens of Japanese blood to be removed from their homes and confined in camps. As events developed, the general's command imposed confinement for three years on most of the people who were evacuated under it.

There are then two basic constitutional problems concealed in the Court's easy dismissal of Ex parte *Milligan*: the arrest, removal and confinement of persons without trial, pending examination of their loyalty; and the indefinite confinement of persons found to be disloyal. On both counts, at least as to citizens, the moral of Ex parte *Milligan* is plain. The *Milligan* case says little about the propriety of a curfew, or perhaps even of the exclusion orders as such. The military necessity of such steps are to be tested independently in the light of all the relevant circumstances. The *Milligan* case does say, however, that arrest and confinement are forms of action which cannot be taken as military necessities while courts are open. For such punitive measures it proposes a clear and forceful rule of thumb: the protection of the individual by normal trial does not under such circumstances interfere with the conduct of war.

Much was made in the Japanese American cases of the analogy of temporary preventive arrest or other restriction, approved for material witnesses, the protection of the public at fires, the detention of typhoid carriers, mentally ill persons, and so on.[90] The analogy has little or no application to the problems presented in these cases, except perhaps for the curfew or conceivably the abstract issue of exclusion, as distinguished from detention. The restrictions involved here were not temporary emergency measures, justified by the breakdown of more orderly facilities for protecting society against espionage and sabotage. As interferences with the liberty of the individual, they go well beyond the minimal forms of precautionary arrest without warrant which were permitted by the statute of 1863, discussed in the *Milligan* case; they are closely comparable to the forms of arbitrary action which were actually presented by the facts of the *Milligan* case, and strongly disapproved by the Court.

As for Japanese aliens, it is orthodox, though not very accurate, to

---

90. For temporary restrictions on access to localities see Warner, *The Model Sabotage Prevention Act* (1941) 54 HARV. L. REV. 602, 611–8; Pressman, Leider and Cammer, *Sabotage and National Defense* (1941) 54 HARV. L. REV. 632, 641. The confinement of alcoholics, psychotic persons, and the like raises different problems. The issue in such cases is not whether persons can be confined in the social interest without trial, but without trial by jury. Ample individual investigation, hearings and other safeguards are required by way of "due" process of law. Minnesota *ex rel.* Pearson v. Probate Court, 309 U. S. 270 (1940); see Hall, *Drunkenness as a Criminal Offense* (1942) 32 J. CRIM. L. & CRIM. 297; Rostow, *The Commitment of Alcoholics to Medical Institutions* (1940) 1 Q. J. OF STUDIES ON ALCOHOL 372. Moreover, the limits to such interferences with individual freedom in the name of protecting society are jealously guarded. Skinner v. Oklahoma, 316 U. S. 535 (1942); see Note (1943) 3 Q. J. OF STUDIES ON ALCOHOL 668.

say that as persons of enemy nationality they are subject only to the Government's will in time of war.[91] But the protection of the Fifth and Sixth Amendments extends generally to aliens.[92] Should arbitrary distinctions be permitted in our policy for enemy aliens, distinctions without reasonable basis? Is it permissible to intern all the Japanese who live on the West Coast, but to allow German and Italian aliens, and Japanese who live elsewhere, general freedom? Lower courts have said they would refuse to review executive action directed at the control of enemy aliens.[93] Such a view is far from necessary. The courts go to great lengths to assure reasonable protection to the property rights of enemy aliens, their privilege of pursuing litigation, and the like. It requires no extension of doctrine to propose that their control and custody in time of war be reasonably equal, and even-handed. So far as accepted notions of international law are concerned, the "single aim" of specialized enemy alien controls is to prevent enemy aliens from aiding the enemy.[94] The present pattern of discriminatory controls bears no relation to the end of safety.

V

These cases represent deep-seated and largely inarticulate responses to the problems they raise. In part they express the justices' reluctance to interfere in any way with the prosecution of the war. In part they stem from widely shared fears and uncertainties about the technical possibilities of new means of warfare. Such fears were strongly felt everywhere on the Allied side after the German victories of 1940 and 1941. It was common then, and still is common, to believe in a vague but positive way that the restoration of mobility in warfare, and the appearance of new weapons, have somehow made all older thought on the subject of war obsolete. We expected fifth-columns and paratroops to drop near San Francisco at any moment. In the panic of the time, it seemed almost rational to lock up Japanese Americans as potential enemy agents.

---

91. See Comment (1942) 51 YALE L. J. 1316, 1317. *Cf.* 3 HYDE, INTERNATIONAL LAW CHIEFLY AS INTERPRETED AND APPLIED IN THE UNITED STATES (2d ed. 1945) §§ 616–617; De Lacey v. United States, 249 Fed. 625 (C. C. A. 9th, 1918).

92. See ALEXANDER, RIGHTS OF ALIENS UNDER THE FEDERAL CONSTITUTION (1931) 127–9; GIBSON, ALIENS AND THE LAW (1940) 151–2, c. 7; Oppenheimer, *The Constitutional Rights of Aliens* (1941) 1 BILL OF RIGHTS REV. 100, 106.

93. *Ex parte* Graber, 247 Fed. 882 (N. D. Ala. 1918); *Ex parte* Gilroy, 257 Fed. 110 (S. D. N. Y. 1919). However, the premise of these cases is hardly compatible with that of *Sterling v. Constantin*, but rather depends on the proposition that the exercise of executive discretion in military and quasi-military matters is not reviewable, except for fraud, mistaken indentity, etc. See also cases cited *supra* note 13. The statute and regulation involved in those cases applies to any persons, not only to citizens or friendly aliens.

94. See HYDE, *loc. cit. supra* note 91. As for the status of enemy aliens in court, see *Ex parte* Kawato, 317 U. S. 69 (1942); as to the property of enemy aliens see Symposium, *Enemy Property* (1945) 11 LAW & CONTEMP. PROB. 1–201.

But the airplane, the tank, and the rocket have not made it necessary to abandon the principles of Ex parte *Milligan.* Whatever the effect of such developments may be on Infantry Field Regulations and the Manual of Arms, they do not compel us to deny suspects the right of trial, to hold people for years in preventive custody, or to substitute military commissions for the civil courts. The need for democratic control of the management of war has not been reduced by advances in the technique of fighting. The accelerated rate of technical advance emphasizes anew the importance of civil control to guard against resistance to novelty, and the other occupational diseases of the higher staffs of all armies. And as warfare becomes more dangerous, and as it embraces more and more of the life of the community, the problem of assuring a sensible choice of war policies, and of preserving democratic social values under conditions of general mobilization, becomes steadily more urgent.

What lies behind Ex parte *Milligan, Mitchell v. Harmony,* and *Sterling v. Constantin* is the principle of responsibility. The war power is the power to wage war successfully, as Chief Justice Hughes once remarked. But it is the power to wage war, not a license to do unnecessary and dictatorial things in the name of the war power. The decision as to where the boundaries of military discretion lie in particular cases has to be made differently in different circumstances. Sometimes the issue will arise in law suits, more often in courts martial, Congressional investigations, reports of the Inspector General, or other law-enforcement procedures. When a court confronts the problem of determining the permissible limit of military discretion, it must test the question by the same methods of judicial inquiry it uses in other cases. There is no special reason why witnesses, depositions, cross-examination and other familiar techniques of investigation are less available in these cases than in others. As *Mitchell v. Harmony* and many other cases indicate, Mr. Justice Jackson is plainly wrong in asserting that judicial control of military discretion is impossible. Mr. Justice Jackson said:

> "The limitation under which courts always will labor in examining the necessity for a military order are illustrated by this case. How does the Court know that these orders have a reasonable basis in necessity? No evidence whatever on that subject has been taken by this or any other court. There is sharp controversy as to the credibility of the DeWitt report. So the Court, having no real evidence before it, has no choice but to accept General DeWitt's own unsworn, self-serving statement, untested by any cross-examination, that what he did was reasonable. And thus it will always be when courts try to look into the reasonableness of a military order." [95]

95. Korematsu v. United States, 323 U. S. 214, 245 (1944). See procedure in *Ex parte*

The Supreme Court had a real alternative in the *Korematsu* case: it could have remanded for trial on the necessity of the orders. The courts have found no special difficulty in investigating such questions, and there is no reason why they should.

The first and greatest anomaly of the *Hirabayashi*, *Korematsu* and *Endo* cases is that they seem to abandon the requirement of a judicial inquiry into the factual justification for General DeWitt's decisions. Despite the careful language of the Chief Justice, these cases treat the decisions of military officials, unlike those of other government officers, as almost immune from ordinary rules of public responsibility. The judges were convinced by the *ipse dixit* of a general, not the factual record of a court proceeding. On this ground alone, the Japanese American cases should be most strenuously reconsidered.

An appropriate procedure for reviewing decisions taken in the name of the war power is an indispensable step towards assuring a sensible result. But the ultimate problem left by these cases is not one of procedure. In these cases the Supreme Court of the United States has upheld a decision to incarcerate 100,000 people for a term of several years. The reason for this action was the extraordinary proposition that all persons of Japanese ancestry are enemies, that the war is not directed at the Japanese state, but at the Japanese "race." General DeWitt's views on this subject are formally presented in his Final Recommendations and his Final Report to the War Department.[96] They are reiterated in his later testimony to a subcommittee of the Naval Affairs Committee. After testifying about soldier delinquency and other problems involving the welfare of his troops, General DeWitt was asked whether he had any suggestions he wanted to leave with the Congressmen. He responded:

> "I haven't any except one—that is the development of a false sentiment on the part of certain individuals and some organizations to get the Japanese back on the west coast. I don't want any of them here. They are a dangerous element. There is no way to determine their loyalty. The west coast contains too many vital installations essential to the defense of the country to allow any Japanese on this coast. There is a feeling developing, I think, in certain sections of the country that the Japanese should be allowed to return. I am opposing it with every proper means at my disposal."
>
> MR. BATES: "I was going to ask—would you base your determined stand on experience as a result of sabotage or racial history or what is it?"

Duncan as described in Frank, *supra* note 84, at 649; General Wilbur has been a witness in the individual exclusion proceedings against one Ochikubo, now pending. See Pacific Citizen, March 17, 1945, p. 2, col. 1.

96. See *supra*, pp. 520–1.

GENERAL DEWITT: "I first of all base it on my responsibility. I have the mission of defending this coast and securing vital installations. The danger of the Japanese was, and is now—if they are permitted to come back—espionage and sabotage. It makes no difference whether he is an American citizen, he is still a Japanese. American citizenship does not necessarily determine loyalty."

MR. BATES: "You draw a distinction then between Japanese and Italians and Germans? We have a great number of Italians and Germans and we think they are fine citizens. There may be exceptions."

GENERAL DEWITT: "You needn't worry about the Italians at all except in certain cases. Also, the same for the Germans except in individual cases. But we must worry about the Japanese all the time until he is wiped off the map. Sabotage and espionage will make problems as long as he is allowed in this area—problems which I don't want to have to worry about."[97]

The Japanese exclusion program thus rests on five propositions of the utmost potential menace: (1) protective custody, extending over three or four years, is a permitted form of imprisonment in the United States; (2) political opinions, not criminal acts, may contain enough clear and present danger to justify such imprisonment; (3) men, women and children of a given ethnic group, both Americans and resident aliens, can be presumed to possess the kind of dangerous ideas which require their imprisonment; (4) in time of war or emergency the military, perhaps without even the concurrence of the legislature, can decide what political opinions require imprisonment, and which ethnic groups are infected with them; and (5) the decision of the military can be carried out without indictment, trial, examination, jury, the confrontation of witnesses, counsel for the defense, the privilege against self-incrimination, or any of the other safeguards of the Bill of Rights.

The idea of punishment only for individual behavior is basic to all systems of civilized law. A great principle was never lost so casually. Mr. Justice Black's comment was weak to the point of impotence: "Hardships are a part of war, and war is an aggregation of hardships." [98] It was an answer in the spirit of cliché: "Don't you know there's a war going on?" It is hard to reconcile with the purposes of his dissent in *Williams v. North Carolina*, where he said that a conviction for bigamy in North Carolina of two people who had been validly divorced and remarried in Nevada "makes of human liberty a very cheap thing—too cheap to be consistent with the principles of free government." [99]

---

97. *Hearings before Subcommittee of House Committee on Naval Affairs on H. R. 30*, 78th Cong., 1st Sess. (1943) 739–40. The text of the testimony is given somewhat differently from current newspaper reports in MCWILLIAMS, PREJUDICE (1944) 116.

98. Korematsu v. United States, 323 U. S. 214, 219 (1944).

99. Williams v. North Carolina, 65 Sup. Ct. 1092, 1116 (U. S. 1945).

That the Supreme Court has upheld imprisonment on such a basis constitutes an expansion of military discretion beyond the limit of tolerance in democratic society. It ignores the rights of citizenship, and the safeguards of trial practice which have been the historical attributes of liberty. Beyond that, it is an injustice, and therefore, like the trials of Sacco, Vanzetti, and Dreyfus, a threat to society, and to all men. We believe that the German people bear a common political responsibility for outrages secretly committed by the Gestapo and the SS. What are we to think of our own part in a program which violates every democratic social value, yet has been approved by the Congress, the President and the Supreme Court?

Three forms of reparation are available, and should be pursued. The first is the inescapable obligation of the Federal Government to protect the civil rights of Japanese Americans against organized and unorganized hooliganism. If local law enforcement fails, prosecutions under the Civil Rights Act should be undertaken.[100] Secondly, generous financial indemnity should be sought, for the Japanese Americans have suffered and will suffer heavy property losses as a consequence of their evacuation. Finally, the basic issues should be presented to the Supreme Court again, in an effort to obtain a reversal of these war-time cases. In the history of the Supreme Court there have been important occasions when the Court itself corrected a decision occasioned by the excitement of a tense and patriotic moment. After the end of the Civil War, Ex parte *Vallandigham*[101] was followed by Ex parte *Milligan.* The *Gobitis* case has recently been overruled by *West Virginia v. Barnette.*[102] Similar public expiation in the case of the internment of Japanese Americans from the West Coast would be good for the Court, and for the country.

---

100. 18 U. S. C. §§ 51, 52 (Criminal Code §§ 19, 20) (1940); Hague v. CIO, 307 U. S. 496 (1939); United States v. Classic, 313 U. S. 299 (1941). *Cf.* Screws v. United States, 65 Sup. Ct. 1031 (U. S. 1945).

101. 1 Wall. 243 (U. S. 1863).

102. Minersville School District v. Gobitis, 310 U. S. 586 (1940); West Virginia State Board of Education v. Barnette, 319 U. S. 624 (1943).

The Case of *Korematsu v. United States*: Could it be Justified Today?

TABLE OF CONTENTS

PREFACE

The bombing of Pearl Harbor on December 7, 1941 abruptly thrust the United States into World War II. The fear of further attacks inflamed the anti-Japanese sentiments already existent among West Coast communities where many Asian immigrants had settled.[1] Looked upon with suspicion, the Japa-

[1] By congressional enactment, the Commission on Wartime Relocation and Internment of Civilians [hereinafter cited as CWRIC] was formed in 1980. CWRIC Act, Pub. L. No. 96-317, 94 Stat. 964 (1980) (codified as amended at 50 U.S.C. app. § 1981 (1976 & Supp. V 1981)). The report issued by the Commission after a two-year investigation presented, among other things, a capsule of the "long and ugly history of West Coast anti-Japanese agitation and legislation." CWRIC, PERSONAL JUSTICE DENIED 4 (1982) [hereinafter cited as CWRIC REPORT].

> Antipathy and hostility toward the ethnic Japanese was a major factor of the public life of the West Coast states for more than forty years before Pearl Harbor. Under pressure from California, immigration from Japan had been severely restricted in 1908 and entirely prohibited in 1924. Japanese immigrants were barred from American citizenship, although their children born here were citizens by birth. California and the other western states prohibited Japanese immigrants from owning land. In part the hostility was economic, emerging in various white American groups who began to feel competition, particularly in agriculture, the principal occupation of the immigrants. The anti-Japanese agitation also fed on racial stereotypes and fears: the "yellow peril" of an unknown Asian culture achieving substantial influence on the Pacific Coast or of a Japanese population alleged to be growing far faster than the white population. This agitation and hostility persisted, even though the ethnic Japanese never exceeded three percent of the population of California, the state of greatest concentration.
>
> The ethnic Japanese, small in number and with no political voice—the citizen generation was just reaching voting age in 1940—had become a convenient target for political demagogues, and over the years all the major parties indulged in anti-Japanese rhetoric and programs. Political bullying was supported by organized interest groups who adopted anti-Japanese agitation as a consistent part of their program: The Native Sons and Daughters of the Golden West, the Joint Immigration Committee, the American Legion, the California State Federation of Labor and the California State Grange.
>
> This agitation attacked a number of ethnic Japanese cultural traits or patterns which were woven into a bogus theory that the ethnic Japanese could not or would not assimilate or become "American." Dual citizenship, Shinto, Japanese language schools, and the education of many ethnic Japanese children in Japan were all used as evidence. But as a matter of fact, Japan's laws on dual citizenship went no further than those of many European countries in claiming the allegiance of the children of its nationals born abroad. Only a small number of ethnic Japanese subscribed to Shinto, which in some forms included veneration of the Emperor. The language schools were not unlike those of other first-generation immigrants, and the return of some children to Japan for education was as

nese, both American citizens and resident aliens alike, were perceived as potential saboteurs and spies who would aid the enemy in another invasion. The government concluded that areas vulnerable to attack by Japanese forces required protection from such subversive threats. President Roosevelt, with the approval of Congress, authorized the military to take measures necessary to defuse this menace.[2] What was deemed "necessary" was the wholesale removal or exclusion of Japanese Americans and aliens from areas considered by the military to be of strategic value.[3] Once removed, the evacuees were shunted to internment camps, where many remained for over two years.[4]

Although most of the internees were residents of the West Coast, about 1,900 of them were also relocated from Hawaii.[5] Among them was George Hoshida. Prior to the attack on Pearl Harbor, Hoshida thought of himself as an ordinary family man. He cherished his wife and three daughters, and was grateful that he could provide them with a home in a pleasant neighborhood in Hilo, Hawaii. He earned a living as a salesman for the Hilo Electric Light Company. As a pastime and as a means of keeping physically fit, he practiced and taught judo, and served as president of an association of judo clubs. He also conducted Sunday school classes at the local Buddhist temple and was an officer of a church-affiliated organization, the United Young Buddhist Association of Hawaii. Although Hoshida continued to observe certain Japanese customs, he nevertheless considered himself a permanent member of the Hawaiian community. He felt fortunate to be among people who were able to retain an identity with their ethnic origins, while living together harmoniously.[6]

> much a reaction to hostile discrimination and an uncertain future as it was a commitment to the mores, much less the political doctrines, of Japan. Nevertheless, in 1942 these popular misconceptions infected the views of a great many West Coast people who viewed the ethnic Japanese as alien and unassimilated.

Id. at 4-5. *See also id.* at 27-46 (presenting in detail the socio-economic atmosphere in California preceding World War II); M. FUKUDA, LEGAL PROBLEMS OF JAPANESE-AMERICANS (1980); F. CHUMAN, THE BAMBOO PEOPLE: THE LAW AND JAPANESE-AMERICANS (1976).

[2] *See infra* text at part I, sec. A.

[3] *Id.*

[4] CWRIC REPORT, *supra* note 1, at 12-15.

[5] *Id.* at 277.

[6] The CWRIC considered cultural harmony to be one of the significant factors in the difference between the fate of the Japanese in Hawaii from those on the West Coast during World War II:

> Hawaii was more ethnically mixed and racially tolerant than the West Coast. Race relations in Hawaii before the war were not infected with the virulent antagonism of 75 years of anti-Asian agitation. While anti-Asian feeling existed in the territory, it did not represent the longtime views of well-organized groups as it did on the West Coast and, without statehood, xenophobia had no effective voice in the Congress.

Id. at 16.

Other factors cited by the Commission were the greater number of ethnic Japanese in Hawaii,

In the aftermath of the Pearl Harbor bombing, Hoshida's self-perception was drastically altered. Those personal traits he once considered unremarkable now set him apart from the rest of the community. His association with the Buddhist church, his judo practice, his respect for Japanese ways, and his alienage suddenly brought loyalty to his adopted country into question. Afraid of courting suspicion, he refrained from conversing in his native language in public; he burned many Japanese books and articles he had collected and disposed of Japanese dolls given to his daughters in observance of Girl's Day, a holiday of the old country.

These attempts to allay mistrust proved to be futile. Hoshida was taken into custody on February 6, 1942 by his own brother-in-law, a detective on the local police force. With other detainees,[7] he was confined under guard at the Kilauea Military Camp, a facility used in peacetime as a vacation resort by armed forces personnel. Uncertain about his future, he worried about his family's ability to cope without him. He learned by letter that his wife was pregnant with their fourth child. In addition, she could no longer rely on his help to care for their eldest daughter Taeko who had been left blind, mute, and partially paralyzed as a result of an automobile accident.

About a month after his seizure, Hoshida faced a panel of military officers and civilians charged with reviewing the loyalty of the detainees. The judgment rendered by the panel would determine Hoshida's fate during the course of the war. He later recounted his thoughts about the confrontation:

> Why didn't I keep my mouth shut? If only I had kept quiet when the examiners told me that judo was pro-Japanese training! Why did I have to argue that judo in itself is not pro-Japanese or pro-any nationality? Why didn't I just nod and say, "Yes, sir, yes, sir?"[8]

Not allowed to notify his family of his departure or to bid them farewell, Hoshida was sent to an internment camp in Lordsburg, New Mexico. In retrospect, "he believed he 'may have gotten out clear' if he had patiently listened to the lecture a panel member gave him instead of disagreeing."[9]

---

the imposition of martial law in Hawaii, which gave the military more effective control over daily affairs, and the difference in the racial attitudes of the military commander of each area. *Id.* at 16-17; *see also id.* at 261-82 (describing in greater detail the Hawaii situation and contrasting it with the West Coast experience).

[7] In an unpublished autogiography, Hoshida noted that those places under initial custody at the military camp were referred to as *detainees*. Those detainees who were subsequently sent to centers on the Mainland after loyalty hearings were classified as *internees*. G. Hoshida, Life of a Japanese Immigrant Boy in Hawaii & America: (The First Half) From Birth Thru WWII 1907-1945 at 256 & 274 (Dec. 14, 1983) (unpublished manuscript).

[8] P. Saiki, Ganbare! An Example of Japanese Spirit 92 (1982).

[9] E. Joesting, Hawaii: An Uncommon History 316 (1972).

Hoshida was virtually a prisoner-of-war in his own country.[10] The internment camp was surrounded by double barbed wire fences and constantly under armed military guard. The ordeal of confinement proved to be intolerable for some. One internee from Hawaii was shot and killed while attempting to escape over the compound's barrier.[11]

In the meantime, the family Hoshida had left behind in Hawaii was also undergoing hardship. His wife Tamae could not work because of her advancing pregnancy. She had to accept welfare benefits to relieve financial burdens despite help from family members. Unable to cope after a nervous collapse, Mrs. Hoshida was forced to institutionalize her severely handicapped daughter.

Eventually, Hoshida's family was also interned. Two months after the birth of their fourth daughter, Mrs. Hoshida left with the children for a relocation camp in Jerome, Arkansas. Before leaving Hawaii, she hastily sold their home. Hoshida wrote in a journal he had kept during his internment, "It seems so unreal that the house we've struggled to acquire should no longer be called our home. It feels as though a bond which has been binding us to Hawaii has snapped. Perhaps we may never return there again."[12]

The anxieties of relocation were lessened for Mrs. Hoshida by the promise of being with her husband once again. But the family would not be reunited for another year as petitions for Hoshida's release made their way through the administrative morass. While still apart, Mrs. Hoshida no longer had the support of the close friends and relatives she had left behind in Hawaii. A few months after their reunion, the family was informed of the death of the eldest girl Taeko. Later, they would learn that she had drowned while left unattended in a bathtub.

The family returned to Hilo in 1945. With no money and no livelihood, Hoshida at the age of thirty-seven, had to rebuild a life for himself and his family:

> He filed claims against the government for the property he had lost, but in the end he was awarded $500, a small part of what he had lost. George Hoshida and the others who had been condemned to live out the war behind barbed wire lost

---

[10] Hoshida's autobiography includes excerpts from a diary of his internment years. His October 28, 1942 entry notes the arrival at the internment camp of about fifty Japanese prisoners-of-war:

> The majority of the POWs were captured adrift in a life boat after the Battle of Midway where the Japanese [navy] suffered a great loss. They were the engine crew of [the] aircraft carrier *Hiryu* sunk at Midway. . . . The group included five officers. . . . Five others were captured in the [Aleutians] and were members of a sub which was sunk. One was taken from a lookout boat near Ogasawara [Island] off Japan.

G. Hoshida, *supra* note 7, at 350.

[11] *Id.* at 320.

[12] *Id.* at 378 (relating journal entry of Jan. 23, 1943).

not only their possessions but also valuable years from their lives.[13]

George Hoshida is my father. I know this gentle man would never have aided the Japanese military—they were his enemy, too. But, because he fit the government's profile of a potential saboteur, he was sent away. He expressed his bewilderment about the experience in an autobiography he is currently writing:

> Although it was war, it seemed so unbelievable that we who had been honest, law-abiding residents of this land, should be seized and thrown into such humiliating situations for the reason that we were the nationals of the enemy country. We had looked upon this country as our home but to be treated as an enemy alien and dangerous to the security of this country and be segregated and confined behind barbed wire under guard, was a rude awakening. It was a great shock to realize that this land which we had considered as home for life was after all not our true home. We would have gladly relinquished our affiliation to the land of our birth and become citizens of this country of adoption but discriminating naturalization laws [had] denied us that privilege.[14]

My own memories of the internment are vague. I was only two years old when my family was ordered to leave Hawaii. Hearing and reading about the experience only increased my own bewilderment: How could such events have taken place at all? This paper was the vehicle I used to understand what happened and to reassure myself that the injustice could not be repeated today. The political, social and legal forces that operated to justify the exclusion and internment of the Japanese Americans are clearer to me now. However, they leave me far less certain that another misunderstood minority might not be similarly mistreated again.

INTRODUCTION

Nearly four decades ago while the country was still embroiled in World War II, the United States Supreme Court decided the case of *Korematsu v. United States.*[15] The highest court of the land upheld the conviction of one Fred Toyosaburo Korematsu, a Japanese American, for failure to vacate a zone de-

---

[13] E. JOESTING, *supra* note 9, at 319. Today, George Hoshida does not seek additional monetary reparations. Despite what happened, he feels a debt to the country that offered him a home and a way of life. Interview with Ray Lovell, newscaster for Channel 2, Honolulu, Hawaii (July 1981). *But cf.* CWRIC, PERSONAL JUSTICE DENIED PART 2: RECOMMENDATIONS 9 (1983) (recommending, among other things, that Congress establish a fund "to provide a one-time per capita compensatory payment of $20,000 to each of the approximately 60,000 surviving persons excluded from their places of residence pursuant to Executive Order No. 9066").

[14] G. Hoshida, *supra* note 7, at 306.

[15] 323 U.S. 214 (1944).

clared off-limits by the military to all those of his ancestry.[16] After the bombing of Pearl Harbor, the ethnic Japanese on the West Coast and in Hawaii were feared to be potential saboteurs and spies who would willingly aid the enemy in another invasion. As a result, their presence was forbidden in those strategic areas believed to be under threat of attack by Japan's armed forces. Because the government believed that the exigencies of the time justified the wholesale restraint of a single ethnic group, the United States Supreme Court upheld the action as a constitutional exercise of the war power.[17]

Today, Fred Korematsu is free of the stigma of the conviction levied against him over forty years ago. In November 1983, United States District Judge Marilyn Patel vacated the 1942 judgment, accepting Korematsu's assertion that the Supreme Court ruling had been attained with evidence falsified by the government.[18] Although the present Justice Department supported the petition to set aside the conviction,[19] the 1944 *Korematsu* decision remains as precedent for justifying restrictive government actions against an ethnic minority on the basis of military necessity.[20]

Does the evolution of caselaw ensure that the Court will not sanction governmental decisions based on "race prejudice, war hysteria and a failure of political leadership?"[21] To attempt an answer, this comment first presents the facts and legal reasoning of the actual decision. The Court's analysis is then examined vis-a-vis the direction certain areas of constitutional law have taken since that landmark case, specifically:

1. Constitutional authorization for exclusion and internment and the permissible interplay between Congress and the Executive in the exercise of the war power;

---

[16] *Id.* at 215-16.

[17] *Id.* at 217-18.

[18] Honolulu Star-Bulletin, Nov. 11, 1983, at A-5, col. 1.

[19] Honolulu Star-Bulletin, Oct. 4, 1983, at A-6, col. 5.

[20] Professor Gerald Gunther of Stanford University is troubled by this: "The unsettling aspects of *Korematsu* are not removed by the fact that 40 years later someone can say that it shouldn't have happened." *Bad Landmark: Righting A Racial Wrong*, TIME, Nov. 21, 1983, at 51 (quoting Professor Gunther). *See also* Justice Jackson's dissent in *Korematsu*:

> But once a judicial opinion rationalizes such [a military] order to show that it conforms to the Constitution, or rather rationalizes the Constitution to show that the Constitution sanctions such an order, the Court for all time has validated the principle of racial discrimination in criminal procedure and of transplanting American citizens. The principle then lies about like a loaded weapon ready for the hand of any authority that can bring forward a plausible claim of an urgent need.

Korematsu, 323 U.S. at 246 (Jackson, J., dissenting).

[21] In its report to the United States Congress, the CWRIC concluded that "[t]he broad historical causes which shaped these decisions [to detain, to end detention, and to end exclusion of the ethnic Japanese] were race prejudice, war hysteria and a failure of political leadership." CWRIC REPORT, *supra* note 1, at 18.

2. Current equal protection law and the deference the Court might accord to military judgment in its decisions today; and
3. Justiciability of the *Korematsu* question.

The paper also identifies some nonlegal factors that may have entered into the judicial decisionmaking of the 1944 Supreme Court bench, and explores the extent to which such variables could be controlled today.

Refinements to the equal protection doctrine may demand that the Court exercise the strictest scrutiny where official actions aimed at a single ethnic group result in serious deprivations of freedom. Certainly, the revelations of history provide guideposts to direct governmental conduct today.[22] Nonetheless, judicial decisions are not calculated from the mechanical insertion of variables into legal formulae. Enough latitude exists in the application of rules and doctrines to permit the influence of nonjudicial factors to enter into the Court's decisions. Its review is also limited by the facts laid before it; it cannot act on evidence withheld from its scrutiny. Thus, historical and legal developments do not serve as guarantees that a case such as *Korematsu* would be decided differently today.

---

[22] *See generally* CWRIC REPORT, *supra* note 1, and P. IRONS, JUSTICE AT WAR (1983) (an exposition of the events leading to the three Japanese American World War II Supreme Court decisions—Hirabayashi v. United States, 320 U.S. 81 (1943), Korematsu v. United States, 323 U.S. 214 (1944), and *Ex parte* Endo, 323 U.S. 283 (1944)).

Specifically, the CWRIC Report presented personal assessments by some of those involved in the events:

Henry L. Stimson, then Secretary of War: "While believing in the context of the time that evacuation was a legitimate exercise of the war powers, [he] recognized that 'to loyal citizens this forced evacuation was a personal injustice.' " CWRIC REPORT, *supra* note 1, at 18.

Francis Biddle, then United States Attorney General: "In his autobiography, [he] reiterated his beliefs at the time: 'the program was ill-advised, unnecessary and unnecessarily cruel.' " *Id.*

Justice William O. Douglas, who had joined the majority opinion in *Korematsu*: "[He] found that the evacuation case 'was ever on my conscience.' " *Id.*

Milton Eisenhower, then director of the War Relocation Authority, the civilian agency created by President Roosevelt to oversee the relocation of civilians: "[He] described the evacuation to the relocation camps as 'an inhuman mistake.' " *Id.*

Chief Justice Earl Warren, who, as Attorney General of California, had promoted the evacuation: " 'I have since deeply regretted the removal order and my own testimony advocating it, because it was not in keeping with our American concept of freedom and the rights of citizens.' " *Id.*

Justice Tom C. Clark, then liaison between the Justice Department and the Western Defense Command: " 'Looking back on it today [the evacuation] was, of course, a mistake.' " *Id.*

But cf. John J. McCloy, then Assistant Secretary of War: "The historic reality is that the wartime Japanese government made the evacuation necessary. . . . The consensus of prudent, responsible officials, without rebuttal from any quarter, was that an attack was possible, accompanied by sabotage by the ethnic Japanese heavily concentrated around vulnerable West Coast defense installations." McCloy, *Payment to Internees Opposed*, Honolulu Star-Bulletin, May 6, 1983, at A-21, col. 1.

I. The *Korematsu* Decision

A. *The Factual Setting*[23]

On the day after the December 7, 1941 bombing of Pearl Harbor, the United States Congress declared war on Japan.[24] Early measures to protect the national security were instituted under the Alien Enemy Act.[25] By authority of presidential proclamations, the United States Attorney General seized and detained mainly those enemy aliens previously identified as dangerous to the national security by civilian and military intelligence services.[26] As a result, ini-

---

[23] Much of the facts and circumstances that were before the Court in *Korematsu* are culled from Hirabayashi v. United States, 320 U.S. 81 (1943). The basis for looking to the earlier decision to reconstruct the record is the reliance by the Court in *Korematsu* on the facts and principles expressed in *Hirabayashi*.

> The Hirabayashi conviction and this one [Korematsu] thus rest on the same 1942 Congressional Act and the same basic executive and military orders, all of which orders were aimed at the twin dangers of espionage and sabotage.
>
> . . . .
>
> In the light of the principles we announced in the *Hirabayashi* case, we are unable to conclude that it was beyond the war powers of Congress and the Executive to exclude those of Japanese ancestry from the West Coast war area at the time they did.

Korematsu, 323 U.S. at 217-18.

In addition, another source is utilized for supplemental background material. Lieutenant General John L. DeWitt, United States Army, who was Military Commander of the Western Defense Command (which included the area that Korematsu had been directed to leave) submitted a report on the evacuation to the United States Army Chief of Staff. Headquarters Western Defense Command and Fourth Army, Final Report [on the] Evacuation of Japanese from the West Coast 1942 (June 5, 1943) [hereinafter cited as Final Report]. The Court in *Korematsu* had the Final Report before it, citing it in a number of instances. *See, e.g.*, *Korematsu*, 323 U.S. at 219 n.2, 236-38 nn.1, 3, & 4-10. The document provides historical information and sociological embellishments presented in this section and in other parts of this paper.

[24] Pub. L. No. 328, 55 Stat. 795 (1941) (codified as 50 U.S.C. app. § 1 note (1976)).

[25] 50 U.S.C. § 21 (1976). Still in force today, the Act provides for the restraint, regulation and removal of alien enemies. Specifically, the statute states, in part:

> Whenever there is a declared war between the United States and any foreign nation or government, or any invasion or predatory incursion is perpetrated, attempted, or threatened against the territory of the United States by any foreign nation or government, and the President makes public proclamation of the event, all natives, citizens, denizens or subjects of the hostile nation or government, being of the age of fourteen years and upward, who shall be within the United States and not actually naturalized, shall be liable to be apprehended, restrained, secured, and removed as alien enemies.

Id.

[26] Final Report, *supra* note 23, at 3. Under the Alien Enemy Act, the President immediately issued proclamations declaring all nationals of the countries with which the United States was at war to be enemy aliens. *Id.* The Attorney General, through the Department of Justice, was directed to implement the provisions of the proclamations. He was given the authority to (1) promulgate administrative procedures, (2) declare zones from which enemy aliens were to be

tially about 2,300 enemy aliens were apprehended: 1,291 Japanese (367 in Hawaii and 924 on the mainland), 857 Germans, and 147 Italians.[27] However, no American citizen of Japanese ancestry was involved in this early stage.[28] At this time, the Attorney General did not exercise any other discretionary options. He did not designate off-limit areas, nor did he require the collection of contraband.[29]

Lieutenant General John L. DeWitt, Commander of the Fourth Army on the West Coast (and later the leader of the Western Defense Command), was "convinced that the military security of the coast required [more assertive] measures."[30] At his urging, representatives of the Justice and War Departments met and eventually agreed to coordinate efforts to exclude enemy aliens from zones surrounding vital installations and to enforce the prohibition against contraband.[31] The joint undertaking, however, led to disagreements between the parties. The Attorney General refused to adopt General DeWitt's recommendation to expand the off-limit areas by including regions not contiguous to important defense installations.[32] The Justice Department official did not wish to proceed without "convincing evidence of the military necessity."[33]

In a February 14, 1942 memorandum to the Secretary of War, General DeWitt presented his own assessment of the military necessity for expanding the exclusion areas and for including Japanese American citizens in the program.[34]

---

excluded, (3) collect from such persons articles described as contraband, and (4) intern such enemy aliens believed to be dangerous to the national security. *Id.*

[27] CWRIC REPORT, *supra* note 1, at 55.

[28] Final Report, *supra* note 23, at 3, 6.

[29] *Id.* at 3.

[30] *Id.* In a memorandum to Assistant Attorney General James Rowe, Jr., General DeWitt expressed his misgivings:

> The developments which have resulted in the current conferences between the Attorney General's representative, and General DeWitt and his staff, have been occasioned by the almost complete absence of action on the part of the Department of Justice over a period of nearly four weeks since promulgation of the [executive] proclamations, toward implementing [necessary measures].
>
> . . . .
>
> . . . [I]t is considered desirable to request advice as to the extent to which the Department of Justice is prepared to assume and to discharge the responsibility of taking whatever steps are necessary for the prevention of sabotage, espionage, and other fifth column activities from enemy alien [quarters].

Memorandum from General DeWitt to Assistant Attorney General James Rowe, Jr., (Jan. 5, 1942), *reprinted in* Final Report, *supra* note 23, at 19, 22 (app. to ch. II).

[31] Final Report, *supra* note 23, at 4.

[32] *Id.* at 7.

[33] *Id.*

[34] Memorandum from General DeWitt to Secretary of War Henry L. Stimson (Feb. 14, 1942), *reprinted in* Final Report, *supra* note 23, at 25, 33-38.

His proposal was the impetus for Executive Order No. 9066, issued on February 19, 1942.[35] The presidential directive declared that "[t]he successful prosecution of the war requires every possible protection against espionage and against sabotage to national-defense material, national-defense premises and national-defense utilities."[36] Accordingly, the order authorized the Secretary of War and his designated military commanders, when they deem it necessary or desirable, "to prescribe military areas . . . from which any or all persons may be excluded. . . . The designation of military areas . . . shall supersede designations of prohibited and restricted areas by the Attorney General . . . and shall supersede the responsibility and authority of the Attorney General . . . in respect of such . . . areas."[37] The day after the order was issued, the Secretary of War named General DeWitt as Military Commander of the Western Defense Command.[38] This appointment was pivotal in expanding the scope of the West Coast defense program beyond that envisioned by the Attorney General.[39]

To implement his earlier proposal, General DeWitt issued a series of proclamations and orders. The first directives, released in early March 1942, established certain military areas within the Western Defense Command.[40] These designations were necessary, according to the official statements, because the Pacific Coast was "particularly subject to attack, to attempted invasion by the armed forces of nations with which the United States [was] at war, and . . . [was] subject to espionage and acts of sabotage."[41] The proclamations also warned that exclusion measures might be taken against "such persons or classes of persons as the situation may require" by subsequent announcements.[42]

Additionally, on March 18, 1942, the President issued the executive order which established the War Relocation Authority, the civilian agency charged with supervision of the evacuees after departure from assembly centers.[43] Although the order did not expressly call for relocation camps, the newly created unit was given wide discretion in deciding the fate of the Japanese who were forced to leave their homes.[44]

---

[35] 7 Fed. Reg. 1407 (1942); Final Report, *supra* note 23, at 25.

[36] Exec. Order No. 9066, 7 Fed. Reg. 1407 (1942).

[37] *Id.*

[38] The Western Defense Command Area comprised the West Coast states and Idaho, Montana, Nevada, Utah, and Arizona as well. Final Report, *supra* note 23, at 16 (figure 1).

[39] P. IRONS, *supra* note 22, at 24. For a detailed account of the clash between the Justice Department and the War Department before the issuance of Executive Order No. 9066, see *id.* at 25-47 (chapter 2: "An American Citizen Is an American Citizen").

[40] Pub. Proclamation No. 1, 7 Fed. Reg. 2320 (1942) (issued on Mar. 2, 1942); Pub. Proclamation No. 2, 7 Fed. Reg. 2405 (1942) (issued on Mar. 16, 1942).

[41] Pub. Proclamation No. 1, 7 Fed. Reg. 2320 (1942).

[42] *Id.*

[43] Exec. Order No. 9102, 7 Fed. Reg. 2165 (1942).

[44] CWRIC REPORT, *supra* note 1, at 107.

Within a month thereafter, General DeWitt imposed a curfew and began to issue exclusion orders for certain areas within his command.[45] These applied to all persons of Japanese ancestry. Another order from the commander prohibited departure from military areas until further instructions were issued.[46] The purpose of this restriction was to "insure the orderly evacuation and resettlement of Japanese voluntarily migrating from Military Area No. 1."[47] Five months after the bombing of Pearl Harbor, in May 1942, General DeWitt issued the civilian exclusion order which Korematsu eventually defied.[48] It required all persons of Japanese ancestry, both aliens and citizens alike, to evacuate a prescribed portion of Military Area No. 1 and to report within five days to assembly centers.[49] Indefinite detention of persons reporting to these centers was decreed in another order by the General, promulgated eleven days before Korematsu was charged.[50]

Those who failed to leave as directed were subject to criminal prosecution.[51] A conviction was punishable by a fine of $5,000, or up to one year of imprisonment, or both.[52] These sanctions were provided by an act of Congress passed before General DeWitt began issuing the curfew and exclusion orders.[53] However, from the legislative history, it is evident that Congress was aware of the

---

[45] *Korematsu*, 323 U.S. at 228.

[46] *Id.*

[47] Pub. Proclamation No. 4, 7 Fed. Reg. 2601 (1942). Korematsu's home was located in Military Area No. 1, which comprised the coastal region of California, Oregon and Washington, as well as southern Arizona. *Hirabayashi*, 320 U.S. at 87.

[48] Civilian Exclusion Order No. 34, 7 Fed. Reg. 3967 (1942) (issued on May 3, 1942; reporting to center required by May 8, 1942); *Korematsu*, 323 U.S. at 216, 229.

[49] *Korematsu*, 323 U.S. at 229; *see supra* note 48.

[50] Civilian Restrictive Order No. 1, 8 Fed. Reg. 982 (1943); *Korematsu*, 323 U.S. at 221.

[51] *Korematsu*, 323 U.S. at 229.

[52] Restrictions in Military Areas and Zones Act, Pub. L. No. 503, ch. 191, 56 Stat. 173 (1942) (codified as amended at 18 U.S.C. § 1383 (1976) (repealed 1976)). The statute, at enactment, read:

> To provide a penalty for violation of restrictions or orders with respect to persons entering, remaining in, leaving, or committing any act in military areas or zones.
>
> *Be it enacted by the Senate and House of Representatives of the United States of America in Congress assembled*, That whoever shall enter, remain in, leave, or commit any act in any military area or military zone prescribed, under the authority of an Executive order of the President, by the Secretary of War, or by any military commander designated by the Secretary of War, contrary to the restrictions applicable to any such area or zone or contrary to the order of the Secretary of War or any such military commander, shall, if it appears that he knew or should have known of the existence and extent of the restrictions or order and that his act was in violation thereof, be guilty of misdemeanor and upon conviction shall be liable to a fine of [sic] not to exceed $5,000 or to imprisonment for not more than one year, or both, for each offense.

[53] *Id.*

Military Commander's plans.[54] It enacted the statute "to provide the means for the enforcement of orders issued under Executive Order No. 9066."[55]

Thus, at the time of his arrest, Korematsu was under compulsion of a military order forbidding him from leaving the area except as authorized. Other orders required that he evacuate the military area in which he resided and report to an assembly center where he could be indefinitely detained. Failure to comply subjected him to criminal penalties under an act of Congress. Yet, Korematsu's personal make-up did not distinguish him from any other Ameri can citizen:

> The record in Korematsu's case showed by stipulation and uncontroverted evidence that he had violated the Exclusion Order applicable to San Leandro, which was the place of his residence. It further showed without contradiction that Korematsu had been born in California; that he had never departed from the continental limits of the United States nor renounced his American citizenship nor did he have any form of allegiance to any country other than the United States; that he was willing to render any service requested of him in the war against Japan and had no sympathy for Japan in the war; that he was a registered voter and was in various respects assimilated into the American community; and that he had remained in his place of residence in violation of the Exclusion Order because of friendly relations with the residents of the locality, particularly with a girl not of Japanese ancestry, and because he considered himself an American.[56]

Korematsu's conviction was thereafter upheld by the United States Supreme Court.

---

[54] *Hirabayashi*, 320 U.S. at 89-91. The decision related some legislative historical background:

> The Chairman of the Senate Military Affairs Committee explained on the floor of the Senate that the purpose of the proposed legislation was to provide means of enforcement of curfew orders and other military orders made pursuant to Executive Order No. 9066. He read General DeWitt's Public Proclamation No. 1, and statements from newspaper reports that "evacuation of the first Japanese aliens and American-born Japanese" was about to begin.

Id. at 90.

[55] *Id.* at 89.

[56] Dembitz, *Racial Discrimination and the Military Judgment: The Supreme Court's Korematsu and Endo Decisions*, 45 COLUM. L. REV. 175, 182 (1945) (extracting information from Brief for the United States at 4-5, Korematsu v. United States, 323 U.S. 214 (1944)). The author, Nanette Dembitz, was a member of the legal staff of the Justice Department during preparations for presentation of the *Korematsu* and *Endo* cases to the Supreme Court. P. IRONS, *supra* note 22, at 119.

B. *The Decision*

Justice Hugo Black delivered the opinion of the Court. He was joined in the majority opinion by Chief Justice Stone and by Justices Reed, Douglas, and Rutledge. Justice Frankfurter concurred in a separate opinion.[57] Justices Roberts, Murphy and Jackson each voiced their dissent in individual opinions.[58]

1. *The Majority Opinion*

Justice Black recognized that "all legal restrictions which curtail the civil rights of a single racial group are immediately suspect."[59] He articulated the oft-cited standard of review that applies even today in equal protection cases addressing racial issues—"the most rigid scrutiny" and the requirement of "[p]ressing public necessity."[60] Reaffirming the principles proclaimed in *Hirabayashi v. United States*,[61] the Court held the exclusion of "a single racial

---

[57] *Korematsu*, 323 U.S. at 224 (Frankfurter, J., concurring).

[58] *Id.* at 225 (Roberts, J., dissenting); *id.* at 233 (Murphy, J., dissenting); *id.* at 242 (Jackson, J., dissenting).

[59] *Korematsu*, 323 U.S. at 216.

[60] *Id. See, e.g.*, Fullilove v. Klutzwick, 448 U.S. 448, 507 (1980) (Powell, J., concurring); Harper v. Virginia Bd. of Elections, 383 U.S. 663, 668 (1966); McLaughlin v. Florida, 379 U.S. 184, 192 (1964); Narenji v. Civiletti, 617 F.2d 745, 754 (1979).

[61] 320 U.S. at 81 (1943). The *Korematsu* decision did not reiterate those principles. From a reading of the *Hirabayashi* decision, they appear to be:

1. *Congress may delegate legislative powers by ratification of measures already promulgated by the executive branch.* Legislative history suggested that Congress, aware of the military plans for the West Coast pursuant to Executive Order No. 9066, had effectively ratified orders such as the curfew restriction by enacting the statute under which Korematsu and Hirabayashi were charged. *Id.* at 90-91. In addition, the Court did not hesitate to look to the military orders themselves to find the standards and policies that must be articulated by Congress when delegating its legislative powers. *Id.* at 103-04.

2. *Congress and the President through joint action possess the constitutional power to impose the curfew restriction as an emergency war measure.* "The war power of the national government is 'the power to wage war successfully.' " *Id.* at 93 (quoting Hughes, *War Powers under the Constitution*, 42 A.B.A. Rep. 232, 238 (1917)). Since the constitutionally granted power must be wielded by Congress and the President within the context of warfare, the Court reasoned that the grant necessarily included "wide scope for the exercise of judgment and discretion in determining the nature and extent of the threatened injury or danger and in the selection of the means for resisting it." *Id.* Accordingly, the judicial branch should not review the advisability of wartime measures employed by its coequal branches nor substitute its judgment for theirs. *Id.* The Court in *Hirabayashi* then held that the curfew was a justifiable governmental restriction in light of (1) the military situation in the Pacific following the Pearl Harbor attack; (2) the military judgment that the West Coast, vulnerable to attack by the Japanese, must be safeguarded against espionage and sabotage; and (3) the reasonableness of imposing the curfew on a few citizens, the

group" to be within the war power of Congress and of the President.[62] The "pressing public necessity" rested on the military judgment that immediate segregation of the disloyal—potential spies and saboteurs who threatened West Coast security—from the loyal was not possible.[63] And, in turn, reliance on the military to decide the best course of action was held to be proper, in light of congressional authorization.[64]

The Court limited its review to the constitutionality of the exclusion order. Refusing to regard the exclusion and subsequent detention as "one and inseparable,"[65] Justice Black found "no reason why violations of these orders, insofar as they were promulgated pursuant to congressional enactment, should not be treated as separate offenses."[66] He surmised that Congress would have *provided* distinct sanctions for violation of each order, if it had chosen to legislate the separate directives into one act.[67] Since Korematsu had been convicted of violating only the exclusion order, the Associate Justice saw no need to address any other issues.[68] He presumed there would be "time enough" later, when the assembly or detention orders were enforced against Korematsu, for the Japanese American to raise constitutional issues relevant to these military restrictions.[69]

2. *The Concurrence*

Justice Frankfurter agreed with the majority that the Court's decision in

---

ethnic Japanese, rather than on the entire West Coast populace.

3. ***Facts and circumstances indicating that an ethnic group poses a danger to others provide a rational basis for the implementation of public safety measures by the government in the face of threatened invasion and warfare.*** The Court held that there was no denial of equal protection in imposing the curfew only upon those of Japanese ancestry. *Id.* at 100-02. The "facts and circumstances" relied on by the Court include: (1) the effectiveness of espionage in the surprise attack on Pearl Harbor; (2) the concentration of persons of Japanese ancestry along the West Coast; (3) the failure of the ethnic Japanese to assimilate into the American culture; (4) the attachment to Japan as evidenced by the continued practice of Japanese customs; (5) the military judgment that segregation of the disloyal from the loyal could not be done with the swiftness demanded by the crisis of threatened attack and by the possibility of espionage and sabotage; and (6) the greater source of danger posed by those residents having ethnic affiliations with the enemy attackers.

Id. at 96-102.

[62] *Korematsu*, 323 U.S. at 217-18.

[63] *Id.* at 218-19.

[64] *Id.* at 223.

[65] *Id.* at 221.

[66] *Id.* at 222.

[67] *Id.*

[68] *Id.*

[69] *Id.*

Hirabayashi sustained the exclusion order at issue in *Korematsu*.[70] Additionally, the Justice appeared to hold a more expansive view of the war power and of the scope of its legitimate exercise. As his premise, he recited former Chief Justice Hughes' charge that "the war power of the Government is 'the power to wage war successfully,' "[71] Frankfurter firmly fixed this power within the bounds of the Constitution. He concluded that military actions, though possibly illegal measures in time of peace, "must be judged wholly in the context of war."[72] The constitutional test he would apply was that the "military order . . . not transcend the *means appropriate for conducting war*."[73]

His standard reflected a degree of deference to the military that the majority opinion apparently was not willing to entertain. Justice Black's constitutional gauge required "apprehension by the proper military authorities of the gravest imminent danger to the public safety" and "a definite and close relationship to the prevention of espionage and sabotage."[74]

The latitude that Justice Frankfurter would accord the legislative and executive branches in the waging of war necessarily limited the judicial role. "To find that the Constitution does not forbid the military measures now complained of does not carry with it approval of that which Congress and the Executive did. *That is their business, not ours*."[75]

3. *Justice Roberts' Dissent*

Justice Roberts' dissent from the majority opinion was based on an elemental difference, the issue before the Court. The governmental intrusion he considered under judicial review was not "keeping people off the streets at night, . . . nor . . . temporary exclusion of a citizen from an area for his own safety or that of the community, nor . . . offering him an opportunity to go temporarily out of an area where his presence might cause danger to himself or to his fellows."[76] Rather, he framed the issue as a "case of convicting a citizen as a punishment for not submitting to imprisonment in a concentration camp, based on his ancestry, and solely because of his ancestry, without evidence or inquiry concerning his loyalty and good disposition towards the United States."[77] So stated, the issue invites the Justice's conclusion that "[c]onstitutional rights have been

---

[70] *Id.* at 224. (Frankfurter, J., dissenting).

[71] *Id.* (citing *Hirabayashi*, 320 U.S. at 93, quoting Hughes, *War Powers Under the Constitution*, 42 A.B.A. Rep. 232, 238 (1917)).

[72] *Id.* at 224.

[73] *Id.* at 225 (emphasis added).

[74] *Id.* at 218.

[75] *Id.* at 225 (emphasis added).

[76] *Id.* at 225-26 (Roberts, J., dissenting).

[77] *Id.* at 226.

violated."[78]

Justice Roberts' position was that "exclusion was but a part of an overall plan for forceable detention."[79] As support, he recited a chronology of the executive and military orders to illustrate progressive governmental restraint that operated to narrow the options available to Korematsu and other ethnic Japanese evacuees.[80] He described Korematsu's situation, prior to his arrest, as a dilemma: the Californian could not remain in his home or leave the area voluntarily without inviting criminal arrest; yet, to avoid punishment he had to submit to military custody at an assembly center.[81]

Justice Roberts could not accept the majority's holding that the principles announced by the Court in *Hirabayashi* applied here—"that exclusion from a given area of danger, while somewhat more sweeping than a curfew regulation, is of the same nature—a temporary expedient made necessary by a sudden emergency."[82] He characterized this as "a substitution of an hypothetical case for the case actually before the court,"[83] and questioned the use of an "artificial situation" instead of "the actualities of the case."[84] He found it untenable that Korematsu would have to submit to detention to escape violation of, in his mind, an unconstitutional statute and to delay until then the benefit of legal remedy by way of a writ of habeas corpus.[85]

4. *Justice Murphy's Dissent*

Justice Murphy vehemently dissented, refusing to uphold "this legalization of racism."[86] His opinion is the only one of the five delivered that delved extensively into General DeWitt's *Final Report*.[87] The dissenter parsed the document to support his contention that the military necessity upon which evacuation rested "resolve[d] itself into a few intimations that certain individuals actively aided the enemy, from which it is inferred that the entire group of Japanese Americans could not be trusted to be or remain loyal to the United States."[88]

Justice Murphy would accord the military a wide scope of discretion in pro-

---

[78] *Id.*
[79] *Id.* at 232.
[80] *Id.* at 226-30.
[81] *Id.* at 230 & 232.
[82] *Id.* at 231; *see also supra* note 61.
[83] *Korematsu*, 323 U.S. at 231 (Roberts, J., dissenting).
[84] *Id.* at 232.
[85] *Id.* at 233.
[86] *Id.* at 242 (Murphy, J., dissenting).
[87] Final Report, *supra* note 23.
[88] *Korematsu*, 323 U.S. at 240 (Murphy, J., dissenting).

tecting the national security.[89] However, in the absence of martial law, he could not find the necessary imminent and immediate public danger to support the evacuation.[90] The assumption that all persons of Japanese ancestry are potential saboteurs or spies he found unsupported by the *Final Report*.[91] He characterized its findings as based on "questionable racial and sociological grounds not ordinarily within the realm of expert military judgment, supplemented by certain semi-military conclusions drawn from an unwarranted use of circumstantial evidence."[92]

Moreover, the Justice could not agree that a sense of urgency existed to justify dispensing with individual loyalty hearings. He noted, instead, a sense of "[l]eisure and deliberation" in the lapsing of eleven months between the Pearl Harbor attack and removal of the last of the evacuees.[93] He found no adequate proof that government intelligence services alone could not control the espionage and sabotage threat, a fear that failed to materialize after Pearl Harbor.[94] The situation, he surmised, would have permitted the holding of loyalty hearings for the "mere 112,000 persons involved, "as in Britain, for approximately 74,000 German and Austrian aliens."[95]

5. *Justice Jackson's Dissent*

Justice Jackson appeared to have doubted the justiciability of military judgments because of the Court's inability to assess adequately the necessity and reasonable basis for wartime decisions.[96] Once confronted with the issue, however, the Justice saw a grave danger in according constitutionality to government action directed against individuals by virtue of their membership in "a race from which there is no way to resign."[97] He warned of the durability of the Court's imprimatur upon racially discriminatory measures; its precedential value could be utilized to justify racial discrimination in future crises.[98] Further, he cautioned against reliance upon a judiciary, inexpert in military matters, to restrain military power.[99] In essence, he advised a nation to choose its generals carefully.

---

[89] *Id*. at 233-34.
[90] *Id*. at 235.
[91] *Id*. at 235-41.
[92] *Id*. at 236-37.
[93] *Id*. at 241.
[94] *Id*.
[95] *Id*. at 241-42 & n.16.
[96] *Id*. at 244-45 (Jackson, J., dissenting).
[97] *Id*. at 243.
[98] *Id*. at 246.
[99] *Id*. at 248.

C. *Summary of the Issues*

As outlined above, the five opinions in *Korematsu* confronted several issues in varying degrees. In addition, the majority opinion incorporated the reasoning in the *Hirabayashi* decision by its reliance on the principles announced in the earlier case.[100] From both decisions, the following issues have been identified for discussion in this comment:

1. Whether the military authority exercised over civilians during wartime was within the constitutional boundaries of the war power of Congress, the President, and the military.
   a. What is the nature of the war power?
   b. What military measures should be subjected to judicial review?
   c. Were the military actions taken under proper authority?
2. Whether the restriction of only those of Japanese ancestry was racially discriminatory in violation of the fifth amendment.[101]
   a. Did the ethnic classification have "a definite and close relationship" to the prevention of espionage and sabotage?
   b. Was there a "pressing public necessity" for the classification?
3. Whether the orders merited judicial deference or, more fundamentally, were not justiciable.

II. The War Power: Constitutional Limits of Exercise

The first issue is whether the military control of civilians of Japanese ancestry during World War II was exercised within the constitutional boundaries of the war power of Congress, the President, and the military. The discussion here focuses on the nature and scope of the military actions taken and the source of authority for the measures instituted by General DeWitt. The equal protection issue raised by the operation of the military orders upon one ethnic group is addressed in Part III.[102]

Writing for the majority in *Korematsu*, Justice Black concluded that it was not "beyond the war power of Congress and the Executive to exclude those of Japanese ancestry from the West Coast war area at the time they did."[103] The task presented is to determine whether that same conclusion can be supported

---

[100] *See supra* note 61.

[101] U.S. Const. amend. V.

[102] The discussion here in part II is limited to whether the Constitution provides for the exercise of the war power through such measures as the exclusion orders. Whether such actions were in fact necessary for the successful conduct of the war is examined in a later section. *See infra* text at part III. The equal protection issue focuses on the asserted military necessity as a compelling governmental interest and on the least restrictive alternative required to protect that interest.

[103] *Korematsu*, 323 U.S. at 217-18.

today. The inquiry proceeds in the following manner: First, as an introduction, the source of the war power and its recent definitions are reviewed. Next, the question of which military measures should be the object of judicial scrutiny is addressed; specifically, is the scope of the *Korematsu* case properly limited to the exclusion measure, or should it include review of the detention and relocation orders? The inquiry then explores the constitutionally rooted mechanisms for apportioning the war power among Congress, the President, and the military.

A. *Nature of the War Power*

The legitimate exercise of the war power does not lend itself to mechanical verification. The difficulty rests, partly, with the lack of specificity in the definition of its boundaries and its proper locus.

1. *Constitutional Definition*

The United States Constitution contains a number of clauses pertinent to the conduct of war, none of which clearly commits this power to either Congress or the President.[104] The lawmaking body is delegated the power "[t]o declare

---

[104] Professor tenBroek compiled a list of those constitutional provisions related to the exercise of war:

> The Congress shall have power to lay and collect taxes, duties, imposts and excises, to . . . provide for the common defense and general welfare of the United States. [U.S. CONST. art. I, § 8, cl. 1].
>
> To borrow money on the credit of the United States [*Id.* at art. I, § 8, cl. 2];
>
> . . . .
>
> To declare war [*Id.* at art. I, § 8, cl. 11]; To raise and support armies, but no appropriation of money to that use shall be for a longer term than two years [*Id.* at art. I, § 8, cl. 12];
>
> To provide and maintain a navy [*Id.* at art. I, § 8, cl. 13];
>
> To make rules for the government and regulation of the land and naval forces [*Id.* at art. I, § 8, cl. 14];
>
> To provide for calling forth the militia to execute the laws of the Union, suppress insurrection and repel invasions [*Id.* at art. I, § 8, cl. 15];
>
> To provide for organizing, arming and disciplining, the militia, and for governing such part of them as may be employed in the service of the United States, reserving to the States respectively the appointment of the officers, and the authority of training the militia according to the discipline prescribed by Congress [*Id.* at art. I, § 8, cl. 16];
>
> . . . To make all laws which shall be necessary and proper for carrying into execution the foregoing powers, and all other powers vested by this Constitution in the government of the United States, or in any department or officer thereof [*Id.* at art. I, § 8, cl. 18];
>
> . . . The executive power shall be vested in a President of the United States of America [*Id.* at art. II, § 1, cl. 1]; . . . The President shall be commander in chief of the army and the navy of the United States, and of the militia of the several states, when called into

war"[105] and "[t]o provide for calling forth the militia to execute the laws of the Union, suppress insurrection and repel invasions."[106] The President, on the other hand, is the "Commander in Chief of the Army and Navy of the United States . . . when called into actual service of the United States."[107] The provisions of the Constitution clearly avoid assigning the war power exclusively to one branch of government. As a result, the scope of authority resting in each branch is not clearly defined. The Constitution itself contains no apparent guidelines to determine the appropriate application of the war power by either branch of government.

Historically, this void has resulted in political and judicial definitions and redefinitions of the constitutionally acceptable exercise of the war power—both in terms of which branch is empowered to act and in terms of the proper scope of its authority.[108] This continual translation has its genesis in the birth of the Constitution. Unbridled military power under the King of England was one of the evils the Founding Fathers sought to eradicate when the Constitution was written. Through the separation of powers to create a system of checks and balances and by the declaration of a bill of rights, they intended the subordination of the military to civilian control.[109]

---

actual service of the United States [*Id.* at art. II, § 2, cl. 1].
TenBroek, *Wartime Powers of the Military Over Citizen Civilians Within the Country*, 41 CALIF. L. REV. 167, 168-69 n. 5 (1953).

[105] U.S. CONST. art I, § 8, cl. 11.

[106] *Id.* at art. I, § 8, cl. 15.

[107] *Id.* at art. II, § 2, cl. 1.

[108] For historial background, *see* SENATE COMM. ON FOREIGN RELATIONS, NATIONAL COMMITMENTS, S. REP. NO. 797, 90th Cong., 1st Sess. (1967).

[109] In a speech given at the New York University Law Center, Chief Justice Earl Warren reviewed the constitutional history:

> I think it desirable to consider for a moment the principle of separation and subordination of the military establishment, for it is this principle that contributes in a vital way to a resolution of the problems engendered by the existence of a military establishment in a free society.
>
> It is significant that in our own hemisphere only our neighbor, Canada, and we ourselves have avoided rule by the military throughout our national existences. This is not merely happenstance. A tradition has been bred into us that the perpetuation of free government depends upon the continued supremacy of the civilian representatives of the people. To maintain this supremacy has always been a preoccupation of all three branches of our government. . . .
>
> . . . [T]he people of the colonies had long been subjected to the intemperance of military power. Among the grievous wrongs of which they complained in the Declaration of Independence were that the King had subordinated the civil power to the military . . . . Our War of the Revolution was, in good measure, fought as a protest against standing armies . . . .
>
> . . . Distrust of a standing army was expressed by many [of the Founding Fathers when they drafted the Constitution] . . . .

2. *Political Definition*

Politically, the war power has migrated toward the executive branch.

> The principal cause of the constitutional imbalance has been the circumstances of American involvement and responsibility in a violent and unstable world. . . .
> . . . Prior to 1940, foreign crises were infrequent and therefore put no lasting strain on our institutions. Since 1940 crisis has been chronic and, coming as something new in our experience, has given rise to a tendency toward anxious expediency in our response to it. The natural expedient—natural because of the real or seeming need for speed—has been the executive action . . . .[110]

The War Powers Resolution, passed in 1973, manifested Congress' concern over this shift and represented an attempt to reintroduce a means for checks and balances.[111] The principal concern appeared to be presidential commitment

---

> Their apprehensions found expression in the diffusion of the war powers granted the Government by the Constitution . . . .
> Despite these safeguards, the people were still troubled by the recollection of the conditions that prompted the charge of the Declaration of Independence that the King had "effected to render the military independent and superior to the civil power." . . . Although there is undoubtedly room for argument based on the frequently conflicting sources of history, it is not unreasonable to believe that our Founders' determination to guarantee the preeminence of civil over military power was an important element that prompted adoption of the Constitutional Amendments we call the Bill of Rights.

Warren, *The Bill of Rights and the Military*, 37 N.Y.U. L. Rev. 181, 183-85 (1962).

[110] Senate Comm. on Foreign Relations, *supra* note 108, at 422. *But cf.* Youngstown Sheet & Tube Co. v. Sawyer, 343 U.S. 579, 630 (1952) (Douglas, J., concurring) ("We . . . cannot decide this case by determining which branch of government can deal most expeditiously with the present crisis [the threat of a nationwide strike of steelworkers in the face of Korean War supply needs]. The answer must depend on the allocation of powers under the Constitution.").

[111] War Powers Resolution, Pub. L. No. 93-148, 87 Stat. 555 (1973) (codified as 50 U.S.C. §§ 1541-48 (1976)). The purpose and policy statement reads:

> It is the purpose of this chapter to fulfill the intent of the framers of the Constitution of the United States and insure that the collective judgment of both the Congress and the President will apply to the introduction of the United States Armed Forces into hostilities, or into situations where imminent involvement in hostilities is clearly indicated by the circumstances, and to the continued use of such forces in hostilities or in such situations.

The legislative history further describes Congress' position:

> The interpretation and application of [the] constitutional grants [of the war power] have varied widely throughout our Nation's history. Testimony received during hearings . . . confirmed the view of many Members of Congress . . . that the constitutional "balance" of authority over warmaking has swung heavily to the President in modern times. To restore the balance provided for and mandated in the Constitution, Congress must now reassert its own prerogatives and responsibilities.
> In shaping legislation to that purpose, the intention was not to reflect criticism on activities of Presidents, past or present, or to take punitive action. Rather, the focus of concern

of troops without congressional approval or adequate consultation with Congress.[112] The joint resolution articulated Congress' elaboration of the President's and the lawmaking body's constitutional war powers.[113] The legislative body recognized that, as Commander in Chief, the President may engage the armed forces in hostilities "only pursuant to (1) a declaration of war, (2) specific statutory authorization, or (3) a national emergency created by attack upon the United States."[114] The resolution requires the Chief Executive "in every possible instance" to consult Congress before and during the commitment of troops to hostilities or to imminent confrontation with hostile forces.[115] He must also report to the Congress on such deployments in the absence of a declaration of war.[116]

The President is required to terminate the use of armed forces within sixty days after the submission of a report, unless Congress (1) has declared war, (2) has enacted specific authorization for such troop commitment, (3) has extended the sixty-day period, or (4) is unable to convene as a result of armed attack upon the nation.[117] Where there is no declaration of war or specific statutory authorization, Congress may, by concurrent resolution, direct the removal of military forces by the President.[118]

> was the appropriate scope and substance of congressional and [p]residential authority in the exercise of the power of war in order that the Congress might fulfill its responsibilities under the Constitution while permitting the President to exercise his responsibilities.

H.R. COMM. ON FOREIGN AFFAIRS, WAR POWERS RESOLUTION OF 1973, H.R. REP. NO. 287, 93d Cong., 1st Sess., *reprinted in* 1973 U.S. CODE CONG. & AD. NEWS 2346, 2349 [hereinafter cited as WAR POWERS RES. LEGISLATIVE HISTORY].

112 *Id.* at 2349.

113 "[T]he resolution is merely intended to elaborate upon the application of the warmaking powers of the Congress and the President mentioned in the Constitution. [It] does not attempt any itemized definition of the war powers." *Id.* at 2350.

> A *joint resolution* is [a] resolution adopted by both houses of [C]ongress or when such a resolution has been approved by the [P]resident or passed with his approval it has the effect of law.
>
> The distinction between a joint resolution and a concurrent resolution of [C]ongress, is that the former requires the approval of the [P]resident while the latter does not.

BLACK'S LAW DICTIONARY 1178 (rev. 5th ed. 1979).

The War Powers Resolution was passed on November 7, 1973, by Congress, overriding President Nixon's veto. Pub. L. No. 93-148, 87 Stat. 555, 557 (1973).

114 War Powers Resolution, 50 U.S.C. §§ 1541-58, § 1541(c) (1976).

115 *Id.* at § 1542.

116 *Id.* at § 1543.

117 *Id.* at § 1544(b).

118 *Id.* at § 1544(c). The use of a concurrent resolution, which does not require executive approval, was intended as a vehicle for legislative veto of presidential troop commitments that met with congressional disapproval. WAR POWERS RES. LEGISLATIVE HISTORY, *supra* note 111, at 2357-58. Its constitutionality was questioned by certain members of Congress. *Id.* at 2358-63. A recent Supreme Court decision, Immigration & Naturalization Serv. v. Chadha, 103 S. Ct. 2764

The resolution's effectiveness remains to be tested both politically and judicially. However, recent presidential troop deployments to Lebanon and to Grenada without prior congressional approval suggest the resolution has not met the lawmaking body's expectations.[119]

3. *Judicial Definition*

The Supreme Court has varied in its perception of the range within which legitimate exercise of the war power falls.[120] Generally, the Court has accorded, with some consistency, much deference to the President and to Congress in matters involving national security vis-a-vis foreign affairs.[121] In contrast, where the assertion of military authority has involved individual rights, the judicial attitude has fluctuated.[122]

The Court's deferential attitude toward broad national security issues reflects

---

(1983), underscores that uncertainty. The Court in *Chadha* struck down as unconstitutional the congressional veto of Attorney General decisions allowing particular deportable aliens to remain in the United States. *Id.* at 2780-88. The veto power of the House of Representatives was held to be contrary to the separation of powers doctrine. Writing for the majority, Chief Justice Burger characterized the veto as legislative action, which therefore required passage by both houses and presentation to the President for his approval. *Id.* The veto capability reserved by the War Powers Resolution may meet similar judicial challenge. However, a concurrent resolution *does* require passage by both houses, but does not submit to presidential review. Additionally, at issue will be the nature of the veto—that is, whether it is legislative in character.

[119] Congress attempted to use the War Powers Resolution to rein in the powers of the President. When President Reagan deployed troops to Lebanon, the lawmaking body adopted a resolution authorizing the United States Marines to remain in that country as part of a multi-national peace-keeping force through March 1985. The Wash. Post, Sept. 24, 1983, at A1, col. 5. Although the President agreed on the 18-month period of Marine presence in Lebanon, he did *not* agree that his deployment decision required congressional authorization or approval. *Id.* at A22, col. 1. *See also id.*, Sept. 25, 1983, at A1, col. 4; *Strengthening the American Hand*, *id.*, Sept. 26, 1983, at A12, col. 1; *id.*, Sept. 28, 1983, at A1, col. 1.

The Grenada invasion by the United States Marines on October 26, 1983, produced similar reactions from both political branches of government. *Id.*, Oct. 28, 1983, at A1, col. 5 & at A3, col. 3. It was anticipated that the 60-day cutoff for unilateral presidential troop deployment under the War Powers Resolution would not be tested because the United States armed forces were expected to depart before that deadline. *Id.* at A3, col. 3.

[120] *See supra* text accompanying note 71 and *infra* text accompanying note 234.

[121] *See, e.g.*, Dames & Moore v. Regan, 453 U.S. 654 (1981) (judicial deference to presidential action freezing Iranian assets in the United States as bargaining chips in negotiations for release of American hostages captured by Iranians); United States v. Curtiss-Wright *Export* Corp., 299 U.S. 304 (1936) (judicial deference to presidential prohibition of arms sales to certain foreign countries under broad delegation of congressional authority).

[122] *See, e.g.*, United States v. Robel, 389 U.S. 258 (1967) (membership in Communist party not sufficient to deny citizen employment at defense plant); Duncan v. Kahanamoku, 327 U.S. 304 (1946) (civilians improperly tried by military tribunal); cases cited *infra* note 360.

judicial wariness in addressing political questions.[123] "Conventional wisdom suggests that in a democracy the judiciary should refrain from deciding 'political questions.' Such issues as initiating, conducting, and terminating war and hostilities require policy decisions that are political rather than legal in character."[124] It has even characterized the war power as:

> [one of the] powers of external sovereignty [that do] not depend upon the affirmative grants of the Constitution. The powers to declare and wage war, to conclude peace, to make treaties, to maintain diplomatic relations with other sovereignties, if they had never been mentioned in the Constitution, would have vested in the federal government as necessary concomitants of nationality.[125]

Where the war power has collided with individual rights, the Court has intervened. However, the judicial results have varied enough to leave uncertain the constitutional boundaries of the war power in relation to personal liberties.[126] Former Chief Justice Warren has postulated that the variations may be due to whether decisions are rendered during times of war or peace:

> War is . . . a pathological condition for our Nation. Military judgments sometimes breed action that, in more stable times, would be regarded as abhorrent. Judges cannot detach themselves from such judgments, although by hindsight, from the vantage point of more tranquil times, they might conclude that some actions advanced in the name of national survival had in fact overridden the strictures of due process.[127]

Thus, as Chief Justice Warren's analysis suggests, the constitutionality of government measures affecting individuals and taken in the name of the war power appears to depend less on theoretical limits in the minds of jurists than on the "winds of war." In *Korematsu*, the federal actions at issue are just such measures. The next section looks at the case to determine exactly ***which*** military orders should be subject to judicial review.

---

[123] *See, e.g.*, Baker v. Carr, 369 U.S. 186, 210-11 (1962) (addressing justiciability of political questions). *See generally* Keynes, *Democracy, Judicial Review, and the War Power*, 8 OHIO N.U.L. REV. 69 (1981) (discussing judiciary's role in review of exercise of war power by the President and Congress).

[124] Keynes, *supra* note 123, at 69.

[125] *Curtiss-Wright Export Corp.*, 299 U.S. at 318.

[126] *See generally* Warren, *supra* note 109, for an exposition of Bill of Rights cases involving the exercise of military authority.

[127] *Id.* at 191-92. *See infra* note 149 for contrasting cases.

B. *War Power Exercise Under Review*

What military measures were at issue in the *Korematsu* case? An answer is prerequisite to determining the larger question—whether Korematsu's conviction was unjustified because the military restraints imposed upon him were outside the constitutional limits of the war power. Obviously, a less restrictive measure is more easily placed within constitutional boundaries than a more intrusive one. Thus, a court's delineation of the object of scrutiny may very well presage the judicial outcome.[128]

The majority opinion in *Korematsu* limited review to the exclusion order and held that it represented a constitutional exercise of the war power.[129] Its reasoning rested on (1) an assumption that Congress would have treated the violation of each order as separate offenses if it had chosen to pass such legislation; (2) the fact that Korematsu had been convicted solely of defying the exclusion order; and (3) he would have time enough to seek legal relief when he was subjected to detention.[130]

Strong arguments were raised in the dissent by Justice Roberts for the review of exclusion, detention and relocation as a "single and indivisible" military plan.[131] His factual appraisal led the Justice to conclude that Korematsu was under the compulsion of all the orders. The Japanese American could not remain at home without courting arrest; to leave as ordered meant inevitable relocation and detention. In Justice Roberts' estimation, limiting judicial review to the exclusion was "a substitution of an [sic] hypothetical case for the case actually before the court."[132]

Judicial resolution of the scope of review in this case must focus on congressional intent in the enactment of the statute under which Korematsu was charged.[133] On its face, the act is devoid of any reference to relocation and detention. It provides unambiguously for the restriction of entry into and exit from military areas, but makes no mention of the authority to relocate or to

---

[128] *See supra* text accompanying notes 76-78 (discussing Justice Roberts' framing of issue in his dissent, contrasting with the question addressed by the majority opinion). *See also* P. IRONS, *supra* note 22, at 325-27 (explaining Justice Black's difficulties, while drafting the majority opinion, over treatment of the orders for exclusion and for detention as separable issues).

[129] 323 U.S. at 221-24. *See also supra* text accompanying notes 65-69.

[130] 323 U.S. at 222.

[131] *Id.* at 226 (Roberts, J., dissenting). *See also supra* text accompanying notes 79-85; *Korematsu*, 323 U.S. at 243 (Jackson, J., dissenting) ("[The orders] were so drawn that the only way Korematsu could avoid violation was to give himself up to the military authority. This meant submission to custody, examination, and the transportation out of the territory, to be followed by indeterminate confinement in detention camps."); *infra* note 216.

[132] 323 U.S. at 231 (Roberts, J., dissenting).

[133] 18 U.S.C. § 1383 (repealed 1976). *See supra* note 52 for text of statute.

detain individuals.[134] Thus, from the statutory language alone, it is not evident whether Congress intended a comprehensive plan of removal and detention—or separate and independent measures, each of which was not necessarily applicable to all evacuees.

The Court in *Hirabayashi* was very likely aware of the danger of reading too much from the sparse language of the statute alone. In order to sustain constitutional delegation of legislative power, the decision readily found that the congressional legislation, Executive Order No. 9066 and General DeWitt's orders "were not to be read in isolation from each other. They were parts of a single program and must be judged as such."[135] The measures were aggregated in order to meet the requirements for constitutional delegation of legislative power—that the legislation contain a standard to which the military orders must conform and that it require findings of necessity for the orders.[136] The Court looked to Executive Order No. 9066 for the requisite standard—the need to protect military facilities from espionage and sabotage—and to the military orders themselves for the findings required to support their promulgation.[137] By combining the orders and legislation into "a single program," the Court in *Hirabayashi* was free to look to all to find the prerequisites for constitutional delegation of legislative powers.

The facts of the case in *Korematsu* support an analogous use of the approach taken by the Court in *Hirabayashi*. The statute under which Korematsu was charged was enacted *after* issuance of Executive Order No. 9066, with knowledge of General DeWitt's plans, and for the purpose of providing sanctions for the violation of the military orders,[138] all of which suggest incorporation of these measures by the legislation. To return to the question—was the legislative intent to provide for a comprehensive and unified program of exclusion, relocation, and detention—or for discrete measures for independent application? Since the statute itself is silent, the implementing orders are examined to determine the nature of the legislative authority delegated.

At the time of arrest, Korematsu was subject to General DeWitt's orders to leave Military Area No. 1, where his home was located—but to leave only as authorized.[139] The "authorized" departure was by way of assembly centers, where evacuees faced indefinite detention before removal to relocation centers.[140] The degree of compulsion under which Korematsu acted is certainly diluted if viewed as solely the command to evacuate the area. If exclusion were

---

[134] *Id.*

[135] 320 U.S. at 103.

[136] *Id.* at 102-03.

[137] *Id.* at 103.

[138] *See supra* notes 51-55 and accompanying text.

[139] *See supra* notes 46-49 and accompanying text.

[140] *Id. See also supra* text accompanying notes 43-44.

the only order Korematsu was expected to obey, although intrusive enough, he would be free to choose his course of action once outside the prohibited area. But that was not the case.

Korematsu was to leave the area to go where the government had directed he should go. Exclusion was but the first step in the military's program to restrict the Japanese Americans and aliens on the West Coast. The government did not intend to do one—to exclude—without the other—to relocate and to detain. Thus, rationally and realistically, exclusion, relocation and detention are inseparable. And *all* require examination by the reviewing court to determine whether the military authority exercised over Korematsu was within constitutionally acceptable limits of the war power.

C. *Mechanisms for Exercise of War Power*

What was the source of authority for the military actions taken by General DeWitt and was that authority assumed or delegated constitutionally? Restricting review to the exclusion orders, without consideration of detention or relocation, the *Korematsu* Court was "unable to conclude that it was beyond the war power of Congress and the Executive to exclude those of Japanese ancestry from the West Coast war area."[141] Justice Black's summary conclusion relied on principles expounded in *Hirabayashi* (left unarticulated in *Korematsu*)[142] and on ready acceptance of congressional and military judgment of the need for exclusion.[143] He could not discard as groundless the belief of "the war-making branches of government" that quick action was necessary to deal with the disloyal, "a menace to the national defense" who could not be readily segregated from the loyal.[144]

To assess Justice Black's appraisal, the inquiry in this section looks first to the conditions under which the military may possess its own authority to exert controls over civilians. Specifically, the question is, did the West Coast situation at the time of General DeWitt's command warrant martial rule? Second, in the absence of any inchoate military power arising from the prevailing circumstances, presidential authority and the vehicle for its delegation are examined as the enabling source. Third, congressional authorization and delegation of legislative power are considered.

---

[141] 323 U.S. at 217-18, 221-23.

[142] *Id.* at 217-18.

[143] *Id.* at 218.

[144] *Id.*

1. *Military Power Over Civilians During Wartime*

What are the conditions under which inchoate military power develops to authorize military control of civilians during wartime?[145] The declaration of martial law, or the conditions that would necessitate its invocation, provides one basis for the wielding of such authority *sua sponte* over civilians.[146]

In *Duncan v. Kahanamoku*,[147] a case arising out of the imposition of martial law in Hawaii during World War II, the Supreme Court could find no precise meaning of martial law either within the Constitution or in any congressional enactment.[148] Writing the plurality opinion,[149] Justice Black[150] reviewed the development of the nation's political institutions and found a consistent theme—fear and opposition to subordination of civilian governance to military

---

[145] *Cf. Duncan*, 327 U.S. at 313-14 & nn.6-10 (enumerating different conditions which lend themselves to military authority—e.g., military jurisdiction over its own personnel or over prisoners of war).

[146] Former Chief Justice Stone expressed one view of the conditions giving rise to martial law:

> [M]artial law is the exercise of the power which resides in the executive branch of the government to preserve order and to insure the public safety in times of emergency, when other branches of the government are unable to function, or their functioning would itself threaten the public safety.

Id. at 335 (Stone, C.J., concurring) (citing Luther v. Borden, 48 U.S. (7 How.) 1, 45 (1849)).

[147] 327 U.S. 304 (1946).

[148] *Id.* at 315. After the Japanese attack on Pearl Harbor, the Governor of Hawaii placed the Territory (now the state of Hawaii) under martial law pursuant to the Hawaii Organic Act, 48 U.S.C. § 532 (1946) (omitted as obsolete 1959). *Id.* at 307. Two civilians, White and Duncan, were arrested by military police while martial rule was still in force. *Id.* at 309-10. Both were convicted by the military tribunals whose jurisdiction they subsequently challenged. *Id.* at 311. Legislative history of the enabling act was deficient in aiding the Court to determine whether the scope of martial law intended by the statute was to include supplanting of courts by military tribunals.

[149] Justice Jackson did not participate in the case. Chief Justice Stone and Justice Murphy each concurred in separate opinions. Justice Burton, with Justice Frankfurter, dissented.

[150] Note that Justice Black authored both the *Duncan* and *Korematsu* opinions. His expansive view of permissive war power exercise in *Korematsu* [*see supra* text accompanying notes 141-44] was not evident in *Duncan*. In the latter case the Justice saw in our nation's history a reflection of the nearly universal abhorrence to complete military dominance over civilian government. As a result, he interpreted the scope of permissible rule narrowly in *Duncan*. *See also* CWRIC REPORT, *supra* note 1, at 280-82 (comparing Justice Black's decision in the two cases). *But cf. id.* at 282 n.92 (citing Fairman, *The Supreme Court on Military Jurisdiction: Martial Rule in Hawaii and the Yamashita Case*, 59 HARV. L. REV. 833 (1946), which reconciled *Duncan* and *Korematsu*); *Duncan*, 327 U.S. at 314 n.9 (where Justice Black appeared to distinguish the two cases: "Nor is this [*Duncan*,] a case where violators of military orders are to be tried by regular courts. *Cf.* Hirabayashi v. United States, 320 U.S. 81.").

As former Chief Justice Warren posited, such differing judicial postures may be due to the context in which decisions are made. *See supra* text accompanying notes 126-27. *Korematsu* was decided when the country was still engaged in World War II; *Duncan* was a post-war decision.

authority.[151] The Court held that the act under which martial law had been declared in Hawaii did not contemplate authorization of military rule so broad as to supplant civilian courts with military tribunals.[152] The decision in *Duncan* rested on the interpretation of statutory intent. It did not need to reach constitutional issues since its reading of legislative purpose was sufficient to decide the case.

In its reasoning, the Court cited *Ex parte Milligan*, a case that did involve constitutional issues raised by the exercise of martial law.[153] This landmark case provides a framework for assessing the constitutionality of military control over civilians.[154] In *Milligan*, a civilian was convicted by a Civil War Military Commission in Indiana. The Supreme Court held that the commission had no jurisdiction to do so, when Indiana had not been invaded or engaged in rebellion and the arrestee was neither a member of the armed forces nor a prisoner of war. The Court reasoned that "[m]artial rule can never exist where the courts are open, and in the proper and unobstructed exercise of their jurisdiction. It is also confined to the locality of actual war."[155]

The Court in *Milligan* delineated the conditions under which martial law may properly be imposed:

> If, in foreign invasion or civil war, the courts are actually closed, and it is impossible to administer criminal justice according to law, *then*, on the theatre of active military operations, where war really prevails, there is a necessity to furnish a substitute for the civil authority, thus overthrown, to preserve the safety of the army and society; and as no power is left but the military, it is allowed to govern by martial rule until the laws can have their free course. As necessity creates the rule, so it limits its duration; for, if this government is continued *after* the courts are reinstated, it is a gross usurpation of power.[156]

The imposition of martial law was not at issue in *Korematsu*. However, it is worth noting that the criteria of valid military control are the *Milligan* factors,

---

[151] 327 U.S. at 319-24.

[152] *Id.* at 324.

[153] 71 U.S. (4 Wall.) 2 (1866), *cited with approval in Duncan*, 327 U.S. at 322, 324. In his concurrence, Justice Murphy applied the constitutional tests developed in *Milligan* and found the trials at issue in *Duncan* "were forbidden by the Bill of Rights." *Duncan*, 327 U.S. at 325, 326 (Murphy, J., concurring).

[154] "Decided in 1866 and arising out of an episode of the Civil War, [*Ex parte Milligan*] has stood as a landmark in our constitutional history on the nature and extent of the war-time power of the military over civilians within the country." tenBroek, *supra* note 104, at 171. The approach utilized in this section relies on TenBroek's treatment of *Ex parte Milligan* in the cited article.

[155] 71 U.S. (4 Wall.) at 127.

[156] *Id.* (emphasis included).

and not the incantation of the words, "martial law," however officially pronounced.[157] The *Milligan* factors are: (1) the occurrence of foreign invasion or civil war; (2) the impossibility of administering criminal law through the civil courts; (3) the presence of the "theatre of active military operations, where war really prevails"; and (4) overthrown civil authority, leaving none but military power.[158]

Applying the factors to the *Korematsu* case, the inescapable conclusion is that the West Coast situation did not warrant the exercise of military control over civilians, or that the declaration of martial law (which would have been the source of authorization for *sua sponte* military control) would not have been justified. Indeed, a foreign invasion had occurred in an area under the aegis of the United States. On December 11, 1941, Army Chief of Staff General George C. Marshall did declare the Pacific Coast to be a "theatre of operations."[159] But it was not the site "where war really prevails."[160] During the period that General DeWitt was in command of the Western Defense area, all civil agencies, including the courts, were open. The conditions then present did not make it "impossible to administer criminal justice according to law."[161] The situation on the West Coast was not so critical as to justify military control over civilians,[162] and, in fact, mimicked the consequences of indiscriminate

---

[157] Although the *Duncan* case involved incidents occurring under officially declared martial law, the court appeared to reject the label of "martial law" as self-justifying or self-defining. *Duncan*, 327 U.S. at 314-24. It thus avoided validating the circular reasoning that: when civil courts are not able to function because of critical wartime conditions, martial law is warranted; therefore, when martial law has been invoked, civil courts ought to be closed and supplanted by military tribunals.

[158] 71 U.S. (4 Wall.) at 127.

[159] F. Chuman, *supra* note 1, at 155.

[160] *But cf. Duncan*, 327 U.S. at 344 n.3 (Burton, J., dissenting) (quoting, from the *Duncan* record, testimony of Lieutenant General Robert C. Richardson, Jr., United States Army, then Commanding General of the Central Pacific Area). A very broad interpretation of the "theatre of operations" was suggested. It would be the area needed for defensive or offensive operations, including administrative agencies which are necessary for the conduct of these operations.

[161] *Milligan*, 71 U.S. (4 Wall.) at 127. In fact, Korematsu himself had been tried and convicted in a federal district court in San Francisco. P. Irons, *supra* note 22, at 151-54.

[162] *See* CWRIC Report, *supra* note 1, at 261-62. The report described the conditions in Hawaii where martial law had been imposed. No wholesale evacuation of persons of Japanese ancestry occurred; yet, the area included the Pearl Harbor bombing site and was certainly closer to the active theatre of war in the Pacific. *See also* F. Chuman, *supra* note 1, at 154-55. In his legal history of the Japanese Americans, attorney Frank Chuman described the role of then Army Provost Marshall General Allen W. Gullion, that service's highest ranking law enforcement officer, in the decision to evacuate the Japanese. In December 1941, after an unsuccessful attempt to have the enemy alien program transferred from the Department of Justice to the War Department, he recommended to General DeWitt that all Japanese, American citizens and aliens alike, be taken into custody. *Id.* at 155. The Provost Marshall told General DeWitt that such action had been urged by a representative of the Los Angeles Chamber of Commerce. *Id.* at 155 and ch.

martial rule that *Milligan* found untenable:

> [W]hen war exists, foreign or domestic, and the country is subdivided into military departments for more convenience, the commander of one of them can, if he chooses, within his limits, on the plea of necessity, with the approval of the Executive, substitute military force for and to the exclusion of the Laws, and punish persons, as he thinks right and proper, without fixed or certain rules.[163]

In *Korematsu*, Justice Black alluded to "conditions of modern warfare [under which] our shores are threatened by hostile forces . . ."[164] and described the West Coast as a "war area."[165] The bombing of Pearl Harbor may have subjected the West Coast to the threat of attack. The prevailing military judgment was that the threat was exacerbated by possible espionage and sabotage by persons of Japanese ancestry living in the area. However, to expand the *Milligan* "theatre of war" concept to include such conditions would effectively remove any protection the doctrine would provide against unrestrained military control over civilians during wartime. Modern concepts of war could turn the entire United States into a "theatre," as Justice Black would suggest, on the basis of threatened invasion, even when civil authority was not rendered powerless by hostilities. In his later opinion in *Duncan*, Justice Black provided his own counterpoint:

> Our system of government clearly is the antithesis of total military rule and the founders of this country are not likely to have contemplated complete military dominance within the limits of a Territory made part of this country and not recently taken from an enemy. They were opposed to governments that placed in the hands of one man the power to make, interpret and enforce the laws.[166]

2. *Executive Power*

If General DeWitt lacked any independent power to exercise control over

---

9, n.25. Yet, in his previous capacity as Judge Advocate General, the Army's highest ranking legal officer, General Gullion's official opinion had been that the "military . . . does not have jurisdiction to participate in the arrest and temporary holding of civilians who are citizens of the United States" outside of any combat zone where civilian courts were still open. *Id.* at 154-55 & ch. 9, n.24 (quoting General Gullion). Ironically, at that time in December 1941, General DeWitt was doubtful of the wisdom of General Gullion's wholesale internment idea and thought that the disloyal could be ferreted out. *Id.* at 155 & ch. 9, n.26.

[163] 71 U.S. (4 Wall.) at 124-25.

[164] 323 U.S. at 220.

[165] *Id.* at 218.

[166] 327 U.S. at 322.

civilians, did he derive authority from the powers of the President? On February 19, 1942, President Roosevelt issued Executive Order No. 9066 authorizing the Secretary of War or his designate "to prescribe military areas" and to exclude "any or all persons" as deemed necessary.[167] A month later, by Executive Order No. 9102,[168] the President established the War Relocation Authority and entrusted it with the removal and relocation of persons excluded under military orders emanating from Executive Order No. 9066. Such presidential statements may have the force and effect of public law when based on constitutional powers or on congressional authorization.[169] There is no statute that specifically defines the term "executive order." Essentially, it is a presidential device used to direct the actions of administrative officials and agencies.[170]

Did the President have the requisite constitutional or congressional authorization to promulgate Executive Orders No. 9066 and 9102? The inquiry will depend in part on whether the Court were to construe the presidential orders alone or as part of a joint congressional-executive plan. If the orders are viewed as independent executive actions, they would require constitutional or statutory bases to be valid.[171] In this case, the declaration of war by Congress arguably could be sufficient authorization. Alternative justification could be sought under the President's constitutional power as Commander in Chief. On the other hand, if specific legislation can be identified, the requisite congressional approval or participation would be less difficult to substantiate. In *Korematsu*, although Congress did pass Public Law No. 503,[172] a statute permitting restrictions in military areas under the authority of executive order, the enactment *followed* the issuance of Executive Order No. 9066. The tardy legislative measure could be characterized as ratification by Congress after the fact.[173]

If assumed to be unilateral actions by the President, do the orders pass judicial muster under his constitutional role as "Commander in Chief of the Army

---

[167] 7 Fed. Reg. 1407 (1942).

[168] 7 Fed. Reg. 2165 (1942).

[169] Jenkins v. Collard, 145 U.S. 546 (1891) (presidential pardon, a constitutional prerogative of the Executive [U.S. CONST. art. II, § 2, cl. 1] made by way of proclamation had force of public law). It should be noted that "[t]he difference between Executive orders and proclamations is more one of form than of substance since in each instance the effective action sought or directed by the document is an exercise of the Executive power under article II of the Constitution and must be based on authority derived from the Constitution or statute." HOUSE COMM. ON GOVERNMENTAL OPERATIONS, 85th Cong., 1st Sess., EXECUTIVE ORDERS AND PROCLAMATIONS: A STUDY OF A USE OF PRESIDENTIAL POWERS 1 (Comm. Print 1957) (footnote omitted).

[170] HOUSE COMM. ON GOVERNMENTAL OPERATIONS, *supra* note 169, at 1.

[171] *Youngstown Sheet & Tube Co.*, 343 U.S. at 585 ("The President's power, if any, to issue the order must stem either from an act of Congress or from the Constitution itself.").

[172] Codified as amended at 18 U.S.C. § 1383 (1976) (repealed 1976).

[173] *See infra* notes 185-90 and accompanying text.

and Navy of the United States?"[174] Immediately after congressional declaration of war, security measures decreed by the President were directed only against suspected enemy aliens.[175] These early presidential proclamations were issued under the authority of the Alien Enemy Act.[176] However, Executive Order No. 9066, not specifically limited in application to enemy aliens, was promulgated "by virtue of the authority vested in [the Chief Executive] as President of the United States, and *Commander in Chief of the Army and Navy*."[177] Executive Order No. 9102, establishing the War Relocation Authority, followed and purportedly did no more than "implement the program authorized by Executive Order No. 9066."[178]

Upon declaration of war by Congress, the legitimacy of the President's conduct would turn on whether it was executive in character (and thereby within his own constitutional powers) or whether it could be action committed in the "zone of twilight in which he and Congress may have concurrent authority or in which its distribution is uncertain."[179] Clearly "executive" actions would be those implementing constitutional or congressional policy.[180] The President encroaches upon the lawmaking function when he "does not direct that a congressional policy be executed in a manner prescribed by Congress—[he is legislating when he] directs that a presidential policy be executed in a manner prescribed by the President."[181]

There very likely would be judicial reluctance to sanction the decisions to exclude and to detain persons of Japanese ancestry as unilateral executive measures. In *Youngstown Sheet & Tube Co.*, the Court appeared to suggest that the exercise of the President's war power as Commander in Chief would not go beyond the broad powers exercised in "day-to-day fighting in a theater of war."[182] The decision recognized that the concept "theater of war" was an expanding one, but held that seizure of private property to prevent production stoppage could not meet constitutional strictures even when taken under the war power of the Commander in Chief.[183] Moreover, in light of the War Pow-

---

[174] U.S. CONST. art. II, § 2, cl. 1.

[175] *See supra* note 26 and accompanying text.

[176] 50 U.S.C. § 21 (1976). *See also supra* notes 25-26 and accompanying text.

[177] Exec. Order No. 9066, 7 Fed. Reg. 1407 (1942) (emphasis added).

[178] *Ex parte* Endo, 323 U.S. 283, 298-301 (1944) (construing Executive Order No. 9102 as based on Executive Order No. 9066 which, in turn, rested for authority on Public Law No. 503, the Act providing penalties for violation of the military orders).

[179] *Youngstown Sheet & Tube Co.*, 343 U.S. at 635-37 (Jackson, J., concurring); *see also id.* at 587-89 (plurality opinion).

[180] *Id.* at 585, 587 (plurality opinion).

[181] *Id.* at 588.

[182] *Id.* at 587.

[183] *Id. See also* letter from Black, J., to Stone, C.J., (Jan. 18, 1946) (responding to criticism of a draft of his *Duncan* opinion) *quoted in* CWRIC REPORT, *supra* note 1, at 282:

ers Resolution, the Court may hesitate to sanction unilateral presidential action if congressional approval is not evident—either because the political nature of the question would caution against justiciability or because specific congressional direction required by the resolution is missing.[184]

However, the Court very likely would not consider the executive actions independent of the legislative authorization of Public Law 503.[185] To do so would indicate a narrow adherence to form.[186] Congress, fully aware of the intent of Executive Order No. 9066, ratified it after the fact. Thus, the inquiry turns to whether such ratification would be recognized as valid. In the *Prize Cases*[187] the Court upheld the seizure of ships transporting goods to Confederate areas in defiance of a presidential blockade instituted before any congressional declaration of war. It noted that Congress had passed subsequent legislation "approving, legalizing, and making valid all the acts, proclamations, and orders of the President . . . as if they had been *issued and done under the previous express authority* and direction of the Congress of the United States."[188] The Court concluded that "it is plain that if the President had in any manner assumed powers which it was necessary should have the authority or sanction of Congress, . . . this ratification has operated to perfectly cure the defect."[189]

I think the Executive is without Constitutional powers to suspend all legislative enactments in loyal, uninvaded states, to substitute executive edicts for those laws, and to provide for their enforcement by agents chosen by and through tribunals set up by the Executive . . . . In other words, the Constitution, as I understand it, so far as civilians in legal uninvaded territory is concerned, empowers the Executive to "execute" a general code of civil laws, not executive edicts.

[184] *See supra* notes 110-19 and accompanying text.

[185] Codified as amended at 18 U.S.C. § 1383 (1976) (repealed 1976).

[186] Legislative history before the Court in *Korematsu* indicated that the congressional act was imposed "to provide the means of enforcement of curfew orders and other military orders made pursuant to Executive Order No. 9066." *Hirabayashi*, 320 U.S. at 90; *see also supra* notes 51-55 and accompanying text.

[187] 67 U.S. (2 Black) 635 (1863); *see also* Haig v. Agee, 453 U.S. 280, 292-300 (1981) (nonstatutory administrative practices taken under original enactment considered adopted by Congress when, in amending statute, there was no evidence of legislative intent to repudiate such practices).

[188] 67 U.S. (2 Black) at 670 (*quoting* Act of Aug. 6, 1861, ch. 63, § 3) (emphasis added).

[189] *Id.* at 671; *see also* Mason Co. v. Tax Comm'n of Wash., 302 U.S. 186, 208 (1937) ("The federal intent in this instance is clearly shown. It is shown not merely by the action of administrative officials, but by the deliberate and ratifying action of Congress, which gives the force of law to the prior official action even if unauthorized when taken."); Swayne & Hoyt, Ltd. v. United States, 300 U.S. 297, 301-02 (1937) ("It is well settled that Congress may, by enactment not otherwise inappropriate, 'ratify . . . acts which it might have authorized,' and given the force of law to official action unauthorized when taken [citation omitted] . . . . The mere fact that the validation is retroactive in its operation is not enough, in the circumstances of this case, to render it ineffective."); Isbrandtsen-Moller Co. v. United States, 300 U.S. 139, 146-48 (1937) ("Whatever doubt may be entertained as to the intent of Congress [that the President

In *Korematsu* Congress did not expressly ratify the executive order in the body of the statute.[190] However, the President was acting in the aftermath of an official declaration of war. Moreover, the Court had information before it indicating that the legislation was enacted in response to a War Department request and that Congress was told of General DeWitt's initial plans.[191] Accordingly, the enactment could be construed to confer after the fact ratification upon the executive order.

3. *Congressional Delegation of Power*

If the exclusion, relocation, and internment orders required ratification by Congress, the legislative authorization must have been constitutionally delegated. A congressional statement of policies and standards is required for presidential exercise of the delegated power.[192] "The degree to which Congress must specify its policies and standards in order that the administrative authority granted may not be an unconstitutional delegation of its own legislative power is not capable of precise definition."[193] Certainly, "the Constitution [must] be scrupulously obeyed."[194] Nevertheless, where the war power is concerned, the Court has shown its customary deference in finding constitutional delegation even when congressional directions have been vaguely stated and leave much to

---

was authorized to transfer specific functions between agencies of the executive branch], . . . Congress appears to have recognized the validity of the transfer and ratified the President's action by [appropriation acts earmarking related salaries to the agency assigned the transferred functions].").

[190] *See supra* note 52 for the text of the Act.

[191] *See supra* notes 23, 51-55, 186 and accompanying text. General DeWitt's Final Report reveals that:

> Immediately upon the promulgation of Executive Order No. 9066, the War Department, with the approval of the President, requested the Congress to enact legislation to provide sanctions for the enforcement of directives issued under the authority of the Executive Order. A draft of proposed legislation for this purpose was transmitted [to Congress] . . . .
>
> . . . .
>
> While the legislation was under consideration, the Secretary of War . . . transmitted another letter to the Congress . . . urging immediate enactment, [which stated in part:] "General DeWitt is strongly of the opinion that the bill, when enacted, should be broad enough to enable the Secretary of War or the appropriate military commander to enforce curfews and other restrictions within the military areas and zones."

Final Report, *supra* note 23, at 29-30.

[192] *See, e.g.*, Greene v. McElroy, 360 U.S. 474, 507 (1959); Kent v. Dulles, 357 U.S. 116, 129 (1958); Lichter v. United States, 334 U.S. 742, 779 (1948); A.L.A. Schechter Poultry Corp. v. United States, 295 U.S. 495, 530 (1935); Panama Ref. Co. v. Ryan, 293 U.S. 388, 421 (1935).

[193] *Lichter*, 334 U.S. at 779.

[194] *Id.*

presidential discretion.[195]

> *A constitutional power implies a power of delegation of authority under it sufficient to effect its purpose.* This power is especially significant in connection with constitutional war powers under which the exercise of broad discretion as to methods to be employed may be essential to an effective use of its war powers by Congress.[196]

The requisite congressional guidance to the President is not evident on the face of the statute[197] identified as the ratification instrument in *Korematsu.* It merely stated as its purpose, "[t]o provide a penalty for violation of restrictions or orders with respect to persons entering, remaining, leaving, or committing any act in military areas or zones."[198] This deficiency may not be fatal, in light of decisions in which the Court has looked to the presidential action itself to find the necessary legislative intent and direction where the war power, national security, or foreign affairs were involved.[199]

This stance could be further supported by the Court's recognition that the actions in question, if not specifically taken "pursuant to an express or implied authorization of Congress,"[200] were at least within "a zone of twilight in which [the President] and Congress may have concurrent authority, or in which its distribution is uncertain."[201] In this zone of shared power, the constitutional question would not turn so much on the specificity of statements of congressional policies and standards. Rather, "any actual test of power is likely to depend on the imperatives of the events and contemporary imponderables rather than on abstract theories of law."[202]

Moreover, the power to wage war has been viewed by the Court as one of the federal government's inherent powers of sovereignty.[203] As such, the political branches of government do not need to look to the Constitution for delineations of such authority.[204] "As a member of the family of nations, the right and power of the United States in that field are equal to the right and power of the other members of the international family. Otherwise, the United States is not

---

[195] *See, e.g., Lichter*, 334 U.S. at 783-86; *cf. Curtiss-Wright Corp.*, 299 U.S. at 322-29 (listing several legislative and judicial decisions confirming the need for broad delegation of legislative powers under which the President defines the appropriate parameters for action).

[196] *Lichter*, 334 U.S. at 778-79 (emphasis included).

[197] 18 U.S.C. § 1383 (1976) (repealed 1976). *See supra* note 52 for text of statute.

[198] *Id.*

[199] *See supra* note 195.

[200] *Youngstown Sheet & Tube Co.*, 343 U.S. at 635 (Jackson, J., concurring).

[201] *Id.* at 637; *see also supra* notes 102-27 and accompanying text.

[202] *Youngstown Sheet & Tube Co.*, 343 U.S. at 637 (Jackson, J., concurring).

[203] *Curtiss-Wright Corp.*, 299 U.S. at 318.

[204] *Id.*

completely sovereign."[205] The effectiveness of federal actions in this area often requires prompt response based on an assessment of conditions best known to the President alone.[206] Congress has thus frequently granted the Chief Executive wide discretion in national security affairs.[207] Especially where the two political branches act in concert, a judgment of "the imperatives of the events" by either entity could be sufficient to provide "the test of power." Accordingly, as in *Hirabayashi*, the executive order could very well serve as the source of the standard by which adherence to congressional policy is measured.[208] In Executive Order No. 9066, the President declared that "[t]he successful prosecution of the war requires every possible protection against espionage and against sabotage to national-defense material, national-defense premises and national-defense utilities."[209]

Against the standard of the need to protect national defense resources against espionage and sabotage, could exclusion, relocation and internment have been measures contemplated by the Congress in its delegation of authority? The Court has certainly recognized that the grant of legislative power cannot anticipate all myriad situations which the executive branch is better structured to meet.[210] "Undoubtedly, legislation must often be adapted to complex conditions involving a host of details with which the national Legislature cannot deal directly."[211] As intrusive as exclusion and relocation appear to be, they are less troubling than internment as measures that could be necessary and proper[212] to protect designated military zones from espionage and sabotage. The prohibition of entry into and the removal of individuals from vital areas to safeguard against those twin dangers appear to be within the realm of necessary and proper military options.

However, if these measures are but initial phases of a comprehensive plan of internment, they take on a more questionable cast. Internment, as carried out in World War II, was a constriction of virtually all freedom of choice in movement—at least until plans for the leave program and Nisei combat teams were formulated in the fall of 1942.[213] The judicial device of severability could per-

---

[205] *Id.*

[206] *Id.* at 320. *See also supra* note 110 and accompanying text.

[207] *Curtiss-Wright Corp.*, 299 U.S. at 321-22.

[208] 320 U.S. at 103-05; *see also supra* note 54.

[209] 7 Fed. Reg. 1407 (1942).

[210] *See, e.g.*, Carlson v. Landon, 342 U.S. 524, 542 (1952); *Lichter*, 334 U.S. at 384-86; *Panama Ref. Co.*, 293 U.S. at 421.

[211] *Panama Ref. Co.*, 293 U.S. at 421.

[212] *Lichter*, 334 U.S. at 781 (legislative power includes all that is necessary and proper for executing that power).

[213] CWRIC REPORT, *supra* note 1, at 185-212. *Nisei* is a Japanese word denoting the "[s]econd generation[,] [p]articularly a person born in the United States of Japanese parents." BLACK'S LAW DICTIONARY, *supra* note 113, at 944.

mit the Court to find exclusion acceptable, while ruling out internment as within the powers delegated, "[u]nless it is evident that the *legislature* would not have enacted those provisions which are within its power, independently of that which is not."[214] This exercise of "concurrent authority," however, relied on the President, and ultimately his military commanders, for definition.[215] Thus, the earlier military orders calling for exclusion and removal were, as discussed previously, part and parcel of a unified program leading inevitably to the internment of evacuees.[216]

The broad authority conferred by Congress upon the President to protect wartime facilities gave him much discretion in defining the scope of the plan necessary to meet that objective. The defense measures taken by General DeWitt, under the authority of Executive Order No. 9066, were designed to move potential saboteurs and spies out of designated military areas and to detain them in internment camps. Exclusion and relocation may arguably be within the scope of actions contemplated by Congress as necessary for the protection of vital resources. The constitutionality of the delegation thus hinges on the internment measure and on whether this restriction of the ethnic Japanese fell within the authorized legislative power. Was such a drastic restriction upon the evacuees' movements necessary, when measured against the standard of protection of military facilities from subversion? The Court in *Ex parte Endo*[217] established the purpose of the prolonged detention to be the protection of the *evacuees* from hostilities perceived to be present in the communities expected to receive the displaced West Coast ethnic Japanese.[218] Justice Douglas could find no support in the statute or executive orders for such a ground for internment.[219] The entire, unified plan of evacuation and internment therefore fails as

---

[214] Buckley v. Valeo, 424 U.S. 1, 108-09 (*quoting* Champlin Ref. Co. v. Corporation Comm'n, 286 U.S. 210, 234 (1932)).

[215] *See supra* text accompanying notes 200-09.

[216] In his autobiography, Justice William Douglas expressed regret over his position in the *Korematsu* decision. He had written a draft of a concurring opinion in which he agreed to the constitutionality of evacuation, but not to evacuation through internment. He was persuaded by Justices Black and Frankfurter to withdraw his opinion, both jurists arguing that the issue of detention was not before the Court. Justice Douglas reflected: "Technically, . . . the question of detention was not presented to us. Yet evacuation via detention camps was before us, and I have always regretted that I bowed to my elders and withdrew my opinion." W. DOUGLAS, THE COURT YEARS: 1939-1975 at 279-80 (1980), *quoted in* CWRIC REPORT, *supra* note 1, at 239 n.112. *See also supra* notes 128-40 and accompanying text.

[217] 323 U.S. 283 (1944).

[218] *Id.* at 301.

[219] *Id.* at 302-04. Unlike the decisions in *Hirabayashi* and *Korematsu*, in *Endo* a citizen's loyalty was important in determining whether the governmental action was appropriate to protect against espionage and sabotage. *Id.* at 302. But the Court was not implying that "detention in connection with no phase of the evacuation program would be lawful. The fact that the Act and the orders are silent on detention does not mean that any power to detain is lacking." *Id.* at 301.

a constitutionally delegated measure.

If the issue of internment were to be addressed squarely by the Court today, there might be even less latitude in finding constitutional delegation of authority to do so. The 1971 repeal of the Emergency Detention of Suspected Security Risks Act (passed in 1950 and thus not before the *Korematsu* Court for consideration) provides legislative history requiring specific congressional authorization for internment.[220] Until its repeal, the enactment permitted the detention of suspected security risks if the President proclaimed a "state of 'Internal Security Emergency' " and "[i]n the event of any one of the following: (1) [i]nvasion of the territory of the United States or its possessions, (2) [d]eclaration of war by Congress, or (3) [i]nsurrection within the United States in aid of a foreign enemy."[221] Its repeal was demanded because the statute was perceived as a means by which internment camps could again be used. "[G]roups of Japanese-American citizens regard[ed] the legislation as permitting a recurrence of the roundups which resulted in the detention of Americans of Japanese ancestry in 1941 and subsequently during World War II."[222]

When it repealed this act, Congress also amended another statute to provide that "[n]o citizen shall be imprisoned or otherwise detained by the United States except pursuant to an Act of Congress."[223] Its purpose was twofold: to remove legislation that contained "concentration camp implications" and to "assure that no detention camps can be established without at least the acquiescence of Congress."[224]

Congress was not satisfied with repeal alone as a means to restrict presidential conduct:

> Repeal alone might leave citizens subject to arbitrary executive action, with no clear demarcation of the limits of executive authority. It has been suggested that repeal alone would leave us where we were prior to 1950. The Committee believes that imprisonment or other detention of citizens should be limited to situations in which a statutory authorization, an Act of Congress, exists.[225]

The legislative history of the repeal and amendment suggests that internment is not necessarily foreclosed. As stated in the passage quoted above, it is emphatic in requiring a congressional command for instituting such measures.

[220] The statute was 50 U.S.C. §§ 811-26 (1970) (repealed 1971). For the legislative history of its repeal see H.R. REP. NO. 116, 92nd Cong., 1st Sess., *reprinted in* 1971 U.S. CODE CONG. & AD. NEWS 1435 [hereinafter cited as H.R. REP. NO. 116].

[221] 50 U.S.C. §§ 811-26 (1970) (repealed 1971).

[222] H.R. REP. NO. 116, *supra* note 220, at 1436.

[223] 18 U.S.C. § 4001(a) (1976).

[224] H.R. REP. NO. 116, *supra* note 220, at 1438.

[225] *Id.*

Thus, "[s]hould drastic measures be called for at some future time, it is inconceivable that this already dated statute [subject of the repeal] would fill the needs of the moment. Almost certainly, ***new and different legislation would be called for, tailored to current needs.***"[226]

No doubt, the 1942 statute under which President Roosevelt acted[227] would be found constitutionally deficient on its face for delegation purposes today. However, if internment were instituted under current laws, the authorizing statute would in all likelihood be drafted more carefully in light of legislative history. The irony is that, by requiring specificity in legislating the command for internment, the amended statute and its history ensures that today the delegation issue would be more apt to withstand constitutional attack.

III. Equal Protection v. Military Necessity

> It should be noted, to begin with, that all legal restrictions which curtail the civil rights of a single racial group are ***immediately suspect.*** That is to say that all such restrictions are unconstitutional. It is to say that courts must subject them to the ***most rigid scrutiny.*** Pressing public necessity may sometime justify the existence of such restrictions; racial antagonism never can.[228]

This seminal passage articulates the tests by which courts today judge the constitutionality of racially classifying governmental actions.[229] Yet the decision from which it sprang has been the target of criticism for not heeding its own command to subject suspect legal restrictions to strict scrutiny.[230] The excerpt is, of course, from *Korematsu v. United States.*[231]

Presently, strict, or rigid, scrutiny is the acknowledged standard of review courts use to assess restrictions upon fundamental rights or upon suspect classifications.[232] Accordingly, in *Korematsu* the military orders singling out the ethnic

---

[226] *Id.* (emphasis added).

[227] 18 U.S.C. § 1383 (1976) (repealed 1976).

[228] *Korematsu,* 323 U.S. at 216.

[229] *See, e.g.,* McLaughlin v. Florida, 379 U.S. 184, 192 (1964); cases cited *supra* note 60.

[230] *See, e.g.,* P. Irons, *supra* note 22, at 337-38; CWRIC Report, *supra* note 1, at 236; TenBroek, *supra* note 104, at 182-83.

[231] 323 U.S. 214 (1944) (majority opinion).

[232] For fundamental rights cases, *see, e.g.,* Planned Parenthood v. Danforth, 428 U.S. 52 (1976) (woman's decision to terminate pregnancy protected by right to privacy); Shapiro v. Thompson, 394 U.S. 618 (1969) (right to travel interstate impaired by one-year residency requirement for welfare qualification; no compelling state interest thereby served); William v. Rhodes, 393 U.S. 23 (1968) (right to vote and to associate impaired by burdensome election requirements for new political parties; no compelling state interest served by burden).

For cases involving race or ethnicity as suspect category, *see, e.g.,* University of Cal. Regents v.

Japanese would be "immediately suspect."[233] By its own directive, the Court should have subjected the exclusion order, if not also the order for internment, to the most rigorous review. In a more recent decision, it has held that "the phrase 'war power' cannot be invoked as a talismanic incantation to support any exercise of congressional power which can be brought within its ambit. '[E]ven the war power does not remove constitutional limitations safeguarding essential liberties.' "[234]

The Court in *Korematsu* recognized that "exclusion from the area in which one's home is located is a *far greater* deprivation than constant confinement to the home from 8 p.m. to 6 a.m. [the curfew measure approved by the court in *Hirabayashi*]."[235] Its conclusion was that "exclusion from a threatened area, *no less than curfew*, has a definite and close relationship to the prevention of espionage and sabotage."[236] The Court relied, as it did in *Hirabayashi*, on the military and congressional judgment that the order was necessary. Thus, the Court apparently rested the justification for the "far greater deprivation" on the rationale for the less intrusive curfew.[237] And that, essentially, was the extent of the "most rigid scrutiny" it undertook.

In this section, the *Korematsu* case is examined, using recent case law on official actions that involve racial classifications. Additionally, decisions revealing the weight given by the Court to military justifications for official actions are

---

Bakke, 438 U.S. 265 (1978); *McLaughlin*, 379 U.S. 184 (1964); Hernandez v. Texas, 347 U.S. 375 (1954); Oyama v. California, 332 U.S. 633 (1948).

[233] Whether a racial, national origin, or ethnic category is the subject of discrimination, the Supreme Court has treated all such classifications as "suspect" for equal protection purposes. *See, e.g., Bakke*, 438 U.S. at 290-91 ("Racial and ethnic classification . . . are subject to stringent examination without regard to [notion that a group must be a 'discrete and insular minority' requiring extraordinary protection from the majoritarian political process]."); *Hernandez*, 347 U.S. at 478 (1954) (reasoning that while race defines easily identifiable groups requiring special protection under the laws, "community prejudices are not static, and from time to time other differences from the community norm may define other groups [such as defendant's Mexican sector] which need the same protection"); *Oyama*, 332 U.S. at 647 (1948) (applying same standards of review as in racial discrimination cases to hold that "the rights of a citizen may not be subordinated merely because of his father's country of origin"); *Hirabayashi*, 320 U.S. at 100 (Court treating synonymously race and ancestry: "Distinctions between citizens solely because of their ancestry are by their nature odious to a free people whose institutions are founded upon the doctrine of equality. For that reason, legislative classifications or discrimination based on race alone has often been held to be a denial of equal protection."); Yick Wo v. Hopkins, 118 U.S. 356, 374 (1886) ("No reason for [ordinance requiring laundries to be operated in brick or stone buildings] is shown, and the conclusion cannot be resisted that no reason for it exists except hostility to the race and nationality [Chinese] to which the petitioners belong.").

[234] United States v. Robel, 389 U.S. 258, 263-64 (1967) (*quoting* Home Bldg. & Loan Ass'n v. Blaisdell, 290 U.S. 398, 426 (1934)).

[235] 323 U.S. at 218 (emphasis added).

[236] *Id.* (emphasis added).

[237] *Id.*

reviewed. The positions of the Court in these two areas are juxtaposed to determine whether the *Korematsu* situation might produce a different result today.

A. *Racial Discrimination*

An equal protection challenge involving a suspect classification may be brought against the federal government through the due process clause of the fifth amendment.[238] Generally, two approaches have emerged. First, if the racial classification is manifest on the face of the official measure, those seeking redress must show that no other purpose is served except antagonistic racial discrimination.[239] If a public interest is proffered by the government as justification for the classification, it must be a compelling one and the means employed must be narrowly tailored toward that end.[240] Secondly, if the measure is facially neutral, a demonstration that a "clear pattern unexplainable on grounds other than race emerges from the . . . action"[241] will serve to invalidate the official act. Otherwise, a showing of an intent to discriminate must be made, either by direct or circumstantial evidence. This may include, but is not necessarily satisfied by, facts or statistics supporting any racially disproportionate impact.[242]

---

[238] *See, e.g., Fullilove*, 448 U.S. at 480; Bolling v. Sharpe, 347 U.S. 497, 499 (1954).

[239] *See, e.g., McLaughlin*, 379 U.S. at 198 (Stewart, J., concurring) (reasoning that state statute conditioning imposition of criminal punishment on the color of an individual's skin is *per se* invidious discrimination); Strauder v. West Virginia, 100 U.S. 303, 308 (1880) (holding that statute denying blacks right to serve on juries "is practically a brand upon them, . . . an assertion of their inferiority, and a stimulant to that race prejudice which is an impediment to securing to individuals of the race that equal justice which the law aims to secure to all others.").

[240] *See, e.g.*, Swain v. Alabama, 380 U.S. 202 (1965) (prosecutor's use of peremptory challenges to remove all six black jurors not unconstitutional; essential nature of peremptory challenge system noted by Court). Cases upholding racially discriminatory action because of a compelling governmental interest are extremely rare.

> Since *Korematsu* the Court has had few occasions to consider the possibility . . . that racial discrimination against minorities might be constitutionally permissible in some circumstances because of some overriding justification.
>
> . . . .
>
> . . . *Swain v. Alabama* . . . seems to be the only case since *Korematsu* in which the Court has written an opinion upholding deliberate official racial discrimination directed against a minority group.

N. Dorsen, P. Bender, B. Neuborne, S. Law, 2 Emerson, Haber and Dorsen's Political and Civil Rights in the United States 67, 68 (4th ed. 1979).

[241] Village of Arlington Heights v. Metropolitan Hous. Dev. Corp., 429 U.S. 252, 266 (1977) (*citing* Gomillion v. Lightfoot, 364 U.S. 339 (1960); Lane v. Wilson, 307 U.S. 268 (1939); Guinn v. United States, 238 U.S. 347 (1915); Yick Wo v. Hopkins, 118 U.S. 356 (1886)).

[242] *See, e.g., Village of Arlington Heights*, 429 U.S. 252 (1977); Washington v. Davis, 426 U.S. 229 (1976).

In *Korematsu* the particular governmental measure that should have been the focus of judicial review is not clear. In previous sections, the argument was presented that the executive and military orders, as well as the authorizing legislation, should be viewed as a comprehensive and unified plan to remove and to intern those of Japanese ancestry.[243] If this argument were accepted, the equal protection analysis should follow the first approach delineated above since General DeWitt's orders, on their face, evidence a suspect racial classification. On the other hand, if each official action is considered separately, the general's orders may arguably be simply evidence of the discriminatory impact of a facially neutral enactment. In this case, the second approach would be the appropriate equal protection analysis. Under either analysis, the orders should be viewed as being unconstitutional.

1. *Manifest Racial Classification*

The authorizing legislation[244] and Executive Order No. 9066,[245] on their face, did not direct restrictions against any particular "suspect classification." General DeWitt's orders, though, contained language clearly specifying all persons of Japanese ancestry in a military area be evacuated and detained at assembly centers.[246] The Court had before it the officer's *Final Report* which narrated the steps leading up to the congressional enactment.[247] The suggested legislative awareness implied adoption of the general's plan by the act's passage.[248] If all the measures did comprise a comprehensive and unified plan, a reviewing court could find the entire package facially discriminatory, based on the provisions of the military orders.

A showing that a suspect category appeared on the face of the legislation may not be sufficient indication of "an invidious racial discrimination forbidden by the Constitution."[249] Many government decisions require recognition of a particular racial or ethnic group, but do not fail constitutionally. In *University of Cal. Regents v. Bakke*,[250] for example, the Court considered the admissions program of a California medical school which had allocated a fixed number of positions for racial minorities. Although the program was struck down, the Court recognized that "a substantial state interest" could validate racial classifi-

---

[243] *See supra* part II, §§ B and C.

[244] Pub. L. No. 503, ch. 191, 56 Stat. 173 (1942) (codified as amended at 18 U.S.C. § 1383 (1976) (repealed 1976)). *See supra* note 52 for text of statute.

[245] 7 Fed. Reg. 1407 (1942).

[246] *See supra* text accompanying notes 45-50.

[247] Final Report, *supra* note 23, at 25-31; *see also supra* notes 51-55 and accompanying text.

[248] *See supra* text accompanying notes 185-91.

[249] Washington v. Davis, 426 U.S. 229, 242 (1976).

[250] 438 U.S. 265 (1978).

cation.[251] Other decisions have upheld the use of racial criteria in a remedial context to correct prior *de jure* school segregation[252] and to ensure that minority businesses receive subcontracts comprising a fixed percent of federal funds granted to public works projects.[253] In these cases, the racial classification was not indicative of antagonism toward a minority.

Similarly, in *Korematsu*, could there have been a reason, other than racial prejudice, for singling out only those of Japanese ancestry for exclusion and internment?[254] The evacuation of designated military zones was prompted by the military judgment that persons associated by ethnic affinity to wartime enemies were security threats. At its earliest stages, the government's evacuation plan included German and Italian aliens, as well as all persons of Japanese ancestry.[255] Later, when by memorandum, the Secretary of War named General DeWitt head of the Western Defense Command, the Secretary expressed his desire that those of Italian descent should not be part of the exclusion scheme. He considered them less dangerous, and, at the same time, a burden on military resources because of their number.[256] No reason was given in the *Final Report* for the later omission of German persons from the exclusion program.

The German enemy armed forces produced far greater actual damage on the East Coast than the Japanese military did on the western shores. American ships were lost to German attacks along the Atlantic coast, as well as on the high seas.[257] "This devastating warfare often came alarmingly close to shore. Sinkings could be watched from Florida resorts and, on June 15, 1942, two American ships were torpedoed in full view of bathers and picknickers [sic] at Virginia Beach."[258] Yet no mass evacuation or detention of German aliens or German Americans occurred. Only those German individuals deemed dangerous by the Justice Department were interned.[259]

---

[251] *Id.* at 294-95.

[252] *See, e.g.*, Milliken v. Bradley, 433 U.S. 267 (1977) (remedial educational programs to restore discrimination victims to educational positions, they would otherwise have attained); Swann v. Charlotte-Mecklenburg Bd. of Educ., 402 U.S. 1 (1972) (racial criteria used in assigning students and teachers to remedy prior *de jure* school segregation).

[253] Fullilove v. Klutznick, 448 U.S. 448 (1980).

[254] In this section, the legislative basis for the classification would have been one that justified selecting only the ethnic Japanese, and not any other. The "compelling governmental interest" as justification for racial discrimination is discussed in section B.

[255] Memorandum from Assistant Secretary of War John J. McCloy to General DeWitt (Feb. 20, 1942), *reprinted in* Final Report, *supra* note 23, at 27-29.

[256] Letter from Secretary of War Henry L. Stimson to General DeWitt (Feb. 20, 1942), *reprinted in* Final Report, *supra* note 23, at 25-26.

[257] CWRIC REPORT, *supra* note 1, at 283; *see infra* note 288.

[258] *Id.*

[259] *Id.* at 284.

> Was there a coherent policy behind treating the German aliens and German Americans on the East Coast differently from the Japanese on the West Coast? If one accepted the Western Defense Command's view that ethnic groups remain loyal to their ancestral nation, and further argued that mass measures were necessary only against Japanese Americans either because the loyal could not be distinguished from the disloyal within Asian groups or because urgency did not permit individual review, one would expect a careful official review of all German Americans in order to detain the disloyal. The government made no such review . . . . The divergent treatment of ethnic Japanese and Germans does not make a logical pattern; one must look elsewhere to understand these events.[260]

In its report, the Commission on Wartime Relocation and Internment of Civilians suggests two factors for the less harsh treatment of the ethnic Germans. First, by sheer numbers, those of German descent would have posed a monumental logistics problem to exclude and, an even greater one, to intern.[261] In 1940 there were 1,237,000 of them living in the United States, "the largest foreign-born ethnic group except for the Italians."[262] Secondly, their numbers also endowed them with political influence, especially in those areas of the country where they represented substantial voting blocs.[263]

Added to these considerations were certain negative factors against the ethnic Japanese on the West Coast. Much pressure for internment was exerted upon cabinet members by California politicians, probably stemming from the atmosphere of pre-war antagonism against the Asians in that area.[264] In addition, the successful farming enterprises of the ethnic Japanese were considered threatening by other produce growers, who collectively became advocates for removal of their competitors.[265]

The obvious dangers to which the East Coast was exposed could hardly justify differential treatment between those of Japanese and German descent, assuming that the reasoning for detention of the ethnic Japanese was valid. Without more, the restrictions aimed solely at nonwhite enemy aliens and similar American citizens would be unconstitutional, in light of the Court's holding in *McLaughlin v. Florida*.[266] That decision involved a Florida statute which, in

---

[260] *Id.* at 288-89.

[261] *Id.* at 289.

[262] *Id.*

[263] *Id.*

[264] P. IRONS, *supra* note 22, at 38-40. For example, California Congressman Leland M. Ford called for placement of all ethnic Japanese in "inland concentration camps," rationalizing that a Japanese American could prove that " 'he is patriotic and is working for us' " by agreeing to be interned. *Id.* at 38 (*quoting* from M. GRODZINS, AMERICANS BETRAYED 65 (1947)). *See also* CWRIC REPORT, *supra* note 1, at 4, 67-86.

[265] P. IRONS, *supra* note 22, at 39-40; CWRIC REPORT, *supra* note 1, at 69.

[266] 379 U.S. 184 (1964).

effect, barred only interracial couples from living together if unmarried. Its purpose was "to prevent . . . breaches of basic concepts of sexual decency."[267] The Court held it to be unconstitutional because of its discriminatory restriction of only interracial couples who are sexually promiscuous.

> That a general evil will be partially corrected may at times, and without more, serve to justify the limited application of a criminal law; but legislative discretion to employ the piecemeal approach stops short of permitting a State to narrow statutory coverage to focus on a racial group.[268]

Enactments offering only partial solutions to problems have been approved by the Court in other cases not involving suspect classifications.[269] In *Korematsu*, those ethnically related to countries at war with the United States were considered military security risks. Yet the harshest measures were directed only against the nonwhite group. If the official action was intended to prevent sabotage and espionage, its discriminatory application solely upon the ethnic Japanese, who comprised only a part of the pool of potential enemy saboteurs, would be a constitutionally unacceptable partial solution to the national security problem.

2. *Facially Neutral Legislation*

If the legislation, the statute and the executive order in the *Korematsu* case were examined and found to be facially neutral, could a "clear pattern" of racial discrimination in its application be found that could not be explained? Otherwise, could direct and circumstantial evidence be presented to demonstrate violation of equal protection rights?

In the first instance, the Court would be looking for an obvious scheme of racial discrimination emerging from execution of the legislative directive. The unexplainable and patent character of the racially discriminatory application of authority would render the legislation unconstitutional. Two classic cases provide illustrations of this pattern—*Yick Wo v. Hopkins*[270] and *Gomillion v. Lightfoot*.[271] In the first case, an ordinance of the city of San Francisco forbade the conduct of laundry businesses in buildings other than those of brick or stone. This effectively barred the Chinese from such enterprises in the area,

---

[267] *Id.* at 193.

[268] *Id.* at 194.

[269] *See, e.g.*, Minnesota v. Clover Leaf Creamery Co., 449 U.S. 896 (1981) (step-by-step environmental protection measure valid); Dandridge v. Williams, 397 U.S. 471 (1970) (limited welfare grants regardless of family size valid under the equal protection clause as step-by-step program to combat poverty).

[270] 118 U.S. 356 (1886).

[271] 364 U.S. 339 (1960).

since nearly all of their laundries were operated in wooden structures. On the other hand, exceptions to the ordinance were granted to all ***non-Chinese*** petitioners. The Court held that "[n]o reason for [this pattern] is shown, and the conclusion cannot be resisted that no reason for it exists except hostility to the [Chinese] . . . which, in the eye of the law, is not justified."[272]

In *Gomillion*, an Alabama redistricting statute changed the shape of the city of Tuskegee "from a square to an uncouth twenty-eight-sided figure."[273] The result excluded almost all black voters from the city limits, while no white voters were affected. The Court viewed the gerrymandering as "tantamount for all practical purposes to a mathematical demonstration, that the legislation is solely concerned with segregating white and colored voters by fencing Negro citizens out of town."[274]

Was there a "clear pattern" of discriminatory application of restrictions levied to protect sensitive military areas in *Korematsu*? The purpose of the legislation, imputed from the executive order, was to guard against the dangers of espionage and sabotage in those military areas vulnerable to attack by enemy forces.[275] In *Hirabayashi*, the principles of which were relied upon in the *Korematsu* decision, the Court posited that "[t]he fact alone that attack on our shores was threatened by Japan rather than another enemy power set these citizens [of Japanese descent] apart from others who have no particular association with Japan."[276] That reasoning, in light of the purpose of the statute, and the greater incidence of German enemy activity along the East Coast,[277] would clearly call for restriction of aliens and citizens of German origin. If a danger to the nation's safety was perceived from those who had ethnic ties with enemy countries, no reason was advanced for limiting exclusion and internment to only the nonwhite group. The emerging picture of an unexplained and "clear pattern" of racial discrimination is accented by the congressional testimony given by the military officer who issued the West Coast orders: "It makes no difference whether he is an American citizen, he is still a Japanese. American citizenship does not necessarily determine loyalty . . . . But we must worry about the Japanese all the time until he is wiped off the map."[278]

If the Court is not persuaded that invidious discrimination was clearly demonstrated from the law's execution, direct and circumstantial evidence could

---

[272] 118 U.S. at 374.

[273] 364 U.S. at 340.

[274] *Id.* at 341.

[275] *See supra* text accompanying notes 36-41 and 197-209.

[276] 320 U.S. at 101.

[277] *See supra* text accompanying notes 257-59.

[278] *Hearing Before the House Naval Affairs Subcommittee to Investigate Congested Areas*, 78th Cong., 1st Sess. 739-40 (1943) (statement of General DeWitt), *quoted in Korematsu*, 323 U.S. at 236 n.2 (Murphy, J., dissenting).

be used to show a racial intent. Some of the criteria developed by the Court to prove intent are: (1) disproportionate impact on a minority;[279] (2) the historical background of the government decision, including any series of actions instituted for invidious purposes, the specific events leading up to the decision, or any departure from routine or traditional procedure;[280] and (3) lack of any neutral justification for the government's conduct.[281] *Washington v. Davis* articulates the approach to be employed: "[A]n invidious discriminatory purpose may often be inferred from the *totality of the relevant facts*, including the fact . . . that the law bears more heavily on one race than another."[282]

A disproportionate impact on one ethnic group, the Japanese, is readily discernible in this case. According to the *Washington* framework, however, this showing alone would not suffice to prove intent. "Disproportionate impact is not irrelevant, but it is not the sole touchstone of an invidious racial discrimination forbidden by the Constitution. Standing alone, it does not trigger the rule . . . that racial classifications are to be subjected to the strictest scrutiny and are justifiable only under the weightiest considerations."[283] However, in *Korematsu*, disproportionate impact was *not* the "sole touchstone." Procedural departures and the absence of neutral justifications for the official actions also provide support for the conclusion that the restrictive measures imposed upon the ethnic Japanese were fomented by invidious racial discrimination.

Turning, then, to the historical background of the government's decisions, does it reveal such irregularities as to implicate "invidious purpose"? The difficulty is in determining the routine or traditional lines along which national defense decisions should be made.[284] In *Korematsu* a procedural departure of sorts was the ratification of presidential action after the fact. The accepted sequence of events is the delegation of authority by Congress, *after* which the president acts under the conferred power. However, this kind of deviation has occurred in the past, with the subsequent approval of the Supreme Court.[285] The turbulent nature of war may require that legislative authorization be forthcoming only after field decisions have been made—as occurred in this case.

There was, however, another inconsistency in the course of the government's decision making process that the Court could have questioned. While the threat

---

[279] *See, e.g., Village of Arlington Heights*, 429 U.S. at 266; Washington v. Davis, 426 U.S. 229, 242 (1976).

[280] *See, e.g., Village of Arlington Heights*, 429 U.S. at 267.

[281] *See, e.g.*, Washington v. Davis, 426 U.S. at 242.

[282] *Id.* (emphasis added).

[283] *Id.* (citation omitted).

[284] This difficulty suggests one of the reasons for the judicial deference accorded federal action during wartime. Note that the examination here focuses on the procedural aspect of the decision, not the substance of it.

[285] *See supra* notes 185-91 and accompanying text.

of enemy attack was greater initially than at the time of Korematsu's arrest, the scope of defensive measures taken by the President and his subordinates was limited to enemy aliens—Japanese, Germans, and Italians alike—who had been identified as potentially dangerous by intelligence agencies before the Pearl Harbor bombing.[286] Yet as the probability of attack upon the West Coast by Japan diminished[287] and American casualties inflicted by the Germans rose in the Atlantic,[288] restrictions upon the Japanese element widened to include all aliens and citizens as well. Concomitantly, detention of Italians ceased and actions against the German sector remained selective and limited.[289] This disparity was never explained in terms of military necessity.[290]

Another factor in the determination of discriminatory intent is the absence of any neutral justification for the official action. General DeWitt highlighted the military activities affecting the West Coast area in his *Final Report*: clashes with Japan's armed forces in the Pacific theatre and reports of enemy attacks on coastal points, with evidence, according to the general, of shore-to-sea communication of information to the enemy.[291] These, perhaps, could be characterized as "neutral justifications" for the exclusion. But they lose their patina of legitimacy when the East Coast situation failed to generate similar defense concerns.

Once removal of potential saboteurs from the threatened area had been effected, however, these military conditions did not justify indefinite detention or internment. By Executive Order No. 9066, the Secretary of War was empowered "to provide for residents of any such [military] area who are excluded

---

[286] *See supra* notes 24-29 and accompanying text.

[287] *But cf.* memorandum from General DeWitt to Secretary of War Stimson, *supra* note 34, *reprinted in* Final Report, *supra* note 23, at 34 ("The very fact that no sabotage has taken place to date [Feb. 14, 1942] is a disturbing and confirming indication that such action will be taken.").

[288] Before the issuance of the exclusion order in March 1942, 93 American vessels were sunk in the Atlantic, a loss of nearly 200,000 tons. In March, 28 ships, totaling over 150,000 tons, were lost along the East Coast. Additionally, 15 more, comprising over 90,000 tons, went down in the Gulf of Mexico and the Caribbean Sea. CWRIC REPORT, *supra* note 1, at 283.

[289] *See supra* text accompanying notes 257-63.

[290] No basis for the noninclusion of the ethnic Germans in the exclusion and detention orders is given by General DeWitt in his Final Report, *supra* note 23. *See also Korematsu*, 323 U.S. at 240 (Murphy, J., dissenting) (pointing out the disparity).

[291] Final Report, *supra* note 23, at 18-19. *But cf.* CWRIC REPORT, *supra* note 1, at 7 ("But the reports of shore-to-ship signaling were investigated by the Federal Communications Commission, the agency with relevant expertise, and no identifiable cases of such signaling were substantiated."). There was indication of General DeWitt's awareness that reports of the alleged radio signalings were false. P. IRONS, *supra* note 22, at 285. In fact, two deputies on the United States Attorney General's staff, John Burling and John Ennis, had inserted a footnote in an early version of the government's *Korematsu* brief, informing the Court that military necessity could not rest on such tenuous grounds. *Id.* at 286. Before the brief was finalized, the language of that crucial footnote was watered down, removing the clear message to the Court that General DeWitt's judgment was not premised on substantiated fact. *Id.* at 292.

therefrom, such transportation, food, shelter, and other accommodations as may be necessary, in the judgment of the Secretary . . . or the [designated] Military Commander."[292] Initially, it was envisioned that evacuation of Japanese American citizens was to be followed by "voluntary internment," with those persons declining such custody to "be excluded from all military areas, and left to their own resources."[293] The plan to provide "shelter" for evacuees from which they could voluntarily leave was displaced by required detention at assembly centers and eventual internment at relocation camps.

Although the involuntary custody of evacuees was accomplished under the delegated authority of both the executive order and legislation, General DeWitt conceded that it was not done to protect against espionage and sabotage. Rather, as he stated in his *Final Report*:

> Essentially, military necessity required only that the Japanese population be removed from the coastal area and dispersed in the interior, where the danger of action in concert during any attempted enemy raids along the coast, or in advance thereof as preparation for a full scale attack, would be eliminated. That the evacuation program necessarily and ultimately developed into one of complete Federal supervision, was due primarily to the fact that the interior states would not accept an uncontrolled Japanese migration.[294]

Confinement of an ethnic minority cannot pass constitutional tests when employed simply to avoid the adverse reactions of communities that would have had to accept evacuees. If a benevolent motive was present, such as protection of the evacuees by placement in internment camps, it produced a result contrary to the dictates of our criminal justice system. That institution envisions that the offender is incarcerated, not the victim. It is ludicrous to suggest that it is legitimate to segregate and to imprison a minority group in anticipation of probable racial clashes. The Court could hardly return now to the doctrine of "separate but equal."[295] Moreover, the conditions awaiting the evacuees at the camps could *not* have been characterized as "equal" to those not under restraint.[296]

---

[292] Exec. Order No. 9066, 7 Fed. Reg. 1407 (1942).

[293] Memorandum from General DeWitt to Secretary of War Stimson, *supra* note 34, *reprinted in* Final Report, *supra* note 23, at 36-37. This document served as the basis for Exec. Order No. 9066, which was issued on Feb. 19, 1942.

[294] Final Report, *supra* note 23, at 43-44. *See also Ex parte* Endo, 323 U.S. at 301-04 (reasoning that community hostility against the ethnic Japanese evacuees could not be used to justify detention, which was prompted by the need to protect against espionage and sabotage).

[295] Plessy v. Ferguson, 163 U.S. 537 (1896).

[296] *See* CWRIC REPORT, *supra* note 1, at 158-84 (describing the harsh conditions under which the internees lived). Although my personal memories of camp life are vague, the stories related to me by my mother, Tamae Hoshida, conjured images of hard times. One, especially, has

In total, the evidence supports an inference of discriminatory intent. Only one minority group was decidedly burdened by the government's actions. The decision making process involved some questionable steps, although the wartime atmosphere may have affected the procedure. The "neutral justifications" fail when the entire program instituted against persons of Japanese ancestry is taken into account. The Court would have a strong basis for finding an intent to discriminate at this point, and for shifting the evidentiary burden to the government.

B. *Military Necessity*

Confronted with an assertion that its actions were racially discriminatory, the government has at least two avenues of defense: (1) the rebuttal of the racial discrimination contention, or (2) a showing that a compelling governmental interest was at stake and that it had been advanced by a narrowly tailored, least intrusive alternative. In *Korematsu*, the Court relied on military justification to address both approaches.

1. *Rebuttal to Racial Discrimination Contention*

The racial discrimination contention was rebutted in *Korematsu* by the same "military imperative" created by the impossibility of immediate segregation of the disloyal that justified the curfew order in *Hirabayashi*.[297] According to the later decision, the refusal by approximately 5,000 Japanese American internees to swear allegiance to the United States and to renounce any ties to the Japanese emperor was further confirmation of disloyalty.[298] The Court in *Hirabayashi* acknowledged that "[d]istinctions between citizens solely because of their ancestry are by their very nature odious to a free people whose institutions are founded upon the doctrine of equality. For that reason, legislative classification or discrimination based on race alone has often been held to be a denial of equal protection."[299] But, the Court countered, "in dealing with the perils of war, Congress and the Executive are [not] wholly precluded from taking into

---

stayed with me—a tiny woman, not five feet tall, trudging along a highway with bucket in hand, searching for bits of coal that might have fallen from passing trucks to use to warm her family.

[297] *Korematsu*, 323 U.S. at 218-19.

[298] *Id.* at 219. *But cf.* CWRIC REPORT, *supra* note 1, at 191-97 (describing the confusion with which the internees confronted the loyalty questionnaire presented to them—whether asserting loyalty to the United States and renouncing Japanese nationality left Japanese aliens "without a country" or would be interpreted as a pledge by the Japanese Americans to volunteer for the military; offering testimony from those who answered in the negative as a protest against their summary internment).

[299] 320 U.S. at 100.

account those facts and circumstances which are relevant to measures for our national defense and for the successful prosecution of the war, and which may in fact *place citizens of one ancestry* in a *different category from others*."[300] The Court's conclusion was that these "facts and circumstances" indicated "that a group of one national extraction may menace [public] safety [in time of war and threatened invasion] more than others."[301]

The "facts and circumstances" the Court delineated were essentially sociological in nature and based on questionable authority.[302] Most of the "data" were derived from a congressional hearing and a House of Representatives report.[303] Yet the Court approved of their use by the political branches of government in wartime decisionmaking. "These are only some of the many considerations which those charged with responsibility for the national defense could take into account in determining the nature and extent of the danger of espionage and sabotage, in the event of invasion or air raid attack."[304]

These kinds of data were indeed taken into account by General DeWitt, and probably formed the primary basis for his military orders.[305] Aside from his concern about the contraband seized and the supposed shore-to-ship signaling by saboteurs,[306] the commander's conclusion that all persons of Japanese ancestry posed threats of espionage and sabotage was based on his collection of sociological information on the Japanese in America. His data, also of questionable validity and reliability, were essentially the same as that reviewed by the Supreme Court in *Hirabayashi*.[307] His reasoning followed this tack:

1. The Japanese alien in the United States maintains close allegiance to his mother country and has failed to assimilate into the American mainstream because of nonacceptance by other Americans and because of strong bonds of tradition and custom. The group conducts pro-Japanese activities through Japanese organizations, religious practices, and native language newspapers.[308]
2. Because of nonassimilation, the Japanese born in America identify with the aliens among them and their traditions. Indoctrination is accom-

---

[300] *Id.* (emphasis added).

[301] *Id.* at 101.

[302] *Id.* at 96-98. *See also supra* note 61.

[303] *Id.* at 97 nn.5 and 6, 98 n.9.

[304] *Id.* at 99.

[305] Most of chapter II, "Need for Military Control and for Evacuation," in General DeWitt's Final Report deals extensively with his version of a sociological profile of the ethnic Japanese. No authoritative citations are offered. Final Report, *supra* note 23, at 7-19. *See also supra* note 291.

[306] Final Report, *supra* note 23, at 8. *But cf.* CWRIC REPORT, *supra* note 1, at 62, 88 (Justice Department reports minimizing significance of material gathered in contraband round-up); *supra* note 291.

[307] *Compare Hirabayashi*, 320 U.S. at 96-98 *with* Final Report, *supra* note 23, at 7-19.

[308] Final Report, *supra* note 23, at 9-12.

plished through Japanese language schools and by the practice of sending American-born children to Japan for their education.[309]

3. The Japanese have located themselves along strategic military points on the West Coast.[310]
4. There are some disloyal persons of Japanese ancestry on the West Coast who pose a threat of espionage and sabotage.[311]
5. Because of their strong cultural identification, loyal Japanese Americans cannot be distinguished from the disloyal.[312]

From this logic, the general concluded that all persons of Japanese ancestry must be excluded from specified military areas and detained at assembly centers.

If the sociological features of a group ethnically related to an enemy nation were of military significance, why were similar analyses not performed of the Italian and German population? Using General DeWitt's criteria of disloyalty, Morton Grodzins[313] developed a profile of the Italians in the United States. His result, astonishing in its similarity to the portrayal of the Japanese, was presented in his 1949 book *Americans Betrayed*:

> Because of their concentration in the fishing industry, Italians if anything were located in more strategic coastal locations than the Japanese. This was especially true of the San Francisco Bay area and adjoining counties.
>
> The Italians had their full quota of language schools and their own churches. They and their children made numerous trips to their home country. The Italian consuls were active and important members of the community, and Fascist propaganda was reflected in a vernacular press which supported Mussolini's domestic and foreign policies. If naturalization were any indication of acculturation, then the single fact that more than half the foreign-born Italians had not become citizens of the United States demonstrated a low degree of Americanization. Educational achievement rates of children of Italian ancestry were lower, and their delinquency rates were higher, in comparison with those of Japanese ancestry. Italians in California had contributed funds to the Italian relief agencies following the conquests of Ethiopia and Albania.[314]

---

[309] *Id.* at 12-14.

[310] *Id.* at 9.

[311] *Id.* at 18.

[312] *Id.* at 9.

[313] Morton Grodzins is professor of political science at the University of Chicago, former dean of the Division of the Social Sciences, and a former member of the Evacuation and Resettlement Study of the University of California. He has participated in the Pugwash Conferences on Science and World Affairs and is the co-author of GOVERNMENT AND HOUSING IN METROPOLITAN AREAS and the author of THE LOYAL AND THE DISLOYAL and AMERICANS BETRAYED.

[314] M. GRODZINS, AMERICANS BETRAYED 172-73 (1949), *quoted in* CWRIC REPORT, *supra* note 1, at 90.

Nearly identical phenomena generated antipathy toward persons of German descent in America during World War I.[315] Although by 1942 the German sector had undergone a deliberate and almost defensive assimilation, remnants of cultural ties to the country of origin were still present in 1942.[316] If sociological data are indeed valid indices of disloyalty, the attacks by German enemy forces along the East Coast certainly warranted a study of ethnic Germans in the United States.

The "proof" of disloyalty among the Japanese group thus loses its force as justification for the military orders. At the same time, the suspicion that a racially motivated reason for the exclusion and internment is reinforced.

2. *Compelling Government Interest: Military Necessity*

If the government is unable to dispel the cast of racial discrimination, can its actions be justified by a "compelling governmental interest"? Further, were the means taken to advance that interest narrowly tailored to their purpose and the least intrusive alternative at hand?[317]

The Court in *Korematsu* upheld the constitutionality of the restrictive actions taken against a single ethnic group because there existed "[n]othing short of apprehension by the proper military authorities of the gravest danger to the public safety."[318] The Court accepted the judgment of the military and Congress that, because prompt identification of the disloyal among the ethnic Japanese was not possible, the exclusion of the entire group from vital areas was a military necessity. In *Hirabayashi*, the curfew case, the Court detailed the military conditions confronting the President and Congress in early 1942. The Pearl Harbor attack and the many incursions by Japanese enemy forces in the Pacific theatre created a critical situation along the West Coast.[319] Thus, the military judgment was that the presence of the disloyal, sympathetic to the nation's enemies, presented a threat of espionage and sabotage to defense installations in the area.[320] And the Court in *Korematsu* concluded that "exclusion . . ., no less than curfew, has a definite and close relationship"[321] to the prevention of fruition of that twin threat.

The military assessment of disloyalty, and the willingness to attribute to the entire ethnic group the subversiveness of a few of its members, raises at least

---

315 CWRIC REPORT, *supra* note 1, at 289-92.
316 *Id.* at 291-92.
317 *See supra* note 240 and accompanying text.
318 323 U.S. at 218.
319 *Hirabayashi*, 320 U.S. at 93-96.
320 *Id.* at 95-96.
321 323 U.S. at 218.

two points that are troubling: (1) that a military officer's sociological analysis could be accepted as the basis for such a military judgment, and (2) that, in accord with that judgment, an entire ethnic group may be labelled as "disloyal," foregoing a less restrictive and narrower means to remove potential subversives from critical areas.

First, the wholesale exclusion measures ultimately rested on the acceptance of General DeWitt's conclusion that all Japanese, whether American citizen or alien, could be disloyal.[322] His judgment was not military in foundation. A civilian sharing his attitude could have made the same conclusion, but his at least would not have been clothed with the color of military authority. The inclusion of all persons of Japanese ancestry, aliens, those with dual citizenship, and Americans alike—within the group to be restricted was overinclusive and not justified by the officer's reasoning. He viewed the loyalty of both aliens and citizens as questionable only because of their ethnic similarity.[323]

The general's "guilt by association" line of argument could meet judicial disapproval today in light of *United States v. Robel*.[324] In that case, Robel's membership in the Communist Party was seen as a possible threat to the national security. As a result, under authority of a statute, the federal government barred him from working in defense facilities. The Court held such governmental prohibitions unacceptable when appellee's guilt is established "by association alone, without any need to establish that an individual's association poses a threat feared by the Government in proscribing it."[325] Korematsu's loyalty was not in dispute.[326] If alienage is to be equated with disloyalty, Korematsu's "guilt" as a possible saboteur or spy was established only by his common ancestral link to Japanese aliens. Since Robel's membership in the Communist party was presumably a voluntary act, and one that he could rescind, the reasoning in *Robel* is even more persuasive than in Korematsu's case.

> [G]uilt is personal and not inheritable. . . . But here is an attempt to make an otherwise innocent act, being present in the state whereof he is a citizen, near the place where he was born, and where all his life he has lived, a crime merely because this prisoner is the son of parents as to whom he had no choice, and belongs to a race from which there is no way to resign.[327]

However, the significance of *Robel* may have been eroded to some extent by

---

[322] *See supra* notes 291 and 306.
[323] *See supra* notes 305-12 and accompanying text.
[324] 389 U.S. 258 (1967).
[325] *Id.* at 265 (footnote omitted).
[326] *Korematsu*, 323 U.S. at 216.
[327] *Id.* at 243 (Jackson, J., dissenting).

Haig v. Agee.[328] That case appeared to have expanded the discretionary powers of the executive branch to restrict the rights of an individual when national security or foreign policy is involved, even without clear and specific delegation of authority by Congress. In that case the Court upheld, as constitutionally delegated legislative power, the Secretary of State's discretionary revocation of an American citizen's passport where "there is a substantial likelihood of 'serious damage' to national security or foreign policy as a result of a passport holder's activities in foreign countries."[329] The Court validated the delegation, despite the absence of specific statutory authorization for revocation. The congressional approval was inferred from Congress' silence over or at least acquiescence to a long standing administrative assertion of the power to revoke passports.[330] Yet the finding of a "long standing" practice relied on no more than a handful of revocations[331] over the life of the Passport Act of 1926.[332] The Court defended the paucity of the evidence by asserting that "[t]he exercise of a power emerges only in relation to a factual situation, and the continued validity of the power is not diluted simply because there is no need to use it."[333] As to the charge by the defendant of discriminatory enforcement, the Court dismissed this claim in a footnote, according to the government entitlement "to concentrate its scarce legal resources on cases involving the most serious damage to national security and foreign policy."[334]

The claim by Agee that his first amendment rights of free speech[335] had been abridged had "no foundation" according to the Court.[336] The reasoning by the Court was that the revocation of Agee's passport " 'is an inhibition of *action*' [his campaign to expose intelligence operations abroad] rather than of speech. Agee is as free to criticize the United States Government as he was when he held a passport."[337] However, in light of the wide latitude shown by

---

[328] 453 U.S. 280 (1981). The defendant Agee, a former Central Intelligence Agency (CIA) employee, had openly criticized the agency and had undertaken a campaign to expose CIA agents abroad. He began these activities in 1974, but was not notified of his passport revocation until a month after the November 4, 1979 seizure of the American embassy in Iran. *Id.* at 286 and n.8. But the holding in *Agee* approving of the Secretary's discretionary revocation of passports does not rely on the declaration of war or national emergency. *Id.* at 288-89 n.14.

[329] *Id.* at 309.

[330] *Id.* at 291-301.

[331] *Id.* at 301-03 (majority opinion); *id.* at 317 and n.7, 318 (Brennan, J., dissenting).

[332] 22 U.S.C. § 211a (1976 & Supp. III, 1979).

[333] *Agee*, 453 U.S. at 302.

[334] *Id.* at 309 n.61.

[335] U.S. CONST. amend. I.

[336] *Agee*, 453 U.S. at 308-09.

[337] *Id.* at 309 (*quoting* Zemel v. Rusk, 381 U.S. 1, 16-17 (1965) (emphasis added by the Court in *Agee*)). The Court in *Agee* also pointed out that the defendant had violated an express term of his employment contract not to disclose CIA matters without prior clearance, but this fact

the Court in its deference both to the executive branch's determination of danger to the national security, without explicit congressional authorization, and to that branch's discretion in allocating its "scarce legal resources," it is conceivable that mere criticism, without more, could be the basis for passport revocation if, in the Secretary of State's estimation, a "substantial likelihood" of damage to the national security exists.

The Court in *Korematsu* and *Hirabayashi* accepted the link of ethnicity to a high probability of subversive behavior. Could *Agee* permit that to happen again? The reach of that decision could be limited to passport revocation cases and to government actions not involving racial overtones. However persuasively that limitation is argued, the Court's own reaction in a wartime or crisis milieu would seem to be the important gauge. In *Agee*, the Court was approving of the Chief Executive's broad exercise of discretion without reliance on a declaration of national emergency.[338] Would the judicial deference to presidential discretion be heightened and broadened under more stressful conditions?[339]

A strong possibility exists that judicial weight will be placed on questionable sociological arguments presented under the guise of military expertise, given the pressures and uncertainties of wartime. The second equally troubling aspect of reliance on the military judgment in *Korematsu* was the government's failure to use less restrictive means of removing potential spies and saboteurs from the military areas. According to General DeWitt, the bypassing of any loyalty

---

did not appear to control the Court's position on Agee's first amendment claim. *Id.* at 284 and nn.4-5. *But cf. id.* at 320 n.10 (Brennan, J., dissenting) (analogizing the majority's assertion that Agee is free to criticize the government despite passport revocation, to the freedom of a prisoner to do the same while serving a 40-year sentence for criticizing the government's food stamp policy).

[338] *See supra* note 328.

[339] How, for example, would the Court decide if confronted with Narenji v. Civiletti, 617 F.2d 745 (D.C. Cir. 1979). In that case, decided during the Iranian crisis, the Circuit Court of the District of Columbia held that the Immigration and Nationality Act (INA), 8 U.S.C. § 1103(a), grants the Attorney General broad enough authority to issue regulations which draw distinctions on the basis of nationality "[s]o long as the distinctions are not wholly irrational." *Narenji*, 617 F.2d at 747. The regulation in question, 8 C.F.R. § 214.5 (1979), provided that all nonimmigrant alien post-secondary school students who are Iranian natives or citizens must meet certain reporting requirements with a local Immigration and Naturalization Service Office. Failure to comply subjected the foreign student to deportation proceedings. The Circuit Court found a reasonable relationship between the Attorney General's action and his duties under the INA. *Narenji*, 617 F.2d at 747. The court had been informed that the regulation was issued as a "fundamental element of the President's effort to resolve the Iranian crisis, the seizure of the American embassy in Iran, and to maintain the safety of the American hostages in Tehran.' " *Id.* (*quoting* an affidavit from the Attorney General). *See also* Note, *Constitutional Limits on the Power to Exclude Aliens*, 82 COLUM. L. REV. 957 (1982) *and* Note, *Aliens—Constitutionality of Discrimination Based on National Origin*—Narenji v. Civiletti, 21 HARV. INT'L L.J. 467 (1980).

screening was necessitated by the exigency of time.[340] Yet the record before the Court showed that the government had conducted itself with "[l]eisure and deliberation," taking nearly eleven months to complete evacuation of the designated areas.[341]

Moreover, the situation in Hawaii and the limited measures taken by the Attorney General immediately after the bombing of Pearl Harbor (before the War Department took over) suggest that other less intrusive alternatives to the DeWitt scheme were available. Following the Attorney General's lead, only enemy aliens could have been summarily removed, with suspected citizens undergoing scrutiny by a panel, as in Hawaii and in England.[342] In the alternative, once all were removed from the area, disloyal evacuees could be weeded out; indefinite custody was not necessary.[343]

To show that the orders were necessary, the commander's sociological appraisal of the Japanese would not be sufficient. Similar ethnic factors present in Hawaii did not trigger large-scale evacuation, detention and internment of the Japanese Americans there. From a military standpoint, it would appear that Hawaii was in a much more strategic position. It was the site of the original attack by the Japanese military. It was in closer vicinity to the Pacific theatre of war. Martial law had been declared. The United States armed forces were stationed in Hawaii, making it an inviting target to the enemy. Despite the Islands' vulnerability, of the 158,000 persons of Japanese ancestry living there at that time, less than 2,000 were interned.[344]

No explanation for the disparate treatment between the ethnic Japanese in Hawaii and those on the West Coast was offered in *Korematsu*. The Commission on Wartime Relocation and Internment of Civilians concluded that four factors contributed to the difference, none of which was of much military significance: (1) the greater ethnic mixture and cultural harmony present in Hawaii; (2) the larger number of ethnic Japanese in Hawaii; (3) the imposition of mar-

---

[340] Final Report, *supra* note 23, at 18-19.

[341] It is asserted merely that the loyalties of this group "were unknown and time was of the essence." Yet nearly four months elapsed after Pearl Harbor before the first exclusion order was issued; nearly eight months went by until the last order was issued; and the last of these "subversive" persons was not actually removed until almost eleven months had elapsed. Leisure and deliberation seem to have been more of the essence than speed. And the fact that conditions were not such as to warrant a declaration of martial law adds strength to the belief that the factors of time and military necessity were not as urgent as they have been represented to be.

Korematsu, 323 U.S. at 241 (Murphy, J., dissenting).

[342] *See supra* notes 25-29 and accompanying text. *See also Korematsu*, 323 U.S. at 241-42 and n.16 (Murphy, J., dissenting) (proposing loyalty hearings and describing those conducted by the British government).

[343] *See supra* text accompanying notes 292-93.

[344] CWRIC Report, *supra* note 1, at 261.

tial law in Hawaii, giving the military greater responsiveness to deal with emergencies; and (4) the difference in racial outlook of the commanders of each area.[345]

The presence of ethnic Japanese in larger numbers in Hawaii was viewed as an economic plus, for they were represented throughout the governmental work force and among the various professions.[346] In contrast, many among the West Coast Japanese population were successful farmers and posed competitive threats to other agricultural enterprises.[347] The economic difference, the greater degree of racial tolerance in Hawaii and the lack of congressional representation from the Territory produced no major movement to exclude the ethnic Japanese from the Islands, unlike the entreaties by California politicians and agricultural organizations.[348]

General Delos Emmons, United States Army, was General DeWitt's equivalent in Hawaii. Just as the social and political environments within the two commands differed, the attitudes of the two officers toward the ethnic Japanese contrasted. General Emmons chose to see the individual Asian as loyal to the United States unless shown otherwise by supporting evidence.[349] Where pressures for mass evacuation were exerted, he defused them with the practical arguments of economic disruption and logistical problems of transportation.[350] His own staff rejected the Western Command's assessment of the difficulty of loyalty screening.[351] Later, in September 1943, when General Emmons replaced DeWitt as head of the Western Command, he sought to reduce the size of the designated military area and to end exclusion of persons who were not shown to be disloyal or dangerous.[352]

The difference in treatment between the ethnic Japanese and those of German and Italian descents is also indicative of the unnecessarily broad sweep of General DeWitt's orders. At the start of the war, only enemy aliens were under suspicion—whether white or nonwhite.[353] Later, the West Coast Japanese group was the only one in which American citizens, as well as aliens, were considered potentially subversive. Throughout World War II, only those among the German and Italian groups deemed to be potentially dangerous were interned.[354]

---

[345] *Id.* at 261-62.

[346] *Id.* at 263.

[347] *See supra* note 1 and note 265 and accompanying text.

[348] CWRIC REPORT, *supra* note 1, at 67-72, 261.

[349] *Id.* at 262.

[350] *Id.*

[351] *Id.* and n.3.

[352] *Id.* at 262.

[353] *See supra* text accompanying notes 24-29.

[354] CWRIC REPORT, *supra* note 1, at 284-85. *See also supra* notes 257-65 and accompanying

Perhaps, if the Court in *Korematsu* had heeded its own admonition that racially directed government restrictions require "the most rigid scrutiny," the nonmilitary character of the justification for exclusion and internment would have become apparent. Justice Murphy apparently scrutinized General DeWitt's report, and found it unacceptable. Although Justices Roberts and Jackson also voiced their dissent, the majority of the Court was persuaded to defer to the military judgment that the presence of the ethnic Japanese along the West Coast was a threat to the national security. It was wartime, our shores were threatened, and there was a need to wage war, and to wage it successfully, no matter what the cost to the constitutional freedoms of one small group of citizens.

IV. JUSTICIABILITY

In a dissenting opinion in *Korematsu*, Justice Jackson expressed doubts about the Court's intervention in matters involving military judgment.

> In the very nature of things, ***military decisions are not susceptible of intelligent judicial appraisal***. They do not pretend to rest on evidence, but are made on information that often would not be admissible and on assumptions that could not be proved. Information in support of an order could not be disclosed to courts without danger that it would reach the enemy. Neither can courts act on communications made in confidence. ***Hence courts can never have any real alternative to accepting the mere declaration of the authority that issued the order that it was reasonably necessary from a military viewpoint.***[355]

Despite his hesitancy about challenging the validity of military declarations, Justice Jackson feared to a greater degree the lasting and durable nature of judicial opinions that sanction unconstitutional military orders.[356] He saw the majority opinion in *Korematsu* as an example of the need for judicial caution.[357] Here, the Court relied on the principles expounded in *Hirabayashi* to validate exclusion and detention. But, according to Justice Jackson's reading of *Hirabayashi*, that decision was intended to be a limited one, a validation of "discrimination on the basis of ancestry for mild and temporary deprivation of liberty [the curfew]."[358] In *Korematsu*, he saw the legitimizing effect of *Hirabayashi* being expanded to include more restrictive military measures and pondered "[h]ow far the principle of this case would be extended before plausible reasons

text.

[355] 323 U.S. at 245 (Jackson, J., dissenting) (emphasis added).

[356] *Id.* at 246.

[357] *Id.* at 246-47.

[358] *Id.* at 247.

would play out."[359]

Justice Jackson's dissent embodies the tension that flows between the legitimate exercise of the war power and the protection of individual rights. To some extent, other decisions of the Court reflect that tension by their often polar recitations. For example, in *Haig v. Agee*, the Court acknowledged the "volatile nature of problems confronting the Executive in foreign policy and national defense."[360] It further characterized these areas as "rarely proper subjects for judicial intervention."[361] On the other hand, the Court in *United States v. Robel*[362] declared that "the phrase 'war power' cannot be invoked as a talismanic incantation to support any exercise of congressional power which can be brought within its ambit. '[E]ven the war power does not remove constitutional limitations safeguarding essential liberties.' "[363]

Where the judiciary has accorded great deference in military and national security affairs to the political branches of government, it has recognized the broad constitutional powers each possesses,[364] and its own lack of competence, in those areas.[365] "Such issues as initiating, conducting, and terminating war and hostilities require policy decisions that are political rather than legal in character."[366] This political character, or question, raises the issue of justiciability in cases involving the exercise of the war power. Writing for the majority in *Baker v. Carr*,[367] Justice Brennan reviewed several "political question" decisions and concluded that "it is the relationship between the judiciary and the coordinate branches of the Federal Government which gives rise to the 'political question.' . . . The nonjusticiability of [such a] question is primarily a function of the separation of powers."[368]

In his survey of cases, Justice Brennan sought to determine "the contours of

---

[359] *Id.*

[360] 453 U.S. 280, 291 (1981) (*citing* United States v. Curtiss-Wright Export Corp., 299 U.S. 304 (1936)). For other cases reflecting judicial deference to military matters, *see, e.g.*, Brown v. Glines, 444 U.S. 348 (1980) (excluding speech likely to interfere with military effectiveness from defense installation); Middendorf v. Henry, 425 U.S. 25 (1976) (need for counsel in summary court martial overriden by military demand for discipline); Greer v. Spock, 424 U.S. 828 (1976) (no generalized constitutional right to make political speeches and distribute leaflets on military base).

[361] 453 U.S. at 292.

[362] 389 U.S. 258 (1967).

[363] *Id.* at 263-64 (*quoting* Home Bldg. & Loan Ass'n v. Blaisdell, 290 U.S. 398, 426 (1934)). For other cases, *see, e.g.*, Duncan v. Kahanamoku, 327 U.S. 304 (1946); *Ex parte* Endo, 323 U.S. 283 (1944).

[364] *See supra* note 360.

[365] *See infra* notes 373-76 and accompanying text.

[366] Keynes, *supra* note 123, at 69.

[367] 369 U.S. 186 (1962).

[368] *Id.* at 210.

the 'political question' doctine."[369] He abstracted several political question formulations:

1. [A] textually demonstrable constitutional commitment of the issue to a coordinate political department; or
2. [A] lack of judicially discoverable and manageable standards for resolving it; or
3. [T]he impossibility of deciding without an initial policy determination of a kind clearly for nonjudicial discretion; or
4. [T]he impossibility of a Court's undertaking independent resolution without expressing lack of the respect due coordinate branches of government; or
5. [A]n unusual need for unquestioning adherence to a political decision already made; or
6. [T]he potentiality of embarrassment from multifarious pronouncements by various departments on one question.[370]

Justice Brennan concluded that one of these formulations must be inherent in the issue presented before the Court would dismiss it for nonjusticiability.

Determining justiciability by applying the formulations requires certain precautions. Particularly concerning the war power, Justice Brennan pointed out that although the Court would defer to political decisions concerning the duration of hostilities:

> [D]eference rests on reason, not habit. The question in a particular case may not seriously implicate considerations of finality [as in the determination of when or whether a war has ended]. Further, clearly definable criteria for decision may be available. In such a case the political question barrier falls away: "[A] Court is not at liberty to shut its eyes to an obvious mistake, when the validity of the law depends upon the truth of what is declared. . . . [It can] inquire whether the exigency still existed upon which the continued operation of the law depended."[371]

Does *Korematsu* present a political question that precludes judicial review? The *Baker* criterion that may raise the issue is the possible "lack of judicially discoverable and manageable standards" for resolving the controversies in the case.[372] In *Korematsu*, governmental restrictions were directed against members

---

[369] *Id.*

[370] *Id.* at 217.

[371] *Id.* at 213-14 (*quoting* Chastleton Corp. v. Sinclair, 264 U.S. 543, 547-48 (1923)).

[372] The other conditions may be present in lesser degrees in *Korematsu*, but would not be "inextricable from the case at bar." *Baker*, 369 U.S. at 217. As discussed earlier, *supra* text accompanying notes 93-98, the Constitution does not explicitly circumscribe the scope of the war power to be exercised by the "coordinate political departments." In addition, although the issues in *Korematsu* may require the Court to confront the validity of certain military judgments, no

of one ethnic group. The need for such actions rested on the military judgment that vulnerable national defense areas required protection against the threat of espionage and sabotage posed by disloyal persons of Japanese ancestry. Removal of all who were of that ethnic background, whether loyal or disloyal, was deemed a military necessity. The exigencies of time and the fear of invasion, according to military assessment, did not permit identification and removal of only the disloyal.

The Court may choose not to substitute its judgment for that of the military in weighing the need for evacuation and detention, because it lacks the knowledge to develop "judicially discoverable and manageable standards." The shortfall in knowledge may result from either the Court's own limited expertise in military matters or from the relative inaccessibility of information necessary for the setting of appropriate evaluation criteria.

The Court often rests its lack of competence on the nature of the military itself. In *Parker v. Levy*,[373] a claim of infringement of the right of free speech was rebuffed with the reasoning that "[w]hile the members of the military are not excluded from the protection granted by the [f]irst [a]mendment, the different character of the military community and of the military mission requires a different application of those protections."[374] The Court reemphasized its perception of the special character of the military in *Rostker v. Goldberg*:[375]

> [I]t is difficult to conceive of an area of governmental activity in which the courts have less competence. The complex, subtle, and professional decisions as to the composition, training, equipping, and control of a military force are essentially professional military judgments, subject *always* to civilian control of the Legislative and Executive branches.[376]

Although these decisions reflect deference to the military decisions affecting its own personnel, in a wartime context, the Court may be even more reluctant

---

policy determination is necessary for its review. Further, if the Court should strike down the exclusion and detention actions of the executive branch, its ruling may "embarrass" the individuals responsible for the political and military decisions, but would not necessarily undermine the nation's credibility vis-a-vis other countries; it is understandable that decisions made in the urgency of wartime may not survive later judicial scrutiny. An adverse ruling would also not be a show of disrespect, especially if it is based on firm constitutional grounds.

[373] 417 U.S. 733 (1974) (officer claiming first and fifth amendment rights infringement, where overly broad and vague provisions of Uniform Code of Military Justice forbade public statements urging blacks not to go to Vietnam).

[374] *Id.* at 758.

[375] 453 U.S. 57 (1981) (holding that draft registration of males and not females does not violate fifth amendment rights).

[376] *Rostker*, 453 U.S. at 65-66 (*quoting* Gilligan v. Morgan, 41 U.S. 1, 10 (1973)) (emphasis included). *See also Agee*, 453 U.S. at 291-92.

to impose its judgment on the soundness of military decisions made to defend against espionage and sabotage. Even the test that Justice Frankfurter appeared to suggest in his concurring opinion in *Korematsu*, that a military order be a "means appropriate for conducting war,"[377] would raise difficulties for the Court. It assumes what the Court, in the first place, is disclaiming—a knowledge of the military mission and its requirements.

However, it is unlikely that the Court would reject as nonjusticiable a case such as *Korematsu*. Where the protection of the rights of individuals are involved, the Court has not hesitated to subject cases to review.[378] Moreover, when an equal protection question exists and involves a suspect class, the Court is equipped with judicially developed standards to guide its strict scrutiny of the case.[379] If a suspect class is singled out and subjected to governmental restraint, the official body must demonstrate that a compelling interest justifies *all* of its actions.[380] This requirement serves to ensure that the alternative it has selected to protect that interest is the least restrictive one and is narrowly tailored to that end. If the Court chooses to defer to military judgment, its scrutiny at a minimum should focus on whether defensive action was indeed prompted by military necessity, and not by any covert invidious motives, and "whether the exigency still existed upon which the continued operation of the law depended."[381]

An effective review depends, of course, on the presentation to the Court of adequate information by the parties involved. The nature of military decisions during wartime necessarily involves confidential information, the secrecy of which must be protected against divulgence to enemy forces. The Court, in *United States v. Nixon*, has recognized the President's need to guard "military, diplomatic, or sensitive national security secrets," despite requiring disclosure of his confidential communications applicable to a criminal prosecution.[382] Whether justified or not, the government may shield information from the Court's review by claim of national security.[383] Additionally, the individual challenging the government's actions may find information important to his arguments beyond his reach because of the same claim of national security.[384]

---

[377] 323 U.S. at 225 (Frankfurter, J., dissenting).

[378] *See, e.g.*, Reid v. Covert, 354 U.S. 1 (1957); *Ex parte* Endo, 323 U.S. 283 (1944); West Virginia Bd. of Educ. v. Barnette, 319 U.S. 624 (1943); Minersville School Dist. v. Gobitis, 310 U.S. 586 (1940).

[379] *See supra* note 232-33 and accompanying text.

[380] *See supra* note 240 and accompanying text.

[381] Keynes, *supra* note 123, at 93 (*quoting* Chastleton Corp. v. Sinclair, 264 U.S. 543, 547-48 (1923)).

[382] United States v. Nixon, 418 U.S. 683, 703, 706 (1974).

[383] In *Korematsu*, critical information had been removed from the government's brief. *See supra* note 291.

[384] *See, e.g.*, Weinberger v. Catholic Action of Hawaii, 454 U.S. 139 (1981) (exempting the

However, when military control is exercised over civilians, the Court has a positive duty to attempt judicial review to protect the personal rights guaranteed by the Constitution. "The self-restraint and constitutional sensitiveness of the generals cannot be relied upon as adequate sources of protection. Because of its organization, mode of selection and function, the military is less likely thus to confine itself than are other agencies of government [that are] subject to judicial surveillance."[385]

Even more critical, an individual in Korematsu's predicament may find himself without any meaningful forum, should the Court fail to review his case. When Congress and the President engage jointly in the exercise of the war power, as in *Korematsu*, the political avenues of redress are effectively closed to an individual whose constitutional rights have been abridged. His last resort is foreclosed if the Court removes itself from review of his cause.

V. Conclusion

What assurances do we have that the strength of history and of legal development will prevent another *Korematsu*? Constitutional case law since the 1944 decision provides strong precedent for reversing the decision made almost forty years ago. The United States Supreme Court could find at the outset that the delegation of power, either from Congress to the President or from the President to the military, was not within constitutional boundaries of specificity. If the validity of the delegation is accepted, it is unlikely that the Court would foreclose judicial review by asserting the political question doctrine. The magnitude of the racial discrimination issue should prompt scrutiny of the case.

The Court also has equal protection bases for invalidating potential evacuation and internment programs. Such measures today would have to meet the strict scrutiny test. There would have to be a compelling governmental interest and the method of implementation would have to be the least intrusive means possible. In *Korematsu*, the federal measures, General DeWitt's orders, were racially discriminatory on their face. The compelling governmental interest of protecting vital defense facilities against the disloyal ethnic Japanese would probably fail as a counterargument. The nonmilitary factors upon which a supposedly military judgment of urgency was made were based substantially on unverified sociological information about the Japanese. Certainly, the selective treatment of German and Italian aliens suggests that the wholesale internment of the Japanese, both aliens and American citizens alike, was an unnecessarily broad and intrusive alternative.

United States Navy from public disclosure of an environmental impact statement which contains classified material dealing with national security).

[385] TenBroek, *supra* note 93, at 207. *See also supra* note 109 and accompanying text.

What of the impact of variables that do not ostensibly enter into the judicial equation? Theoretically, the legal system is structured to provide *constitutional* safeguards. But the principal actors within that arena are fallible human beings. Although their roles are usually guided by the highest of ideals, they are often the target of forces that seem at times beyond their control or detection. In *Korematsu*, the pressures of wartime, of politicial interests and of racial prejudice may have combined to influence the conduct of usually better-thinking persons.

Professor Arthur Miller has asserted that the role of the courts as guardians of civil liberties may be eroded by the political forces that influence judicial decisons.[386] If this is true, Justice Jackson's words are worth heeding:

> If the people ever let command of the war power fall into irresponsible and unscrupulous hands, the courts wield no power equal to its restraint. The chief restraint upon those who command the physical forces of the country, in the future as in the past, must be their responsibility to the political judgments of their contemporaries and to the moral judgments of history.[387]

To ignore his warning is to risk hearing the echo of Korematsu's plaintive words when he was shown, nearly four decades later, the evidence withheld by the government that would have supported his case: "They did me a great wrong."[388]

Sandra Takahata

---

[386] [T]he Supreme Court—courts generally—have never been the staunch bastions for protection of American rights and liberties that many consider them to be.

The lesson is clear and unmistakable: civil rights and liberties receive judicial protection when the interests of the state, speaking through government, are not considered jeopardized—as determined by those in public office.

. . . .

. . . [G]overnment in the United States, both in war and in peace, has always been relative to circumstances—and the Supreme Court is usually a willing ally of those who wield effective political power in the nation.

Miller, *Courts Haven't Always Been Guardians of Rights*, The Sunday Star-Bulletin & Advertiser, Mar. 13, 1983, at I-3, col. 1.

[387] *Korematsu*, 323 U.S. at 248 (Jackson, J., dissenting).

[388] P. IRONS, *supra* note 22, at 367.

California Law Review

VOL. 41 SUMMER, 1953 No. 2

Wartime Power of the Military Over Citizen Civilians Within the Country

*Jacobus tenBroek**

"WHEN an area is so beset," said Justice Jackson in his *Korematsu* dissent, "that it must be put under military control at all, the paramount consideration is that its measures be successful, rather than legal. . . . No court can require such a commander in such circumstances to act as a reasonable man; . . . He issues orders, and they may have a certain authority as military commands, although they may be very bad as constitutional law.

"But if we cannot confine military expedients by the Constitution, neither would I distort the Constitution to approve all that the military may deem expedient."[1]

"I should hold that a civil court cannot be made to enforce an order which violates constitutional limitations even if it is a reasonable exercise of military authority."[2]

To this, Justice Frankfurter, concurring with the majority, took vigorous exception.

"To talk about a military order that expresses an allowable judgment of war needs by those entrusted with the duty of conducting war as 'an unconstitutional order' is to suffuse a part of the Constitution with an atmosphere of unconstitutionality . . ." The Constitution, he said, explicitly granted the war power "for safeguarding the national life by prosecuting war effectively . . ." Hence, "If a military order . . . does not transcend the means appropriate for conducting war, such action by the military is as constitutional as would be any authorized action by the Interstate Commerce Commission. . . ."[3]

This Jackson-Frankfurter exchange highlighted and placed in modern context one of the oldest, most crucial and well-nigh insoluble problems of

---

* University of California at Berkeley. This article will appear in an altered form as a chapter in *The Japanese American Evacuation*, by tenBroek, Barnhart and Matson.

[1] Korematsu v. United States, 323 U.S. 214, 244 (1944).

[2] *Id.* at 247.

[3] *Id.* at 224-225.

constitutional democracy, viz., how to reconcile the conflicting demands of unfettered military power necessary to preserve the State in times of crises with the system of constitutional limitations and individual rights. Can this sort of military power be brought within the confines of the Constitution, subjected to rule, circumscribed by limitation—or is it governed only by necessity as uninhibited and elemental as national self-preservation?[4] Must we, as Justice Jackson suggests, regard unconstitutionality as an inevitable concomitant of the exercise of military authority even when it is "reasonable?"

The evacuation of American citizens of Japanese ancestry from the west coast, and their subsequent detention, carried out during World War II by the military under a plea or pretext of military necessity, provided the occasion for the Jackson-Frankfurter exchange and placed the perennial issue in its most critical modern form.[5] The circumstances contained at

---

[4] This question has been raised not only by modern total war but by the total war in years gone by. Said Abraham Lincoln after the presidential election of 1864 which some sober-minded citizens would have postponed because of the war: "It has long been a grave question whether any government not too strong for the liberties of its people, can be strong enough to maintain its existence in great emergencies. On this point the present rebellion brought our government to a severe test, and a presidential election occurring in regular course during the rebellion, added not a little to this strain . . . ."Speech of November 10, 1864, quoted in CORWIN, TOTAL WAR AND THE CONSTITUTION 132 (1947).

[5] KONVITZ, THE ALIEN AND THE ASIATIC IN AMERICAN LAW 241-279 (1946); Alexandre, *The Nisei—A Casualty of World War II*, 28 CORNELL L. Q. 385 (1943); Corwin, *op. cit. supra* note 4, at 91-100; Dembitz, *Racial Discrimination and the Military Judgment: The Supreme Court's* Korematsu *and* Endo *Decisions*, 45 COL. L. REV. 175 (1945); Fairman, *The Law of Martial Rule and the National Emergency*, 55 HARV. L. REV. 1253 (1942); and THE LAW OF MARTIAL RULE 157-167, 255-261 (2d ed. 1943); Freeman, *Genesis, Exodus, and Leviticus. Genealogy, Evacuation, and Law*, 28 CORNELL L. Q. 414 (1943); Graham, *Martial Law in California*, 31 CALIF. L. REV. 6 (1942); Rostow, *The Japanese American Cases—A Disaster*, 54 YALE L. J. 489 (1945); Watson, *The Japanese Evacuation and Litigation Arising Therefrom*, 22 ORE. L. REV. 46 (1942); Wolfson, *Legal Doctrine, War Power, and Japanese Evacuation*, 32 KY. L.J. 328 (1944). The war power of the national government is variously described as a single inherent power and as an aggregate of all the specifically delegated constitutional powers having a bearing upon the conduct of war. The most famous recent expression of the first of these theories is the statement of Justice Sutherland in United States v. Curtiss-Wright Corp., 299 U.S. 304, 316 (1936): "A political society cannot endure without a supreme will somewhere. Sovereignty is never held in suspense. When, therefore, the external soverignty of Great Britain in respect of the colonies ceased, it immediately passed to the Union . . . ."

"It results that the investment of the federal government with the powers of external sovereignty did not depend upon the affirmative grants of the Constitution. The powers to declare and wage war, to conclude peace, to make treaties, to maintain diplomatic relations with other sovereignties, if they had never been mentioned in the Constitution, would have vested in the federal government as necessary concomitants of nationality." *Id.* at 318.

A list of the specifically delegated powers having a bearing on the conduct of war might include: "The Congress shall have power to lay and collect taxes, duties, imposts and excises, to . . . provide for the common defense and general welfare of the United States" (U. S. CONST. Art. I, § 8, clause 1);

"To borrow money on the credit of the United States; . . . (clause 2); To declare war . . . (clause 11);

least two complicating factors. The first was the discriminatory character of the action taken. Had that factor not been present—had the question simply been the validity of a universally applied curfew or of the evacuation of all persons from a limited area threatened even remotely by invasion—it is doubtful that there would have been so much question about the power of the military to impose the restriction, though a basic constitutional issue of fact would still have had to be decided. But the curfew reached citizens of Japanese ancestry only, and mass evacuation and detention were applied exclusively to persons of that ancestry.[6]

Second, the problem was complicated by the fact that it involved military control over civilians within the country and thus opened to view the baffling lay and professional chaos surrounding martial law.[7] Had a civilian

---

"To raise and support armies, but no appropriation of money to that use shall be for a longer term than two years (clause 12);

"To provide and maintain a navy" (clause 13);

"To make rules for the government and regulation of the land and naval forces (clause 14);

"To provide for calling forth the militia to execute the laws of the Union, suppress insurrection and repel invasions (clause 15);

"To provide for organizing, arming and disciplining, the militia, and for governing such part of them as may be employed in the service of the United States, reserving to the States respectively the appointment of the officers, and the authority of training the militia according to the discipline prescribed by Congress (clause 16);

". . . To make all laws which shall be necessary and proper for carrying into execution the foregoing powers, and all other powers vested by this Constitution in the government of the United States, or in any department or officer thereof (clause 18);

". . . The executive power shall be vested in a President of the United States of America (Art. II, § 1, clause 1) . . . The President shall be commander in chief of the army and navy of the United States, and of the militia of the several states, when called into the actual service of the United States . . . ." (Art. 2, § 2, clause 1).

[6] In a strict sense, the separate classification of all Japanese Americans was based on ancestry and not on race, since all Orientals belong to the same race and only the Japanese Americans were evacuated. Yet basically, the classifying trait was race since those Americans having an ethnic affinity with our Asiatic enemy alone were excluded and imprisoned; those Americans having an ethnic affinity with our white European enemies were not subjected to similar treatment.

[7] Compare the following statements, for example: ". . . there are occasions when martial rule can be properly applied. If, in foreign invasion or civil war, the courts are actually closed, and it is impossible to administer criminal justice according to law, then, on the theatre of active military operations, where war really prevails, there is a necessity to furnish a substitute for the civil authority, thus overthrown, to preserve the safety of the army and society; and as no power is left but the military, it is allowed to govern by martial rule until the laws can have their free course. As necessity creates the rule, so it limits its duration; for, if this government is continued after the courts are reinstated, it is a gross usurpation of power. Martial rule can never exist where the courts are open, and in the proper and unobstructed exercise of their jurisdiction. It is also confined to the locality of actual war." *Ex parte* Milligan, 4 Wall. 2 (U.S. 1866).

"Martial law is the public law of necessity. Necessity calls it forth, necessity justifies its exercise, and necessity measures the extent and degree to which it may be employed." WIENER, A PRACTICAL MANUAL OF MARTIAL LAW 16 (1940).

"Martial law—or better, martial rule— . . . consists in the total or partial exercise by

agency alone handled the evacuation no martial law questions would have been raised to confuse the war powers problem.

The Japanese American cases required the Court to answer the following fundamental questions: (1) In time of war, what are the tests of the war power of the national government, vested by the Constitution in Congress and the Executive as the war-waging branches of government, especially over citizen civilians within the country? (2) How far does the power of the military extend over civilians in time of war and within the country? (3) Does such power arise from necessity and exist independently of the Constitution or is it constitutionally authorized and granted? (4) No matter what its sources, is such power subject to the limitations of the Constitution? (5) To what extent, if at all, is this power dependent upon the concept of martial law? (6) Is the military judgment of the military neces-

---

military authority, of governmental functions over our own people in our own territory; . . . embeded in the very fiber of the Constitution, we find not only the authority for martial rule but the occasions which require and justify it, and as well the limits of its operations . . . the power delegated under Public Law 503 [and Executive Order 9066] to the Secretary of War and the military commanders named by him is enormous, and places under military jurisdiction, every man, woman and child on the Pacific Coast. And, in my judgment, except as it implies the trial by jury in federal courts of civilian violators, it authorizes martial rule without limit or restraint." Graham, *supra* note 5, at 7, 9, 14.

"Those exercises of the war power which involve the most drastic changes from governmental customs in times of peace have been, but need not be, termed exercises of martial law." "The test for the validity of the exercise of martial law is identical to the test for the propriety of any exercise of the war power . . .

"The fact that many exercises of the war power, such as the peace time maintenance of military establishments, require no particular formal invocation of the war power and the fact that martial law customarily is formally declared either by the Chief Executive or by a military commander supplies no basis for a distinction between the war power and martial law, because the proclamation of martial law is not necessary to its exercise but 'must be regarded as the statement of an existing fact, rather than the legal creation of that fact . . . '

"It is, therefore submitted that the fundamental question in the case at bar is whether the evacuation program is, under all the circumstances, a proper exercise of the war power. It is unnecessary legally, and it may be undesirable generally, to decide the subsidiary question of terminology whether this particular exercise of the war power falls within that part of the scope of the war power which is popularly called martial law. On the one hand it is a substantial exercise of the war power over personal liberty and for that reason might possibly be called martial law. On the other hand it is not the type of control usually associated with martial law . . . and might be exercised by civilian authorities as is contemplated under the English statute and regulations . . . without making any reference to martial law. Indeed, in view of the fact that in popular opinion greater military control than here involved and suspension of the civil government including the courts is associated with the state of martial law, it might be more useful not to label the exercise of the war power here involved as martial law." Brief for the United States, pp. 80-81, Yasui v. United States, 320 U.S. 115 (1943).

"Probably the problem will only be confused by talking about martial law. The President had made no such proclamation and if he did his constitutional powers would not be increased one whit. The question in every case of military control would still be, can the action complained of be justified as apparently reasonable and appropriate under the circumstances, to the defense of the nation and the prosecution of the war?" San Francisco Chronicle, March 4, 1942, p. 14.

sity for controls over civilians final? Does the responsibility rest primarily with Congress to determine the existence of conditions justifying the establishment of partial or total military government, and of establishing it? To what extent, and by what tests, if any, are such judgments, by the military or by Congress, subject to judicial review and control? (7) To what degree, if any, do the methods and character of modern warfare require that we relax our democratically indispensable doctrines of the civil control of the military and the responsibility of the latter for its acts?

The Milligan Case and the Alternatives

Merely to mention these questions raises immediately the character and scope of the decision in Ex Parte Milligan.[8] Decided in 1866 and arising out of an episode of the Civil War,[9] that case has stood as a landmark in our constitutional history on the nature and extent of the war-time power of the military over civilians within the country. It was a "brooding omnipresense" in the Japanese American cases, albeit undiscussed, unanalyzed and all but unmentioned. Like many another historical landmark, it has been held to stand for various and often conflicting propositions. In recent years, the majority opinion in the *Milligan* case has been sharply criticized as importing "into the Constitution a mechanical test" which, though a "salutary restraint upon the tyranny of the Stuarts," is not "an appropriate limit on the powers of both executive and legislature in the highly responsible national government."[10] Just as stoutly, the majority opinion has been defended as "a monument in the democratic tradition" which "should be the animating force of this branch of our law."[11]

The facts of the *Milligan* case were not exactly like those of the Japanese American cases. In October, 1864, Milligan was arrested by military order, tried, found guilty and sentenced to hang by a military commission. The offense charged was not disloyalty or suspicion of disloyalty but conspiracy to overthrow the government, a crime under the laws of Congress, punishable in the civil courts, as disloyalty or suspicion of disloyalty is not. In May, 1865, Milligan petitioned the United States Circuit Court to be discharged from unlawful imprisonment. The two-judge circuit court divided and certified the question to the United States Supreme Court.

By Act of March 3, 1863, Congress had authorized the President to suspend the privilege of the writ of habeas corpus "whenever, in his judgment, the public safety may require it;" and the President had acted by

---

[8] *Ex parte* Milligan, 4 Wall. 2 (U.S. 1866).

[9] Stampp, *The Milligan Case and the Election of 1864 in Indiana,* 31 MISS. VALLEY HIST. REV. 41 (1944) explodes the belief traditionally held by constitutional writers that the conspiracy with which Milligan was charged was a serious and substantial one.

[10] Fairman, *supra* note 5, at 1286.

[11] Rostow, *supra* note 5, at 524.

proclamation doing so. Under the statute, however, the privilege of the writ was not to be suspended if the person;

> was detained in custody by the order of the President, otherwise than as a prisoner of war; if he was a citizen . . . and had never been in the military or naval service, and the grand jury of the district had met, after he had been arrested, for a period of twenty days, and adjourned without taking any proceedings against him. . . .

Milligan was not a prisoner of war and had not been in the military or naval service. Moreover, a Federal Grand Jury had convened in the district and failed to indict him. The privilege of the writ therefore had not been suspended as to him. So the United States Supreme Court unanimously held. The case being settled on these narrow statutory grounds, the Court might have rested. Unlike their successors eighty years later, however, these judges were not seeking opportunities to evade the underlying grave constitutional issues.

Justice Davis, for a majority of the Court, went on to enunciate the doctrine which is our principal heritage from the *Milligan* case. Trial by military commission, he maintained, violated the Third Article of the Constitution vesting the judicial power in courts ordained and established by Congress and composed of judges appointed during good behavior. It also violated the jury trial guarantees in the original Constitution and in the Fifth and Sixth Amendments and the search and seizure provisions of the Fourth Amendment. The Fathers, said Justice Davis:

> secured the inheritance they had fought to maintain, by incorporating in a written constitution the safeguards which *time* had proved were essential to its preservation. Not one of these safeguards can the President, or Congress, or the Judiciary disturb, except the one concerning the writ of *habeas corpus*.[12]

The Fathers "limited the suspension to one great right, and left the rest to remain forever inviolable."[13] "No doctrine, involving more pernicious consequences, was ever invented by the wit of man than that any of its [the Constitution] provisions can be suspended during any of the great exigencies of government."[14] It did not follow, said Justice Davis, that the nation was helpless in the face of a war crisis, barring the allowance of a theory of power by necessity; "for the government, within the Constitution, has all the powers granted to it, which are necessary to preserve its existence; . . ."[15]

This, then, was the system of the *Milligan* majority: That the power

[12] *Supra* note 8, at 125.
[13] *Id.* at 126.
[14] *Id.* at 121.
[15] *Ibid.*

to wage war and to wage it successfully, the power to preserve the existence of the nation, is granted by the Constitution; but at no point and in no crisis will its exercise justify the suspension of the constitutional guarantees and limitations, with the single exception of the privilege of the writ of habeas corpus. Moreover, the privilege of the writ of habeas corpus could be suspended only to the extent of warranting detention of individuals, in an hour of emergency, without trial; not to the extent of supplying unconstitutional trials. This aspect of the *Milligan* majority opinion, showing the power of the military to be constitutionally derived and constitutionally limited, is often ignored in the preoccupation of courts and commentators with Justice Davis' remark about martial law.

This position of the *Milligan* majority is sustained and amplified by the claim that it rejected. That claim was that:

> in a time of war the commander of an armed force (if in his opinion the exigencies of the country demand it, and of which he is to judge) has the power, within the lines of his military district, to suspend all civil rights and their remedies, and subject citizens as well as soldiers to the rule of *his will*; and in the exercise of his lawful authority cannot be restrained, except by his superior officer or the President of the United States.

Justice Davis answered:

> If this position is sound to the extent claimed, then when war exists, foreign or domestic, and the country is subdivided into military departments for mere convenience, the commander of one of them can, if he chooses, within his limits, on the plea of necessity, with the approval of the Executive, substitute military force for and to the exclusion of the laws, and punish all persons, as he thinks right and proper, without fixed or certain rules.
>
> The statement of this proposition shows its importance; for, if true, republican government is a failure, and there is an end of liberty regulated by law. Martial law, established on such a basis, destroys every guarantee of the Constitution, and effectually renders the "military independent of and superior to the civil power"—the attempt to do which by the King of Great Britain was deemed by our fathers such an offence, that they assigned it to the world as one of the causes which impelled them to declare their independence. Civil liberty and this kind of martial law cannot endure together; the antagonism is irreconcilable; and, in the conflict, one or the other must perish.[16]

The Court's opinion in the *Milligan* case thus makes clear that military necessity, even when approved by the executive, is not a self-justifying plea; that the military judgment that military necessity exists, cannot be allowed to stand alone; that the commander is responsible not only to his superiors and the President but to the law and the courts; that the military must act, at least when dealing with citizens within the country, within the

[16] *Id.* at 124-125.

confines of the Constitution and subject to civil, that is, judicial control.

All of this is true, said the *Milligan* majority, with one exception; and it is the scope of that exception upon which critics of the opinion have concentrated. Justice Davis said:

> . . . there are occasions when martial rule can be properly applied. If, in foreign invasion or civil war, the courts are actually closed, and it is impossible to administer criminal justice according to law, *then*, on the theatre of active military operations, where war really prevails, there is a necessity to furnish a substitute for the civil authority, thus overthrown, to preserve the safety of the army and society; and as no power is left but the military, it is allowed to govern by martial rule until the laws can have their free course. As necessity creates the rule, so it limits its duration; for, if this government is continued *after* the courts are reinstated, it is a gross usurpation of power. Martial rule can never exist where the courts are open, and in the proper and unobstructed exercise of their jurisdiction. It is also confined to the locality of actual war.[17]

> . . . Martial law cannot arise from a *threatened* invasion. The necessity must be actual and present; the invasion real, such as effectually closes the courts and deposes the civil administration.[17a]

Thus in these conditions, when warfare is actually raging in the community and has knocked out civil government, the military is in control by default and necessity. Its power arises from these facts and not from the Constitution, which does not limit it. But even here, the facts establishing the necessity are not finally determined by the military. The Supreme Court will have to be satisfied that the courts are actually closed and civil administration deposed.

Chief Justice Chase, joined by Justices Wayne, Swayne and Miller, expressed basic disagreement. In their view, ". . . there is no law for the government of the citizens, the armies or the navy of the United States, within American jurisdiction, which is not contained in or derived from the Constitution."[18] Thus, presumably even in a civil government vacuum, the military does not derive its power from extra-constitutional necessity. "The Constitution itself," said the Chief Justice, "provides for military government as well as for civil government."[19] Military government is derived from the constitutional powers to declare and wage war, to raise and support armies and perhaps from the power to provide for the government of the national forces. Moreover, continued the Chief Justice, military government is not confined to the area and circumstance of actual military conflict. Though the courts are not closed and the civil officials deposed, mili-

17 *Supra* note 8 at 127.

17a *Ibid.*

18 *Id.* at 141.

19 *Id.* at 137.

tary government may constitutionally be established "in time of insurrection or invasion, or of civil or foreign war, within districts or localities where ordinary law no longer adequately secures public safety and private rights."[20] The Chief Justice's description of the situation prevailing in Indiana at the time of Milligan's arrest, though historically incorrect, still supplies an example of particular circumstances included within this general formula:

> . . . the state was a military district, was the theatre of military operations, has been actually invaded, and was constantly threatened with invasion. . . . a powerful secret association, composed of citizens and others, existed within the state, under military organization, conspiring against the draft, and plotting insurrection, the liberation of prisoners of war at various depots, the seizure of the state and national arsenals, armed cooperation with the enemy, and war against the national government.[21]

What branch or agency of government was authorized to establish military government and determine the circumstances warranting it? The answers can be guessed from the depository of the constitutional powers involved. The Chief Justice said:

> . . . it is within the power of Congress to determine in what states or districts such great and imminent public danger exists as justifies the authorization of military tribunals . . . .[22]

> MARTIAL LAW PROPER . . . is called into action by Congress, or temporarily, when the action of Congress cannot be invited, and in the case of justifying or excusing peril, by the President . . . .[23]

Finally, to military government, when operating within its proper sphere, "the civil safeguards of the Constitution"[24] have no application. Citizens may thus be tried by military commissions instead of constitutional courts and constitutionally guaranteed juries.

With respect to the questions above formulated and presented to the Court in the Japanese American cases: the majority and minority in the *Milligan* case agreed that the wartime power of the military over civilians within the country, in most circumstances, is only such as is authorized by Congress and the Executive in the exercise of their constitutional powers to wage war and is subject to all limitations, guarantees and civil controls of the Constitution. The majority and minority agreed moreover that there were circumstances in which the military could displace, alter or substitute for civil government and could operate unrestrained by constitutional limi-

[20] *Id.* at 142.
[21] *Id.* at 140.
[22] *Ibid.*
[23] *Id.* at 142.
[24] *Id.* at 137.

tations. They disagreed as to the character of those circumstances: the majority finding them present only when civil government was knocked out and there existed a governmental vacuum; the minority finding them present more broadly "where ordinary law no longer adequately secures public safety and private rights." The majority and minority agreed that when those circumstances were present the military is not limited by the civil rights and guarantees of the Constitution. They disagreed as to the source of the power of the military; the majority found it outside the Constitution in necessity, "usages of war," or "martial law"; the minority, in the war powers vested by the Constitution—the term "martial law" merely describing the circumstances in which the war powers might be exercised uninhibited by the Constitution. The majority thus regarded martial law as a "generating source of power"; the minority, though its use of the terms was not careful, seems to have regarded it as an explanatory rubric. The majority and minority disagreed as to the agency of government responsible for finally determining the existence of the conditions freeing the military from constitutional restraints, though they agreed the military judgment was not final. The majority thought this a function of the judiciary and in this very case reversed the conclusion of the military on that point. The minority spoke mainly in terms of a congressional determination or, "temporarily . . . in the case of justifying or excusing peril," a presidential determination. The minority said nothing as to whether the judges would review the factual determination of Congress or the President.[25] But it must be remembered both that discussion on this point was not necessary in the opinion, and that in 1866, the methods and character of judicial review, as presently conceived, were still in their infancy. Neither the majority nor the minority commented on the provision of the Act of 1863 authorizing the President to suspend the privilege of the writ "whenever, in his judgment, the public safety may require it."

In brief, the difference between the majority and the minority in the *Milligan* case comes down to this: the minority maintained that, short of conditions of actual warfare and civil government vacuum, there is an area in which the military may operate with respect to civilians within the country, unrestrained by the civil guarantees of the Constitution except that it must act pursuant to congressional or executive findings and authorization. This the majority denied, holding rather that in that area constitutional restraints as well as constitutional powers apply.

Against the background of the *Milligan* case, the Court in the Japanese American cases had a number of alternatives:

---

[25] Contrary to Professor Fairman's presumption there is nothing in Chief Justice Chase's opinion to indicate that the judiciary would subject the legislative or executive action to normal constitutional tests as to the legitimacy of the end and the appropriateness of the means.

(1) Strict application of the rule of the *Milligan* majority. This would confine the constitutionally unhindered wartime powers of the military over civilians within the country to those areas in which battle is raging and civil government is unable to function. Since neither of these conditions was present on the west coast, the rule would require testing the validity of the curfew, evacuation and detention program within the ordinary framework of the constitutionally granted and constitutionally limited executive and legislative power to wage war.

(2) Adoption of the position of the *Milligan* minority—that Congress can declare martial law short of a breakdown of civil government and in areas removed from actual combat; that accordingly Congress can authorize military government over selected subjects or activities in the presence of civil authority duly and regularly functioning in all other respects; that it is up to Congress to decide in what "districts such great and imminent public danger exists" as to justify the military measures authorized; and that once the military power is thus properly invoked constitutional safeguards of civil and individual rights are irrelevant.

(3) Accept the martial rule doctrine in the *Milligan* majority but expand the area and liberalize the conditions of its application. This might be done under the guise of interpreting and applying that doctrine in the light of the conditions and methods of modern warfare. As stated by the Attorney-General of California:

> In a total global war not confined to the actual scene of hostilities but waged swiftly and violently and at long range upon civilians, factories and fields far beyond the front line and conducted by sabotage, espionage and propaganda everywhere, the army must undertake certain precautionary and preventive measures in areas not directly under the siege guns of the enemy, the object of which is the protection of the civilian population and the successful prosecution of the war . . . . The touchstone by which these preventive measures are justified is the military necessity [for them].[26]

> The view of the majority [in the *Milligan* case] that martial law must be confined to the locality of actual war does not require a change of this phase of the test of necessity but merely a new and realistic conception of the type of warfare being waged today.[27]

While this might be called an interpretation and application of the martial-law-rule-of-necessity dictum of the *Milligan* majority, it would be a basic repudiation of the whole spirit and tendency of the *Milligan* majority opinion, which sought to restrict the area of martial law to the narrowest confines of absolute necessity. It would in effect be tantamount to an adop-

[26] Brief for California as *Amicus Curiae*, pp. 6-7, Korematsu v. United States, 323 U.S. 214 (1944).

[27] *Id.* at 11.

tion of the inherent war powers theory classically stated by John Quincy Adams in 1831:

> Sir, in the authority given to Congress by the Constitution of the United States to declare war, all the powers incidental to war are, by necessary implication, conferred upon the government of the United States. Now, the powers incidental to war are derived, not from the internal municipal sources, but the laws and usages of nations . . . . There are, then, in the authority of Congress and in the Executive, two classes of powers altogether different in their nature and often incompatible with each other—war power and peace power. The peace power is limited by regulations and restricted by provisions in the Constitution itself. The war power is only limited by the usages of nations. This power is tremendous. It is strictly constitutional, but it breaks down every barrier so anxiously erected for the protection of liberty and of life.[28]

The Answers of the Japanese Cases

The *Hirabayashi* case stands generally upon the first of these alternatives; the *Korematsu* case upon the third; and the *Endo* case, though the judges there evaded all constitutional alternatives, stands as a disparagement of the spirit and rule of *Milligan*.

In *Hirabayashi* v. *United States*,[29] a Japanese American, born in the United States of Japanese immigrant parents, reared in the United States and never having visited Japan, a member of the Quaker faith, educated in our public schools and at the time a senior at the University of Washington, was criminally prosecuted for violation of the curfew order, tried by jury, convicted and sentenced to three months imprisonment. The United States Supreme Court upheld the conviction and sentence. The legal foundation for the prosecution rested on Executive Order 9066, issued February 19, 1942, Public Law 503, adopted March 21, 1942, and Public Proclamation No. 3 of the Western Defense Command promulgated March 24, 1942. In Executive Order 9066, the President, after declaring that:

> the successful prosecution of the war requires every possible protection against espionage and against sabotage to national defense material, . . . premises and . . . utilities

authorized and directed the Secretary of War or any military commander designated by him

> to prescribe military areas . . . from which any or all persons may be excluded, and with respect to which, the right of any person to enter, remain in, or leave shall be subject to whatever restrictions the Secretary of War or appropriate Military Commander may impose in his discretion.[30]

---

[28] Register of Debates xii, pp. 4037-38 quoted by Corwin, *op. cit. supra* note 4, at 78.
[29] Hirabayashi v. United States, 320 U.S. 81 (1943).
[30] 7 Fed. Reg. 1407 (Feb. 19, 1942).

Public Law 503 provided:

> That whoever shall enter, remain in, leave, or commit any act in any military area or military zone prescribed, under the authority of an executive order of the President, by the Secretary of War, or by any military commander designated by the Secretary of War, contrary to the restrictions applicable to any such area or zone or contrary to the order of the Secretary of War, or any such military commander, shall, if it appears that he knew or should have known of the existence and extent of the restrictions or order and that his act was in violation thereof, be guilty of a misdemeanor and upon conviction shall be liable [to fine or imprisonment or both].[31]

Public Proclamation No. 3, issued by General DeWitt, proclaimed that "military necessity" required "the establishment of certain regulations pertaining to all enemy aliens and all persons of Japanese ancestry" within Military Area No. 1, prescribed by earlier proclamations. Accordingly, Public Proclamation No. 3 ordered that:

> all alien Japanese, all alien Germans, all alien Italians, and all persons of Japanese ancestry residing or being within the geographical limits of Military Area No. 1 . . . shall be within their place of residence between the hours of 8 p.m. and 6 a.m. . . .[32]

Chief Justice Stone's opinion for the Court is divided into three main areas. In the first, he considers the extent of the national war power and whether prevention of espionage and sabotage falls within it. In the second, he considers whether the military peril, that is, danger of "air attack and invasion," actually existed, whether there was a likelihood of attempts to commit acts of espionage and sabotage and their bearing, if committed, on air attacks and invasion, and, finally, whether curfew was a device reasonably adapted to the prevention of these dangers. In the third, the Chief Justice specifically analyzes the issue of whether the curfew "unconstitutionally discriminated between citizens of Japanese ancestry and those of other ancestries in violation of the Fifth Amendment" stated at the outset of the opinion to be "the question for our decision. . . ."[33]

Thus, the Court in the *Hirabayashi* case, invokes for the military curfew the tests which are traditionally announced for all other constitutional powers: (1) was the end sought to be achieved within the war powers granted by the Constitution; (2) were the means employed appropriate to the achievement of that end; (3) were the means chosen selected with due recognition of the substantive and comparative guarantees of the Fifth and other Amendments?

---

[31] 18 U.S.C. § 97a (1946).

[32] 7 Fed. Reg. 2543 (March 24, 1942).

[33] Hirabayashi v. United States, *supra* note 29 at 83.

Whatever quarrel one may have with the Court's actual decision on these points, there can be little doubt but that the Court regarded the exercise of military power as subject to these limitations.

Justices Murphy and Rutledge wrote concurring opinions. Justice Murphy especially asserted that "the mere existence of a state of war" does not "suspend" the broad guarantees of the Bill of Rights and other provisions of the Constitution protecting essential liberties." ". . . the war power like the other great substantive powers of government, is subject to the limitations of the Constitution."[34] Justices Murphy and Rutledge apparently agreed that drastic wartime invasion of constitutional rights could only be permitted if martial law were declared and in circumstances warranting martial law.

Like that of Hirabayashi, Korematsu's conviction was based on violation of orders issued by the Western Defense Command under the authority of the President and Congress granted in Executive Order 9066 and Public Law 503. At the time of his conviction, there were three such military orders outstanding and applicable to him. The first of these was Public Proclamation No. 4 of March 27, 1942,[35] terminating the original exclusion program which had left free choice of route and destination to the evacuee and instituting a system of rigid controls on movement. It forbade persons of Japanese ancestry to leave the area except as authorized and directed by the Western Defense Command. The second outstanding military order applicable to Korematsu was Civilian Exclusion Order No. 34, issued on May 3, 1942.[36] It provided that after midnight, May 28, all Japanese Americans were to be excluded from a given portion of Military Area No. 1 which included San Leandro, Alameda County, California, the place of Korematsu's residence. Civilian Exclusion Order No. 34 required a responsible member of each family and each individual living alone to report to a civil control station for instructions to go to an assembly center. By its terms, the order did not apply to Japanese Americans who were within the area and in assembly centers. The effect of these two orders, therefore, was to direct all Japanese Americans to proceed to assembly centers according to instructions to be received from a civil control station. The third outstanding military order was Civilian Restrictive Order No. 1, of May 19, 1942.[37] It provided for the indefinite detention of persons of Japanese ancestry in assembly or relocation centers.

Korematsu was born in Oakland and graduated from high school. He was classified as 4-F because of stomach ulcers. After graduation he worked

[34] Hirabayashi v. United States, *supra* note 29 at 110.

[35] 7 Fed. Reg. 2601 (1942).

[36] 7 Fed. Reg. 3967 (1942).

[37] 8 Fed. Reg. 982 (1943).

in a shipyard as a welder until, after the outbreak of war, the Boiler Makers Union cancelled his membership because of his race. He could not read or write Japanese, had never been outside of the United States, and was not a dual citizen. Romance rather than disloyalty was his undoing. The evacuation orders disrupted his plans to marry a Caucasian girl, so he decided to evade them and remain behind. He hoped to escape detection by having a plastic surgical operation performed on his nose and by changing his name. The ruses failed and the F.B.I. seized him.

Korematsu was prosecuted in the Federal District Court under Public Law 503 for knowingly remaining within the forbidden territory contrary to Civilian Exclusion Order No. 34, that is, for remaining within the forbidden territory but not in an assembly center.

> The court set the bail at $1,000 which was posted by the American Civil Liberties Union. But when served with the court order for Korematsu's release, the gaoler telephoned Military Police, who, without any warrants or writ whatever, placed Korematsu in military custody and took him to the Tanforan Assembly Center. . . . Judge Welch subsequently sanctioned this action and refused to release the bail despite the fact that the defendant was in the hands of the Government. Instead, the bail was raised to $2,500 when Korematsu declined to sign the useless bail bond. The unsigned bond was exonerated and the defendant was ordered into the custody of the United States Marshal and again placed in the county jail.[38]

On the trial, Korematsu was found guilty but his sentence was suspended and he was placed on probation for five years.

At no stage of the proceeding had Korematsu's loyalty to the United States been put in issue.

Confronted with the situation of a Japanese American still at large in the prohibited area after the sanctions provided by the law of Congress had been applied, the Western Defense Command again resorted to its own devices. Once more Korematsu was picked up by soldiers and lodged in the assembly center.

The Supreme Court's opinion in the *Korematsu* case is an attempt to occupy both areas marked out and imperatively separated by the *Milligan* majority: the rule of necessity, and the rule of the Constitution. It is a futile attempt to reconcile the irreconcilable. The opinion, as a result, is a muddled hodge-podge of conflicting and barely articulate doctrine. On the one hand, Justice Black explicitly builds upon the *Hirabayashi* case and treats the evacution decision as largely settled by the principles announced there. On the other hand, he lays such heavy stress upon the emergency character of the military action as in effect to take the position of the *Milligan* majority martial law dictum that the existence of dire emergency re-

38 American Civil Liberties Union News, July 1942, p. 1.

sults in substituting untrammeled military judgment for constitutional limitations. Justice Black said:

> We upheld the curfew order as an exercise of the power of the government to take steps necessary to prevent espionage and sabotage in an area threatened Japanese attack.[39]
> ... exclusion from threatened area, no less than curfew, has a definite and close relationship to the prevention of espionage and sabotage.[40]
> The Hirabayashi conviction and this one thus rest on the same 1942 Congressional Act and the same basic executive and military orders, ...
> * * *
> In the light of the principles we announced in the Hirabayashi case, we are unable to conclude that it was beyond the war power of Congress and the Executive to exclude those of Japanese ancestry from the West Coast war area at the time they did.[41]

To this explicit assimilation of the *Korematsu* and *Hirabayashi* cases must be added the Court's express denial that Korematsu was evacuated "because of hostility to him or his race."[42] This at least suggests that the military decision is subject to some judicial review, however perfunctory, in terms of the legitimacy of its purpose.

Side by side with these remarks and implications, however, are others far more reminiscent of the *Milligan* dictum than of *Hirabayashi*. For example:

> Compulsory exclusion of large groups of citizens from their homes, except under circumstances of direst emergency and peril, is inconsistent with our basic governmental institutions. But when under conditions of modern warfare our shores are threatened by hostile forces, the power to protect must be commensurate with the threatened danger.[43]

And look again at this passage:

> He [Korematsu] *was* excluded because we are at war with the Japanese Empire, because the properly constituted military authorities feared an invasion of our West Coast and felt constrained to take proper security measures, because they decided that the military urgency of the situation demanded that all citizens of Japanese ancestry be segregated from the West Coast temporarily ....[44]

This language of military necessity is augmented by other statements in the opinion. The courts, said Justice Black, must give "rigid scrutiny" to "legal restrictions which curtail the civil rights of a single racial group.... "[45] Just such restrictions were before the Court. Yet nothing like

---

[39] *Supra* note 1 at 217.

[40] *Id.* at 218.

[41] *Id.* at 217-18.

[42] *Id.* at 223.

[43] *Id.* at 219-220.

[44] *Id.* at 223.

[45] *Id.* at 216.

"rigid scrutiny" was given. Nor could it be in the presence of the latitude allowed to the military. But, in any event, legal restrictions of this order, that is, legal restrictions abrogating constitutional rights, can be justified, said Justice Black by "pressing public necessity."[43] No doubt they can be, but this is hardly a constitutional test.

In determining whether the stress on "pressing public necessity" as the justifying basis of the *Korematsu* decision should be regarded, on the one hand, as a mere shift in emphasis or rhetorical deviation, or, on the other hand, as a real departure from the reasonable relation formula of the *Hirabayashi* case, several factors must be borne in mind. It must be remembered, in the first place, that the Court which had reached a unanimous holding in the *Hirabayashi* case was now bitterly and publicly divided. Not only did the dissenters feel that the hardship of evacuation was quite out of the class of that involved in curfew but they denied that it was justified by the military situation at the time and raised serious questions about the extent to which non-military considerations entered into the decision. General DeWitt's "Final Report" which had come out prior to the *Korematsu* decision, in addition to casting a curious light on some of the General's beliefs, gave plausibility, if not something more, to the points of the dissenters. In these circumstances it was certainly easier for the Court to sustain the evacuation as an emergency action within the allowable discretion of the military than to attempt to justify it as reasonable in detail. In the second place, if one rejects this interpretation of the *Korematsu* case and assumes that the reliance on the *Hirabayashi* case implies a basic acceptance of the formula there stated, the question must be answered whether the application of the formula was not so feeble and uncertain as to amount to no application at all and thus in effect to leave the Constitution suspended. The later day, the diminished emergency, the incomparably more drastic invasion of individual rights, the application of the evacuation order exclusively to Japanese Americans (whereas German and Italian enemy aliens had also been included in the curfew), the evidence that racism played a part in the final decision—all demanded, if not a rigid and minute scrutiny, at least a substantial re-evaluation of ends and means. If the reasonable relation test was to mean anything in these circumstances, one would think that it required something more than the cursory and almost imperceptible application given it by Justice Black.

In view of these factors, it is difficult to avoid the conclusion that the *Korematsu* decision is the exact antithesis of the spirit and decision of the *Milligan* majority. In fact, a more complete rejection of the entire *Milligan* case, majority and minority, is hard to imagine.

---

43 *Ibid.*

The principle doctrine of the *Milligan* majority, that the exigencies of government, even the exigencies of war—which the war-waging power is constitutionally granted to meet—do not suspend the limitations and guarantees of the Constitution, is spurned.

Korematsu sustained far-reaching military controls over the lives and property of citizen civilians within the country on the theory that "pressing public necessity" or "dire emergency" justified them and were their sources of power and sole limitations, thus in effect, suspending the Constitution.

The *Milligan* majority dictum about martial law is also distorted out of its contemplated proportions. In the sense of the Attorney-General of California that dictum was interpreted and applied in *Korematsu.* It was simply given "a new and realistic conception" in the light of "the type of warfare being waged today." That conception, however, so interpreted and applied, is of such a character as to destroy the conception of the *Milligan* majority. *Korematsu* vastly enlarged the sphere in which the military is freed of the Constitution by necessity. It permitted the institution of constitutionally unrestrained military rule in an area which was not actually invaded, or even subjected to a sizable threat of invasion. In *Korematsu,* the threat of hit-and-run raids is apparently sufficient, for, after Midway, that was all that was possible. Could an historical landmark such as the *Milligan* case, in which military authority had been boldly resisted and constitutional guarantees firmly vindicated, ever come to a more ignominious end than that a dictum there uttered about a rigidly confined area in which the military might prevail over the Constitution should now be expanded to consume the principle which made the case a landmark? When to this doctrine is added automatic acceptance by the Court of the military judgment that the course of the war required mass evacuation of the Japanese Americans, without the Court itself investigating the factual relationship of the course of the war to the evacuation, or requiring the military to make any showing on that point; and when, in addition, in the presence of an open avowal by the commander that he acted on and justified the program by beliefs about race, the Court attributed to the military the judgment that the course of the war necessitated the program, without evidence or proof that the military held such judgment; one can see how far the *Korematsu* case relaxes judicial control of the military, how doubtful the standards of military responsibility have become, and the degree to which the *Milligan* dictum has been interpreted to the point of extinction. The *Milligan* majority was emphatic that, in the end, the courts—not the military—would have to decide whether the facts were such as to justify substitution of martial rule for the civil government and guarantees of the Constitution.

Finally, by the *Korematsu* ruling, even the *Milligan* minority is left far in the dust. The minority there contemplated martial rule in connection with the suspension of the privilege of the writ of habeas corpus. No such suspension was involved in *Korematsu;* but the martial rule was tolerated. Further, according to the *Milligan* minority, it was up to Congress—or "temporarily, . . . in the case of justifying or excusing peril . . . when the action of Congress cannot be invited," up to the President—to determine whether the circumstances exist justifying martial law and to establish it. The *Korematsu* case leaves this to the military. Lastly, since under Public Law 503 the ordinary courts were to enforce the military decisions, the *Korematsu* case far outstrips the *Milligan* minority in permitting Congress "to throw over martial law the sanctifying aegis of civil authority."[47]

Justice Jackson dissented in the *Korematsu* case, but on the ground that the civil courts could not be required to enforce the military evacuation orders. As to the unconstitutional source of the power of the military and the absence of constitutional limitations, he agreed with the Court's opinion. He merely stated it with the irresponsible clarity of a dissenter. Though he wrote a concurring opinion, the real dissenter in the *Korematsu* case, on the war powers issue, was Justice Frankfurter. He thought that the military power exercised in the evacuation was merely a part of the national government's power, constitutionally derived and constitutionally limited, as any other power granted by the Constitution. The real debate on this phase of the *Korematsu* case was between Justices Jackson and Frankfurter, not between the dissenter and the majority. And the division was not along the lines of the *Milligan* majority and minority. The Jackson-Frankfurter points of view represented tendencies in the *Milligan* majority opinion, Frankfurter invoking the rules applicable short of extra-constitutional martial rule, Jackson, the rules of constitutionally unfounded and unfettered power.[48]

Thus, as to the answers to the questions posed at the outset of this section, the *Hirabayashi* and *Korematsu* cases are poles apart. The *Hirabaya-*

---

[47] CORWIN, *op. cit. supra* note 4, at 98.

[48] Justice Roberts dissented on the ground that "the indisputable facts exhibit a clear violation of Constitutional rights . . . it is the case of convicting a citizen as a punishment for not submitting to imprisonment in a concentration camp, based on his ancestry, and solely because of his ancestry, without evidence or inquiry concerning his loyalty and good disposition towards the United States." Korematsu v. United States, *supra* note 1, at 225-26.

"The opinion refers to the *Hirabayashi* case, to show that this court has sustained the validity of a curfew order in an emergency. The argument then is that exclusion from a given area of danger, while somewhat more sweeping than a curfew regulation, is of the same nature, —a temporary expedient made necessary by a sudden emergency. This, I think, is a substitution of an hypothetical case for the case actually before the court." *Id.* at 231. "No pronouncement of the commanding officer can, in my view, preclude judicial inquiry and determination whether an emergency ever existed and whether, if so, it remained, at the date of the restraint out of which the litigation arose." *Id.* at 231, n.8.

shi case, at least as to the doctrine announced, treated the wartime power of the military over civilians within the country as part of the war-waging powers of the national government. It is accordingly both constitutionally derived and eonstitutionally limited. It may be exercised under the authority of Congress and the President when it is directed to an end within the war powers, when it is reasonably appropriate to the achievement of that end, and when it does not violate the civil and individual guarantees of the Constitution. The techniques of warfare do not alter these constitutional rules or the need for civil control. They do, however, affect the circumstances which are determinative of appropriateness. The *Korematsu* case uses *Hirabayashi* as a justifying foundation but shunts its doctrine aside. It turns to necessity, rather than to the Constitution, as the source and sole limitation of the authority of the military; and, in view of the character and conditions of modern warfare, apparently would apply this rule whenever the nation is at war. Neither case discusses martial law, but in *Korematsu* it is apparently the basic conception underlying the outcome.

The Alternative Answers Evaluated

As against the doctrine of necessity, time has, if anything, sustained and strengthened the wisdom of the doctrine of the *Milligan* majority. This results from nothing so much as the very reason given for abandoning or modifying that doctrine: namely, the changed conditions of modern warfare. Total war, we are told, implies a degree of military control over civilians as a military matter of waging war successfully not hitherto imaginable.[49] The need for the increased participation of the military in the regu-

---

[49] Note for example the language of the Attorney General in the case of the Nazi saboteurs: ". . . war today is so swift and so sudden and so universal that it would be absurd to apply a doctrine like the doctrine in the *Milligan* case, where they said that Indiana during the Civil War had not recently been invaded. The facts existing in 1863 do not today exist, and a bomber may drop a bomb tomorrow in Chicago. Can it be said that there is no area of warfare, no area of military operations in Chicago under those circumstances? I think not." Transcript of Oral Argument quoted by FAIRMAN, *op. cit. supra* note 5, at 199. Ochikubo v. Bonesteel, 60 Fed. Supp. 916, 932 (S.D. Cal. 1945): "The contention fails to take into consideration the patent fact, so awful in its consequence, that modern wars are not limited to clashes at arms on particular fields of battle of comparatively insignificant area, but are exertions of the complete strength and ingenuity of all our people and all our resources wherever located, against the complete strength and ingenuity of the enemy and all their resources wherever located."

"Judge Lloyd J. Black delivered an oral opinion in *Ex parte* Ventura, 44 Fed. Supp. 520, 522 (W.D. Wash. 1942): "In the Civil War . . . no invasion could have been expected into Indiana except after much prior notice and weary weeks of slow and tedious gains by a slowly advancing army. They then never imagined the possibility of flying lethal engines hurtling through the air . . . They never visioned the possibility of far distant forces dispatching an air armada that would rain destroying parachutists . . . and invade and capture far distant territory overnight. They never had to think then of fifth columnists far, far from the forces of the enemy successfully pretending loyalty to the land where they were born, who, in fact, would forthwith guide or join any such invaders. The past few months in the Philippines . . . established that apparently peaceful residents may become enemy soldiers overnight. The

lation and administration of wide areas of national life, however, increases rather than decreases the necessity to retain normal judicial safeguards in the management of war and preparations for war. To allow this expansion of military activity under an expanded martial law doctrine, with its lack of constitutional bases and limitations—justified, if at all, only under a conception severely limited to facts that overwhelm—is to create a breach in the Constitution, relax the constitutionally and democratically required judicial control of the military and cut down immeasurably the operative sphere of civil rights. It is virtually to surrender the government to the military in time of war, and perhaps even in time of international stress and national preparedness. Standards that make sense when the allowable area of military absolutism is very small and confined to an extreme exigency, make little sense indeed when the scope of military authority is extended to embrace areas related to the war effort by the nature of modern warfare. Martial rule accepted and applied in these circumstances represents the sort of military accretion of power apprehended and reprobated by Justice Davis. It is based upon the doctrine characterized by him as the most "pernicious . . . ever invented by the wit of man." The Japanese American program itself stands as a most convincing modern example that Justice Davis' conception and words have not been outmoded by time.

Lying between the *Korematsu* surrender to the military and the *Milligan* majority's insistence on the application of constitutional limitations (but far nearer to the former than to the latter) is the ground taken by the *Milligan* minority. The most effective and forceful exponent of this position in recent years is Professor Charles Fairman. In a much quoted article, Professor Fairman has stated the case thus:

> When one considers certain characteristics of modern war—mobility on land, surprise from the air, sabotage and the preparation of fifth columns [the depth and dispersion of the Army,]—it must be apparent that the dictum that "martial rule cannot arise from a threatened invasion" is not an adequate definition of the extent of the war power of the United States. . . . It does not take an actual bombing of Pearl Harbor or a shelling of Santa Barbara to unchain the hands of the commander on the spot. Facts of this sort prove the reality of the danger, but the courts should be prepared to sustain vigilant precautions without waiting for such proof. A commander should not be put in a worse position legally because he has contrived to keep disaster at arm's length.

---

orders and commands of our President and the military forces, as well as the laws of Congress, must, if we secure that victory that this country intends to win, be made and applied with realistic regard for the speed and hazards of lightning war."

Judge Denman, concurring in Korematsu v. United States, 140 F.2d 289, 296 (1943), reached the conclusion that: "A threatened air invasion, directed by saboteur signals, which in an hour's time could destroy every federal court house in California presents the necessity for the substitute [sic] of military action against such sabotage for that of civil courts."

. . . The war power, distributed between Congress and the President, comprehends all that is requisite to wage war successfully.[50]

In modern dress and in apter phrase, this is the position of the Milligan minority. Its essential feature is not that the war power is traced to a constitutional source, for the *Milligan* majority does that. Rather, its essential feature is the broad field in which the suspension of constitutional guarantees is allowed. That can be the only point to allowing martial law in cases of threatened invasion; the *Milligan* majority thought that ample power existed to prepare against and resist threats of invasion, but maintained that within that range, the war power was subject to constitutional limitations.

There can be no quarrel with the proposition that the war power "comprehends all that is requisite to wage war successfully." That is a virtual tautology. The critical question is, does waging war successfully, in the circumstances envisioned, necessarily involve removing from the hands of the commander on the spot the chains of the Constitution? Under the *Milligan* rule, is such a commander "put in a worse position legally because he has contrived to keep disaster at arm's length?" To what extent is such a military commander hindered in his military function by the constitutional machinery and requirements of criminal justice; by the due process and equal protection requirements of fair dealing and reasonable classification? Certainly, such a commander is not "put in a worse position legally" if he has "contrived to keep disaster at arm's length" by the preparation of shore defenses, the maintenance of air and naval patrols, the disposition and maneuver of the men and machines of war. It is only if the commander seeks to protect Pearl Harbor and Santa Barbara from the threat of air delivered bombs and submarine delivered shells by abolishing jury trial and the civil courts, by doing away with confrontation of witnesses, immunity from self-incrimination, counsel for the defense, and the other guarantees of the Bill of Rights that the commander is "put in a worse position legally;" and these are not likely contrivances to keep that sort of disaster at arm's length. Moreover, even when the bombs and shells are actually falling rather than merely threatened, it does not automatically follow that closing the courts or suspending constitutional rights will be a helpful military measure or one calculated to improve the military situation. If the bombs that are falling are atomic, they may blast whole districts into perdition and along with them all civilian agencies. Such facts as these raise constitutional questions about the meaning of "invasion" amid the new methods of producing a civil government vacuum, much more than they justify overruling the dictum about threatened invasions. Whatever one

[50] Fairman, *supra* note 5 at 1287, 1288.

might say about Pearl Harbor, the shelling of Santa Barbara and the threat of other such acts along the coast, for instance, left the courts open and in the unobstructed exercise of their jurisdiction, and left the civilian agencies in the unhindered performance of their duties. What in these conditions would justify military trial of citizen civilians or patently unreasonable and discriminatory classification? Can we not be mindful of Professor Fairman's admonition that we ought not legally to place at a disadvantage the commander who by his vigilant precautions has kept disaster from our shores without, at the same time, automatically concluding that everything the military does in a period of possibly threatened invasion must be permitted, at the expense of personal rights and civil liberties? Would all of the proper functions of the military be impaired by a constitutional rule of reasonableness applied in a context of constitutional limitations and rights?

What the *Milligan* minority says in effect is that when Congress establishes martial law in the broad area in which it is permitted to do so, constitutional rights lose their special weight. They are then of the order of all other rights and interests. Military measures are to be judged exclusively in the light of their military appropriateness. Their constitutional validity does not depend upon whether they result in an invasion of the most precious and basic rights of men or some mere statutory privilege.[51] By this view, in the area mentioned, a military order which will accomplish an end within the war power will be sustained no matter what rights of civilians are destroyed thereby. Different degrees of military necessity will not be required to sustain curfew, evacuation and detention since distinctions based on the character of the rights invaded by the military order are immaterial.

Though this position does not surrender to the military with the abandon of *Korematsu*, one must still ask whether it too does not grant to the military more than is necessary to win wars.

Some phases of the *Milligan* majority doctrine, in the opinion of the present writer, require modification. These phases, however, do not directly involve the basic proposition of the normal subjection of the military to the Constitution. Nor do they involve directly the dictum about threatened

[51] Professor Fairman does not associate himself with this phase of the *Milligan* minority position. Instead, he suggests an adaptation of the rule of appropriateness: "The nature and proximity of the danger," he argued, "must, of course, have a bearing on the type of control which the military authorities may reasonably enforce. The *Milligan* case had to do with an attempt to inflict the extreme penalty, death, for an offense known to the law and triable . . . by judges. . . . The removal of civilians, for cause or suspicion, from areas of military importance, interferes with interests of a much lower order. Measures of prevention, such as curfew, may be appropriate where no reason could be offered for assuming the functions of the courts of law." *Ibid.*

invasions, so vigorously attacked by Professor Fairman.[52] They do not even directly involve the techniques and character of modern warfare. They turn rather upon the mechanical appearance of the test applied and the rigid conception of constitutional rights and guarantees apparently entertained. Both must be modified to adjust to the modern view of the function of the Court as judgematic rather than automatic and to the modern view of constitutional rights and guarantees as relative rather than absolute.

Whether the martial rule test of the *Milligan* majority is "mechanical" is of course a matter of interpretation. So long as the test is to be applied by the courts, its rigidity or flexibility will depend upon the attitude of the court in each particular case rather than upon anything controlling in the nature of the test itself. Certainly, its use as an absolute command unrelated to a specific situation could not be defended. Properly construed, the test presents a matter of fact, not of fiction; of substance, not of form. The question is, are the courts open in the sense that they are able to function substantially in their usual way and with their usual degree of efficiency, not are the courts formally closed. "Are the courts open and in the proper and unobstructed exercise of their jurisdiction?" In the words of Charles Evans Hughes in his 1917 war powers address:

> Certainly, the test should not be a mere physical one, nor should substance be sacrificed to form. The majority [in the *Milligan* case] recognized "a necessity to furnish a substitute for the civil authority," when overthrown, in order "to preserve the safety of the army and society." If this necessity actually exists it cannot be doubted that the power of the Nation is adequate to meet it, but the rights of the citizen may not be impaired by an arbitrary legislative declaration. Outside the actual theatre of war, and if, in a true sense, the administration of justice remains unobstructed, the right of the citizen to normal judicial procedure is secure.[53]

In present day judicial discussion, constitutional rights are seldom described as fixed and immutable. They are viewed rather as conditional guarantees, dependent on time, place and circumstance for their meaning and substance. The rights of life, liberty and property, for example, guaranteed by the due process clause of the Fifth and Fourteenth Amendments, and mainly at stake in the curfew, evacuation and detention situations, are nothing in our law, if not governed by the rule of reason. In fact, the due process clauses are held to impose upon government little more than a rule of judicially determined reasonableness. Life, liberty and property may all be taken away, singly or collectively, if the courts find an adequate or reasonable factual justification.

The First Amendment's rights, too, are conditional and dependent upon

---

[52] See also 3 WILLOUGHBY, CONSTITUTIONAL LAW 1602 (2d ed. 1943) and GLENN, THE ARMY AND THE LAW 188-90 (1943).

[53] Hughes, *War Powers Under the Constitution*, 42 A.B.A. REP. 232, 245-46 (1917).

circumstances, as anyone may see who reads the line of cases from *Schenk*[54] to *Dennis*.[55] Even the relatively fixed criminal law procedural safeguards of the Fifth and Sixth Amendments and, for that matter, the Third Article's vestiture of the judicial power in constitutional courts, are not absolutely applied. Still other rights, of course, are stated in qualified terms. Only "unreasonable searches and seizures" are forbidden by the Fourth Amendment.

The civil rights and guarantees of the Constitution are treated as a flexible but weighty set of values which must be thrown into the scale of every relevant decision. They may never be ignored, though they may be outbalanced. They should not and need not be withdrawn from consideration altogether because the times are militarily troubled and many programs are administered by uniformed men.

But if the restraints are not absolute, neither is the war power plenary; and certainly it is not self-contained and self-defining. It "comprehends all that is requisite to wage war successfully." But that is all. It is not a grant of authority to do all manner of irrelevant things. Actions taken under the war power may not be so arbitrary as to violate the due process requirement. The situation may not be so lacking in clear and present danger that the actions violate First Amendment guarantees of speech, press and assembly. Without emergency justification, they may not be so sweeping as to treat dissimilar things alike or so uneven as to treat similar things differently without violating the constitutional guarantee of equality. They may not interfere with the processes and jurisdiction of the civil courts in the absence of battlefield conditions.

Understood as thus modified to encompass a conception of constitutional rights and guarantees governed by the rule of reason and as expressive of a test governed by substance rather than form, the *Milligan-Hirabayashi* doctrine, now as in the past—precisely because of the nature of total war and the need for an omnipresent military, and notwithstanding modern techniques of fighting—is the only hopeful path to the adjustment of military power and constitutional guarantees, to allowing the military such freedom of action as is necessary to preserve the nation and at the same time retaining democratic controls. That doctrine would subject the

---

54 Schenck v. United States, 249 U.S. 47, 52 (1919): "The character of every act depends upon the circumstances in which it is done. . . . The most stringent protection of free speech would not protect a man in falsely shouting fire in a theatre and causing a panic. It does not even protect a man from an injunction against uttering words that may have all the effect of force. . . . When a nation is at war many things that might be said in time of peace are such a hindrance to its effort that their utterance will not be endured so long as men fight and that no Court could regard them as protected by any constitutional right."

55 Dennis v. United States, 341 U.S. 494 (1951).

wartime power of the military to the same rule as a peace-time exertion of the war powers or any other power.

The *Milligan-Hirabayashi* doctrine might be restated thus: Even in time of global and total war, when the nation is straining at every sinew fighting on world-wide battle fronts, when invasion is threatened and island outposts have fallen or been crippled, when the waters off-shore are infested with enemy submarines, the coast line itself shelled and remote parts of the mainland bombed—in circumstances such as these, the military, though its authority increases as military necessity increases, still must act within the confines of the Constitution and the safeguards of individual rights, at least with respect to civilians within the country. The military may take drastic measures, it may impose dimouts, brownouts, blackouts. It may restrict travel of all persons within given areas. It may evacuate all inhabitants from prospective landing beaches. It may take precautionary steps to protect power and water supplies and other public utilities or productive facilities. But these measures must be both appropriate to meet the emergency and reasonable in the circumstances. One part of the circumstances with respect to which reasonableness must be determined consists of the rights invaded and the interests disturbed. Consequently, a measure thoroughly appropriate to meet the emergency when only the military features are considered, may not be reasonable in the circumstances when the constitutional rights of citizens are thrown into the balance. The effect of applying constitutional limitations and guarantees to wartime exertions of the war power is to add a determinative factor in selecting among measures all reasonably appropriate to meet a given military danger; it is to forbid the choice of militarily appropriate measures which displace civil government or invade civil rights if less civilly drastic but militarily adequate alternatives are available. Curfew may not be imposed if dimouts, with their less drastic invasions of the liberty of the citizen civilians, will serve the same military purpose. Exclusion of an individual from an entire military district may not be ordered if terminating his employment in war plants and access to military secrets will end his threat of harm. Evacuation of a portion of the population is forbidden if curfew is adequate to meet the danger, or, given the time, if individual hearing will identify the disloyal. Evacuation will not be tolerated at all if the basis of selection is a trait unrelated to the war objectives.

Of course, all or any of these measures will not be tolerated if there is no military peril, or if there is a military peril but the measures are not appropriate to meet it. Even in the absence of the specific limitations of the Constitution, the war power is, after all, only the power to wage war. Military necessity and military fiat are not necessarily identical. If they turn out not to be, the fiat is null.

Because the war power is a grant within the framework of a constitutional system in which power not granted is not possessed, because the war power in addition is subject to the amendments and guarantees provided elsewhere in the Constitution, and because in our system the judges have become the final enforcers of constitutional allocations and limitations—the courts will sit in judgment on the military requirements of the hour, at least to the extent of determining whether the military judgment as to the existence of the emergency, the appropriateness of the means, and the necessity to invade civil rights which makes that invasion reasonable, have a substantial basis.

The programs for the curfew and evacuation of Japanese Americans during World War II—the very programs before the Court in the *Hirabayashi* and *Korematsu* cases—especially reveal the urgent reasons supplied by total war for applying the *Milligan-Hirabayashi* doctrine.

For the first two-and-a-half months after Pearl Harbor, dangers of espionage, sabotage and fifth column activity thought to arise out of the presence of Japanese on the west coast were within the jurisdiction of the Justice Department. Under a plan prepared in advance, sizable numbers of Japanese American aliens, believed on some evidence to be dangerous or disloyal, were apprehended and detained. Enemy aliens, including Japanese, were subjected to travel restrictions, were required to surrender as contraband firearms and other weapons and certain radio and camera equipment. Enemy aliens were registered. In fact, the department, on recommendation of the Western Defense Command, established the first prohibited zones from which all enemy aliens were excluded. These surrounded airports and airfields, hydro-electric dams, pumping and power plants, gas and electric works, harbor areas and military installations. In addition, restricted zones were established along the coast within which enemy aliens were curtailed as to travel and were curfewed. Executive Order 9066, signed on February 19, 1942, turned the whole matter over to the War Department. This was done with the concurrence of the Justice Department because, first, the Justice Department thought that further restrictive measures were unnecessary; because, second, it was unwilling to extend restrictions to citizens on a discriminatory basis, and, finally, because it lacked the staff and facilities to carry out any more far-reaching project. The first two of these reasons justify adamant resistance to the proposal rather than concurrence in it; and the mere possession of facilities by the War Department is hardly an argument for undertaking the plan.

On March 24, 1942, the Western Defense Command issued a curfew order covering all persons of Japanese ancestry, and German and Italian aliens.[56] On March 27, it announced, and on March 30, actually commenced

[56] Curfew applied to such individuals in Military Area No. 1 and in the 1,132 small zones in the other parts of the Western Defense Command.

the compulsory uprooting and removal of all Japanese Americans, aliens and citizens alike, herding them first into assembly and then relocation centers.[57] Once the removal was accomplished, custodianship was given, at least in part, to the War Relocation Authority, a civilian agency. The actual process of removal could have been carried on as well by the Justice Department as by the Army if the former had had the manpower, the disposition and the barbed-wire enclosures. There was nothing peculiarly military about the function. In performing it, the Western Defense Command and the War Department were acting as an extended arm of the F.B.I.

Why should not the action of the military in such cases be measured by the same "conventional tests of constitutionality" which are applied to civil government when doing the identical tasks? Why is it "impracticable and dangerous idealism" to insist that tasks appropriately performed by the military in time of war and when it has much of the manpower and many of the facilities, but which are not peculiarly military in character, conform to the standards exacted of civilians? There is certainly little justification for the view that a military commander, even in carrying out such civilian tasks, should not, in the words of Justice Jackson, be required to act like a "reasonable man," or at least as like a reasonable man as the civilians are.

Milligan, Endo and Detention

Under the doctrine of the *Milligan* majority, curfew and exclusion would stand or fall upon their individual merits as particular exertions of the war power, to be judged in the light of all relevant circumstances. If there was a danger of invasion by the forces of Japan, if ethnic affiliation with the Japanese people determined the loyalty of American citizens, if the circumstances were such that persons loyal to Japan could and were likely to do acts helpful to Japan and harmful to us, if curfew or exclusion as the chosen method of prevention was appropriate to achieve that end, and, if, finally, there were no available alternative methods of prevention which would accomplish the military objective, and, at the same time, be more consistent with the individual and civil guarantees of the Constitution, then curfew and exclusion were constitutionally authorized exercises of the national war power. If these conditions did not obtain, curfew and exclusion were unconstitutional both as going beyond the granted war power and as transcending the guarantees and prohibitions of the Constitution.

The same test must be applied to detention. If detention was an appropriate means for meeting an existing or imminently threatening peril and could not reasonably have been replaced by some other means less de-

[57] This proclamation applied only to Military Area No. 1 of course.

structive of individual rights and civil guarantees, it too was a constitutional exercise of the nation's war power. With respect to the existence of danger and the likelihood that Japanese Americans would assist the forces of Imperial Japan, exactly the same factual inquiry was necessary for detention as for evacuation. But, with respect to the appropriateness of the means, detention was far more difficult to justify than evacuation, just as evacuation in turn had been far more difficult than curfew. Not only was the deprivation of the rights of the victim group far more drastic but as a method of preventing Japanese American collaboration with submarine-landed saboteurs, hit-and-run air raids, or with invasion itself, detention added little to evacuation. If exclusion was not only constitutional but also successful in keeping the Japanese Americans out of the coastal area, detention was entirely unnecessary for that purpose.

If detention had been intended as a device to facilitate the separation of the disloyal from the loyal, or if, though not itself helpful in the sifting process, it had been instituted as an intermediate make-shift pending sorting, the case for the constitutionality of detention would be placed on its strongest grounds. In that event, however, the duration of the incarceration would be a determinative factor. Detention for a few weeks, or, considering the size of the group, for two or three months, might have been held administratively necessary; but hardly the two-and-a-half years which was the period of confinement for most excludees. In that event, also, unconditional release should immediately have followed a determination of loyalty, instead of the continued detention and conditional leave procedure actually enforced. However, it is clear from the facts that the plan for detention originated with the War Relocation Authority and not with the military, that so-called voluntary migration was tried before the program of detention was initiated, and that no plan for the separation of the disloyal from the loyal was undertaken until about four months after assembly center detention had begun. It is clear from these facts that incarceration was not originally intended as a step in or an aid to a process of sorting. The historical fact is that segregation of the disloyal from the loyal came almost as an afterthought.

Whatever may be said of the military's participation in its legal authorization and in its execution, the program for the wartime detention of the Japanese American population resulted not from a judgment of military necessity made by the military, but from a judgment of social desirability made by civilians.

A clearcut statement in General DeWitt's *Final Report* bears upon the attitude of the Western Defense Command toward the introduction of detention and the non-military reasons for it:

> Essentially, military necessity required only that the Japanese population be removed from the coastal area and dispersed in the interior, where the danger of action in concert during any attempted enemy raids along the coast, or in advance thereof as preparation for a full scale attack, would be eliminated. That the evacuation program necessarily and ultimately developed into one of complete Federal supervision, was due primarily to the fact that the interior states would not accept an uncontrolled Japanese migration.[58]

The reasoning of the WRA is plainly set forth in a remarkable pamphlet prepared by WRA lawyers and published over the signature of the Director of the WRA and the Secretary of the Interior.

> Detention was a policy which the responsible officers of WRA decided upon reluctantly, out of a conviction that no other course was administratively feasible or genuinely open to them. The agitation for mass evacuation had repeatedly asserted that west coast residents of Japanese ancestry were of uncertain loyalty. The Government's later decision to evacuate was widely interpreted as proof of the truth of that assertion. Hence, a widespread demand sprang up immediately after the evacuation that the evacuees be kept under guard, or at the very least, that they be sorted and that the dangerous ones among them be watched and kept from doing harm. In these circumstances it was almost inescapable that the program administrators should come to the conclusion that if the right of free movement throughout the United States was to be purchased for any substantial number of the evacuees, the price for such purchase would have to be the detention of all the evacuees while they were sorted and classified, and then the continued detention of those found potentially dangerous to internal security. The detention policy of WRA was born out of a decision that this price would have to be paid, that it was better to pay this price than to keep all the evacuees in indefinite detention, and that to refuse to pay this price would almost certainly mean that the prevailing fear and distrust could not be reasoned with and could not be allayed.[59]

Speaking of the leave program, the pamphlet continued:

> These conditions to departure—that the evacuee shall have been found to be nondangerous to internal security, that he shall have a job or some other means of support, that there shall be "community acceptance" at his point of destination, and that he shall keep the Authority notified of his changes of address—represented, in fact, the heart of the relocation program. They were designed to make planned and orderly what must otherwise have been helter-skelter and spasmodic.[60]

Again,

> If the constitutionality of the evacuation itself be assumed, the situation that was inevitably created by the evacuation does of itself give rise to new problems which Government must undertake to solve by appropriate means. Thus, the conditions attached to departure from the centers enabled a

[58] *Final Report: Japanese Evacuation from the West Coast 1942* 43-44 (1943).

[59] *Legal and Constitutional Phases of the War Relocation Authority Program* 11 (1946).

[60] *Id.* at 12.

sifting of a possibly questionable minority from the wholesome majority whose relocation it became the principal object of WRA to achieve. These restrictions enabled WRA to prepare public opinion in the communities to which the evacuees wished to go for settlement, so as to avoid violent incidents, public furor, possible retaliation against Americans in Japanese hands, and other evil consequences. The leave regulations "stemmed the flow"; they converted what might otherwise be a dangerously disordered flood of unwanted people into unprepared communities into a steady, orderly, planned migration into communities that gave every promise of being able to amalgamate the newcomers without incidents, and to their mutual advantage. The detention, in other words, was regarded as a necessary incident to this vital social planning.[61]

Even as to the disloyal, detention was not justified as a means of preventing them from committing acts harmful to the war effort.

WRA took the position that it sought to detain those deemed ineligible to leave until after all those deemed eligible had been relocated. Such detention, it maintained, was necessary to build upon public acceptance of those found eligible to relocate. The detention was thus regarded as an essential step in the accomplishment of the relocation objective. Since the war ended before relocation of the eligibles had been completed, the Government never had to face the question of whether it could or would attempt to detain those deemed ineligible after the relocation objective had been fully achieved.[62]

---

[61] *Id.* at 12-13. The Justice Department made a similar justification for relocation center detention in its argument for the constitutionality of that program before the Supreme Court. It said: "It is clear from the facts stated above that the program and procedures of the War Relocation Authority have been undertaken as a result of the situation arising from the evacuation of persons of Japanese ancestry from the West Coast area. If the evacuation is upheld by this Court [Korematsu v. United States] as a measure validly undertaken in the prosecution of the war, conditions occasioned by it may properly be dealt with under the war power by reasonable means. A relocation problem was inescapable as a result of the evacuation. By the removal of 110,000 persons from the West Coast area the Government assumed moral and political responsibility for their fate and for the wise handling of conditions which their relocation might precipitate." Brief for the United States, p. 75, *Ex parte* Endo v. United States, 323 U.S. 283 (1944).

"The purpose of the relocation program has been to minimize the sufferings of the evacuated population. This purpose has entailed a restriction of the liberty of the individuals affected—liberty to go and come, to seek out opportunity wherever they might choose, to meet with such failure or success in the world at large as fortune and individual capacity might yield. This restriction is each day becoming less as additional persons are granted leave. The principle of restoration of the citizen's liberty has been kept constantly in mind. We do not contend that under any set of circumstances less unique or less definitely a product of an extreme war measure the Government might bestow advantage, as viewed by officials, at the price of the elementary personal freedom of individuals *sui juris*. We suggest, however, that the issue here involved must be judged in the light of its origin in a measure adopted in the course of a declared war, under a threat of invasion to which it was related—a measure fraught with the gravest human consequences, which the Government has striven to render as little productive of permanent harm as the forces with which it has had to cope permitted." *Id.* at 81-82.

[62] *Supra* note 59 at 15. The WRA centers were analogized to temporary refuges, set up by the government following a flood or other natural disaster to provide shelter for the victims, and as to which the government "would have . . . to regulate the entries and departures." *Ibid.* It was admitted that the regulations with respect to departure at least were somewhat more stringent in the case of the Japanese Americans than in the case of flood victims.

Detention, the only course "genuinely open" or "administratively feasible"; the purchase price for "the right of free movement"; "a method of allaying popular fear and distrust"; "a necessary incident to . . . vital social planning"; "an essential step in the accomplishment of the relocation objective"—these are hardly the categorical imperatives of military necessity. They are the social desiderata of welfare planners.

In the *Korematsu* case the Court passed upon the constitutionality of evacuation after declaring it to be separable from assembly center detention. The Court indicated, however, that if detention were not separable from evacuation, if detention was the means for executing evacuation, the detention would be constitutional. Said Justice Black for the majority:

> The Assembly Center was conceived as a part of the machinery for group evacuation. The power to exclude includes the power to do it by force if necessary. And any forcible measure must necessarily entail some degree of detention or restraint whatever method of removal is selected.[63]

No doubt assembly centers might reasonably be instituted "as a part of the machinery for group evacuation." As such, they would serve as control points or check stations to make certain that the excludees were actually departing from the area. Once having the excludees in the assembly centers for this purpose, the government might reasonably convert the centers into permanent shelters for those who, whatever the cause, did not wish to disperse in the interior; and the government might lay down appropriate rules of notice and sign-out prior to departure from the shelters. But control-point processing would have warranted compulsory confinement for a few days, or, at the most, a few weeks. Assembly center detention lasted months and did not end in dispersal in the interior or a voluntary decision to remain in a government provided shelter. The assembly centers were in fact prisons and could never be properly analogized to refuges for, say, flood victims, as to which the government might reasonably regulate the entry and departure. Moreover, they were not so much "part of the machinery for group evacuation" as they were part of the machinery for further detention. Confinement in the assembly centers was simply a prelude to more confinement in the relocation centers. The Court's analysis in the *Korematsu* case, while perhaps a reasonable statement of the general principle, had little relevance to the facts there presented.

The issue of relocation center detention was squarely and unavoidably presented to the Court in the *Endo* case.

Mitsuye Endo was an American citizen of Japanese ancestry, 22 years old, who prior to the evacuation was a civil service employee of the State of California. Her brother Kunio was serving in the U.S. Army. On May 14,

---

63 *Supra* note 1 at 223.

1942, she was ordered to an assembly center and then to the Tule Lake Relocation Center. On July 13, 1942, she petitioned for a writ of habeas corpus to seek freedom from Tule Lake. The United States District Court signed a brief order declaring (1) that Endo was not entitled to the writ and (2) that she had not exhausted her administrative remedies because she had not sought to secure her release under WRA regulations. On February 19, 1943, she appealed from the District Court decision. Later in the same month the WRA granted leave clearance but Endo made no application for indefinite leave. The Circuit Court certified four questions to the Supreme Court on April 22, 1944:

1. May an American citizen be held in a concentration camp without the right to a hearing which has all the elements of due process merely because such citizen is of Japanese ancestry?

2. May a loyal citizen be so confined until she satisfies the WRA that she can support herself and receive assistance in the community where she desires to live?

3. May such issues of self-support and community acceptance be decided by the WRA without a hearing at which the citizen enjoys all the elements of due process?

4. May the WRA in addition require that she report after she has left the camp?

The Supreme Court held that Endo must be given her liberty. The WRA, the Court said, had "no authority to subject citizens who are concededly loyal to its leave procedure"[64] or to detain them or release them conditionally. Any power of detention possessed by the WRA, the Court argued, would have had to be received by redelegation from General DeWitt of powers conferred on him under Executive Order 9066 and Public Law 503. Executive Order 9102, establishing the WRA, was issued only to implement the measures already authorized by Executive Order 9066. These two executive orders, the Act of Congress and the pertinent legislative history do not "use the language of detention."[65] Hence, the authority to detain, if it existed, must be implied. And "[If] there is to be the greatest possible accommodation of the liberties of the citizen with this war measure, any such implied power must be narrowly confined to the precise purpose of the evacuation program."[66]

The "single aim" of Executive Orders 9066, 9102 and Public Law 503, said Justice Douglas for the Court, was "to protect the war effort against espionage and sabotage, detention which has no relationship to that objective is unauthorized . . . ."[66a]

---

[64] *Ex parte* Endo, 323 U.S. 283, 297 (1944).
[65] *Id.* at 300.
[66] *Id.* at 301.
[66a] *Id.* at 302.

> A citizen who is concededly loyal presents no problem of espionage or sabotage....[67]
> Nor may the power to detain an admittedly loyal citizen or to grant him a conditional release be implied as a useful or convenient step in the evacuation program, whatever authority might be implied in case of those whose loyalty was not conceded or established.... Community hostility even to loyal evacuees may have been (and perhaps still is) a serious problem. But if authority for their custody and supervision is to be sought on that ground, the Act of March 21, 1942, Executive Order 9066 and Executive Order 9102, offer no support.... To read them that broadly would be to assume that the Congress and the President intended that this discriminatory action should be taken against these people wholly on account of their ancestry even though the government conceded their loyalty to this country.[68]

Thus, in the *Endo* case, the Supreme Court invalidated relocation center detention for persons whose loyalty was granted and who therefore were clearly held in confinement or subjected to leave procedures and conditional release for social rather than for military reasons. The ground for the action was that WRA and the Western Defense Command lacked authority to make such detentions under the pertinent executive orders and congressional legislation. The majority of the Court steadfastly declined to place its holding upon a constitutional basis, though some of the reasons given for confining the executive orders and legislation to a narrow scope were equally, if not more, compulsive of a constitutional negative on the program. The Court, however, was content to refer to the relevant constitutional provisions:

> not to stir the constitutional issues which have been argued at the bar but to indicate the approach which we think should be made to an Act of Congress or an order of the Chief Executive that touches the sensitive area of rights specifically guaranteed by the Constitution.[69]

If the broad doctrine enunciated in the *Milligan* case and subsequently repeated but not followed in the *Hirabayashi* case were to be applied, it is difficult to see how the detention of a concededly loyal citizen, not charged and convicted of crime, could escape constitutional condemnation. Certainly, such detention cannot constitutionally be justified as "a necessary incident" of "vital social planning" or as the purchase price "of the right of free movement throughout the United States . . . for any substantial number of the evacuees." Rendering "planned and orderly what must otherwise [be] helter-skelter and spasmodic" is not a power conferred by the Constitution on the national government; and the purchase price of the

67 *Ex parte* Endo, *supra* note 64 at 302.
68 *Id.* at 302-304.
69 *Id.* at 299.

right of free movement of citizens was paid a long time ago, when the right was embodied in the Constitution. To exact a new purchase price, in these circumstances, would ordinarily go by the name of extortion. Imprisonment of a hapless racial minority as an element in "vital social planning" for the improvement of race relations hardly seems a promising approach either to the Constitution or to the problem of race relations. Mass and racially discriminatory incarceration of well over 100,000 persons, as a result of "vital social planning" to protect them aginst community hostility, is the compulsory acceptance of an unwanted benefit which can be constitutionally justified, if at all, not in terms of the good done for the victims but in terms of the interests of society. In the words of Miss Dembitz:

> . . . the theme of benefaction which runs through the utterances of the military as well as, subsequently, of the War Relocation Authority, may have given the officials involved a feeling of satisfaction, it does not make the deprivations and restraints imposed on the donees any more constitutional.[70]

Detention of the entire group while they were being sorted and classified as to loyalty, justified not on military grounds but as a means of reasoning with and allaying "prevailing popular fear and distrust"; detention of disloyals, not in order to prevent activities harmful to the war efforts, but "as necessary to build public acceptance of those found eligible to relocate," and thus "as an essential step in the accomplishment of the relocation objective"—these are merely variants of the self-same protective custody argument that "it was all for their own good." As such, they stand upon the same constitutional footing as the detention of concededly loyal citizens pending efforts by the WRA "to prepare public opinion in the communities to which the evacuees wished to go . . . ." Detaining disloyal persons for the reasons given, however, could not even be rested on the spurious ground that it was for their own benefit since their detention was for the benefit of the members of the group who were eligible for leave.

If detention is to be sustained against a claim of unconstitutionality on any of the grounds advanced by the Western Defense Command or the WRA, it must be on the ground that protective custody is a reasonable method for preventing disturbances of such a character or extent as to interfere with the prosecution of the war. Evaluated in these terms, the program would have had to hurdle a number of serious obstacles. What was the likelihood of such disturbances in the interior? General DeWitt's *Final Report* contains no assessment. The WRA produced none supported by fact. Could such threat as existed have been subdued by reasonably vigorous precautionary action on the part of local police backed up by firm

---

70 Dembitz, *supra* note 5, at 202-203.

statements from the War Department that the measures taken by it were adequate to prevent espionage and sabotage and that lawless acts against the migrants would be punished? Normally, of course, efforts to prevent violence center about those likely to perpetrate it rather than about the prospective victims of it. Also bearing upon the existence of the danger and its degree is the fact that all of the evidence tends to show that, had assembly and relocation centers been shelters rather than prisons, there would still have been no "dangerously disordered flood of unwanted people into unprepared communities." The Japanese Americans themselves would understandably not have been anxious to rush into hostile communities. Having been completely uprooted, many would prefer to sit out the war in government supplied refuges. Still others, possessed of more initiative and adjustive capacity, would lack the means to go and establish themselves in new communities. All but a very few, if they left the shelters at all, would in all likelihood have been anxious to avail themselves of government proffered assistance to assure "migration into communities that gave every promise of being able to amalgamate the newcomers without incidents and to their mutual advantage." In that event, the movement, though not proceeding out of compulsory confinement, would, because of the self-interest of the group affected, have been planned and orderly. That detention was necessary to the accomplishment of these objectives, even assuming the constitutionality of the objectives, is consequently far from established.

The narrow rule or holding of the *Milligan* case also stands squarely against the constitutionality of the detention imposed on the Japanese Americans.

Justice Davis' opinion, in substance, was that the judicial and jury trials prescribed by the Constitution do not unreasonably interfere with the prosecution of war and their abolition is not reasonably appropriate to its conduct, unless the area is a battlefield where the military is supreme, not by the Constitution but by compelling fact. Accordingly, except in such battlefield conditions, the military may not arrest, try and punish citizen civilians within the country.

Suppose the military arrests and confines without trial. Suppose, further, that the military is aided in accomplishing detention by a civilian agency. Suppose, finally, that the persons so arrested and imprisoned are not guilty or even charged with being guilty of any crime at the common law, by the statutes of the country, or by the laws of war. In all three of these respects—true of the mass detention of Japanese Americans—the situation differed from that presented to the Court in the *Milligan* case. Would these be grounds for a different decision or would they merely make the unconstitutionality of the action more apparent?

Remarkable as it may seem, the Court in the *Endo* case relied on the absence of a military trial and the presence of civil participation and sanctions as bases for distinguishing and refusing to apply the *Milligan* rule. Justice Douglas, speaking for the Court, said:

> . . . we do not have here a question such as was presented in Ex parte Milligan, . . . or in Ex parte Quirin, . . . where the jurisdiction of military tribunals to try persons according to the law of war was challenged in *habeas corpus* proceedings. Mitsuye Endo is detained by a civilian agency, the War Relocation Authority, not by the military. Moreover, the evacuation program was not left exclusively to the military; the Authority was given a large measure of responsibility for its execution and Congress made its enforcement subject to civil penalties by the Act of March 21, 1942. Accordingly, no questions of military law are involved.[71]

To Justice Douglas' list of the civil elements in the detention program should be added other items, in addition to the administrative role of the War Relocation Authority and the civil sanctions under Public Law 503. As we have already pointed out, there was nothing peculiarly military about the operation of the whole program. That the decision to undertake the program itself was not strictly military is plain from General DeWitt's *Final Report*. Moreover, such steps as were taken in the total process by the Army were taken under the supervision and approval of the civilian heads of the War Department. They in turn acted under the authority of Executive Order 9066 and Public Law 503. The WRA itself and the detention camps were established by Executive Order 9102 and the civilian head of the WRA was appointed by the President. Finally, the WRA did not justify detention on grounds of military necessity, but on grounds of desirable social policy.

All of these civilian elements, however, though they existed and need to be given their due weight, do not warrant Justice Douglas' subordination of the part actually played by the military authorities. The original decision to evacuate was made by military authorities. They made the decision first to carry it out on a voluntary basis, that is, leaving to the individual the choice of route, means and destination. They made the decision later to change to a compulsory and controlled exodus. They issued all of the various 108 civilian exclusion orders which marked the step by step progress of the evacuation. These were published in the form of and were widely publicized and understood to be military commands. Transporta-

[71] *Supra* note 64 at 297. In *Ex parte* Quirin, 317 U.S. 1 (1942) the case involving the Nazi saboteurs who entered the country secretly from German submarines as embodied elements of the German Armed Forces, the Court was at pains to point out that the military might try Haupt—the only American citizen among the saboteurs—on the ground that "Citizens who associate themselves with the military arm of the enemy government . . . are enemy belligerents" under the laws of war, and to be treated as such. *Id.* at 37-38.

tion of the victims from gathering places to assembly centers was under armed military guard. The assembly centers were run by the military and armed military personnel were abundantly in evidence. Conveyance from assembly centers to relocation centers again was under military auspices and guard. At the relocation centers, the administration was civilian and some of the guards had different uniforms. Law and order were maintained within the camps in ordinary circumstances by WRA personnel. Military police, however, were at the gates to prevent unauthorized ingress and egress. At Camp Gila in the summer of 1943 they shot and killed a Nisei who strayed outside the barbed wire. Everybody understood that disturbances which could not be quelled by center authorities would bring the troops. At Tule Lake, that is exactly what happened. Following riot or near riot conditions, the army took over and actually ran the center from November 4, 1943 to January 15, 1944.[72] It was the military which suppressed the Manzanar riot of early December 1942. The army stood ready to intervene in the Poston disturbance of November 18, 1942, but the project director thought such intervention unnecessary. All of the relocation centers, including that in far-off and unindustrialized Arkansas were declared by the military to be military areas.[73] This made the civil sanctions of Public Law 503 applicable, which could only be done, under Executive Order 9066, by the Secretary of War or by commanding generals, not by the WRA. By a series of orders issued by General DeWitt and by the War Department for the four camps outside of the Western Defense Command, evacuees were forbidden to leave their relocation centers or their work sites except with written permission.[74] Evacuees granted indefinite or other leaves remain "in the constructive custody of the military commander . . . ."[75] The Joint Army-Navy Board, from January 1943 to April 1944, investigated individuals in the camps, held hearings and made recommendations to the civil administration concerning leave clearances. The Western Defense Command alone could rescind the civilian exclusion orders and proclamations. A word from it or the War Department could have brought the whole program, including the detention, to an end conditionally or unconditionally. And that is what happened finally.

Thus, the total program of exclusion and detention began with a military order, continued during the pleasure of the military, and was terminable at the will of the military. It included the physical removal of the victims by the military. It proceeded with their incarceration for a time in

---

72 THOMAS AND NISHIMOTO, THE SPOILAGE c. 6 (1945).

73 Public Proclamation No. 8, 7 FED. REG. 8346 (June 27, 1942); Public Proclamation W. D. 1, 7 FED. REG. 6593 (August 13, 1942).

74 *War Relocation Authority Administrative Instruction No. 22,* ¶ 9 (June 20, 1942).

75 *Final Report, supra* note 58, at 242.

camps run by the military. It ended, for thousands of American citizens, in three years of imprisonment, the prisons being located in places designated military areas in order to assure adequate military authority and controls, and being manned by administrators and jailers, maintained in their positions at times by the active presence of troops and always with troops in the background. Throughout the process, the sounds and trappings of the military were to be heard and seen.

Yet, said Justice Douglas, there are here "no questions of military law" and the *Milligan* case is not in point! Military arrest and military confinement, says the *Milligan* case, cannot be tolerated while the courts are open. In the Japanese American cases, there was such military arrest. There was also militarily enforced and militarily terminable confinement. Surely the ground will not be taken that military arrest and imprisonment which flout the Constitution when connected with a military trial, become constitutional when executed without such a trial. Surely, also, the ground will not be taken that such arrest and imprisonment lose their military character by having behind them the added sanction of civil imprisonment or when the jail is run by civilians.

The differences between *Milligan* and *Endo* advanced by Justice Douglas serve no purpose quite so much as to make clearer the unconstitutionality of the *Endo* detention. To say otherwise is to say that in this context a military order has a different constitutional significance from the action of a military tribunal; or that imprisonment by civilians stands in a better constitutional position than imprisonment by the military which, after all, is permissible in some circumstances.

Conclusion

The Japanese American cases—*Hirabayashi, Korematsu* and *Endo*—though shrouded in great confusion of rhetoric, and despite the careful statement of doctrine by Chief Justice Stone in *Hirabayashi,* which the Chief Justice failed to apply, represent a constitutional yielding to the awe inspired in all men by total war and the new weapons of warfare. They disclose a judicial unwillingness to interfere with, or even to look upon, the actions of the military taken in time of global war, even to the extent of determining whether those actions are substantially or somehow connected with the prosecution of the war. That the actions were directed to and drastically affected citizen civilians within the country and involved decisions, policies and administration dominantly civil rather than military in character were facts that were hardly noticed, let alone assigned their proper significance. In this context, the Japanese American cases diminish and render uncertain the public responsibility of the military, and relax democratic and judicial controls. In these cases, the historically established balance between the military and the civil—constitutionally sanctified in the United States by the classic majority opinion in *Ex parte Milligan*—has been shifted dangerously to the side of the military by the known and unknown terrors of total war and by a quiescent and irresolute judiciary. In them, the *Milligan* rule of subordination of the military to the Constitution except in battlefield conditions is abandoned. Instead, the national war powers, though explicitly conferred by the Constitution and not exempted from its limitations, are founded on and circumscribed by a military estimate of military necessity. Citizens, on a mass basis, were allowed to be uprooted, removed and imprisoned by the military without trial, without attribution of guilt, without the institutional or individual procedural guarantees of Article III and Amendments V and VI, and without regard to the individual guarantees of Amendments I, IV, V, and others. The military action was taken upon a mere suspicion of disloyalty arising from racial affinity with the enemy, and was applied discriminatorily to one race only. During most of the period of evacuation and detention, there was not even a threat of invasion. The *Milligan* dissenters do not go nearly so far. Can circumstances short of battlefield conditions justify this kind of surrender of the Constitution to the generals? Does the winning of total war require so much—that the military be immune from review in its civil, sociological and anthropological judgment; that the military be allowed to do militarily irrelevant things; that the military be permitted arbitrarily and unnecessarily to invade individual and civil rights? One may insist with Charles E. Hughes that

> the power to wage war is the power to wage war successfully[76] [and] that power, explicitly conferred and absolutely essential to the safety of the Nation is not destroyed or impaired by any later provision of the Constitution or by any one of the amendments.[77]

One may insist on all that and yet at the same time not deviate from the basic proposition—equally plain if not equally explicit in the Constitution and "absolutely essential" to the perpetuation of the Republic—that the war power, when exerted in the military government of citizen civilians within the country, does not exist in the absence of a grave military peril, does not exceed measures reasonably appropriate to cope with that peril, and does not comprehend violations of civil and individual guarantees of the Constitution in the presence of a militarily adequate alternative.

The war powers of the national government traverse a wide area, ranging from economic regulations such as were embodied in the Emergency Price Control Act of 1942 to the command of troops in battle. So far as

---

[76] HUGHES, *supra* note 53, at 238.
[77] *Id.* at 248.

those powers are exerted in the government of citizen civilians within the country, in a place not actually a battlefield, whether administered by the military or civilian executive, the Court should exercise its constitutional and historical function of review and impose the above listed standards of public responsibility. This is not to say that the judges should not candidly appreciate their own fallability. Nor is it to say that the rigor of judicial scrutiny will not or should not vary as the circumstances vary. The military, when making strictly military decisions, must be allowed a reasonable latitude of military error. Yet keeping the military within the confines of the Constitution, at least when it acts with respect to citizen civilians within the country, is a civil imperative if the Republic is to persist. The self-restraint and constitutional sensitiveness of the generals cannot be relied upon as adequate sources of protection. Because of its organization, mode of selection and function, the military is less likely thus to confine itself than are other agencies of government which, despite their representative and responsible character, have traditionally been subjected to judicial surveillance. The techniques and instruments of judicial review are, on the whole, not less applicable or efficient in the case of the military. The courts are hardly the agency to subdue a rebellious general; but in most contexts, where the military touches the civilian in the country, the courts can render it less "vagrant," less "heedless of the individual," and less a "threat to liberty." The expectation of judicial review, or the mere continuing possibility of it, will make most generals more careful.

In the history of the United States, rebellious generals have not been a substantial source of military danger to civil institutions. The real source of that danger is found in other quarters. This point has been well made by Dean Louis Smith:

> Militarism is more than a formal system of thought. It is a type of public opinion and as such is present to some degree in every society. As is true in regard to other such questions, public opinion relative to military doctrines constantly fluctuates in response to various psychological and environmental conditions. The chief danger to states in which militarism is currently in a minor position is that under stress of chronic anxiety over military insecurity, aggravated by a percussive train of war crises, each stopping short of actual conflict but tending ever nearer to it with inescapable indications of its inevitability, militaristic opinion may spread until it captures the minds of almost the whole people.
>
> These facts indicating how the garrison state may possibly come into existence among a hitherto free people are worth pondering. This state may come, not by willful usurpation by the military but by successive adaptations for defense having support of public opinion. It may be ushered in, not by conspiracy, but by plebiscite. It may come into power, not over the wreckage of the civil organs traditionally expected to repress it, but with their active support. It may arrive, not through violence, but by influence,

> an influence born of the demand of the masses that they not be exposed to annihilation from hostile attack by the omission of any security factor. . . . Thus, among peoples obsessed with deepening anxiety regarding imminent warfare and their slender chances of survival in it, by a process which in a less sinister context has been called the "inevitability of gradualism," the garrison state may come into full power.[78]

In this context, the role of the Court is only in part to maintain the proper relationship between the military and civil authority. It is only in part to see that the military does not become master where it should be servant. Beyond that, the function of the Court is to determine the nature of the Constitution itself—the scope of the national war powers exerted by the military and by the civil officials who direct the military vis-a-vis the rights of civilians. Military usurpation may be less a danger to constitutional limitations established to protect the individual than an expansion of the war powers carried out by civil consent and popular insistence. It is the function of the Court to see that "hyperlegality does not impair security and that the shibboleth of 'military necessity' does not justify unnecessary destruction of the rights of the people and bring about an improper impairment of democratic processes of government."[79] When the balance between safety and liberty is struck, it may be the hazardous and often immediately thankless task of the Court to safeguard the people from itself. The performance of this stern task may call, doubtless does call, for statesmanlike self-restraint; it does not call for resignation.

Justice Jackson's worry that wartime review of military action will tend to distort the Constitution and find unhappy application in peacetime cases is certainly legitimate. Yet, it must be remembered that the Supreme Court is often the willing, sometimes the reluctant, but never the helpless victim of its own precedents. The Court also always has the alternative of avoiding bad precedents by not making them. This danger, in any event, must be measured against the danger of constitutionally and judicially unfettered military power.

---

[78] Smith, American Democracy and Military Power 8-9 (1951).
[79] *Id.* at 303.

Collins versus the World: The Fight to Restore Citizenship to Japanese American Renunciants of World War II

John Christgau

The author, a resident of Belmont, California, has completed a book-length manuscript on the U.S. alien enemy internment program during World War II which will be published shortly by Iowa State University Press.

IN LATE 1944, thousands of Japanese Americans renounced their United States citizenship in the Tule Lake Segregation Center in California. The subsequent lawsuit to restore their citizenship was twenty-three years in the courts and ultimately involved over five thousand plaintiffs. Yet with the exception of a dissertation, no paper has focused on the U.S. District Court legal battle between the government and Wayne Collins, attorney for the plaintiffs. In his recent book, *Justice At War*, Peter Irons delineates the earlier, often ineffective, efforts of Wayne Collins on behalf of Fred Korematsu, one of four Japanese Americans who challenged the government's evacuation program. But in 1945 Collins had become even more convinced of the injustices of relocation when he discovered that purely administrative procedures were being used to determine the loyalty of Japanese Americans and to continue their internment. He then hunkered down for a good legal fight. The result was *Abo* v. *Clark*, "an issue without precedent, and unique in the annals of American jurisprudence."[1]

I

By virtue of his birth on American soil in 1911, Tadayasu Abo became a citizen of the United States. Meanwhile, Abo's alien parents, denied the privileges of naturalization because they were Asians, returned to Japan with their young son in 1917. For ten years, Abo received a Japanese education, after which he returned to the United States. Considered now *Kibei* (American born, Japanese educated), Abo married, had a son, and worked in the oyster beds in Olympia, Washington.

Abo registered for the draft as required in 1940. Then, after the Japanese attack on Pearl Harbor, he and his *Nisei* wife Yukiko, whose alien parents also were in Japan, watched with apprehension as plans were formulated for the eventual relocation of 110,000 West Coast persons of Japanese ancestry. Despite the impending relocation, and the humiliation his wife suffered when she was struck one day by an angry white woman, Abo swore his allegiance to the United States.

By the spring of 1942, the Abo family had been relocated to one of the War Relocation Authority's centers. There, Abo received notification that, because of his Japanese ancestry, he had been officially classified 4C, which made him ineligible for the draft as an "enemy alien," even though he was a citizen. Then, in early 1943, the government proceeded with plans to create a Japanese American combat unit. To be eligible for the unit, all male evacuees were required to swear "unqualified allegiance to the United States," while forswearing any form of allegiance or obedience to the Japanese emperor. Furthermore, the potential recruit had to be willing to serve on combat duty wherever ordered.[2]

Despite his previous draft registration and his professed loy-

[1] *Abo* v. *Clark*, 77 F.Supp., 806 (1948); Donald Collins, "Disloyalty and Renunciation of U.S. Citizenship by Japanese Americans During World War II" (Ph.D. dissertation, University of Georgia, 1975); Peter Irons, *Justice at War* (New York, 1983). This paper rests on extensive use of unpublished materials not utilized in earlier treatments of the topic.

[2] Female evacuees were not candidates for a combat unit, but they were required to swear they would serve with the WACS or the U.S. Army Nurse Corps.

alty, relocation bitterness and the prejudicial nature of the questionnaire (German Americans and Italian Americans were not required to forswear allegiance) now prompted Abo to answer "no" to questions twenty-seven and twenty-eight on the government questionnaire. Consequently, when the WRA commenced plans in the summer of 1943 to transfer these so-called disloyal evacuees to a single camp, Tadayasu Abo was among thousands of evacuees transferred to Tule Lake Segregation Center in northern California.

Almost immediately, WRA officials discovered that their segregation camp of some 18,000 evacuees was a smoldering bomb. Organized groups of bitter, pro-Japanese evacuees, determined to be repatriated to Japan, embarked upon a program of threats and beatings to recruit more repatriates. Abo and his wife, who delivered a second child that fall, witnessed gang beatings by young toughs. Shortly, rumors circulated that only those who had requested repatriation would be permitted to remain in the relocation centers rather than be resettled in hostile white communities.

In early November, four days of turmoil broke out at Tule Lake while WRA Director Dillon Myer happened to be inspecting the site. After a clash between WRA security forces and evacuees, the U.S. Army was finally called in to restore order. Still, by late 1944, the pro-Japanese evacuees in camp had organized themselves into groups called the *Sokuji Kikoku Hoshi-dan* (Organization to Return Immediately to the Homeland to Serve) and the *Hokoku Seinen-dan* (Young Men's Organization to Serve the Mother Country). The Hokoku young men, wearing sweatshirts emblazoned with the rising sun and sporting military "bozu" haircuts, began a program of early morning calisthenics which before long grew into close-order drill to the shrill tempo of bugles.[3]

[3]Dillon S. Myer, *Uprooted Americans* (Tucson, 1971); Morton Grodzins, *Americans Betrayed* (Chicago, 1949); William Hosokawa, *Nisei: The Quiet Americans* (New York, 1969), 423–432; Daisuke Kitigawa, *Issei and Nisei: The Internment Years* (New York, 1967); Edward Miyakawa, *Tule Lake* (Waldport, Ore., 1979); D.S. Thomas and R.S. Nishimoto, *The Spoilage* (Berkeley, 1946), 333–361; Audrie Girdner and Anne Loftis, *The Great Betrayal* (New York, 1969); Rosalie Hankey Wax, *Doing Fieldwork* (Chicago, 1971); Michi Weglyn, *Years Of Infamy* (New York, 1976), 249–265; John L. DeWitt, *Final Report: Japanese Evacuation from the West Coast, 1942* (Washington, D.C., 1943); Commission on Wartime Relocation of Civilians (hereafter cited as CWRIC), *Personal Justice Denied* (Washington, D.C., 1982), 245–252; War Relocation Authority, Community Analysis Section, *Tule Lake Inci-*

While the situation worsened at Tule Lake, Justice Department and WRA officials worried over what to do about the whole controversial relocation program. From the very start of that program, the Justice Department had objected to army arguments in favor of wholesale evacuation of aliens from the West Coast.[4] As far back as early 1942, when Attorney General Francis Biddle had hosted a meeting at his home in Washington, D.C., Biddle's assistant, James Rowe, and Edward Ennis, director of the department's Alien Enemy Control Unit, had protested vigorously against any such plan. Then, when U.S. Army Provost Marshal General Allen Gullion had reached into his pocket and pulled out a draft of a presidential order to relocate aliens *and* citizens, Rowe had laughed and called the plan ridiculous. Ennis had agreed and recalled that it was the same Gullion who had earlier proposed registering *all* Americans as a wartime security measure. Still, Ennis recognized now in Biddle's living room that Gullion was serious about interning Japanese Americans, and despite a last minute appeal from Ennis, who looked as if he was about to cry,[5] Executive Order 9066 directing the evacuation of "all persons of Japanese ancestry" from the West Coast was signed by President Franklin D. Roosevelt a week later.

In mid-1943, after a year of hectoring by civil libertarians who argued that the relocation centers were unconstitutional, the WRA began planning to close its camps. The Justice Department was eager to hasten the termination of an internment program it had neither approved of nor administered. The problem was those trouble-making, pro-Japanese demonstrators at Tule Lake. By the fall of 1944, they numbered nearly a thousand young men, and news of their pro-Japanese activities was being used as evidence that on the whole the evacuees were disloyal and should not be released. It occurred then to Edward Ennis and his assistant,

dent (Washington, D.C., 1944); U.S. Congress, House, Special Committee on Un-American Activities, "Investigation of Un-American Propaganda Acts in the U.S.," *Report of Special Subcommittee on the Tule Lake Riot*, 78 Cong., 2 sess. (1944), 939, 1039; Dillon Myer, *Statement Regarding the Events that Occurred Between November 1–4 at Tule Lake Center in Northern California, 1943* (Washington, D.C., 1943).

[4]Edward Ennis, *Written Statement of ACLU Before CWRIC* (Washington, D.C., 1981).

[5]The best description of this critical meeting is in Grodzins, *Americans Betrayed*, 265–267. Grodzins interviewed several of the participants, including Ennis, not too long after the meeting occurred.

John Burling, that if they could get the thousand publicly pro-Japanese young men removed from Tule Lake to alien enemy camps, run by the Department of Justice, it could subsequently be declared that the WRA centers were filled with loyal people who should be free to leave.[6]

There were two significant catches to the plan. First, the alien enemy internment camps[7] were filled with thousands of German, Italian, and Japanese *aliens* who were considered dangerous. However, most of the young demonstrators at Tule Lake were American citizens who could not be legally removed to alien enemy camps. The answer seemed to be to permit the Japanese Americans to renounce their American citizenship. They could then be legally classified as enemy aliens and removed from Tule Lake.

It was renunciation that presented the second catch. Existing provisions of the Nationality Act of 1940 did not permit renunciation of American citizenship while within the U.S. Application had to be made outside the U.S. to a foreign consul. Consequently, Ennis's office originated a proposal to have Congress amend the Nationality Act to permit renunciation while within the U.S. Convinced it would mean at most a thousand applications, Attorney General Biddle went before the House Committee on Un-American Activities and explained the problem, recommending legislation to permit continental renunciation. Congress, which between October 1943 and February 1944 had already considered five bills to deport or denaturalize Japanese Americans,[8] was eager to cooperate in the renunciation plan, and on July 1, 1944, the voluntary renunciation provisions were enacted into Public Law 405. At Tule Lake, news that renunciation of American citizenship was now possible produced only a few hundred applications, mainly from the active Hokoku leaders.

Then on December 19, 1944, a mimeographed announcement by Dillon Myer declared that the WRA would close its relocation centers by the end of 1945. "Coast Ban Lifted," the headlines read

---

[6]Edward Ennis, *Oral Testimony Before CWRIC* (Washington, D.C., Nov. 2, 1981), 156.

[7]There were eight internment camps in operation off and on during the war years. For the duration, the Alien Enemy Internment Program interned 31,275 enemy aliens. (Records of World War II Internment Centers, Record Group 85, Washington National Records Center, Suitland, Maryland (hereafter cited as WNRC-RG 85).

[8]Donald Collins, "Disloyalty and Renunciation," 90.

in the *Newell Star,* the internee newspaper. Tule Lake residents would once again be "able to move freely anywhere in the U.S.," the paper stated.[9] Tule Lake immediately swarmed with rumors that *Nisei* (second generation) children would be separated from their *Issei* (first generation) parents, who were fearful of deportation as aliens. Or worse, it was feared, entire families would be forced to relocate into hostile white communities.[10] Now Hokoku leaders argued that renunciation of citizenship was the only way to remain together as families in the security of a relocation camp.

Meanwhile, a team of twenty army officers began to hold hearings for male residents of Tule Lake, to determine who would be permitted to return to their homes. A number of officers asked those called for hearings, "Do you want to go out, or do you want to renounce your citizenship?"[11] It bred the conviction that "in order to be permitted to stay in the security of the Tule Lake Center, renunciation was necessary."[12] And on Tuesday, December 26, the Justice Department was deluged with over 2000 applications for renunciation. The next night, December 27, seventy Hokoku leaders were arrested at Tule Lake as enemy aliens and shipped off to the alien enemy internment camp at Santa Fe, New Mexico.

Still, with the start of the new year, applications for renunciation continued to pour in. In Washington, D.C., Ennis and Burling were stunned, and they asked Attorney General Biddle to suspend the program.[13] However, Biddle decided they could not legally or even practically shelve a program they had initiated.[14] Instead, it was decided to send Burling back to Tule Lake, with a team of four Justice Department lawyers this time, to administer individual hearings. Charged with exploring the possibility that

[9] *Newell Star,* Dec. 19, 1944, file 78/177C, carton 3, Wayne Collins Papers, Bancroft Library, University of California, Berkeley.

[10] "Affidavit of Tadayasu Abo," Aug. 9, 1956, box 472, *Abo* v. *Clark,* U.S. District Court, Northern District of California, Civil Case Files, 1938–1953, Record Group 21, Federal Archives and Records Center, San Bruno, Calif. (hereafter cited as FARC-RG 21); CWRIC, *Personal Justice Denied,* 250. See also *Newell Star,* May 18, 1945, which reported "New Acts of Terrorism" on the outside, including fifteen shooting attacks, attempted dynamiting, and three arson cases against released internees. Carton 3, Collins Papers.

[11] "Outline of Events Leading to Renunciation of Citizenship," Jan. 1954, carton 3, Collins Papers.

[12] *Ibid.*

[13] "John Burling Affidavit," Nov. 12, 1946, box 467, *Abo* v. *Clark,* FARC-RG 21.

[14] Ennis, *Oral Testimony,* 158.

applicants had been coerced, Burling told the lawyers on the trip across country that "the entire evacuation has been a tragic mistake."[15] They were therefore to make sure that, despite the renunciant's bitterness, he understood what he was doing.

Arriving back at Tule Lake on January 11, 1945, Burling and his team found a whole new cadre of Hokoku leaders had replaced those sent off to New Mexico. A thousand more Hokoku members and leaders were subsequently arrested and entrained to alien enemy camps in New Mexico and North Dakota. Still, 3400 additional renunciation applications were made at Tule Lake in January. Meanwhile, Burling and his fellow hearing officers listened to renunciants who "acted very strangely at [the] hearings. Many gave wild, distorted and fantastic answers and statements as to the reasons why they were renouncing citizenship."[16] When Burling talked with officials at Tule Lake to learn the cause of the peculiar behavior and the continued rush to renounce citizenship, none of the officials suggested coercion. Rather, it was a joke among some Tule Lake staff that hysteria had run rampant, and "the center might properly be . . . run as a species of mental institution."[17]

II

Tadayasu Abo applied for renunciation of his American citizenship in December 1944. Since he was now one of over 5000 such applicants, it was not until August 26, 1945, that he learned from the Justice Department that his application had been approved and he was "no longer a citizen of the United States of America."[18] In the meantime, the Japanese had surrendered and the new President, Harry Truman, had signed Executive Proclamation 2655, making "all dangerous alien enemies [who] adhered to enemy governments . . . subject to removal from the U.S."[19]

---

[15]"John Burling Affidavit," Nov. 12, 1946, box 467, *Abo* v. *Clark*, FARC-RG 21.
[16]"Outline of Events," Jan. 1954, carton 3, Collins Papers.
[17]"John Burling Affidavit," Nov. 12, 1946, *Abo* v. *Clark*, FARC-RG 21.
[18]Copy of form letter, "Notice of Approval of Renunciation of U.S. Nationality," signed by Herbert Wechsler, Assistant Attorney General, Records of World War II Internment Centers, Ft. Lincoln, North Dakota, file 42179, WNRC-RG 85.
[19]Donald Collins, "Disloyalty and Renunciation," 146.

The news set off another round of panic at Tule Lake. Thousands of renunciants, faced with the very family separation they thought renunciation would prevent, came forward to protest that their renunciations had not been freely given. Others explained that they had renounced out of fear of punishment by militant Hokoku. In the *Newell Star,* Tule Lake Project Director Ray Best acknowledged that he was being swamped with letters from Hokoku members wanting to resign "because they had been misinformed or misled by the leaders."[20] Inside the Tule Lake compound, plans were undertaken to form a legal defense committee which would ultimately carry the title of the Tule Lake Defense Committee (TLDC). However, efforts to recruit legal help came to no avail. Shortly, the government began to register, fingerprint, and photograph the potential deportees. Despite the flurry of activity to halt deportation, it seemed inevitable. And Tadayasu Abo resigned himself to involuntary exile.

III

It was into this atmosphere of fatalism and desperation that a San Francisco attorney named Wayne Collins stepped. He was no stranger to Tule Lake, or the entire relocation program for that matter. In June 1942, as counsel for the northern California branch of the American Civil Liberties Union and attorney representing Fred Korematsu, he had argued before the District Court for the Northern District of California, in *Korematsu* v. *the United States,* that the government had no authority to detain and intern loyal American citizens.[21] Then in August 1944, he threatened to file application for writs of habeas corpus on behalf of a number of Tule Lake residents who were being held in the camp stockade.[22] WRA officials yielded, releasing the prisoners and closing the stockade. However, less than a year later, the stockade was reopened and a number of Tule Lake residents were "arrested and placed in the stockade without charges being filed against them."[23]

---

[20] *Newell Star,* Feb. 22, 1945, carton 3, Collins Papers.
[21] *Korematsu* v. *U.S.* 323 U.S. 214 (1944).
[22] "Statement of Wayne Collins," July 10, 1958, p. 4, carton 12, Collins Papers.
[23] *Ibid.,* 6.

Collins set out for Tule Lake immediately. Lou Noyes, Project Attorney for Tule Lake, met Collins at Klamath Falls and drove him to camp headquarters. There, Project Director Best joined Collins and Noyes, and Collins "upbraided them for having incarcerated citizens therein without preferring charges." Confronted again with the feisty Collins, the WRA consented to release the prisoners, who were then brought before Collins so that he could "verify the fact of their release."[24]

The matter settled, Collins was preparing to leave Tule Lake when he heard that a number of camp residents had learned of his presence and wished to see him. A private room was provided, and Collins met with a large group of parents of younger internees. From the worried parents, Collins learned of the renunciations and the threatened deportations. He quickly concluded that renunciation of nationality during detention was unconstitutional. His first advice to the gathering was that the renunciants write the Attorney General and assert that their renunciations were "void for being the direct and proximate result of government duress," along with Hokoku pressure. Collins even hastily wrote out "several identical sample forms of such a cancellation letter in pencil or ink," and urged that the letters be sent immediately.[25]

Collins then returned to San Francisco and began drafting a long letter to Attorney General Tom Clark. In the nineteen-page protest, which Collins mailed to Clark on November 1, he objected to the Justice Department's theory that the renunciants possessed dual citizenship. Renunciation did not automatically make them alien enemy Japanese. If anything, Collins argued, the renunciants were now stateless. Besides, many of them were minors and the Attorney General had no authority to fix eighteen instead of twenty-one as the age of maturity for renunciation. Then Collins presented for the first time what would be a litany of criticism against the entire renunciation program. Those who had renounced, he claimed, were victims of coercion, fraud, and duress by the government and organized gangs at Tule Lake.

---

[24] *Ibid.*, 9.
[25] *Ibid.*, 11–12.

> These renunciants whom I represent have submitted to gross indignities and suffered greater loss of rights and liberties than any other group of persons during the entire history of the nation, all without good cause or reason. They have been misunderstood, slandered, abused and long have been held up to public shame and contempt . . . and now these internees, faced with a loss of citizenship rights, are confronted with a threatened involuntary deportation to Japan. . . .[26]

Appended to Collins's letter were the hastily collected, unalphabetized names of over 900 renunciants which the TLDC had so far gathered, among them Tadayasu Abo, who wished to cancel their renunciations. If the Attorney General did not act promptly, Collins threatened, he could expect legal action immediately to cancel renunciation, to prevent deportation, and to obtain the release of those interned.

Collins delivered on his promise twelve days later and filed suit in the District Court for the Northern District of California on November 13, 1945, two days before the threatened deportation sailings. His action was entitled "Complaint to Rescind Renunciation of Nationality, to Declare Nationality, for Declaratory Judgement, and for Injunction." In simple terms, he wanted the court to cancel the renunciations, declare his clients were American citizens, and stop the threatened deportations. Collins's class action was on behalf of 987 named plaintiffs who were, because of "common questions of fact and law," united in the suit for convenience and order. They were the same renunciants who had appealed to the Attorney General on November 1 for cancellation of their renunciations. Now, however, Collins had had time to alphabetize the list, an arrangement which singled out the first man. It was now "Tadayasu Abo *et al.*, Plaintiffs," pitted against "Attorney General Tom Clark, U.S. Attorney Frank Hennessy, Secretary of State James Byrnes, Secretary of the Treasury Fred Vinson, INS Director Ugo Carusi, S.F. Immigration Chief Irving F. Wixon, Secretary of Interior Harold Ickes, WRA Director Dillon Myer, Project Director Ray Best of Tule Lake, and Ivan Williams of the INS."[27]

Collins also filed a companion action habeas corpus petition the same day demanding release of 397 adults, including Tada-

[26] Collins to Tom Clark, Nov. 1, 1945, box 467, *Abo* v. *Clark*, FARC-RG 21.

[27] "Complaint to Rescind Renunciation of Nationality, to Declare Nationality, for Declaratory Judgment and for Injunction," Nov. 13, 1945, *ibid.*

yasu Abo, who were still being interned at Tule Lake as enemy aliens.[28] In both actions, Tadayasu Abo was the named litigant. In fact, it was Wayne Collins who seemed to be taking on the world, and his twenty-three-page complaint listed three causes for his action. First, he insisted the renunciations were illegal and unconstitutional because the hearings afforded each renunciant by Burling and his team were only "pseudo-hearings" with no opportunity for individual counsel or witnesses. Besides, Collins argued, the congressional amendment to the Nationality Act which permitted renunciation was unconstitutional because it gave "improper legislative and judicial power to the Attorney General."[29] Then Collins asked rhetorically: How could the renunciants legally be deported as "enemy aliens" when the U.S. was at peace with the world?[30]

The second cause of action was an elaboration upon the duress and coercion theme he had raised in his November letter. Now, however, he spelled out how he felt the renunciants had been intimidated by a government which "still holds them in a continual mental state of terror simply because of Japanese ancestry."[31] Collins went on to explain that Japanese parents, led by government actions to fear deportation as aliens, had pressured their American children to renounce in order to keep families together. "[T]he signing of the renunciation application was a trap," Collins charged, "designed by government . . . to cause and result in involuntary deportation . . . and thus result in mass banishment of persons of Japanese ancestry from the U.S."[32] Furthermore, Collins maintained, the government permitted his clients to be subjected to "organized terrorist groups and gangs . . . who were fanatically pro-Japanese."[33] Citing assaults and blacklisting by

---

[28]Since the WRA had already begun closing its relocation centers, it had released the less militant renunciants with the stipulation that they appear when summoned for deportation.

[29]"Complaint to Rescind," Nov. 13, 1945, box 467, *Abo* v. *Clark*, FARC-RG 21.

[30]The so-called Alien Enemy Act (Title 50, USC, chap. 3, sec. 21) spells out that "whenever there is a declared state of war between the U.S. and any foreign nation," citizens of that hostile nation shall be subject to apprehension and removal as enemy aliens of the U.S. The government's position, reaffirmed in the U.S. Circuit Court of Appeals decision of May 2, 1946 (*Citizens Protective League* v. *Tom Clark*, 155 F.2d 290 [1946]) was that the act would remain in full force until a formal peace treaty had been signed with Japan.

[31]"Complaint to Rescind," Nov. 13, 1945, box 467, *Abo* v. *Clark*, FARC-RG 21.

[32]*Ibid.*

[33]*Ibid.*

pro-Japanese goon squads against evacuees who even *talked* to Caucasian personnel in the relocation centers, Collins wrote that "Many loyal and innocent internees were driven into becoming nominal but inactive members of [the Hokoku] to save themselves from danger, physical violence and probable loss of life."[34] The government knew of these group activities and did nothing to suppress them. The result was a "wave of terror" at Tule Lake which led to the renunciations.

Collins's third cause of action alleged that many of the plaintiffs were under twenty-one when they signed their renunciation application.[35] For that reason alone, Collins felt, the renunciations were legally void.

As a result of the habeas corpus petition on November 13, seventy-eight-year-old District Court Judge A. F. St. Sure, to whom *Abo* v. *Clark* was assigned, set a date for Ivan Williams of the INS to appear before the court to show cause why the petitioners still interned at Tule Lake should not be set free. St. Sure sat down with Collins and Assistant U.S. Attorney Robert McMillan to inform them that "he did not desire to hear oral arguments in the case ... and that he desired the motions submitted in briefs."[36] On November 25, repatriation ships began a series of sailings which, during November-December 1945, would remove 7,100 repatriates to Japan.[37] However, Collins's suit had thrown what he characterized as a "monkey wrench" into government plans for involuntary deportation of the renunciants.[38] Collins informed the TLDC by telegram that "the deportation program had been converted from an involuntary to a voluntary basis."[39] At the same time the Justice Department announced that it would hold "mitigation hearings" at Tule Lake and its alien enemy camps. These mitigation hearings were designed to give the renunciants an opportunity to show cause why they should not be removed to Japan. Commenced the week of January 4, 1946, and continued throughout that spring, mitigation hearings were accorded 3,186

---

[34] *Ibid.*

[35] This was not true for Tadayasu Abo; he was thirty-three.

[36] Collins to Iwao Shimizu, Nov. 8, 1946, carton 16, Collins Papers.

[37] *Newell Star,* Jan. 31, 1946, carton 3, *ibid.*

[38] Abstract of phone call, Collins to Larry Takaoka, TLDC, Nov. 30, 1945, carton 16, Collins Papers.

[39] Collins to Takeo Yamamoto, TLDC, Dec. 3, 1945, *ibid.*

renunciants at Tule Lake, on the basis of which 2,280 were granted a stay of deportation orders.

Still, Collins had already stipulated that the mitigation hearings in no way would cancel or replace the legal proceedings for restoration of citizenship he had undertaken on behalf of the renunciants. In fact, at Collins's prompting, the renunciants granted mitigation hearings individually protested to Attorney General Tom Clark that they had had no counsel at the brief hearings, were not given time to prepare, could not subpoena witnesses, and had been confronted with new evidence gathered secretly by the hearing officers.

Throughout that spring, Collins began adding additional parties as plaintiffs in the suit, eventually including over 5,000 in the category of *et al.* Some of these additions were evacuees whom he and his assistant Theodore Tamba personally interviewed on trips during January 1946 to Ft. Lincoln, North Dakota, and Santa Fe, New Mexico, where the "hard core" alien enemies were being held pending deportation. However, at both camps, Collins and Tamba were denied an opportunity to consult privately with their clients. Consequently, on March 4, after returning to San Francisco, Collins filed a twelve-page "Supplement and Amendment" to his original complaint. After his customary recital of Hokoku terrors and duress at Tule Lake, Collins went on to protest that those terrors were continuing at the Ft. Lincoln, Santa Fe, and Crystal City, Texas, internment camps. He also alleged continued government duress in the form of phone taps of his calls to Tule Lake, and eavesdropping on his lawyer-client conferences at Ft. Lincoln and Santa Fe. Then he attacked the mitigation hearings themselves as arbitrary pseudohearings in which new evidence was presented against subjects who had no opportunity to call witnesses or receive legal counsel. In mid-March, he wrote Tom Clark that he "appreciated the present plight of these internees was not your doing," but rather had been inherited by Clark from his predecessor, Francis Biddle, and the WRA. Collins then asked that the renunciants still being detained be released pending the outcome of the case. On March 27, Clark responded, "I have carefully considered your request . . . [and] I am unable to accede to it."[40]

[40]Collins to Tom Clark, March 18, 1946, *ibid.;* Tom Clark to Collins, March 27, 1946, carton 14, *ibid.*

Attached to Collins's "Supplement and Amendment" of March 14 was a letter which would spark one of the key controversies in Collins's complaint. Dated August 6, 1945, the letter was from then Under Secretary of Interior Abe Fortas to the ACLU's Ernest Besig. Besig had protested the existence of coercive, terroristic organizations at Tule Lake. In response to Besig's letter, Fortas wrote, "We have completed our investigations and in this letter I shall report rather fully our findings and conclusions." It was Fortas's conclusion that insofar as the renunciants were concerned, many of them were "in the grip of emotional hysteria created by these organizations, or actually acting under fears of violence." Fortas went on to say that "It was primarily due to the pressure of these organizations that over eighty percent of the citizens eligible to do so applied for the renunciation of citizenship."[41]

Just how official was Fortas's damaging letter to a private citizen became a key issue in the government's first legal response to Collins's complaint. In a "Motion to Strike" filed on April 15, 1946, the government claimed that Collins's original complaint and the amendment to it were full of "redundant, immaterial and impertinent matter."[42] Then, in a brief filed on July 5 in support of its strike motion, the government elaborated that Fortas's letter to a private citizen did not constitute an "official finding." Moreover, it was incompetent evidence because it was "opinion based on hearsay." Fortas had no firsthand knowledge of duress at Tule Lake. And his contention that eighty percent of the renunciations were prompted by duress did not establish which *specific* petitioners were involved.

The government's argument was presented mainly by Frank J. Hennessy, U.S. Attorney for Northern California, and Thomas Cooley, who had become director of the Alien Enemy Control Unit. Hennessy and Cooley conceded many of the relocation particulars spelled out in Collins's complaint, but they argued that the question of whether duress had prompted renunciation would require an investigation into the mind of each individual petitioner. Meanwhile, they wanted to strike the entire complaint because it was an "inextricable congeries of assertions of ultimate fact, evidentiary matter, legal conclusions and irrelevant material."[43]

---

[41] Abe Fortas to Ernest Besig, Aug. 6, 1945, box 467, *Abo* v. *Clark*, FARC-RG 21.
[42] "Motion to Strike," April 15, 1946, *ibid.*
[43] "Memorandum Brief in Support of Motion to Strike," July 5, 1946, *ibid.*

The issue of whether to strike the original complaint was argued in Judge St. Sure's courtroom on July 5, 1946. Cooley spoke for the government, maintaining that the strike motion was not a dilatory tactic. "A proper determination of the very important issues requires that they be drawn as sharp and clear as possible." If the issues were not clear, Cooley warned, the government would insist on trying all the cases separately.

Collins answered, "I might say, your honor, I disagree with Mr. Cooley absolutely.... The issues are specifically set forth!"

St. Sure reminded Collins that, specific or not, "the burden of Mr. Cooley's argument is that the issues should be made clear."

"The issues are very clear, your honor," Collins shot back. "The trouble with the government is [it] would like to avoid the declarations, statements and reports of its own servants which admit the duress." It was, of course, the Fortas letter that Collins claimed the government wanted to avoid. "[T]he mere fact it was sent to a member of the public does not convert it to mere opinion," Collins felt.

Collins then went on to argue that detention by the government was in fact duress. St. Sure responded with surprise: "Don't you think the government was justified in detaining these persons?"

"No, your honor." Collins insisted. He explained that hearings were not even granted the evacuees.

"I know what we should have done," St. Sure interrupted.

Cooley then moved to contradict Collins's assertions that government actions had induced the renunciations. "[N]ow, obviously, the government is not going to concede any such thing," Cooley protested.

Collins answered that he was not asking the government to concede anything. "We merely ask you to answer the allegations."

Cooley insisted that if Collins continued to charge that government duress was an issue, "it will be necessary to split these cases into individual cases and try each one.... Duress is an individual matter."[44]

Despite Collins's belief that the issues were clearly and specifically laid out, on July 11, 1946, Judge St. Sure granted the government's motion to strike the entire original complaint, including the controversial Fortas letter. Collins was given twenty days within which to file an amended complaint. St. Sure granted Col-

[44]From court transcripts for July 5, 1946, *ibid.*

lins an extension to August 19, and just four days before that deadline the determined Collins filed his "Amended Complaint." It was still twenty-three pages long and contained many of the same "inextricable congeries" of material which had previously been ordered stricken. However, Collins now alleged in his assertions of government duress that there were twenty occasions between October 1945 and March 1946 when the government had tapped his phone calls to Tule Lake. And rather than strike the entire Fortas letter, Collins insisted the document was official and he again quoted Fortas's finding that eighty percent of the renunciants had renounced because of group pressure.

IV

The government gave its "Answer" to Collins's amended complaint on September 23, 1946. While refusing to respond to allegations in the stricken Fortas letter, the ten-page answer did admit the facts of Collins's narrative description of evacuation and internment. It also admitted that pro-Japanese groups like the Hokoku had been permitted to operate at Tule Lake. These groups, the government conceded, "engaged in misrepresentations with respect to the purposes and effects of the deportation and renunciation laws ... and engaged in propaganda campaigns."[45] Yes, the government agreed, parental persuasion was present. And some of the plaintiffs, but not all, were minors. However, the government asserted that these confessions did not amount to an admission that duress or coercion had caused the renunciations. There was no influence by the government. Renunciation was not necessary to preserve the family unit. The renunciation hearings were fair, and the subsequent renunciations voluntarily offered. It was not until the fact of the A-bomb and Japanese surrender that any of the plaintiffs had moved to rescind the action. Therefore, the government still intended to deport them to Japan as enemy aliens, and it moved to dismiss the complaint.

Collins promptly filed a "Motion to Strike" the government's answer. "[T]he denials are vague, indefinite, uncertain and evasive," and the entire answer was "sham, false, frivolous, impertinent and evasive."[46] Clearly, Collins was particularly annoyed by the alleged

[45] "Answer," Sept. 23, 1946, *ibid.*
[46] "Motion to Strike," Oct. 10, 1946, *ibid.*

evasiveness. It was, he felt, a stalling tactic until such time as a formal peace declaration was signed with Japan. That would make moot the issue of his clients being alien enemies, and the government would thereby "save face," Collins charged.[47] On October 14, Collins urged St. Sure to deliver his judgment in the proceedings. The issues of fact and law in his complaint were either undenied or admitted in the government's answer, Collins insisted.

The government spent November 1946 filing its own cross-motions for judgment, along with a fifty-one page brief to accompany its answer of October. Complaining to St. Sure that Collins's amended complaint was "substantially indistinguishable" from the one ordered stricken, Frank Hennessy argued that the question of whether minors could legally renounce their citizenship was a difficult one. But regardless of that question, Hennessy insisted that the government had not coerced the renunciations, and therefore the renunciants were alien enemies. To substantiate both claims, Hennessy included affidavits from "authorities" who could speak to the issues. First, John Burling swore in a long, forty-six-page affidavit that "at no hearing which I personally conducted was there any evidence of coercion or duress." Nor were there any reports or claims made to him to that effect. And as far as group pressures to renounce were concerned, they may have only caused "the crystallization of an idea." That was a far cry from coercion, however.[48]

To speak to the legality of the enemy alien label that the government applied to the renunciants, Hennessy secured an affidavit from Thomas Blakemore, a career foreign service officer familiar with Japanese nationality law. Blakemore's affidavit attempted to buttress the government's previous contention that most of the renunciants had held dual nationality. The government's point was that they *automatically* became Japanese citizens, and therefore enemy aliens, with their renunciations of American citizenship.

Collins moved swiftly to enter counterclaims to the Burling and Blakemore affidavits, and on December 11, 1946, he filed a 109-page brief in support of his amended complaint. In a preliminary statement, which again insisted that the government was culpa-

---

[47]Collins to Iwao Shimizu, undated, carton 16, Collins Papers.
[48]"John Burling Affidavit," Nov. 12, 1946, box 467, *Abo* v. *Clark*, FARC-RG 21.

ble, Collins observed that "Herr Hitler was guilty of abusing segments of [his] own citizenry for racial reasons. We were inured, however, to a like abuse of our own citizens by our own government."[49] Government racism was obvious, Collins felt, since German and Italian aliens who were arrested were given prompt, individual hearings,[50] "and all but a few were released."[51]

Burling's affidavit submitted by the government had maintained "it was not legally relevant to determine the ultimate motivation which led to the renunciation."[52] Collins now took issue with that government position, which addressed itself only to threats or intimidation in the examining room at the "precise moment of renunciation."[53] Collins believed that Burling and his fellow hearing officers should not have been looking for revolvers held at renunciants' temples; rather, it was the state of mind induced in a person by forces outside the room they should have been sensitive to. And to elaborate on what those forces were, and the fear induced by them, Collins introduced several affidavits which he felt spoke to the point with more authority than the government's affiants.

The first was a statement from Tetsujiro Nakamura, a twenty-nine-year-old Nisei who had owned a retail food store in Burbank before his internment. Characterizing life at Tule Lake as full of bewilderment and fear, Nakamura testified, "I do not know of a single person who renounced of his own free will."[54] Burling's

[49] "Brief for Plaintiffs," Dec. 11, 1946, box 468, *ibid.*

[50] This was not accurate. In a letter dated August 9, 1948, from W. F. Kelly, assistant commissioner for Alien Control, to A. Vulliet of the YMCA, Kelly totalled the alien enemy figures: 13,183 German and Italian aliens were arrested during the war. Approximately half were interned; the other half were parolled. Many German and Italian seamen, arrested as illegal aliens before the war started, received no hearings for years. And in a memo dated February 7, 1944, from Attorney General Biddle to Earl Harrison, Immigration and Naturalization commissioner, Biddle wrote that Latin-American aliens had been arrested and brought to this country under "unusual circumstances." Therefore, it was "undesirable" to afford them a hearing before an Alien Enemy Hearing Board. (Alien Enemy Internment Camp Records, Central Office File 56163–84, INS, Washington, D.C. These records were in the process of being transferred to the National Archives during the author's research on the complete story of the World War II Alien Enemy Internment Program. It is presumed they are now located in RG 21, Washington National Records Center, Suitland, Md.

[51] "Brief for Plaintiffs," Dec. 11, 1946, box 468, *Abo* v. *Clark*, FARC-RG 21.

[52] "John Burling Affidavit," Nov. 12, 1946, box 467, *ibid.*

[53] "Brief for Plaintiffs," Dec. 11, 1946, box 468, *ibid.*

[54] "Nakamura Affidavit," Dec. 11, 1946, *ibid.*

knowledge of events at Tule Lake was superficial, Nakamura held, and the allegations in the Fortas letter were all true. To support Nakamura's claims, Collins attached an affidavit from Masami Sasaki, a fifty-seven-year-old Issei, who maintained that "the whole camp was terrorized by the pressure groups."[55] He also knew of no one who renounced freely. Then, Collins included an eighteen-page affidavit from Ernest Besig, who cited murder, secrecy, slave labor, and the case of an interned woman who "hammered one of her children to death" as evidence of the atmosphere of fear and terror out of which renunciation emerged.[56] Next, Reverend Thomas Grubbs, whose Tule Lake church was across the compound road from the Japanese Language School, asserted in his affidavit that the school engaged in coercion, intimidation, and terror to encourage renunciation. Immediately after the renunciation hearings were over, Grubbs testified, a number of the renunciants came to him secretly to confess that they had not wanted to cancel their citizenship. Grubbs's testimony was Collins's rebuttal to the government's argument that only the A-bomb and Japanese surrender had prompted the changes of heart.

Collins's final and most powerful affidavit was from Ann Ray, who during late January 1946 had sat in on twenty mitigation hearings as an observer for the Northern California ACLU. Collins had already characterized the mitigation hearings as "star-chamber proceedings . . . [and] nothing but a face-saving device designed to lend the appearance of fairness and legality to proceedings."[57] Now Ann Ray seconded his charge. The renunciants, she stated, were "fearful, grave, serious, nervous . . . [and] incoherent in their speech because of anxiety and fear." When barred from observing the hearing of one internee, Ray complained to Tule Lake officials, only to be told, "Well, it doesn't make any difference. . . . Everything is decided in advance." Ray's affidavit then described the activities of Marvin Opler, a WRA official at Tule Lake who ran "rumor experiments" to gather scientific data on how rumors were disseminated. Ray testified further that she

[55]"Sasaki Affidavit," Dec. 11, 1946, *ibid.*
[56]"Besig Affidavit," Dec. 11, 1946, *ibid.*
[57]"Brief for Plaintiffs," Dec. 11, 1946, *ibid.*

had talked with two hundred renunciants, all of whom had renounced because they had been "prevented from doing otherwise" out of fear of the Hokoku. Finally, she recounted a dinner for the Caucasian staff at the Tule Lake Recreation Club, when a security policeman at her table had remarked, "If there is anything I hate it is a Jap. All Japs should be taken out and shot."[58]

V

A week after filing his brief in December, Collins took legal objection to the government's affidavits from Burling and his fellow hearing officers. The affidavits were full of opinions and hearsay, Collins charged, and they were "self-serving, not binding, sham, evasive[,] .... immaterial."[59] Furthermore, Blakemore's affidavit on Japanese nationality law was irrelevant. The laws of Japan did not affect U.S. citizens, Collins objected.

With Collins's introduction of testimony by Nakamura, Sasaki, Besig, and especially Ann Ray, the government responded by filing a "Supplemental Brief" on January 27, 1947. Charging that Collins's witnesses had made Tule Lake seem like a "horror camp of the type recently made familiar by War Crimes Trials in Germany," the government called Ann Ray's affidavit "plainly worthless hearsay." The background circumstances at Tule Lake could not "vitiate renunciation," the government insisted.[60] Despite that position, the government now included an additional affiant of its own, who spoke to background conditions at Tule Lake. She was Rosalie Hankey, a thirty-one-year-old graduate student in anthropology from the University of California, Berkeley, who had spent a year at Tule Lake as part of the "Evacuation and Resettlement Study" being done by some researchers at the university. Whatever the government intended by the inclusion of the forty-seven-page Hankey affidavit, it seemed to speak more to Collins's arguments than the government's. "It is my opinion," Hankey wrote, "that members of the [Hokoku] who . . . refused to apply for renunciation of citizenship . . . stood in danger of physical violence." Moreover, "a great many citizens made false statements at

[58]"Ann Ray Affidavit," Dec. 11, 1946, *ibid.*
[59]"Objections and Exceptions to Affidavits Filed by Defendants . . . ," Dec. 18, 1946, *ibid.*
[60]"Supplemental Brief for Respondents," Jan. 27, 1947, *ibid.*

their hearings regarding their loyalty to Japan and to the Japanese emperor." Finally, Hankey asserted, the pro-Japanese pressure group had "within its body a group of terrorists who repeatedly assaulted residents" without fear of arrest.[61] Despite such observations, Hankey's perhaps contradictory conclusion was that threats of physical violence were relatively minor incidents.

In its supplemental brief, the government also included thumbnail sketches of 112 renunciants. Based on information the subjects had delivered at renunciation or mitigation hearings, the sketches were included to show that the renunciations had been voluntarily offered by "dangerous" enemy aliens. One profile contained an admission from a renunciant who swore that if he was forced to settle out of the internment camp, he intended to become a member of a Japanese suicide squadron, bent on destroying the U.S. war effort. Another renunciant had confessed at his mitigation hearing that he had been loyal to Japan from 1943 until VJ Day and had wanted Japan to win the war. Others had vowed to return to Japan and serve the war effort. Still another admitted that if he were freed in the U.S., he would wreck factory equipment and kill workers in plants.[62]

It took Wayne Collins only two days to file a motion to suppress the thumbnail sketches as "evidence illegally obtained."[63] Statements by renunciants at either their renunciation or mitigation hearings were obviously coerced statements, Collins argued. Furthermore, the inclusion of these statements was tantamount to compelling a witness to testify against himself. As for the testimony of Rosalie Hankey, it ought to be stricken because it was incompetent and irrelevant. Collins continued his attack on Rosalie Hankey in a "Supplemental Memo" filed on February 11, 1947. She was naive, Collins insisted, if she thought she had gotten the full story of the Tule Lake renunciations.[64] And to contradict her conclusion that physical threats of violence were rela-

---

[61]"Rosalie Hankey Affidavit," Jan. 27, 1947, *ibid.*

[62]"Supplemental Brief for Respondents," Jan. 27, 1947, *ibid.*

[63]"Motion to Suppress Evidence Illegally Obtained," Jan. 29, 1947, *ibid.*

[64]In her book, *Doing Fieldwork* (1971), published some twenty-seven years after her experiences at Tule Lake, Rosalie Hankey nee Wax wrote that by April 1944 she had a good working relationship with more than a dozen of the segregant residents of Tule Lake. Since there were 18,000 residents at Tule Lake then, she was perhaps as naive as Collins charged.

tively minor incidents, Collins cited from *The Spoilage,* ironically the very book on evacuation and resettlement to which Hankey had contributed as a graduate student. "The most prevalent explanation [for the renunciations] dealt with the pressure tactics of *Hokoku* and *Hoshi-dan* in forcing decisions to renounce upon members and unwilling nonmembers alike," stated the book.[65] Unfortunately, Collins noted, these were points that Hankey had conveniently forgotten in her affidavit for the government.

VI

On February 20, 1947, because aging Judge St. Sure had become ill,[66] it was necessary to reassign *Abo* v. *Clark* to Judge Louis B. Goodman. Appointed to the District Court by FDR, it was Goodman who would see the case through the next fourteen years of motions, stipulations, appeals, notices, and orders. Goodman immediately set about the arduous task of familiarizing himself with the volumes of written evidence in the form of memoranda, briefs, and affidavits, along with the various legal points and authorities to which both sides had referred. Not until June 30, 1947, was Goodman prepared to take his first action in the proceedings, granting Collins his habeas corpus petition. It was Goodman's judgment that the petitioners could not be declared enemy aliens by mere renunciation; consequently, their imprisonment was illegal. The government immediately appealed the decision to the Ninth Circuit Court in San Francisco. In the meantime, however, 302 internees still under detention at Crystal City, Texas, or on "relaxed" internment at Seabrook Farms, New Jersey, were ordered paroled into the custody of Collins.[67]

Then, on September 5, even more good news was forthcoming when the Central Division of the Southern California District Court announced its decision in *Inouye* v. *Clark.* The court held that the renunciations of citizenship by four Japanese American women at Tule Lake were invalid. "The modern view and true doctrine of duress and coercion," the court noted, was fear of ac-

[65]"Plaintiffs' Supplemental Memorandum," Feb. 11, 1947, box 468, *Abo* v. *Clark,* FARC-RG 21; Thomas and Nishimoto, *The Spoilage,* 353.

[66]St. Sure died two years later at age eighty.

[67]Collins to Iwao Shimizu, Nov. 5, 1946; and Oct. 3, 1946, carton 16, Collins Papers.

tual or apparent physical violence. Therefore, the "state of mind of the injured party" was critical.[68]

This decision seemed to echo Collins's arguments in San Francisco. It was the fearful *state of mind* of each of his plaintiffs when they had been in the renunciation room which was the crux of the matter. But Attorney General Tom Clark had already objected to Collins that "the record is abundantly clear that at the time they renounced, they did so entirely voluntarily." Furthermore, Clark had argued, he was not empowered to "speculate as to their remote motivation for doing so."[69] Now, however, Collins wrote Judge Goodman to call his attention to the southern California court's decision, which seemed to contradict Clark.

Then, on October 13, both Collins and the government stipulated that, since the merits of the case were now before Goodman in written evidence and affidavits, they were ready for his decision. He could order the production of additional evidence on any particular plaintiff if he wanted. But he could also declare the evidence invalid if it so struck him. It was a stipulation which appeared to favor Collins, who hoped to avoid a lengthy, case-by-case determination of the state of mind of each of the nearly 2,300 plaintiffs whom he now represented. Meanwhile, despite the developments favoring Collins, the government continued to insist that the plaintiffs were deportable Japanese aliens.

The issue was temporarily settled on April 29, 1948, when Judge Goodman delivered a fifteen-page opinion in *Abo* v. *Clark.* Goodman first carefully pointed out that he was not judging the constitutionality of the renunciation statute (which Congress had since repealed), and that he had not considered the mass of documentary evidence both sides had submitted. Moreover, he had ignored the Fortas letter, despite Collins's repeated attempts to introduce it. Goodman agreed with the government that some of the renunciants had acted freely, but they were "not the renunciants here seeking restoration of their citizenship." As for the latter, their actions were due to group pressure, parental pressure, fear of being returned to hostile white communities, fear of de-

[68]*Inouye* v. *Clark,* 73 F.Supp. 1000 (1947).
[69]Tom Clark to Collins, June 25, 1946, carton 16, Collins Papers.

portation, and mass hysteria. "I am convinced," Goodman wrote, "that such factors, singly or in combination, cast a taint of incompetency upon any act of renunciation." Justice Department officials were not alone to blame, he noted, for "the people of the United States, acting through their representatives in Congress, as well as executive and administrative officers of government," were equally culpable. "It is incumbent upon the U.S. to now effect and properly correct the evils. . . . The government need not sheepishly confess error," but rather "must be stalwart and forthright in admitting the wrong done."[70]

The TLDC wired an exhausted Collins: "Wish to thank you for untiring efforts in winning the case."[71] Along with that praise, however, came a letter which read, "The Jap sympathizers seem to be still on the job. . . . You know and I know that that bunch of perverted fanatics can't make good citizens."[72] His convictions stronger than ever, Collins snarled back at the writer, "I suggest sometime that if you are literate enough, you read the Constitution of the United States."[73]

VII

On September 27, 1948, Goodman transformed his earlier opinion into an "Interlocutory Order," cancelling the renunciations on the grounds that they were the result of duress, menace, coercion, and intimidation. He found the plaintiffs to be native-born citizens of the U.S., and thus not subject to detention or deportation. Collins might have whooped, except there was a catch in Goodman's actions. The government would have four months to designate those individuals about whom it could introduce evidence that they "were not affected by the duress at all, but had renounced freely and voluntarily."[74] After that, it would be too late. A final order would be entered.[75]

[70]*Abo* v. *Clark*, 77 F.Supp. 806 (1948).

[71]TLDC to Collins, May 6, 1948, carton 15, Collins Papers.

[72]M. S. Parker to Collins, Aug. 24, 1948, carton 14, *ibid.*

[73]Collins to M. S. Parker, Aug. 30, 1948, *ibid.*

[74]Form letter sent to all renunciants from Wayne Collins, March 19, 1951, carton 10, *ibid.*

[75]Collins later charged that the blame for Goodman's catch lay with the Los Angeles ACLU attorneys who had represented four individual renunciants on August 27, 1944, in the Los Angeles District Court. According to Collins, that case, *Acheson* v. *Murakami*, 176 F.2d 953, branded "a large number of internees as being disloyal to the U.S." (Form letter sent to

At a conference with Judge Goodman on January 25, 1949, just a few days before the four months were up, Hennessy requested an extension of time to file names of the so-called "designations." If his request were not granted, he threatened to "designate" all the plaintiffs for individual trials. At the conference, Collins objected vigorously, claiming that the court had enough evidence to decide every case. But Judge Goodman, eager to avoid Hennessy's threat of a jammed court, felt the "interests of justice would be served" by a month's extension.[76]

On February 25, the government presented to Goodman a "Designation of Plaintiffs" whom it planned to try individually. Despite the government's previous declarations that the extension would expedite matters, Goodman now found the government singling out nearly 3,500 plaintiffs. The names were arranged in twenty-two categories, according to common circumstances which the government considered critical: Were they Kibei? Hokoku leaders? Had they already been voluntarily deported? The largest of these categories included the names of 1,087 Kibei, including Tadayasu Abo, who had applied for renunciation and then renounced, but who were *not* under deportation orders.[77]

Three days after the lists were filed, Collins moved to strike them. Back in October, he charged, Thomas Cooley had told him

---

all renunciants from Wayne Collins, March 19, 1951, p. 5, carton 10, Collins Papers.) It was mostly Kibei whom the court singled out, and the *Murakami* decision established that federal courts "must be more vigilant than ever that the massing of 4,315 plaintiffs" concealed no "enemy minded renunciants" (p. 6, *ibid.*). Collins argued that had that case *not* been pressed and decided before his, Goodman would have had no precedent for his catch.

[76] "Order Extending Defendant's Time in Which to File Designations," Jan. 25, 1949, box 468, *Abo* v. *Clark*, FARC-RG 21.

[77] Of the twenty-two categories, those containing the most renunciants were:

IV. Non-Kibei, Hokoku members who had voluntarily returned to Japan (387).

V. Kibei, non-Hokoku who had renounced and returned to Japan (284).

VI. Non-Kibei, non-Hokoku, who had voluntarily returned to Japan (292).

VIII. Kibei, who had applied for repatriation, renounced, and were under orders to be deported (216).

XIII. Kibei, who had applied for expatriation and renounced, but were not presently under removal orders (1,087).

XIV. Renunciants, who had applied for expatriation prior to renunciation, but were not under removal orders (1,049).

XIX. Non-Kibei, who were not leaders, had not applied for repatriation, and were not under removal orders, but whose renunciations were voluntary (277). (Box 468, *Abo* v. *Clark*, FARC-RG 21.)

that the government had no further evidence to introduce. Now here were these lists incorporating the names of nearly every plaintiff. The lists were a "sham," Collins charged, and a violation of his previous stipulation with the government that the case could be decided on its merits.

Goodman agreed with Collins and ordered the government to show cause why the "designations" should not be stricken. On March 7, Hennessy insisted the lists were legitimate. He also included an affidavit from Cooley, who conceded that he *may* have once said it looked like the introduction of additional evidence had been "foreclosed," but he had never told Collins the government would not, *under any circumstances,* introduce additional evidence. Since there was no explanation of the "circumstances" which now prompted the government to designate 3,500 plaintiffs for trial, Collins concluded that the lists were a tactic by the government to avoid the final order against it. Whatever the motive, the designations raised the threat of the lengthy, case-by-case trials which Goodman had been eager to avoid. On March 23, 1949, Goodman held that "none of the classified lists of plaintiffs . . . has any competency, relevancy or materiality."[78] He then granted Collins his motion to strike the lists and requested that both sides file their motions of fact and law before he issued a final order.

Collins responded with the charges of fear, hysteria, and duress that he had been reciting for over three years, and he pleaded for the restoration of his clients' citizenship. The government responded by labelling Collins's argument as "so patently objectionable as to require no protracted discussion."[79] Still, the government acknowledged that parental pressure, mass hysteria, and fear of deportation had created "neuroses" at Tule Lake. On the other hand, the government insisted that it had not sought to "entrap" the renunciants.

Four days later, on April 12, Goodman issued his "Final Order," holding that there was "complete lack of Constitutional authority [for the U.S. government] . . . to detain and imprison American *Nisei* citizens . . . when they were not charged with criminality."

[78] "Order Striking Defendant's Designations of Plaintiffs," March 23, 1949, *ibid.*
[79] "Defendant's Memorandum of Objections," April 7, 1949, *ibid.*

Conditions at Tule Lake, including government duress, had produced "neuroses built on fear, anxiety, resentment, uncertainty, hopelessness, and despair of eventual rehabilitation."[80] Goodman ordered all effected or pending renunciations set aside. Each plaintiff was a native-born U.S. citizen, and the government could not detain or intern any of them.

The decision was a powerful affirmation of nearly every point Collins had argued since 1945, yet he hardly had time to celebrate his victory before the government on April 26 appealed to the Ninth Circuit Court. In the fall of 1949 the government filed its brief, insisting that Goodman had erred in ruling that the "designated" lists were immaterial. The government had offered to introduce documentary transcripts of hearings, which would have shown that some of the renunciations were voluntary, but Goodman had refused to admit the evidence. Therefore, the question of which renunciations were voluntary still needed to be determined.

Collins responded by accusing the government of "characteristic perversity" for "continuing those deprivations which now are extended into the 8th year since evacuation." Not until the District Court had ruled against the government, Collins explained, had the government filed its "fictitious designations."[81] It had been a dilatory tactic, and Goodman's judgment and orders should be affirmed.

Chief Justice William Denman delivered the court's opinion on January 17, 1951. By that time, 4,315 plaintiffs were party to the suit. The court agreed with Goodman that the regulations which permitted 1,004 plaintiffs to renounce as minors (under age twenty-one) were unconstitutional. On the other hand, the court held that the remaining adult plaintiffs had renounced only after the Hiroshima and Nagasaki bombings had made it clear that Japan had lost the war. Their actions, the court reasoned, can be presumed to be the result of "simple expediency and crass material considerations, not duress."[82] The "burden of proof" to the

---

[80]"Final Order, Judgment and Decree," April 12, 1949, *ibid.*

[81]"Brief for Appellees," March 6, 1950, box 159a, *Abo* v. *McGrath*, Ninth U.S. Circuit Court, FARC-RG 276.

[82]"Opinion," Jan. 17, 1951, *ibid.*; *Abo* v. *McGrath*, 186 F.2nd 766 (1950).

contrary lay with such renunciants. The court noted that the government had previously agreed to accept individual affidavits from each renunciant in lieu of oral testimony. Though the issue would put a heavy burden on the courts, it was remanded to the District Court for further proceedings.

The effect of the partial reversal of Goodman's order was to saddle Collins with the time-consuming task of soliciting individual affidavits from the 3,300 remaining adult plaintiffs. Collins was convinced that the decision was unfair because it "gave the Attorney General another chance to produce evidence" he had failed to produce in four years.[83] Collins petitioned the Circuit Court a month later for a rehearing, arguing that the District Court had never deemed it necessary to order the production of additional evidence about any particular plaintiff. The facts of duress which he had presented had applied to *all* plaintiffs, he insisted.

The petition for a rehearing was denied,[84] and in October 1951, Goodman ordered Collins to begin submitting written affidavits from individual plaintiffs, in lieu of oral testimony, on the involuntary nature of their renunciations. Collins and the government agreed to a standard form which each renunciant would submit.[85] Once the affidavit had been received by the court, if the Justice Department had nothing in its files to contradict the written testimony of the renunciant, it would not oppose a final court order restoring to the plaintiff his citizenship.

VIII

Tadayasu Abo had since returned to Olympia, Washington, with his wife Yukiko and their two children. The government continued to consider him an enemy alien until the war with Japan was officially ended on April 28, 1952. After that, he was still considered a deportable alien who was required to register his status with the government. Availing himself of a form protest letter supplied by Wayne Collins, Abo protested by registered mail to the Justice Department in January 1953: "I deny that I am an alien."[86]

---

[83]Form letter sent to all renunciants from Wayne Collins, March 19, 1951, carton 10, Collins Papers.

[84]Collins's application to the Supreme Court for a writ of certiorari was also denied on October 8, 1951.

[85]Collins negotiated mainly with Warren Burger, then an assistant attorney general.

[86]Abo to U.S. Dept. of Justice, postmarked Jan. 9, 1953, carton 12, Collins Papers.

In the summer of 1956, Abo received from Collins his blank affidavit form. Collins prefaced the form which he sent to each plaintiff with a cover letter entitled "Outline of Events Leading to Renunciation." In it, he refreshed in each recipient's mind the sequence of events, now more than a decade past, which had led to the renunciations. He cautioned the affiants that they were to explain their renunciations in their own words. When Abo submitted his affidavit, he attributed his actions to fear of relocation: "I was afraid that we would be forced to relocate with small children where people would be unfriendly toward us." With respect to the controversial loyalty questions on the WRA questionnaire in 1943, Abo wrote that he had refused to serve in the U.S. Army and refused to forswear allegiance to Japan "because I was fearful for the safety of my alien parents who were in Japan, and I believed they would get in trouble with the Japanese government if I showed a willingness to serve in the U.S. forces." Abo explained that he *had* registered for the draft before relocation. "I would have been willing to swear unqualified allegiance to the U.S.," he insisted, "if the government had assured me and my wife and child that we would be free and safe like U.S. citizens." But, he continued, "I felt that the government was not concerned about us and didn't care what happened to us."

I was afraid for myself and my wife and two children of the gangs at camp .... I heard of beatings ... and I saw a man being beaten ... [and] it frightened my wife to know such things were going on.... My wife witnessed a man being beaten by several others one evening and my wife was constantly worried about family safety .... I thought always and was in fear for the safety of my family, and I was fearful that if my renunciation was not accepted that my family and myself would be in danger....[87]

After reviewing Abo's affidavit, Collins promptly forwarded it to the Department of Justice, where it was received on August 20, 1956. He then cautioned Abo that it "may take several months before a final decision is reached," so "wait patiently."[88] After six months of patient waiting, Abo finally received the good news. "You have been successful in the administrative remedy pursued in your case," Collins told him. The Justice Department agreed

[87]"Abo Affidavit," Aug. 9, 1956, box 472, *Abo* v. *Clark*, FARC-RG 21.
[88]Undated form letter from Collins to all plaintiffs, carton 7, Collins Papers.

that he had renounced "because of fear, coercion and duress."[89] On February 7, 1957, Judge Goodman issued a "Final Order" with respect to Tadayasu Abo. It had been thirteen years since Abo had renounced his citizenship. Goodman found that throughout that time, Abo had remained a citizen of the United States of America. "You need not be ashamed of the fact that you once renounced your citizenship," Collins wrote in a cover letter to Goodman's order which he forwarded to Abo. "You did so because the government took advantage of you while it held you in duress and deprived you of practically all the rights of citizenship."[90]

IX

For Abo, Goodman's order marked the end of a long ordeal. But for Wayne Collins, it would be eleven more years of soliciting affidavits (even from renunciants who had returned to Japan) and then waiting for the final court order in each case before he finished with *Abo* v. *Clark*. Most of the affidavits from Collins's plaintiffs were accepted. By the spring of 1959, the tally was as follows: 5,766 Japanese Americans had renounced. Of those, 5,409 had applied for restoration of their citizenship. And 4,987 were successful.[91]

It had taken decades for racial prejudices to work their way into regulations which relocated Japanese Americans and then sought to strip some of them of their citizenship. And it took Collins decades to undo the constitutional damage that had been done. The final victory was one not only for Japanese Americans, but also for all citizens: Not even the hysterics and exigencies of war, Goodman had warned in his opinion, excused the government for the egregious constitutional wrongs it had committed by imprisoning citizens not charged with a crime.

Judge Goodman died in 1961, and subsequent District Court orders were issued by Judge Alphonse Zirpoli. One after another through the years, new U.S. attorneys general were listed as defendants on the affidavits and orders processed by the court. Over the twenty-three years that the *Abo* case remained active, Collins battled and outlasted eight different attorneys general:

---

[89] *Ibid.*

[90] *Ibid.*

[91] "Statement of Attorney General Wm. Rogers," May 20, 1959, carton 12, *ibid.*

first Tom Clark, then J. Howard McGrath, James F. McGranery, Herbert Brownell, William Rogers, Robert F. Kennedy, and Nicholas Katzenbach. Finally, in an ironic ending to the long episode, Attorney General Ramsey Clark, son of Tom Clark, appeared as a defendant in the final court action of the case, dated March 6, 1968. That action, ordered by Judge Zirpoli, was the "Withdrawal and Dismissal" of the last plaintiff.

Appropriately enough, Zirpoli let Collins have the last word in the case: "This brings to a conclusion these consolidated mass class equity proceedings instituted on November 13, 1945," Collins's statement to the court explained.[92] "The abusive treatment of these citizens was halted by the commencement of these . . . proceedings in this Court. In the course of time, those who had been interned were liberated from internment and returned to their homes. With few exceptions, the purported renunciations of citizenship were ordered cancelled . . . . The episode which constituted an infamous chapter in our history has come to a close."[93]

---

[92]"Withdrawal . . . and Dismissal of last Parties Plaintiff . . . and Statement of Counsel for Plaintiffs' Concluding Cases," March 6, 1968, box 467, *Abo* v. *Clark*, FARC-RG 21.
[93]*Ibid.*

JAPANESE RELOCATION AND REDRESS IN NORTH AMERICA: *A Comparative View*

ROGER DANIELS

Forty years ago the governments of both great North American democracies executed their separate but similar plans for incarcerating their tiny minorities of Japanese origin: just over 125,000 persons in the continental United States and fewer than 25,000 in Canada. There is a large and growing literature devoted to what happened in each country, but very little has been published of a comparative nature.[1] I am going to make the comparison in a very schematic and abbreviated way and then sketch, equally roughly, the comparative rehabilitation and steps taken toward a kind of redress in each country since the Great Pacific War.

On either side of the 49th parallel Japanese immigrants, or Issei, began to arrive in the latter part of the nineteenth century. In each country they followed Chinese as migrants from Asia and they settled almost exclusively in the farms, villages and cities of the Pacific coastal plain: by the census years of 1940-41 more than 88 percent of Japanese Americans lived in California, Washington and Oregon, while more than 95 percent of Japanese Canadians lived in British Columbia. In neither nation was the incidence of persons of Japanese ethnicity high: they represented fewer than one tenth of one percent (.001) of the total population in each country and even on the West Coast their incidence was low: 1.2 percent of the three American states' population and 2.7 percent of the Canadian province's. In each country Japanese had inherited already developed prejudices against Chinese, and these prejudices were exacerbated by the growing strength

Roger Daniels is Professor of History at the University of Cincinnati and an expert in the field of Japanese relocation studies. Among his many works on the subject are such publications as The Decision to Relocate the Japanese Americans, Concentration Camps USA: Japanese Americans and World War II *and numerous shorter articles in popular and scholarly journals.*

and aggressiveness of the Japanese Empire. Nativist anti-Japanese sentiments and movements flourished on both sides of the border and the rhetoric of leaders such as V.S. McClatchy of Sacramento and Miller Freeman of Seattle, American publishers and publicists, or A.W. Neill, Canadian politican from Comox-Alberni, British Columbia, is almost interchangeable.

But the legal and constitutional positions of the immigrants and their descendants were significantly different. In Canada, where naturalization was controlled by the government in Ottawa, Japanese immigrants were eligible for naturalization on the same terms as other immigrants. But in Canada suffrage requirements were determined by the provinces and British Columbia denied the ballot and a long list of other civil rights to Canadian citizens of Asian ancestry, whether naturalized or native born. When war came to Canada, in 1939, Prime Minister William Lyon Mackenzie King's government would not accept Japanese Canadian volunteers because "war with Japan was possible."[2]

In the U.S.A. Asians, except under special circumstances, could not become naturalized citizens because the naturalization statute of 1870 restricted that process to "white persons and persons of African descent."[3] Immigrants from Japan, and other Asians, were thus in a special category: they were "aliens ineligible to citizenship," and, as such, were severely discriminated against by all levels of American government. Their native-born children, the Nisei, fell under the protective umbrella of the Fourteenth Amendment to the Constitution: "All persons born in the United States . . . are citizens of the United States and of the state wherein they reside." Since in the U.S suffrage and other civil rights inhere in citizenship, the legal position of the American Nisei was distinctly superior to that of their Canadian cousins. And, in 1940, when the United States instituted its first peacetime draft, persons of Japanese origin incurred equal liability.

The outbreak of war in the Pacific in December 1941, saw strikingly similar responses by the two governments. In the U.S. the drafting or enlistment of Japanese was stopped and many, but not all, of those already in the military services were discharged. In both countries special restraints were placed on Japanese populations, citizen and alien: fishing boats were seized, bank accounts of aliens or accounts in alien banks were frozen, some enemy aliens were interned, and firearms, explosives, short wave radios and certain photographic equipment were confiscated. The overwhelming majority of the ethnic Japanese population, however, was left in nervous liberty. The never-dormant anti-Japanese movements got new leases on life and from December 7 on, more and more voices clamored for incarceration, or worse, of all Japanese.

Neither government had anything resembling a plan for dealing with domestic Japanese although in the fall of 1940 the Permanent Joint Board on Defense recommended that "in the event of hostilities with Japan, the two governments should follow policies of a similar character" in regard to Japanese residents. That recommendation seems largely to have been ignored.[4] By Feb. 19, 1942, following the advice of Secretary of War Henry L. Stimson and much pressure from assorted west coast groups, President Franklin D. Roosevelt issued Executive Order 9066 which gave the Army the authority to remove Japanese from the West Coast. Five days later, on Feb. 24, Mackenzie King, yielding to British Columbia pressures and his own inclinations and consciously following the American example, issued Order in Council, P.C. 1486, giving similar authority to Minister of Justice Louis S. St. Laurent. In each country special civilian bodies were created to effect the evacuation: the War Relocation Authority in the U.S. and the federal British Columbia Security Commission in Canada. Those few Japanese in either country who lived east of the coastal mountain ranges remained largely undisturbed, and in the Territory of Hawaii, where Japanese comprised about a third of the population —

Courtesy, The Bancroft Library

The American evacuation program for mainland Japanese Americans attempted to keep families together and to provide elements of outside life for internees. In the picture above, an evening English language class is conducted for those wishing to attend at the Heart Mountain, Wyoming, camp. Below, the high school class of 1943 graduates from the Granada Center in Amache, Colorado.

Courtesy, The Bancroft Library

there were more than 150,000, roughly equal to the number in all of continental North America — they were almost all left at liberty. The American evacuation was uncompromisingly racist, insisting, in one instance, on the evacuation of a half-Japanese infant from a Maryknoll orphanage in Los Angeles, while Canada made exceptions for a few Eurasians and Japanese married to persons of other ethnic groups.

The methods of incarceration were quite different. The U.S. placed its Japanese in ten sand and cactus concentration camps — euphemistically called relocation centers — hastily erected in desolate parts of the interior from eastern California to western Arkansas. These were girt with barbed wire and guarded by soldiers although administered by civilians. Canada had a more varied — and less expensive — system. A little over half were sent to what the government called "Interior Housing Centres" in eastern British Columbia which were largely depopulated mining towns, although some "new towns" were constructed for them. They were neither guarded nor wired in, the government assuming, as the Czarist goverment in Russia used to, that climate and geography would keep them in place. Some 2,000 adult males were sent to work camps in British Columbia and Ontario which were loosely guarded, while nearly 1,000 were incarcerated, under close confinement, in northern Ontario. Some 6,000 were allowed to "relocate" to sugar beet work in Alberta and Manitoba and to self-supporting projects elsewhere in the interior. Canadian relocatees were allowed only limited liberty: in Alberta, for example, special permission was needed for a Japanese to enter a city. The Canadian plan originally called for sexual segregation — men to work camps and women and children to interior centers. This provoked a mass protest by Canadian Nisei in April, 1942. Although the leaders were incarcerated and sent to the northern Ontario camps for their pains, the work camps proved ineffective and most Canadian Japanese families were reunited within a year.

The American evacuation was, from the first, largely a family affair and, except for some 2,000 male aliens interned, family units were kept together. The American process, although more total and harsher in its immediate format, had some ameliorating features from the outset. Run largely by sympathetic if paternalistic civilians, the U.S. War Relocation Authority always had the goal of moving "loyal" Japanese Americans into the mainstream of society as soon as possible. Hundreds of Nisei college students were allowed to resume their educations in interior and eastern campuses as early as fall, 1942. Thousands of other Japanese Americans, largely Nisei, were released for short-term agricultural work and for "permanent" relocation in urban centers from Salt Lake City to Boston. Once released from WRA custody relocatees had relative freedom of movement so long as they did not return to the West Coast.

Starting in 1943 Nisei were allowed to enlist, in many cases right from concentration camps, into what became the celebrated 442nd Regimental Combat Team and into Military Intelligence. In 1944 selective service was reinstituted for Japanese Americans, in and out of camp. But to effect all this rehabilitation — the government was primarily concerned with aiding the war effort — it was first felt necessary to determine loyalty. Initially this was done rather casually: anyone who wasn't in the F.B.I.'s bad books and could get testimonials of reliability from Caucasian citizens could be a candidate for what the W.R.A. called "leave clearance." But eventually a mass loyalty test was administered — by questionnaire! The crucial questions — which had originally been framed by the U.S. Army for prospective Japanese American recruits and draftees but which the W.R.A. mindlessly administered to all adults — were those numbered 27 and 28 on the questionnaire. They read:

Courtesy, The Bancroft Library

The U.S. War Relocation Authority, responsible for implementing and supervising the American evacuation, had the final goal of moving "loyal" Japanese back into the mainstream of American society as quickly as possible. The couple pictured here held professional jobs before the evacuation and were among the first internees considered eligible for indefinite leave from the Heart Mountain center.

> 27. Are you willing to serve in the armed forces of the United States on combat duty, wherever ordered?
> 28. Will you swear unqualified allegiance to the United States of America and faithfully defend the United States from any and all attack by foreign or domestic forces, and foreswear any form of allegiance or obedience to the Japanese emperor, to any other foreign government, power or organization?

What aged Issei or women of any age were to make of the first is hard to say, but the second was worse, at least for Issei who could not become citizens. It was asking them voluntarily to assume a stateless status, and, incidently, a clear violation of the Geneva convention. This loyalty program eventually led to some 18,000 non-jurors being labeled "disloyal" and segregated at Tule Lake, California, for repatriation or expatriation to Japan. All told 4,724 Japanese Americans actually left for Japan after the war: of these 1,659 were aliens, 1,949 citizens accompanying repatriating parents, and 1,116 were adult Nisei.

For Japanese Canadians there was no real wartime rehabilitation program: hardly any were enrolled in colleges and perhaps 100 were accepted for military service at the very end of the war. The government pursued an aggressive and successful program of encouraging Japanese Canadians to apply for postwar passage to Japan. Almost half the prewar population — 10,813 persons — applied for "repatriation." Counting adults only, about 42 percent were Japanese nationals, 21 percent naturalized citizens, and 36 percent native-born Canadians. Eventually almost 4,000 did go. That departures for Japan from the U.S. and Canada were almost equal even though Japanese Americans outnumbered Japanese Canadians by more than five to one speaks volumes about their respective visions of the future.

During the war both governments unambiguously expressed the wish to prevent any postwar reestablishment of Japanese enclaves on the Pacific Coast. In the United States the general ban on returning to the Pacific Coast was, in effect, lifted by the U.S. Supreme Court in the famous *Endo* case of December, 1944, which restored domestic freedom of movement to Japanese American citizens. While not all of the exiled Japanese Americans returned to the West Coast, large numbers of them did. As we have seen, 88.5 percent of Japanese Americans lived there in 1940. This fell to 58.2 percent by 1950, but by 1960 was back up to about two-thirds where it has remained. In Canada, where no court could exercise meaningful jurisdiction, the bar on returning to coastal British Columbia stayed in effect until April 1, 1949, more than three and a half years after the end of the Pacific war. British Columbia, which had more than 95 percent of Japanese Canadians in 1941, had just a third in 1951, and the figure has not changed significantly since. Today, there are more Japanese Canadians in Ontario than in B.C., more in Toronto than in Vancouver.

TABLE I

COMPARATIVE POPULATION FIGURES 1941 and 1971 FOR CANADA, THE CONTIGUOUS UNITED STATES AND HAWAII

Japanese in Canada, 1941 and 1971

Province	1941	%	1971	%
Ontario	234	1.0	15,600	41.9
British Columbia	22,096	95.5	13,585	36.5
Alberta	578	2.5	4,460	12.0
Others	241	1.1	3,615	9.6
Total	23,149	100.0	37,260	100.0
Population of Canada	11,506,655		21,568,315	
Japanese as a %	0.2		0.17	

Japanese in the Contiguous United States, 1940 and 1970

Region	1940	%	1970	%
Pacific Coast (CA, OR, WA)	112,353	88.5	239,678	64.6
All other	14,595	11.5	131,471	35.4
Total	126,948	100.0	371,149	100.0
Population of Lower 48	131,699,275		202,163,212	
Japanese as a %	0.1		0.18	

Japanese in Hawaii, 1940 and 1970

	1940	% of HA pop	1970	% of HA pop
Hawaii	157,905	37.3	217,175	28.2

N.B. The preliminary estimated number of persons of Japanese ethnicity in the United States for 1980 (including AL and HA) was 700,747. The comparable 1970 figure was 591,290.

Another method used by the WRA to return Japanese Americans to American society was that of outside employment. This couple preparing to leave the Heart Mountain center were "permanently" relocated in a non-coastal urban center.

In addition to their liberty, Japanese in both countries lost large amounts of real and personal property. Farms, fishing boats, businesses, automobiles, refrigerators and other possessions which had to be abandoned at the time of evacuation were involved in forced sales at a mere fraction of value. Quite often the victims of the relocations were, in current parlance, "ripped off," and usually ripped off to the advantage of some opportunistic Caucasian. In addition, of course, in each country Japanese were largely denied the opportunity to participate fully in the wartime prosperity which finally ended the decade-long depression that Canadians call the "dirty thirties."

The relocations were four decades ago; the war has been over for almost that long. What has happened to the North American Japanese in that time? Their status has changed dramatically. From a pariah group which many wanted to expel they have become, in the words of one American sociologist, a "model minority." Each nation has made a formal apology: in 1964 Prime Minister Lester B. Pearson admitted that the relocation had been a "black mark" against what he claimed was "Canada's traditional fairness and devotion to the principles of human rights." Twelve years later President Gerald R. Ford in a bicentennial year executive order both apologized — "We know now what we should have known then — not only was the evacuation wrong, but Japanese Americans were and are loyal Americans" — and revoked Executive Order 9066. These pronouncements, however symbolic, underline the changed status and situations of the North American Japanese.

In the 1950s legal discrimination was removed from the statute books of each nation.

On February 19, 1976, President Gerald R. Ford formally rescinded Executive Order 9066 in the presence of members of the Japanese American Citizens League and Japanese American members of Congress. On President Ford's immediate left stand Congresswoman Patsy Mink and Senator Daniel Inouye.

Japanese Canadian citizens could vote anywhere and in the U.S. Japanese as well as other Asians became eligible for naturalization. While a few in the majority communities kept wartime memories ever fresh, the postwar Japanese economic miracle and the rise of a threatening People's Republic of China made the public image of Japan an increasingly favorable one. On the American side, the belated admission of Hawaii to statehood in 1959 meant that one state had an Asian American majority and thus there were senators and representatives of Japanese ethnicity in Washington. This new political clout is epitomized by the career of Democratic Senator Daniel Inouye. A much-decorated infantry Captain who had lost an arm fighting in Italy, Inouye was called upon to be the keynote speaker at the 1968 Democratic National Convention. That a Japanese American could fulfill such a sensitive symbolic role less than 27 years after Pearl Harbor says something about the fluidity of American institutions. Six years later, as a member of the Senate Select Committee investigating what became known as Watergate, Inouye helped bring down a President. Polls taken at the time showed him with the highest positive rating of any committee member. Later that year California sent the first Japanese American to Congress from the mainland: Norman Minetta represented not an ethnic community — Japanese are a tiny minority in his San Jose district — but a moderately liberal one. Less than 30 years before his election some of the Japanese returning to San Jose had shots fired into their homes. A second mainland Japanese American, Robert Matsui, now represents Sacramento.

No Japanese Canadian has received national political recognition, although one

Canadian-born Japanese, S.I. Hayakawa, a kind of conservative culture hero, was elected to the U.S. Senate from California in 1976. While some Japanese Canadians have achieved significant positions within the civil service, the Japanese Canadian best known outside of his own ethnic community is Professor David Suzuki, a geneticist at the University of British Columbia, who has done science shows on television.

Despite the not surprising lack of Japanese Canadian political personalities, the contemporary situation of both communities of North American Japanese is strikingly similar. Each group has benefited from ethnic and racial tensions which have served to lessen the once apparently polar differences between persons of Asian and European ancestry. In Canada two kinds of tensions have served to promote the social position of Japanese Canadians. First is the centuries-long conflict between French and English Canada in which language loyalty and acculturation to either an English or a French norm is seen as crucial by both sides. In this conflict, Japanese, by their all but universal adoption of the English language, are seen as allies of anglophone Canadians. Second, immigration from less developed parts of the Commonwealth — Asians from Uganda, blacks from the Caribbean — is seen as a much greater threat by many Canadians than the steady trickle from Japan which has run, in recent years, between 500 and 1,000 annually.

Japanese Americans have benefited similarly from apparent contrasts with other racial and ethnic groups seen as more threatening to the norms of society. Their increasingly middle-class status and outlook makes them truly seem a "model minority" in comparison, say, to blacks or chicanos. The most recent comparative income data (1970) show that more than 90% of Japanese American families were above the poverty line.[5] Although there are no equivalent Canadian data, there is every reason to believe they would be comparable. The numbers demonstrate that Japanese Americans earn considerably more than other minority groups (including other Asian Americans) but considerably less than white Americans, even though Japanese Americans have, on average, more education than the latter. This disparity is most noticeable at each end of the economic spectrum. Despite their greater amount of schooling, a larger proportion of Japanese Americans exist in poverty than their educational attainments would predict, a fact ignored by those who speak of an unalloyed Japanese American success story. Similarly, given their education and the extremely high cost of living where most of them live (California and Hawaii), the percentage of those in the $25,000 and above bracket should have been about twice what it is. This is further evidence of subtle but continuing economic discrimination. Quite often this discrimination takes the form not of exclusion from employment, but rather of a failure to promote Japanese Americans (and other Asian Americans) into higher managerial positions. Many of them are thus conspicuously "over-qualified" for the jobs they hold. There is often a distinct reluctance of employers to put Japanese Americans and other non-whites into positions in which they can hire and fire whites. I suspect, but cannot demonstrate, that similar conditions prevail in Canada.

In both countries the existence of exogamus marriage — once a relative rarity — has become quite pronounced. Although there are no definitive statistics or even large scale studies, research about American Sansei (third generation Japanese Americans) by several scholars show that more than half of their marriages are to non-Japanese, and a study of Canadian Sansei by Gordon Hirabayashi indicates that the rate there may be well over 75 percent.[6] This later figure has caused at least a few Japanese Canadians to question publicly the continued survival of their group. Although this could conceivably happen in a generation or two in Canada, where the ethnic population is quite small and relative-

Courtesy of the Japanese American Citizens League

U.S. Representative Robert Matsui of Sacramento, California is one of several Japanese Americans to attain national political office in the last two decades.

ly dispersed, a Japanese American community of some size seems assured for the foreseeable future.

The role of ethnic Japanese organizations in years to come is also conjectural.[7] Since the 1960s it has been clear to most close observers of the Japanese communities in North America that the winds of change stirring other ethnic groups are not without their effect on the Nisei and particularly upon their children, the Sansei. The rise of ethnic consciousness in the United States, largely in response to the black liberation movement, and the development of multiculturalism as an official policy in Canada, have each affected young Japanese. A distinct but articulate minority of them, particularly those in college or of college age, have become "radical" and "militant" to a degree that shocks their elders even more perhaps than does similar behavior by their white or black peers. The most radical see themselves as allies of lower-class minorities, and on campuses on both American coasts, have coalesced with them in such organizations as the Third World Liberation Front. Much more characteristic is an increased awareness of their ethnic heritage on both sides of the Pacific, with the result that courses in Asian American history and culture and Asian American Study Centers have arisen in many U.S. colleges and universities. In Canada, not surprisingly, such movements have been slower to get under way. As the 1980s began no such courses existed in any Canadian university, but interest in Japanese Canadian history has been greatly increased by a growing number of books, films, conferences and a traveling exhibit celebrating the 1977 centennial of the putative first Japanese arrival in Canada. In addition, newly formed scholarly associations for the interdisciplinary study of Asian North Americans now exist in each country. In the United States the mimetic effect of the black, brown and red power movements has produced a half-serious, half-comic yellow power movement. Its rhetoric, too, is imitative: a popular third generation joke styles accommodationist Japanese as "bananas," that is, "yellow on the outside but white on the inside."

National organizations have changed as well. The Japanese American Citizens League, the one national Japanese American organization, is a case in point. Founded as a defensive device for an oppressed minority, the organization and its membership are not at all sure of their present function since the most obvious oppressions have vanished. In the post-World War II era the JACL, whose actual membership has never amounted to as much as a tenth of the mainland Japanese American population, became one of the most effective minority lobbying groups under the direction of its Washington representative, Mike Masaoka. Exploiting the guilt felt about the evacuation by many Americans, it played an important role in the passage of much legislation affecting the Japanese American community, from the Japanese American Claims Act of 1948 to the symbolic repeal of the Emergency Detention Act in 1970. By the late 1970s, as its leadership was shifting from Nisei to Sansei, from a generation that experienced the wartime ordeals to one for whom they are, at best, a vague memory, the JACL asked, formally, for some kind of "redress" for the wrongs done to the Japanese American people during and after the war. In an argument that pointed out that the West German government was in the process of paying between $35 and $40 billion to Jews and Jewish institutions, but without suggesting an appropriate price tag for their own injuries, the JACL insisted that:

> Redress for the injustices of 1942-1946 is not just an isolated Japanese American issue: it is an issue of concern for all Americans. Restitution does not put a price tag on freedom or justice. The issue is not to recover what cannot be recovered. The issue is to acknowledge the mistake by providing proper redress for the victims of injustice, and thereby make such injustices less likely to occur.[8]

As a result of that campaign a bill was enacted into law in late 1980 creating the presidential Commision on the Wartime Relocation and Internment of Civilians to inquire whether "any wrong" was done pursuant to Executive Order 9066, and should it so find, to recommend appropriate remedies to the Congress within 18 months. That commission has held formal hearings and should make formal recommendations to Congress sometime in 1982.

In Canada the National Association of Japanese Canadians (formerly the Japanese Canadian Citizens Association) has concentrated its energies on a different form of redress. It is concerned chiefly with the proposed Canadian constitution and what it sees as deficiencies in it which might permit a reoccurrence of the events of 1942-1949. As its president, Gordon R. Kadota, put it recently:[9]

> On the current proposal to entrench the Charter of Rights into the Constitution [we] can endorse only the intent to entrench but not the proposal in its entirety. Specifically, we point out that although a number of crucial clauses do cover the protection of rights [a number of] phrases permit Parliament to override these sections.
>
> . . .
>
> [They] provide Parliament with the power to override these very rights by such means as the War Measures Act. The proposal as it stands, *does not* prevent a repetition of the injustices suffered by the Japanese Canadians. We also wish to add that the aboriginal rights of Canada's native peoples should be reflected in the Constitution.

* * *

Although, as E.H. Carr delights in pointing out, history is usually written by the winners, quite often it is also true that history is remembered better by the losers. The relocations of the North American Japanese occupy, at best, a line or two in most textbooks and are completely ignored in many. For the Japanese communities of this continent, however, that experience remains the central event in their history.

FOOTNOTES:

1. The earliest specific scholarly comparison is Robert W. O'Brien, "Evacuation of the Japanese from the Pacific Coast: Canadian and American Contrasts," *Research Studies State College of Washington*, Vol. 14 (1946), pp. 113-120, while Norman Thomas, "Fate of the Japanese in North America and Hawaii," *Pacific Affairs*, Vol. 16 (1943), pp. 92-95 is an acute contemporary comment. For a comparison of the whole experience see Roger Daniels, "The Japanese Experience in North America: An Essay in Comparative Racism," *Canadian Ethnic Studies*, Vol. 9, No. 2 (1977), pp. 91-100. For literature on Asians in North America generally, see Roger Daniels, "American Historians and East Asian Immigrants," *Pacific Historical Review*, Vol. 43 (1974), pp. 449-472 and "North American Scholarship and Asian Immigrants, 1974-1979," *Immigration History Newsletter*, Vol. 9, No. 1 (1979), pp. 8-11. Two new works expand our knowledge: Roger Daniels, *Concentration Camps, North America: Japanese in the United States and Canada During World War II* (Malabar: Krieger Press, Florida, 1981) makes specific comparisons while Ann G. Sunahara, *The Politics of Racism* (Toronto: 1981) is the best treatment of the entire Canadian process.

2. J.W. Pickersgill, *The Mackenzie King Record, Vol.* 1 (Toronto: University of Toronto Press, 1960), p. 149.

3. For a forlorn attempt to get a color-blind naturalization statute made by Charles Sumner see *Cong. Globe*, 41st Congress, 2nd Session, 1870, pp. 4275-79, 5121-77.

4. See Roger Daniels, "The Decisions to Relocate the North American Japanese: A Note" *Pacific Historical Review*, Vol. 51 (February 1982), pp. 71-77.

5. Japanese American income data for families:

Below $4,000	9.5%
$4,000-$7,999	15.0%
$8,000-$11,999	22.5%
$12,000-$14,999	17.4%
$15,000-$24,999	28.3%
$25,000 +	7.3%
	100.0%

Computed from U.S. Dept. of Commerce, Bureau of the Census, *1970 Census of Population: Japanese, Chinese and Filipinos in the United States* (Washington, D.C., 1973), Table 9, p. 72.

6. Harry H.L. Kitano and Akemi Kikumura, "Interracial Marriage: A Picture of the Japanese American," *Journal of Social Issues*, Vol. 29, No. 2 (1973), pp. 67-81 and Gordon Hirabayashi, "From 'Enemy Alien' to Intermarriage in One Third of a Century," *Rikka*, Vol. 3, No. 3 (1976), pp. 3-8. See also Kitano's essay on "Japanese" at p. 570 and David Herr's essay on "Intermarriage" at p. 519 in *Harvard Encyclopedia of American Ethnic Groups* (Cambridge: Harvard University Press, 1980).

7. For further comments on leadership and organizations see Roger Daniels, "The Japanese," in John Higham, ed., *Ethnic Leadership in America* (Baltimore: Johns Hopkins University Press, 1978), pp. 36-63.

8. National Committee for Redress, Japanese Americans Citizens League, *The Japanese American Incarceration: A Case for Redress* (San Francisco: 1978), p. 24.

9. National Association of Japanese Canadians, press release, November 25, 1980.

Redress Achieved, 1983–1990

Roger Daniels

On October 9, 1990, forty-eight years after the incarceration of Japanese Americans began, seven years after the Commission on Wartime Relocation and Internment of Civilians (CWRIC) made its report and recommendations, and two years after Congress passed the Civil Liberties Act of 1988,[1] the American government finally began to make redress payments. Although his immediate predecessors had opposed redress and the whole reopening of the Japanese American incarceration issue, Attorney General Richard Thornburgh hailed the redress process and those who struggled to achieve it. At a special ceremony in which he presented checks to a few of the oldest survivors, the attorney general told them:

> Your struggle for Redress and the events that led to today are the finest examples of what our country is about, and of what I have pledged to protect and defend, for your efforts have strengthened the nation's Constitution by reaffirming the inalienability of our civil rights.[2]

This essay will recount the course of events during the period from the CWRIC's report to the beginning of redress payments.

The CWRIC issued five recommendations for redress to Congress in June 1983. First, it called for a joint congressional resolution acknowledging and apologizing for the wrongs done in 1942. Second, it recommended a presidential pardon for persons who had been convicted of violating the several statutes establishing and enforcing the evacuation and incarceration. Third, it urged Congress to direct various parts of the government to deal liberally with applicants for restitution of status and entitlements lost because of wartime prejudice and discrimination. It gave as an example the less than honorable discharges that were given to many Japanese American soldiers in the weeks after Pearl Harbor which meant that they had no status as war veterans. Fourth, it recommended that Congress appropriate money to establish a special foundation to "sponsor research and public educational activities . . . so that the causes and circumstances of this and similar events may be illuminated." And fifth, in its most important recommendation, it called upon Congress to make a one-time, tax-free, per capita compensation of $20,000 to each survivor who had been incarcerated. An educated guess was that there were perhaps 60,000 such persons alive in 1983, so that the presumed cost was $1.2 billion.[3]

Although the CWRIC report, *Personal Justice Denied*, had been adopted unanimously, one commissioner, Rep. Daniel Lundgren (R-CA), dissented from all recommendations which involved the payment of money. He thought that an apology was enough. He argued that "it is inappropriate that present day taxpayers should be held accountable for actions that occurred 40 years ago. Should we pay monetary redress for the abhorrent practice of slavery or the inhumane treatment of Indians 100 years ago?"[4] As the bill provided redress for survivors only, Lundgren's arguments were a chain of non-sequiturs. Humorist Mark Russell put it nicely by noting that "one congressman is worried that the Chinese will want back pay for building the railroads in the last century. They'll be beefing up security in Washington for another Bonus March by angry Chinese railroad workers."[5]

There was, not surprisingly, significant public reaction against monetary redress, including frequent assertions, in editorials and letters to the editor, that since Japan had bombed Pearl Harbor the now rich

REDRESS AT LAST! In a ceremony at the Department of Justice on October 9, 1990, Attorney General Richard Thornburgh makes ceremonial presentations to a group of the oldest survivors, five of whom were over 100 years of age: Rev. Mamoru Eto, 107 (Gila River); Haru Dairiki, 102 (Tule Lake); Kisa Iseri, 102 (Tule Lake); Hisano Fujimoto, 101 (Minidoka); Senkichi Yuga, 101 (Rohwer); and Sugi Kiriyama, 100 (Manzanar). Courtesy JACL.

Japanese government should make any payments that were due. More frequent were arguments that the United States couldn't afford it or that the Japanese didn't need the money. Often, the objections were accompanied by statements of regret that mass incarceration had occurred. An influential example of this was the *New York Times*. While noting that the incarceration was "a cruel and pointless surrender to panic"—something it had not said during World War II—and that the "lasting resentment" of Japanese Americans was "legitimate," the paper's editorial board found monetary redress inappropriate and suggested a symbolic "gesture of atonement" such as a "fund offering scholarships."[6]

Although clear majorities in both houses of Congress were publicly committed to support of monetary redress, and bills enacting the CWRIC's recommendations into law were introduced in both the Ninety-eighth and Ninety-ninth congresses, none passed. The most significant factor in that failure was the steady opposition to monetary redress by the Reagan administration. Not surprisingly, this led to a great deal of cynicism and doubt within the Japanese American community: one writer spoke for many when he asked, "Will we live [long enough] to collect redress?"[7]

During this period both the *coram nobis* cases discussed by Dale Minami (*infra* pp. 200–202) and the class action suit discussed by William Hohri (*infra* pp. 196–99) were wending their sometimes tortuous paths through the federal courts. Following Fred Korematsu's victory in San Francisco, the federal district court in Portland granted the government's motion to vacate Minoru Yasui's conviction in January 1984, but refused to make the findings of government misconduct that his suit had requested. Judge Robert C. Belloni, unlike Judge Marilyn Hall Patel in Korematsu's case, declined to make such a ruling "forty years after the events took place" and insisted that "courts should not engage in that kind of activity."

Yasui appealed to the Ninth Circuit Court of Appeals, but died in November 1986, before that court had ruled. Under the doctrine of "mootness" his case died with him. A third outcome ensued in Gordon Hirabayashi's case, the last to be reheard. In Seattle federal court, after an evidentiary hearing, Judge Donald S. Voorhees made the same distinction that Supreme Court Justice Owen J. Roberts had made forty-four years before: like the other two judges he vacated the conviction for refusing to be evacuated, but upheld the conviction for curfew violation. He insisted that a curfew that applied only to one ethnic group was a "relatively mild" constraint on those who were restricted by it and ignored the strain that it put on the Constitution. An appeal to the Ninth Circuit resulted in an instruction to Voorhees to vacate the conviction for curfew violation as well, which he did in January 1988.

In this case, and in Korematsu's, the government, as the loser, could have appealed to the United States Supreme Court, but it declined to do so, thus frustrating one of the goals of the *coram nobis* suits: a rehearing of the Japanese American cases before the highest court. The class action suit was never really heard. It was dismissed by a District of Columbia federal judge in May 1983, on the technical grounds of untimely filing. This finding was reversed by the Court of Appeals for the District of Columbia. In this instance, the government was willing to appeal to the highest court, which ruled, without opinion, in June 1987, that the suit was barred by both sovereign immunity and the statute of limitations, and, in addition, had been filed in the wrong federal appellate court. A lower court again dismissed the suit in March 1988, and it lingered in a kind of legal twilight until October 1988, when the Supreme Court denied the last writ of *certiorari*.[8]

By that time, Congress had finally acted. What became the Civil Rights Act of 1988, which wrote into law all five of the CWRIC's recommendations, was taken up by the House in early 1987. It was introduced on the first day of the session by the then majority leader, Tom Foley (D-WA), with an eventual co-sponsorship of at least 140 other House members, twenty of whom were Republicans.[9] It was assigned the symbolic number H.R. 442, for the 442nd Regimental Combat Team, the heroic Nisei unit of World War II. The Reagan administration continued to oppose redress. Its spokesperson, Assistant Attorney General Richard K. Willard, testified at the House hearings:

> It may be that the [CWRIC] is correct in concluding that the assumptions on which the exclusion and detention programs were based were erroneous. It is a long unsubstantiated further step, however, to brand those actions as a product of 'racial prejudice, or hysteria, and a failure of political leadership' [sic]. . . . Moreover, some of the [CWRIC's] conclusions are suspect. These matters are best left to historical and scholarly analysis rather than debated by Congress.

Not surprisingly, he also challenged the idea of cash payment, arguing that the Japanese American Claims Act of 1948 had furnished "a comprehensive statutory scheme which provided a reasonable and balanced contemporaneous remedy to affected individuals." This, of course, was lawyerly nonsense, as the 1948 act provided compensation only for losses of real property and the Commission's recommendations were meant to acknowledge the damage done to the civil rights of individuals, a matter Willard did not mention.[10]

Most testimony at the hearings was favorable. The American Jewish Committee, for example, not only supported payments "of sufficient magnitude at least to create public awareness of the blatant violations of [Japanese Americans'] constitutional rights" but also pointed to the billions of dollars in reparations paid by the West German government to Jewish victims of Nazi oppression as a notable precedent for redress.[11] Prominent among the witnesses were Japanese American congresspersons, many of whom had been active in the almost decade-long fight for redress in Congress. As Norman Mineta, a Democrat from California who had been incarcerated as a child, put the case:

> "I realize that there are some who say that these payments are inappropriate. Liberty is priceless, they say, and you cannot put a price on freedom.
>
> That's an easy statement when you have your freedom. But to say that because constitutional rights are priceless they really have no value at all is to turn the argument on its head.
>
> Would I sell my civil and constitutional rights for $20,000? No. But having had those rights ripped away from me, do I think I am entitled to compensation? Absolutely.
>
> We are not talking here about the wartime sacrifices that we all made to support and defend our nation. At issue here is the wholesale violation, based on race, of those very legal principles we were fighting to defend."[12]

That view prevailed both in the committee and in the full House. The bill passed, on September 17, 1987, by a vote of 243 to 141.[13] Ominously, there were more than enough negative votes to sustain a veto. The noes came from fiscal conservatives in both parties. The vote in the Senate did not come until more than seven months later. On April 20, 1988, the bill passed by a vote of 69–27 and was presumably

THE WHITE HOUSE

WASHINGTON

A monetary sum and words alone cannot restore lost years or erase painful memories; neither can they fully convey our Nation's resolve to rectify injustice and to uphold the rights of individuals. We can never fully right the wrongs of the past. But we can take a clear stand for justice and recognize that serious injustices were done to Japanese Americans during World War II.

In enacting a law calling for restitution and offering a sincere apology, your fellow Americans have, in a very real sense, renewed their traditional commitment to the ideals of freedom, equality, and justice. You and your family have our best wishes for the future.

Sincerely,

Copies of this letter from President Bush accompanied the redress checks. The White House staff did not personalize them with individual names and addresses although that would have meant much to many of the recipients.

veto proof.[14] There were minor differences between the Senate and House versions, and the resulting conference report was agreed to on July 27 in the Senate and on August 4 in the House.[15] Nineteen eighty-eight, of course, was a presidential election year, and, on August 1, just before the final House vote, Ronald Reagan reversed the administration's position and, in a letter to Speaker of the House Wright, endorsed the bill and urged its passage.[16]

The signing ceremony took place on August 10 in the Old Executive Office Building with the late recruit to the redress cause, Ronald Reagan, taking center stage and reading his lines as if he had been a committed supporter all along. He even dredged up an old scene from 1945 by talking about his appearance as a "young actor" in a ceremony at which General Joseph W. Stilwell awarded a posthumous medal to the brother of a Nisei who had been killed fighting with the 442nd Regimental Combat Team. What the president did not point out was that he was a commissioned aide to Stilwell and was following a script written by either the Office of War Information or the Army's public relations people.[17]

The American action on redress triggered an almost immediate response in Canada. (For the experience of Japanese Canadians, see *supra.*, pp. 139–41). There the Mulroney government had been resisting redress payments although negotiations about them had been going on between the government and Japanese Canadian groups for years. Just forty-three days after the American redress bill was signed, the Canadian government and the community representatives reached a stunningly similar agreement, which was announced in Parliament with the support of all three political parties, on September 22, 1988. Canada agreed to pay, "as expeditiously as possible," $21,000 to each survivor (there were presumed to be from 12,000 to 14,000), which could amount to almost $300 million.[18] Since the Canadian dollar was then worth about 85 percent of the American dollar, the Canadian per capita payment was actually smaller, although it seemed larger. All observers agreed that the American precedent was crucial in the volte-face of the Canadian government. It also did not hurt that a national election was approaching.

In Canada the government began to make payments to survivors almost immediately, but in the United States authorization and appropriation are two distinct processes. No payments were authorized in 1988, and in 1989, in the midst of partisan and ideological battles over the budget, nothing was appropriated either. During this period, of course, Issei and older Nisei were dying without being paid. The heirs of those who died *before* August 10, 1988—the day Reagan signed the bill into law—were entitled to nothing as the bill was restricted to survivors. The heirs of those who died on or after that date were entitled to the full $20,000. In late 1989, as part of the budget reconciliation that year, Senator Daniel K. Inouye (D-HA) proposed adding a provision to the 1990 appropriations bill for the Commerce, State, and Justice departments which would make the already authorized redress payments an "entitlement program" beginning in the 1991 fiscal year (Oct. 1, 1990–Sep. 30, 1991) and thus not subject to sequestration or the provisions of the Gramm-Rudman-Hollings Deficit Control Act of 1985. Despite objections from Jesse Helms (R-NC) and some other conservatives, Inouye's provision passed by 74–22, a wider margin than redress itself had passed two years earlier. Warren Rudman (R-NH), an author of the 1985 act and a supporter of redress, helped make the case: "there is a time when one whose name is part of the Deficit Control Act of 1985 believes the Budget Act ought to be waived, and this is one of those times." He urged his colleagues to give "overwhelming support to waive the Budget Act to redress finally for the now elderly [Japanese] Americans, the injustice that money will never recompense."[19]

The House accepted the Senate amendment and

President Bush signed the appropriations bill on November 21, 1989. Although the entitlement designation could have been reversed by subsequent congressional action, the maneuver in late 1989 paved the way for the ceremony and payments that began to be made on October 9, 1990.

• • •

The beginning of redress payments—roughly $500,000,000 a year will be expended with the oldest survivors getting their checks first—was greeted with elation by most of those who had struggled for it over the years. They were entitled to their elation: it was a famous victory. But it was a victory that at least half of the victims did not live to see. Warren Rudman was certainly correct when he noted that no amount of money could properly recompense the injustices of the war years. Perhaps even more important than the money, in the long run, was the way in which the redress struggle has burned the injustices of the 1940s into the minds of most Americans. As recently as the 1970s one could lecture about the relocation and get shocked reactions from college students who had obviously never heard of it. By the late 1980s almost every freshman I taught had some awareness of the way in which the American government, supported by an overwhelming majority of its people, betrayed our highest principles as part of a struggle for human freedom.

NOTES

1. Public Law 100–383, Aug. 10, 1988, 102 *Stat.* 903. It was H.R. 442, 100th Congress. Its legislative history may be traced in: *House Reports* 100–278 (Comm. on the Judiciary) and 100–785 (Comm. of Conference); *Congressional Record*, vol. 133 (1987): Sep. 17, considered and passed the House; vol. 134 (1988): Apr. 20, considered and passed the Senate, amended; July 27, Senate agreed to conference report; Aug. 4, House agreed to conference report. *Weekly Compilation of Presidential Documents*, vol. 24 (1988): Aug. 10, 1988, presidential remarks. The account here ignores those parts of PL 100–383 that pertain to Aleuts and Aleut property.

2. "JUSTICE ISSUES FIRST REDRESS PAYMENTS TO JAPANESE-AMERICANS," press release, Department of Justice, October 9, 1990.

3. The recommendations may be found in CWRIC, "Press Release," 17 pp. (n.d. [June 15, 1983]). By October 1990, the Justice Department's Office of Redress Administration (ORA) was estimating that there were 65,000 survivors alive at that time which seems to me too high.

4. "Testimony of Congressman Daniel E. Lundgren, 42nd District, California, before the Senate Judiciary Committee Subcommittee on Administrative Practice and Procedure, July 27, 1983," 6 pp., mimeographed.

5. *Cincinnati Post*, Aug. 8, 1983.

6. *New York Times*, August 4, 1981.

7. See, for example, Tom M. Nakoji, "Will We Live to Receive Redress?", *Hokubei Manichi* (San Francisco), March 29, 1985.

8. A good summary of these legal proceedings is Peter Irons, ed. *Justice Delayed: The Record of the Japanese American Internment Cases*. Middletown, CT: Wesleyan University Press, 1989, pp. 3–46. Compare, however, Irons' conclusion about the cases' significance at p. 46.

9. Sponsorship data from Japanese American Citizens League Legislative Education Committee, "*ACTION ALERT #4*," June 5, 1987. The Senate bill eventually had 73 cosponsors.

10. "Statement of Richard K. Willard, Assistant Attorney General, Civil Division, U.S. Department of Justice, before the Subcommittee on Administrative Law and Government Relations of the United States House of Representatives, Committee on the Judiciary, concerning H.R. 442 and H.R. 1631 on April 29, 1987," 21 pp., mimeographed. Quotation at p. 8.

11. "Statement of the American Jewish Committee on H.R. 442, Civil Liberties Act of 1987 . . . to the Subcommittee on Administrative Law and Governmental Relations, Committee on the Judiciary, House of Representatives, April 29, 1987," 6 pp., mimeographed. Quotation at p. 6. For the West German so-called *wiedergutmachung*, see pp. 203–9, *supra*.

12. "Testimony in support of H.R. 442 by Norman Y. Mineta, M.C., Subcommittee on Administrative Law, House of Representatives, April 29, 1987," 4 pp., mimeographed. Quotation at p. 3.

13. *Congressional Record* (Sep. 17, 1987), p. H7595. On a prior vote on an amendment by Rep. Lundgren which would have eliminated redress payments to individuals, 162 congresspersons voted yea. *Ibid.* pp. H7587–88.

14. *Congressional Record* (Apr. 20, 1988), p. S4415.

15. *Congressional Record* (July 27, 1988), p. S10153 without recorded vote in the Senate; (Aug. 4, 1988), pp. H6314–15 for the House, where the vote was 257 yeas, 159 nays, and 18 not voting. The long debate converted at least one conservative to the redress cause: see the remarks of Rep. Bill Frenzel (R-MN) at p. H6130.

16. The text of the letter is in *Congressional Record* (Aug. 4, 1988), p. H6312.

17. "Remarks on Signing the Bill Providing Restitution for Wartime Relocation and Internment of Civilians, August 10, 1988." *Weekly Compilation of Presidential Documents*, vol. 24, pp. 1034–35. For the rules and regulations established by the Department of Justice to govern redress, see "Redress Provisions for Persons of Japanese Ancestry," *Federal Register*, vol. 54 (August 18, 1989), pp. 34157–68.

18. A good account can be found in the monthly newspaper *Nikkei Voice*, Scarborough, Ontario, vol. 2, no. 5, Oct., 1988. There are detailed stories in Canadian afternoon newspapers of 22 September 1988 and morning papers of the next day.

19. *Congressional Record* (Sep. 29, 1989). Vote at p. S12225; Rudman quote at p. S12219.

THE JAPANESE AMERICAN CORAM NOBIS CASES: EXPOSING THE MYTH OF DISLOYALTY

INTRODUCTION

During World War II, Japanese Americans fought for the United States in Europe and the Pacific[1] while the United States imprisoned their families in concentration camps. One hundred and twenty thousand persons of Japanese descent were forcibly removed from the West Coast.[2] They suffered income and property losses of between 810 million and 2 billion in 1983 dollars.[3] The Supreme Court upheld the internment, basing its decision on the government's claim that harsh measures were necessary to protect the national security from espionage and sabotage by the ethnic Japanese.[4] The Supreme Court's acceptance of the government's proffered justification for the internment thus imposed upon the Japanese American community a stigma of disloyalty that no other group of Americans has faced.[5]

---

1. Composed mostly of volunteers from the concentration camps, the all-Japanese American (except for high ranking officers) 442nd Regimental Combat Team fought in the battlefields of Europe to become the most decorated combat unit for its size and length of service in American military history. Commission on Wartime Relocation and Internment of Civilians, Personal Justice Denied 253-60 (1982) [hereinafter CWRIC]. Japanese American intelligence units shortened the war in the Pacific by at least two years and prevented hundreds of thousands of American casualties. Id. at 256; F. Chuman, The Bamboo People: The Law and Japanese-Americans 180-81 (1976).

2. The most authoritative account of the circumstances leading to the internment, and of the internment itself, is found in the report of the CWRIC, note 1 supra. For a brief discussion concerning the establishment of the Commission, see note 37 infra. See also R. Daniels, Concentration Camps USA: Japanese Americans and World War II (1972); R. Daniels, The Politics of Prejudice: The Anti-Japanese Movement in California and the Struggle for Japanese Exclusion (1962); M. Grodzins, Americans Betrayed: Politics and the Japanese Evacuation (1949); J. tenBroek, E. Barnhart & F. Matson, Prejudice, War and the Constitution (1954); M. Weglyn, Years of Infamy: The Untold Story of America's Concentration Camps (1976).

3. CWRIC, Study Estimates Economic Loss of Japanese Americans, Resident Aliens During World War II at $810 Million to $2 Billion in 1983 Dollars (June 15, 1983)(press release)(on file at New York University Review of Law & Social Change).

4. Korematsu v. United States, 323 U.S. 214 (1944); Yasui v. United States, 320 U.S. 115 (1943); Hirabayashi v. United States, 320 U.S. 81 (1943).

The government's official military historian of the internment, Stetson Conn, published an account of the internment in which he found "little support for the argument that military necessity required a mass evacuation." S. Conn, R. Engelman & B. Fairchild, The United States Army in World War II, The Western Hemisphere: Guarding the United States and Its Outposts 147 (1964).

John J. McCloy, Assistant Secretary of War during World War II, has stated that the internment had been adopted "in the way of retribution for the attack that was made on Pearl Harbor." P. Irons, Justice at War 353 (1983).

5. In comparison, the United States government took an individualized approach to determining the loyalty of Italian and German Americans. This approach was applied to Germans "despite visible, active pro-Nazi operations among German Americans before the outbreak of war." CWRIC, supra note 1, at 283-93.

Today this seemingly intangible injury fuels the Japanese American community's drive for redress.[6] The search for justice by the Japanese American community has taken a variety of forms, including coram nobis lawsuits which seek to vacate the World War II convictions of three Japanese Americans who defied the military orders that mandated the mass incarceration.[7] Though narrowly stated in terms of the interests of three individuals, these suits aim to destroy the myth of disloyalty that has maligned not only these individuals but the entire Japanese American community. Thus, the metamorphosis of emotion to legal action cannot be accurately stated without accounting for the interests of all Japanese Americans. This Note will describe the Japanese American community's interest in removing the stigma of disloyalty and suggest a basis for viewing stigma as a legally recognizable injury.

I
Historical Background

This unfinished chapter in American history had its genesis in the long and ugly history of anti-Japanese agitation and legislation on the West Coast.[8] Japanese immigrants were barred from becoming naturalized citizens,[9] alien land laws in many states forbade them from purchasing land,[10] and the Japanese Exclusion Act of 1924[11] put an end to Japanese immigration altogether.[12]

In a nation which showed itself to be predisposed to discriminate against persons of Japanese ancestry, the internment of Japanese Americans during World War II is probably more correctly viewed as an historical consistency rather than an aberration. This pervasive racial bias affected the decisionmaking of General John L. DeWitt, Commanding Officer of the Western Defense Command, who was responsible for the security of the West Coast and whose military orders directed the internment of the Japanese American community. In recommending to the Secretary of War that Japanese Americans be interned, General DeWitt said:

> The Japanese race is an enemy race and while many second and third generation Japanese born on United States soil, possessed of United States citizenship, have become "Americanized," the racial strains are undiluted. To conclude otherwise is to expect that chil-

---

6. See text accompanying notes 30-46 infra.

7. Korematsu v. United States, 584 F. Supp. 1406 (N.D. Cal. 1984); Yasui v. United States, Civil No. 83-151-BE, slip op. (D. Or. Jan. 26, 1984); Hirabayashi v. United States, No. C83-122V (W.D. Wash. filed Jan. 31, 1983). For a comprehensive analysis of the factual basis for these coram nobis lawsuits, see P. Irons, supra note 4.

8. See note 38 infra.

9. Ozawa v. United States, 260 U.S. 178 (1922). Japanese immigrants were not allowed to become naturalized citizens until passage of the Walter-McCarran Act in 1952. See generally F. Chuman, supra note 1, at 308-13.

10. See generally F. Chuman, supra note 1, at 73-89. Not until 1952 was the California Alien Land Law declared unconstitutional. Id. at 89.

11. Quota Immigration Law, 43 Stat. 153 (1924).

12. See generally F. Chuman, supra note 1, at 91-103.

> dren born of white parents on Japanese soil sever all racial affinity and become loyal Japanese subjects, ready to fight and, if necessary, to die for Japan in a war against the nation of their parents . . . . It, therefore, follows that along the vital Pacific Coast over 112,000 potential enemies, of Japanese extraction, are at large today.[13]

Facing mounting pressure from General DeWitt, officials of the War Department, and West Coast politicians and pressure groups,[14] President Franklin D. Roosevelt issued Executive Order 9066,[15] authorizing the military to exclude persons endangering the national security from military areas. Under the authority of Executive Order 9066, General DeWitt issued a series of orders designating all of California, Oregon, Washington, and Arizona as military areas.[16] Although these military orders did not refer to any specific ethnic group, Japanese Americans were warned that they would be required to leave these military areas, and were encouraged to "voluntarily" relocate.[17]

To strengthen Executive Order 9066, Congress enacted Public Law 503,[18] imposing criminal sanctions upon any person who knowingly violated military orders made pursuant to Executive Order 9066. With compliance to his military directives assured through the threat of criminal penalties, General DeWitt promulgated a step-by-step program of military orders aimed at the eventual removal and internment of persons of Japanese ancestry.

One of the first such orders imposed a curfew restricting the movements of Japanese Americans residing in military areas.[19] The next series of orders directed Japanese Americans to leave their homes and report to assembly centers established within the military areas.[20] At the same time, other orders

13. J. DeWitt, Final Report: Japanese Evacuation from the West Coast, 1942, 33-34 (1943).

14. See generally CWRIC, supra note 1, at 47-92.

15. Exec. Order No. 9066, 7 Fed. Reg. 1407 (1942), provided in pertinent part:

> [T]he successful prosecution of the war requires every possible protection against espionage and against sabotage to national defense . . . NOW, THEREFORE, by virtue of the authority vested in me as President of the United States, and Commander in Chief of the Army and Navy, I hereby authorize and direct the Secretary of War, and the Military Commanders whom he may from time to time designate, whenever he or any designated Commander deems such action necessary or desirable, to prescribe military areas . . . from which any or all persons may be excluded, and with respect to which the right of any person to enter, remain in, or leave shall be subject to whatever restrictions the Secretary of War or the appropriate Military Commander may impose in his discretion.

16. P. Irons, supra note 4, at 65.

17. Id.

18. Pub. L. No. 503, 56 Stat. 173 (1942), provided in pertinent part:

> [W]hoever shall enter, remain in, leave or commit any act in any military area . . . contrary to the restrictions applicable to any such area . . . shall, if it appears that he knew or should have known of the existence and extent of the restrictions or order and that his act was in violation thereof, be guilty of a misdemeanor and upon conviction shall be liable to a fine of not to exceed $5,000 or to imprisonment for not more than one year, or both, for each offense.

19. P. Irons, supra note 4, at 70.

20. Id.

forbade Japanese Americans from leaving military areas without permission.[21] Consequently, Japanese Americans could not leave the military area in which they resided and could not remain in such areas unless they were in an assembly center.[22] Finally, the military orders called for their mass imprisonment in concentration camps scattered throughout the deserts of the western United States.[23]

Three United States citizens of Japanese descent, Fred Korematsu, Minoru Yasui, and Gordon Hirabayashi, resisted, choosing to violate various aspects of DeWitt's program which would have eventually led to their imprisonment in concentration camps. Each was arrested and convicted of violating Public Law 503 for disobeying military orders made under the authority of Executive Order 9066.[24] The personal loyalty of these individuals to the United States was never the issue; the Supreme Court sustained their convictions based on a perceived threat from the ethnic group in general.[25] In support of its military orders, the government argued that there existed disloyal citizens of Japanese descent, whose number and strength could not be readily ascertained.[26] Furthermore, the government contended that this disloyal element was prone to engage in sabotage and espionage, and thus constituted a menace to the national defense.[27] In light of the facts and circumstances before it, the Supreme Court upheld the military orders mandating the evacuation of all Japanese Americans and resident Japanese aliens[28] from the West Coast. By upholding the military orders, the nation's highest arbiter of justice condoned and ratified the mass imprisonment of an entire ethnic group.[29]

II
Redress Campaigns

Soon after their release in 1944, Japanese Americans began their efforts to obtain redress for the wrongs done them during the war. These efforts gave rise to tangible but inadequate results in 1948 with the passage of the Evacua-

21. Id.
22. *Korematsu*, 584 F. Supp. at 1409.
23. P. Irons, supra note 4, at 73.
24. *Korematsu*, 323 U.S. 214; *Yasui*, 320 U.S. 115; *Hirabayashi*, 320 U.S. 81. For a full discussion of these cases, see generally P. Irons, supra note 4.
25. *Korematsu*, 323 U.S. at 218-19; *Yasui*, 320 U.S. at 116; *Hirabayashi*, 320 U.S. at 96-98.
26. See note 46 infra.
27. *Korematsu*, 323 U.S. at 218; *Yasui*, 320 U.S. at 116; *Hirabayashi*, 320 U.S. at 99.
28. See note 9 supra.
29. *Korematsu* and *Hirabayashi* were cited repeatedly in defense of Nazi war criminals at the Nuremberg Tribunal. The Nazi defendants claimed "military necessity" in the "evacuation" of the Jews. M. Weglyn, supra note 2, at 75.

Critical analyses of the Supreme Court's Japanese American internment decisions can be found in Dembitz, Racial Discrimination and the Military Judgment: The Supreme Court's Korematsu and Endo Decisions, 45 Col. L. Rev. 175 (1945); Rostow, The Japanese American Cases—A Disaster, 54 Yale L.J. 489 (1945); J. tenBroek, E. Barnhart & F. Matson, supra note 2, at 211-321.

tion Claims Act.[30] Symbolic forms of redress, through the repeal of Title II of the Internal Security Act of 1950 and of Executive Order 9066 itself, were obtained in 1971[31] and 1976.[32]

Recently, redress campaigns on the West Coast have successfully obtained monetary and symbolic forms of redress at the local and state levels for some former internees. Several state and local governments, for example, have enacted legislation which compensates Japanese Americans who were dismissed from their government jobs during World War II because of their ancestry.[33] Other communities have placed memorials at the sites of the concentration camps and assembly centers or have taken other steps to commemorate this chapter in American history.[34]

The national redress effort focuses on obtaining remedies through Congress and the federal judiciary. The Japanese American community, through

30. American-Japanese Evacuation Claims Act, 62 Stat. 1231 (1948)(codified at 50 U.S.C. app. § 1981-1987 (1982)). The Act gave persons the right to claim from the government "damages to or loss of real or personal property" which occurred as a "reasonable and natural consequence of the evacuation or exclusion." The Act did not compensate for stigma, psychological impact, deprivation of liberty, personal inconvenience, physical hardship or injury, or loss of anticipated profits or earnings. CWRIC, supra note 1, at 118.

Claims amounting to 148 million dollars were filed under the Act; a total amount of 37 million was paid by the government, an amount which did not fairly compensate for economic hardship and suffering. Id. at 118, 121. It is estimated that uncompensated income and property losses alone fall between 810 million and 2 billion in 1983 dollars. See note 3 supra.

31. In 1971, President Nixon signed into law a bill, 85 Stat. 347 (1971), which repealed Title II of the Internal Security Act of 1950, 64 Stat. 987 (1950). That Act, using the Japanese internment cases as precedent, had authorized the President to apprehend or detain persons who might engage in espionage or sabotage during an "Internal Security Emergency." Such persons were to be "confined in such places as may be prescribed." Japanese Americans had lobbied vigorously for the repeal of Title II. Chuman, supra note 1, at 327-31.

32. In 1976, lobbying efforts led to President Ford's signing of Proclamation No. 4417, which formally rescinded Executive Order 9066. 41 Fed. Reg. 7741 (1976).

33. Each bill calls for the payment of up to $5,000 to each employee who had been dismissed during World War II. The typical bill provided for payment of $1,250 each year for four years. The Washington state bill, ESSB 3163, paid $2,500 each year for two years. The city of Los Angeles paid a lump sum of $5,000. Generally, the bills compensated only the individual who had actually lost his or her job, not the surviving spouse. The Los Angeles ordinance was the most liberal; it provided for payment to the surviving spouse and heirs. See Seattle to Compensate Employees, Pacific Citizen, Mar. 16, 1984, at 1, col. 1; City of Los Angeles to Compensate WW2 Nisei Employees, Pacific Citizen, Feb. 24, 1984, at 1, col. 3; Washington Governor Signs State Worker Reparations Bill, Pacific Citizen, May 27, 1983, at 1, col. 1; S.F. Mayor Feinstein Signs Redress Bill for City Workers, Pacific Citizen, Feb. 4, 1983, at 1, col. 3; Payment for Fired Nikkei L.A. County Workers Passes, Pacific Citizen, Nov. 19, 1982, at 1, col. 5; California Legislature Passes Compensation Bill for Nisei, Pacific Citizen, Aug. 20, 1982, at 1, col. 5.

34. Memorials have been placed at the sites of some of the concentration camps and assembly centers. See, e.g., $8,000 Raised for Amache Camp Shrine, Pacific Citizen, Apr. 1, 1983, at 1, col. 3; Puyallup Fair Will Get Monument, Pacific Citizen, Mar. 18, 1983, at 2, col. 3. "Day of Remembrance" legislation has been passed by some local governments to commemorate February 19, 1942, the day President Roosevelt signed Executive Order 9066. See, e.g., California Legislature Issues Day of Remembrance Resolution, Pacific Citizen, Mar. 9, 1984, at 2, col. 1. Other local governments have adopted resolutions to commemorate Public Law 503. See Nine Local Governments Adopt Resolution to Remember PL 503, Pacific Citizen, Mar. 30, 1984 at 2, col. 1.

the efforts of the Japanese American Citizens League (JACL),[35] the National Coalition for Redress/Reparations (NCRR)[36] and other organizations, is pressing for legislation which generally conforms to the recommendations made by the Commission on Wartime Relocation and Internment of Civilians (CWRIC)[37] in 1984. After an eighteen-month study, CWRIC concluded that the internment was the result of "race prejudice, wartime hysteria and a failure of political leadership."[38] It recommended, among other things, a national apology and monetary compensation.[39]

In addition, a class action was filed in the District Court for the District of Columbia by the National Council for Japanese American Redress

---

35. The Japanese American Citizens League, the oldest and largest national Japanese American organization, has been a major force in the redress movement. The JACL lobbied for the establishment of a congressional commission which would investigate the reasons for the internment and publicize its findings. The JACL believed it was necessary for public education to precede its redress campaign in Congress because Congress, largely ignorant of the internment issue, would have summarily dismissed any petition for redress. JACL Position Statement: Merit of Commission Approach to Redress Shows Strong Promise, Pacific Citizen, December 21-28, 1979, at 60, col. 1.

36. The National Coalition for Redress/ Reparations was formed in 1979 to facilitate the redress movement.

37. CWRIC was established by Congress in 1980 to:

> [1] review the facts and circumstances surrounding Executive Order 9066, issued February 19, 1942, and the impact of such Executive Order on American citizens and permanent resident aliens;
> [2] review directives of United States military forces requiring the relocation and, in some cases, detention in internment camps of American citizens, including Aleut civilians, and permanent resident aliens of the Aleutian and Pribiloff Islands; and
> [3] recommend appropriate remedies.

Commission on Wartime Relocation and Internment of Civilians Act, Pub. L. No. 96-317, 94 Stat. 964 (codified at 50 U.S.C. app. § 1981 (1982)). CWRIC, supra note 1, at 18.

38. According to the CWRIC report, the internment culminated a long history of West Coast anti-Japanese agitation and legislation. CWRIC, supra note 1, at 27-46. See text accompanying notes 8-12 supra.

Anti-Japanese sentiment fed upon economic competition, CWRIC, supra note 1, at 4, 42-44; race prejudice, id. at 36-42; J. tenBroek, E. Barnhart & F. Matson, supra note 2, at 11-68; R. Daniels, The Politics of Prejudice, supra note 2, at 65-78; and the fear of a militaristic Japan, CWRIC, supra note 1, at 37. The public also feared sabotage and espionage by Japanese Americans in Hawaii and on the West Coast. False rumors of sabotage and espionage committed during the Pearl Harbor attack by Japanese Americans in Hawaii had wide circulation. The government knew that these rumors were false, but took no effective measures to disabuse the public of the belief that disloyalty by Japanese Americans had contributed to American losses at Pearl Harbor. CWRIC, supra note 1, at 55-60. The CWRIC concluded that the "hostile reception and treatment of Japanese immigrants on the West Coast are the historical prelude to the exclusion and evacuation." Id. at 28.

39. CWRIC, Federal Commission Recommends $1.5 Billion, Formal Apology to Redress Injustice to Japanese Americans in World War II (undated press release)(on file at New York University Review of Law & Social Change). Professor Alan Dershowitz of Harvard Law School has stated that a "substantial" amount of money should be paid to each internee, but like payments by the German government to Jewish Holocaust survivors, any money must be considered "token." Professor Lawrence Sager of New York University School of Law agreed on the need for monetary compensation: "We must vaccinate ourselves to prevent a recurrence of this tragedy." Statements made at hearing before the CWRIC in Cambridge, Massachusetts, Dec. 9, 1982. Legal Experts Discuss Redress for Japanese American Internees, Pacific Citizen, Jan. 1-8, 1982, at l, col. 3.

(NCJAR)[40] on behalf of interned Japanese and their descendants.[41] The lawsuit sought almost three billion dollars in damages for numerous constitutional and tortious injuries suffered during the evacuation and incarceration. The court dismissed the action as time barred, though it expressed doubt about the government's "military necessity" justification for the incarceration.[42] NCJAR has appealed this decision.[43]

A second major litigation effort aims to erase the stigma of suspected disloyalty which has plagued the Japanese American community ever since the federal judiciary held that the internment was necessary to protect national security. The effort consists of three separate actions which seek the extraordinary writ of coram nobis to vacate the convictions of Japanese Americans who were convicted of violating military curfew and exclusion orders that paved the way for the internment.[44] This campaign is based on newly-discovered evidence which undermines the government's justification for the internment.[45] The crucial evidence casts an unfavorable light upon the methods used by the Department of Justice in defending the position that "military necessity" warranted the internment.[46]

---

40. The National Council for Japanese American Redress (NCJAR) was formed in 1980 by members of the Japanese American community who felt that it was unnecessary to conduct a study to determine whether wrongs had been committed. As an alternative to the Commission approach, see note 37 supra, NCJAR supported legislation which would provide monetary compensation to internees and their heirs. After Congress chose the alternative approach of fact finding by the Commission over immediate monetary compensation, NCJAR decided to bring this class action lawsuit.

41. Hohri v. United States, 586 F. Supp. 769 (D.D.C. 1984).

42. Judge Louis F. Oberdorfer ruled that the evidence which indicates that the internment was not justified by military necessity was available in the late 1940's and, therefore, the case was barred by a six-year statute of limitations. The evidence consisted of intelligence reports that contradicted the government's allegations of Japanese American disloyalty. Although Judge Oberdorfer did not reach the question of whether the internment was required by "military necessity," he said that the new information which relates to a conspiracy to deprive plaintiffs of their rights "fully justif[ies] the condemnation of the wartime Department of Justice . . . voiced by the Commission and the plaintiffs." Id. at 790.

43. Hohri v. United States, 586 F. Supp. 769 (D.D.C. 1984), appeal docketed, No. 84-5460 (filed July 24, 1984).

44. *Korematsu*, 584 F. Supp. 1406; *Yasui*, Civil No. 83-151-BE; *Hirabayashi*, No. C83-122V. The definitive account of this effort is found in P. Irons, supra note 4.

45. This campaign rests heavily upon the CWRIC's conclusion that the "record does not permit the conclusion that military necessity warranted the exclusion of ethnic Japanese from the West Coast." CWRIC, supra note 1, at 8.

46. Internal memoranda of the Justice Department indicate that the Justice Department and General John L. DeWitt knew that the claim of "military necessity" was without foundation. P. Irons, supra note 4, at 186-218.

The Justice Department eventually inserted a footnote in its *Korematsu* brief discrediting the facts in DeWitt's Final Report, supra note 13, which allegedly justified the evacuation. However, under pressure from the War Department, the Justice Department eliminated this footnote and requested the Court to take judicial notice of facts it knew to be false. *Korematsu*, 584 F. Supp. at 1417. See also P. Irons, supra note 4, at 284-92.

III
Writ of Coram Nobis

The writ of coram nobis calls belated attention of the court to facts which, for valid reasons, were not before it in the original trial.[47] "[One] purpose of coram nobis is to allow a defendant to attack a conviction notwithstanding the fact that he has completed his sentence."[48] In the foremost coram nobis case, *United States v. Morgan*,[49] the defendant had been convicted of a violation of federal law and had completed his sentence. Subsequently he was convicted on a state charge and sentenced to a longer term because of his prior federal conviction. He then attacked the federal conviction on the ground that he had lacked counsel. The Supreme Court held that:

> [a]lthough the term has been served, the results of the conviction may persist. Subsequent convictions may carry heavier penalties; civil rights may be affected. As the power to remedy an invalid sentence exists, we think, respondent is entitled to an opportunity to attempt to show that his conviction was invalid.[50]

The Court, however, cautioned that coram nobis relief is to be granted sparingly: "[c]ontinuation of litigation after final judgment and exhaustion or waiver of any statutory right of review should be allowed through this extraordinary remedy only under circumstances compelling such action to achieve justice."[51] A mere technical error will not suffice; the error must be so fundamental as to render the legal proceeding invalid.[52] On review of the defendant's claim that he had been tried and convicted without competent and intelligent waiver of counsel, the Court affirmed the Second Circuit's decision to remand the case for an evidentiary hearing to clarify the record with respect

---

47. See generally Annot., 38 A.L.R. Fed. 617 (1978). The power of the federal courts to grant the writ comes from the All Writs Act, 28 U.S.C. § 1651 (1982). The Act provides in pertinent part: "The Supreme Court and all courts established by Act of Congress may issue all writs necessary or appropriate in aid of their respective jurisdictions and agreeable to the usages and principles of law." 28 U.S.C. § 1651(a) (1982). Although Federal Rule of Civil Procedure 60(b) abolishes the writ for civil proceedings, the writ is available in criminal actions. United States v. Morgan, 346 U.S. 502, 505 n.4, 506 (1954). Unlike a petition for habeas corpus, which is the beginning of a separate civil proceeding and where relief is sought in a separate case and record, coram nobis is a step in the original criminal case. *Morgan*, 346 U.S. at 505 n.4.

The petition for coram nobis is heard in the court that rendered the original conviction. Sinclair v. Louisiana, 679 F.2d 513, 514-15 (5th Cir. 1982); Hensley v. Municipal Court, 453 F.2d 1250, 1252 n.2 (9th Cir. 1972), rev'd on other grounds, 411 U.S. 345 (1973). Appellate leave is not required for the trial court to correct error occurring before it. *Korematsu*, 584 F. Supp. at 1412. The petitioner must overcome the presumption that the original proceedings were correct. *Morgan*, 346 U.S. at 512; Ybarra v. United States, 461 F.2d 1195, 1198 (9th Cir. 1972).

48. Holloway v. United States, 393 F.2d 731, 732 (9th Cir. 1968)(citing *Morgan*, 346 U.S. at 512).

49. 346 U.S. 502.

50. Id. at 512-13.

51. Id. at 511.

52. Id. at 512 (citing United States v. Mayer, 235 U.S. 55, 69 (1914)). See also United States v. Addonizio, 442 U.S. 178, 186 (1979).

to the waiver issue. In searching beyond the deficient trial record, the Court demonstrated its commitment to act to secure justice on a "record that makes plain a right to relief."[53]

The Ninth Circuit has demonstrated a similar willingness to search beyond the trial record in order to scrutinize charges of governmental misconduct. In *United States v. Taylor*,[54] the defendant, convicted of wire fraud, challenged the basis for his conviction with facts assailing the veracity of the government's allegations. The court noted that "prosecutorial misconduct may so pollute a criminal prosecution as to require a new trial, especially when the taint in the proceedings seriously prejudices the accused."[55] Moreover, the court noted that coram nobis is not limited to instances of great prejudice to the petitioner, but "is designed to maintain also public confidence in the administration of justice."[56]

The defects alleged in coram nobis petitions must be of constitutional proportion.[57] Therefore, the criminal justice system's interest in maintaining public confidence warrants special consideration in such actions. It is disturbing for law-abiding citizens to find that the outcome of a criminal prosecution may depend upon the extent to which government law enforcement officials misbehave. When misconduct results in a conviction which abridges a fellow citizen's right to a fair trial, the public's confidence is undermined.

However, defects which do not completely undermine the reliability of a judgment of conviction may be tolerated.[58] For example, a district court in California held that a finding of a fundamental error would not disturb the conviction of one whose guilt was conclusively established.[59] In another case, the government had suppressed evidence during petitioner's trial, but the court refused to grant relief because this alleged act of prosecutorial misconduct, in light of the total record, failed to undermine the validity of the original judgment of conviction.[60]

Prosecutorial misconduct impairs the integrity of the judicial process and

53. 346 U.S. at 505.
54. 648 F.2d 565 (9th Cir. 1981).
55. Id. at 571.
56. Id.
57. Byrnes v. United States, 408 F.2d 599, 602 (9th Cir. 1969). Coram nobis petitions generally allege fundamental errors of fact from which constitutional violations arise. See, e.g., Napue v. Illinois, 360 U.S. 264, 265 (1959) (knowing use of false testimony violates fourteenth amendment). District courts in the Ninth Circuit, however, have shown a willingness to use coram nobis to correct "legal errors of a constitutional or fundamental proportion." United States v. Wickham, 474 F. Supp. 113, 116 (C.D. Cal. 1979)(misapplication of federal statute resulting in conviction may be fundamental error); United States v. Danks, 357 F. Supp. 193 (D. Hawaii 1973)(conviction for not answering 1970 census violates first amendment). Contra, *Korematsu*, 584 F. Supp. at 1420 (coram nobis used only to correct errors of fact).
58. See United States v. Scherer, 673 F.2d 176, 178 (7th Cir. 1982), cert. denied, 457 U.S. 1120 (1982); Willis v. United States, 654 F.2d 23, 24 (8th Cir. 1981); United States v. Keogh, 391 F.2d 138, 148 (2d Cir. 1968); Bateman v. United States, 277 F.2d 65, 68 (8th Cir. 1960).
59. *Wickham*, 474 F. Supp. at 116.
60. In re Alger Hiss, 542 F. Supp. 973 (S.D.N.Y. 1982), cert. denied, 104 S. Ct. 232 (1983).

provides grounds for vacating a conviction through the use of coram nobis, though it does not automatically do so.[61] The Supreme Court has declared that the "touchstone of due process analysis in cases of alleged prosecutorial misconduct is the fairness of the trial, not the culpability of the prosecutor."[62] Where, for example, the prosecutor suppresses evidence at trial, fairness is determined by the character of the suppressed evidence.[63] Thus, there is no due process right to all the evidence in the government's possession which might conceivably assist in the preparation of the defendant's defense, although there is a right to all exculpatory evidence known only to the government.[64]

The petitioner has the burden of proof with respect to the materiality of the suppressed evidence which she has specifically requested.[65] This burden is satisfied if the evidence "might" have affected the outcome.[66] In view of petitioner's light burden, the government's failure to respond to a specific request for evidence is rarely excusable.[67]

Absent a specific request, however, the prosecutor has a duty to voluntarily disclose evidence "so clearly supportive of a claim of innocence" that it gives her "notice of a duty to produce."[68] Failure to disclose such evidence constitutes fundamental error where the suppressed evidence, evaluated in light of the total record, "creates a reasonable doubt that did not otherwise exist."[69] If the verdict is already of questionable validity, additional evidence of relatively minor importance might be sufficient to create a reasonable doubt.[70]

A second category of prosecutorial misconduct concerns the prosecutor's use of false evidence at trial. A conviction may be vacated if the prosecutor presented either testimony known to be perjured[71] or false testimony bearing on the credibility of a key witness,[72] or if she failed to correct unsolicited false testimony.[73] As in the suppression of evidence case, the test for fundamental error is the fairness of the trial, determined by looking at the character of the

61. *Keogh*, 391 F.2d at 148.

62. Smith v. Phillips, 455 U.S. 209, 219 (1982).

63. United States v. Agurs, 427 U.S. 97, 110 (1976).

64. United States v. LeRoy, 687 F.2d 610, 619 (2d Cir.), cert. denied, 459 U.S. 1174 (1982).

65. See *Agurs*, 427 U.S. at 106.

66. Id. at 104 (citing Brady v. Maryland, 373 U.S. 83 (1963)).

67. Id. at 106; United States v. Blasco, 702 F.2d 1315, 1317 (11th Cir. 1983).

68. *Agurs*, 427 U.S. at 107, 110.

69. Id. at 112, 113. See *Keogh*, 391 F.2d at 148 (no duty to disclose unrequested evidence which is not likely to have changed the verdict).

70. Id. at 113; Cannon v. Alabama, 558 F.2d 1211, 1216 (5th Cir. 1977), cert. denied, 434 U.S. 1087 (1978); United States v. Stassi, 544 F.2d 579, 584 (2d Cir. 1976).

71. Mooney v. Holohan, 294 U.S. 103, 112-13 (1934).

72. *Napue*, 360 U.S. at 269.

73. Id. at 265, 269; see United States v. Giglio, 405 U.S. 150, 153 (1972).

evidence, not the culpability of the prosecutor.[74] A conviction must be vacated upon the establishment of a "reasonable likelihood" that the false evidence or testimony could have affected the outcome of the trial.[75]

IV
THE JAPANESE AMERICAN CORAM NOBIS CASES

More than forty years after their convictions were affirmed by the Supreme Court, the three American citizens of Japanese ancestry who were convicted in 1942 uncovered new evidence indicating that they had been the victims of unfair trials. They had been tried for violating the military orders which led to the incarceration of all Japanese Americans on the West Coast. Specifically, they charged that the allegations of disloyalty and treason by the ethnic Japanese were intentional falsehoods designed to mislead the courts into believing that "military necessity" justified the racially discriminatory military orders. Invoking the All Writs Act,[76] each individual filed an identical petition[77] for a writ of coram nobis[78] to vacate their convictions and to have the court make findings of governmental misconduct. The petitions were identical because the legal and constitutional issues raised in the three cases were virtually identical; the newly uncovered evidence was relevant to each case; and the interrelated pattern of alleged governmental misconduct impacted on each petitioner's case.[79]

A. *Korematsu v. United States*

On May 30, 1942, Fred Korematsu was arrested for being in a place where no persons of Japanese descent were allowed, pursuant to Civilian Exclusion Order No. 34 issued by General DeWitt.[80] On September 8, 1942, Korematsu was convicted for violating Public Law 503, which imposed criminal sanctions upon any person who knowingly violated any military order made under the authority of Executive Order 9066. His conviction was affirmed by the United States Supreme Court.[81]

More than forty years after his conviction in the District Court for the Northern District of California, Korematsu filed a petition with that same court for a writ of coram nobis to overturn his conviction and to make findings of governmental misconduct.[82] Petitioner alleged, inter alia, that officials of

74. Id. (citing Savvides v. New York, 1 N.Y.2d 554, 557, 136 N.E.2d 853, 854-855, 154 N.Y.S.2d 885, 887 (1956)).
75. *Agurs*, 427 U.S. at 103.
76. 28 U.S.C. § 1651 (1982). See note 47 supra.
77. Petition for Writ of Error Coram Nobis 3, Korematsu v. United States, 584 F. Supp. 1406 (N.D. Cal. 1984) [hereinafter Petition for Writ of Error, *Korematsu*].
78. See text accompanying notes 47-75 supra.
79. Petition for Writ of Error, *Korematsu*, at 4.
80. Id. at 13.
81. *Korematsu*, 323 U.S. 214.
82. *Korematsu*, 584 F. Supp. at 1409.

the War Department and the Department of Justice suppressed evidence indicating the loyalty of Japanese Americans and the absence of any acts of espionage carried out by them.[83] In addition, Korematsu alleged that government officials failed to advise the Supreme Court of the falsity of the government's allegations that Japanese Americans were disloyal and that they committed acts of espionage.[84] Finally, the petitioner claimed that he was entitled to relief on the ground that his conviction was based upon governmental orders that violate current constitutional standards.[85]

The government requested and was granted an extension on its reply to the coram nobis petition, pending the release of the report of the Commission on Wartime Relocation and Internment of Civilians.[86] This report was the product of an eighteen-month investigation into the facts and circumstances surrounding the evacuation and incarceration, and the government believed that it would have a substantial impact on the proceedings.[87]

Following the issuance of the CWRIC report, the government filed a countermotion under Federal Rule of Civil Procedure 48(a), to vacate the conviction and to dismiss the underlying indictment.[88] The government's objective, however, was not to make an admission of wrongdoing, but rather to avoid an independent assessment of the factual and constitutional issues raised in the petition. In essence, the government wanted the court to set aside Korematsu's conviction without consideration of the record in an effort to forget "this unfortunate episode in our country's history."[89]

Thus the court had before it an unusual posturing of opposing parties: both wanted to vacate petitioner's conviction, but for different reasons. The government's rationale was that the military orders under which Korematsu had been convicted were simply a "mistake" which should be forgotten.[90] On the other hand, the petitioner argued that he had been subjected to a deliberate and calculated plan to exclude and imprison him and other Japanese Ameri-

83. Petition for Writ of Error, *Korematsu*, at ii.

84. Id.

85. Id. at iii.

86. *Korematsu*, 584 F. Supp. at 1410.

87. For a discussion of the CWRIC and its findings, see text accompanying notes 35-39 supra.

88. *Korematsu*, 584 F. Supp. at 1410.

89. Id. at 1413.

90. Transcript of Hearing on Motion to Vacate Conviction and Dismiss Indictment of Fred T. Korematsu 17, Korematsu v. United States, 584 F. Supp. 1406 (N.D. Cal. 1984) [hereinafter Transcript of Hearing, *Korematsu*]. In opposing the government's motion, petitioner's counsel, Dale Minami, said:

> The government's approach turns the idea of public interest on its head. The government, in effect, is advocating letting the guilty go free and keeping the innocent imprisoned in the shame and suffering they endured for [forty] years.
>
> It is advocating keeping the public imprisoned in the ignorant notion that this was an "unfortunate" incident as the government describes.
>
> Even the government's motion to vacate indicates an unwillingness to face the facts and the constitutional issues.

Id. at 24-25.

cans.[91] Therefore a reconsideration of his case required a judicial declaration of the grounds for vacating the conviction.[92] The court decided to focus less on whether the conviction should be vacated than on whether the appropriate ground for relief should be the government's countermotion or petitioner's coram nobis petition.[93]

The court first considered the government's countermotion. It noted that Rule 48(a) allows the prosecutor to terminate, with leave, an indictment, information or complaint. The prosecutor may not, however, invoke this rule after a person has been subject to conviction, final judgment, or exhaustion of all appeals. Nor may the prosecutor invoke Rule 48(a) if judgment has been imposed and the sentence served, at which point "there is no longer any prosecution to be terminated."[94] Since this point had been reached long ago in Korematsu's case, the court held that the government's motion was unavailable.[95]

The court next turned to petitioner's coram nobis petition, finding it to be the appropriate vehicle for relief.[96] After examining the evidence before it, the court determined that the CWRIC report provided ample support for its decision that coram nobis must be granted to prevent manifest injustice and to serve the public interest.[97]

The court then determined whether due process warranted vacating Korematsu's conviction. It addressed the question of whether the judiciary in 1944 had before it all the facts known by the government at that time.[98] Based on the evidence of newly-discovered internal government documents and letters, the court answered in the negative, finding that the government had deliberately withheld and falsified relevant evidence in its papers before the court:

> The facts which the government represented it relied on and provided to the courts were those contained in a report entitled "Final Report, Japanese Evacuation from the West Coast" (1942), prepared by General DeWitt. His evaluation and version of the facts informed the courts' opinions.
>
> . . .
>
> Omitted from the reports presented to the courts was information possessed by the Federal Communications Commission, the Department of the Navy, and the Justice Department which directly con-

91. Id. at 25-26.
92. Id.
93. *Korematsu*, 584 F. Supp. at 1410.
94. Id. at 1411.
95. Id.
96. Id. at 1411-12.
97. Id. at 1419. The government neither opposed any of the allegations made in the coram nobis petition, nor did it confess error. Noting the government's position, the court said: "The government has . . . while not confessing error, taken a position tantamount to a confession of error." Id. at 1413.
98. Id. at 1419.

tradicted General DeWitt's statements. Thus, the court had before it a selective record.[99]

The court concluded that the pervasive nature of the government's misconduct seriously impaired the judicial process, and thus satisfied the compelling circumstances requirement contemplated by the Supreme Court in *Morgan*. Although there was no question that Korematsu had violated the law, the court vacated his conviction in an attempt to restore public confidence in the administration of justice.[100]

After having undermined the factual basis of the military orders, the court expressly declined to go one step further by questioning the law under which Korematsu had been convicted. The court explained that it had no power to correct legal errors because coram nobis is used only to correct errors of fact.[101] Although it is not mentioned in the opinion, even if the district court had found that it did have the power to correct legal errors, the fact that the Supreme Court had upheld the exclusion and internment on the basis of "military necessity" would have rendered dubious any attempt by the district court to declare it otherwise.

In limiting its inquiry to the correction of factual errors, the district court removed the factual justification for the incarceration, but not the legal precedent.[102] Thus, the danger that the doctrines of "military necessity" or national emergency will be used again to legitimize racism looms as a future possibility. Courts today, however, might be more likely to scrutinize the factual basis for such claims in order to safeguard against the recurrence of fraud.[103]

---

99. Id. at 1418-19.

100. Id. at 1419-20.

101. Id. at 1420.

102. The CWRIC believes that "today the decision in *Korematsu* lies overruled in the court of history." CWRIC, supra note 1, at 238. Judge Patel said: "As a legal precedent it is now recognized as having very limited application." *Korematsu*, 584 F. Supp. at 1420. But see note 103 infra.

103. Ironically, although the Supreme Court in *Korematsu* held the incarceration constitutional, the case is cited as the Court's first explicit reference to race as a "suspect" criterion. L. Tribe, American Constitutional Law, §§ 14-16 at 1013 (1978). Since *Korematsu*, the legal doctrine has evolved so that laws based on racial classifications trigger a rigorous standard of judicial review that almost always invalidates racially discriminatory laws. Gunther, The Supreme Court, 1971 Term—Foreward: In Search of Evolving Doctrine in a Changing Court: A Model for a Newer Equal Protection, 86 Harv. L. Rev. 1, 8 (1972).

In reference to this legal development, the CWRIC has suggested that *Korematsu* has been implicitly overruled in the court of history, and is therefore not a precedent on questions of racial discrimination. CWRIC, supra note 1, at 238-39.

Other authorities point out that without the development of a body of constitutional principles protective of minority rights during periods of crisis, the *Korematsu* decision's groundless reference to military necessity could be invoked to imprison an entire racial group during a national emergency. See P. Irons, supra note 4, at 365-66; testimony of Professors Alan Dershowitz and Lawrence Sager, note 39 supra.

Justice Jackson's dissent from the majority opinion in *Korematsu* prophetically characterized that decision's legalization of racial discrimination as a "loaded weapon, ready for the hand of any authority that can bring forward a plausible claim of an urgent need." 323 U.S. at 246.

Although the court restricted its inquiry to the examination of factual errors, the outcome of the case represents judicial recognition that the injustice foisted upon Japanese Americans in 1944 has not faded with the passage of time. It was not too late for the court to try to relieve Fred Korematsu, and the Japanese American community, of some of the stigma of suspected disloyalty which they had carried for four decades.

B. Yasui v. United States

On March 23, 1942, Minoru Yasui was arrested for violating Public Proclamation No. 3, a curfew order which General DeWitt issued against persons of Japanese descent.[104] Yasui was subsequently convicted for a violation of Public Law 503, as Korematsu had been. Yasui's conviction was affirmed by the Supreme Court.[105]

Several months after Korematsu had filed his petition for a writ of coram nobis, Yasui filed his with the district court in Oregon.[106] Yasui's petition requested the court to vacate his conviction, dismiss the underlying indictment, make findings of governmental misconduct, and declare unconstitutional the military orders under which he had been convicted.[107] The government's opposition to this petition conformed to its position in *Korematsu*: it asked the court to vacate the conviction and dismiss the petition without determining the merits of the claims of governmental misconduct.[108]

Although the issues presented in *Yasui* were identical to those decided in *Korematsu*, the *Yasui* decision produced a curiously different outcome. Judge Belloni vacated Yasui's conviction, but granted the government's motion to dismiss the coram nobis petition.[109] In a tersely worded opinion, the judge stated:

> I decline to make such findings [of fact] forty years after the events took place. There is no case nor controversy since both sides are asking for the same relief but for different reasons. The Petitioner would have the court engage in fact finding which would have no legal consequences. Courts should not engage in that kind of activity.[110]

The court apparently agreed with the government's argument that any finding of fact could only be of historical or legislative interest, without any

104. *Yasui*, 320 U.S. 115.

105. Id.

106. The *Hirabayashi* coram nobis case will not be discussed in this Note, as it is not scheduled for an evidentiary hearing until the summer of 1985.

107. *Yasui*, Civil No. 83-151-BE, at 2.

108. Id. at 1. At oral argument, the government said that it was not clear that misconduct had occurred. Transcript of Motions Before the Honorable Robert C. Belloni 15, Yasui v. United States, No. 83-151-BE (D. Or., Jan. 26, 1984) [hereinafter Transcript of Motions, *Yasui*].

109. *Yasui*, Civil No. 83-151-BE, at 2.

110. Id.

legal consequences.[111] Noting that several Presidents had hoped that there would never be another internment, the government agreed to vacate the conviction and dismiss the indictment because of the symbolic importance of such an act.[112] The government, however, declined to admit that it had violated any of Yasui's rights.[113] In effect, the government viewed its concession as an act of forgiveness. A cloud of suspected disloyalty would thus still linger over Yasui, and, by implication, over the Japanese American community.

The court's conclusion that fact finding would have no legal consequences may be attributed to an erroneous view of the case as involving only the interests of one individual. This view is tenable only if one ignores the severe stigma of suspected disloyalty that has attached to Japanese Americans ever since the wartime *Yasui*, *Korematsu*, and *Hirabayashi* decisions. As will be explained in Section V the entire Japanese American community has a profound interest in obtaining a judicial opinion that would correct the record with respect to their loyalty to the United States during World War II. A case or controversy beyond the vacating of petitioner's conviction exists, and its resolution requires fact finding concerning the allegations of governmental misconduct.

V
Social Stigma

Japanese Americans regard the judicial finding of governmental misconduct to be as significant as the vacating of Korematsu's conviction. This broad concern of the Japanese American community developed in the aftermath of the 1944 Japanese internment decisions, which had lent credence to the claim that the national security required the internment of all persons of Japanese ancestry. The entire community has a stake in the coram nobis lawsuits since favorable opinions would diminish the stigma attached to it by the suspicion of wartime disloyalty.

In a real sense, the coram nobis lawsuits are analogous to a class action waged on behalf of all former internees. Viewed is such terms, the relief sought required more than the vacating of one person's conviction. The simple vacating of a conviction, without reference to the government's misconduct, could be erroneously interpreted to mean that a kind-hearted government had forgiven a blameworthy Japanese American community.[114] Thus the vacating of the convictions, without more, would provide insufficient relief from the stigma which affects the entire community. A meaningful remedy requires no less than written opinions explicitly stating that the gov-

---

111. Transcript of Motions, *Yasui*, at 9-10.

112. Id. at 13-14, 16.

113. Id. at 15.

114. Similarly, the CWRIC recommendation, that those who violated the 1942 military orders be "forgiven" or pardoned by the government, suffers from the same ambiguity of interpretation. See CWRIC, supra note 39, at 10.

ernment's misconduct, not its benevolence, compelled the courts to vacate the convictions. For this reason, each Japanese American *coram nobis* case presents the judiciary with a case or controversy that extends beyond the mere vacating of the convictions. Justice requires the judicial declaration of facts which would set the record straight with respect to the loyalty of Japanese Americans to the United States during World War II.

A. The Traditional View

Article III of the Constitution limits the subject matter jurisdiction of the federal courts to cases which present a "case or controversy."[115] In order for there to be a case or controversy sufficient to avoid dismissal of a coram nobis action on grounds of mootness, the petitioner must have a substantial stake in the judgment of conviction which survives satisfaction of the sentence imposed on her.[116] In this context, mootness may be "stated in terms of whether petitioner, who has already fully served his sentence, suffers any collateral consequences such that he should be permitted to apply for a writ of coram nobis."[117]

The Supreme Court in *Lane v. Williams*[118] noted that a "criminal case is moot only if it is shown that there is no possibility that any collateral legal consequences will be imposed on the basis of the challenged conviction."[119] Examples of collateral legal consequences held to present a "case or controversy" in a post-conviction proceeding include the loss of the right to vote and the loss of the right to engage in certain businesses.[120] Since courts recognize as "an obvious fact of life" that collateral consequences flow from a criminal conviction,[121] the burden of proof with respect to showing the existence of such consequences is minimal.

In the Ninth Circuit, the situs of the coram nobis cases, the resolution of mootness questions in post-conviction proceedings gives considerable weight to the merits of petitioner's claim. In *Holloway v. United States*,[122] for example, the petitioner, after serving his sentence for two felony convictions, petitioned for coram nobis relief, alleging that his guilty plea was induced by a bargain made and subsequently broken by the United States Attorney. By granting coram nobis relief, the Court of Appeals reversed the District Court's judgment that there was no "case or controversy" and that it lacked jurisdiction because the sentences had been completed. The court said that "[c]oram nobis must be kept available as a post-conviction remedy to prevent 'manifest injustice' even where the removal of a prior conviction will have little present

115. U.S. Const. art. III, § 2.
116. *Korematsu*, 584 F. Supp. at 1419.
117. Id.
118. 455 U.S. 624 (1982).
119. Id. at 632 (quoting with approval Sibron v. New York, 392 U.S. 40, 57 (1968)).
120. See Carafas v. LaVallee, 391 U.S. 234, 237 (1968).
121. See, e.g., *Sibron*, 392 U.S. at 55.
122. 393 F.2d 731 (9th Cir. 1968).

effect on the petitioner."[123]

This prevention of manifest injustice approach prevailed in *United States v. Danks*,[124] where the petitioner had already paid a fine for refusing to answer questions for the 1970 census. The District Court for the District of Hawaii presumed the existence of collateral legal consequences following from petitioner's misdemeanor conviction. By acknowledging the availability of coram nobis as a post-conviction remedy,[125] the court in effect rejected the government's mootness argument and assumed jurisdiction over petitioner's first amendment claim.

B. A More Expansive View

Generally, the petitioner's interest in removing the social stigma that results from a criminal conviction does not give rise to a case or controversy in the absence of an unrebutted presumption of collateral legal disabilities.[126] However, fundamental concerns of justice should elevate social stigma to the status of a legal disability where, as in the case of the Japanese Americans, individuals have been branded as potential traitors based upon their membership in an ostracized minority group.[127] This should hold particularly true where the label results in the banishment of the branded group by the state. The odious nature of such a stigma should present a case or controversy sufficient to avoid dismissal on grounds of mootness.

The dissenting opinions in *Parker v. Ellis*[128] recognized the concrete and substantial stake of the petitioner in a post-conviction proceeding to remove the taint upon her reputation in society. The *Parker* majority dismissed, on

---

123. Id. at 732.

124. 357 F. Supp. 193 (D. Hawaii 1973).

125. Id. at 195.

126. See The Supreme Court 1967 Term, 82 Harv. L. Rev. 63, 297-98 (1968), and cases cited therein.

127. The CWRIC report concludes that the most devastating loss due to the incarceration was:

> the loss of liberty and the personal stigma of suspected disloyalty for thousands of people who knew themselves to be devoted to their country's cause and to its ideals but whose repeated protestations of loyalty were discounted . . . . The wounds of the exclusion and detention have healed in some respects, but the scars of that experience remain painfully real in the minds of those who lived through the suffering and deprivation of the camps.

CWRIC, supra note 1, at 3.

128. 362 U.S. 574 (1960)(per curiam)(overruled in Carafas v. LaVallee, 391 U.S. 234, 240 (1968)(holding that under the federal habeas corpus statute, once the federal jurisdiction has attached in the District Court, it is not defeated by release of the petitioner prior to completion of proceedings on such application)).

In noting the substantial stake the petitioner retained in pursuing his habeas corpus application even after his release from prison, thus rendering his case not moot, the *Carafas* court spoke of "disabilities or burdens" flowing from his conviction, such as being prohibited by New York law from engaging in certain businesses, serving as a labor union official for a specified period, voting in any election held in New York State, and serving as a juror. Although the *Carafas* opinion did not mention damage to reputation as a consequence of conviction, in general it cited approvingly the reasoning of the *Parker* dissents. Id. at 238-39.

the ground of mootness, a writ of certiorari to review dismissal of the petitioner's habeas corpus application because the petitioner, who had been convicted of a felony, was released from prison prior to the Supreme Court's hearing on the matter. The four dissenting Justices criticized the resolution of mootness questions based upon the existence or nonexistence of a legal disability.[129] The Justices favored an expansive interpretation of what constitutes a case or controversy:

> Conviction of a felony imposes a *status* upon a person which not only makes him vulnerable to future sanctions . . . but which also seriously affects his reputation and economic opportunities . . . . There is, after all, such a thing as rehabilitation and reintegration into the life of a community . . . . Five years of law-abiding life in a new community give [petitioner] a significant enough stake in the outcome of this adjudication to preclude a finding of mootness.[130]

C. *Reputation for Loyalty*

In limited situations, the judiciary may treat social stigma as a legal disability. The determination hinges upon the nature of the stigma. The stigma of real or suspected disloyalty to one's country has drawn special consideration from the courts.[131]

In particular, the judiciary has acknowledged the damaging effect a court martial conviction or dishonorable discharge from the military has upon the reputation of an individual. Such an individual is effectively branded by the government as disloyal, and, by implication, less of an American than other citizens. In *McAliley v. Birdsong*,[132] for example, the petitioner, who had been undesirably discharged "for the good of the service,"[133] sought to have his record expunged and his discharge upgraded to conscientious objector. The court concluded that petitioner's claim was not moot. It stated that while the disability from an "actual conviction in civilian or military court may be more pronounced than those accruing with an undesirable discharge from the Army, it is common knowledge that a discharge . . . can seriously jeopardize an individual's prospects for future employment as well as his general reputation."[134]

In light of the debilitating social and psychological impact upon one branded as disloyal, it is not surprising that Japanese American soldiers court-martialed in 1944 for protesting the internment of their families successfully argued in 1983 that the Army Board for the Correction of Military Records

129. *Parker*, 362 U.S. at 592-94.

130. Id. at 593-94.

131. See Homcy v. Resor, 455 F.2d 1345, 1349 (D.C. Cir. 1971); Grubb v. Birdsong, 452 F.2d 516, 517-18 (6th Cir. 1971); Bland v. Connally, 293 F.2d 852 (D.C. Cir. 1961).

132. 451 F.2d 1244 (6th Cir. 1971).

133. Id. at 1246.

134. Id.

should expunge their dishonorable discharges and the years of imprisonment from their records.[135] Clearly, dishonorable discharges and court-martial convictions adversely impact upon the reputation and psychological well-being of the individual. The enduring stigma of dishonorable discharges and court-martial convictions is evidenced by the persistent efforts that an individual will take to vindicate herself in remedial proceedings long after the sentence has been served.

The stigma associated with the crime of treason is even more lasting in effect than the stigma resulting from a court-martial conviction. Treason is the most serious crime that may be committed against the United States.[136] An act of treason is perpetrated when one "owing allegiance to the United States, levies war against them or adheres to their enemies, giving them aid and comfort."[137] The magnitude of this crime may be inferred from the fact that it is the only offense defined by the Constitution.[138] Treason may be punishable by death.[139]

One stamped with the label of "traitor" has an even greater stake in the outcome of a proceeding which could cleanse her reputation among her peers than someone labeled as disloyal. Therefore, a person convicted of treason retains a strong interest, even after serving her sentence, in showing that her conviction was invalid. Thus, an individual's petition for a post-conviction proceeding to cleanse her name from an undeserved stain of treason presents the judiciary with a "case or controversy" and precludes dismissal on grounds of mootness.

D. Racial Segregation

As with challenges to an individual's national loyalty, governmental discrimination on the basis of race has a severely damaging effect on one's reputation. A minority group alleged to have a propensity for committing acts of treason must live cautiously among neighbors who may not trust them during a national emergency. The accused group is subject to the whims of society's prejudices and fears; thus restrictions of liberty based on suspicion of un-American activities, disloyalty, or inferiority are easily justified.

When prejudice forms the basis for state-enforced racial exclusion, the federal judiciary has responsibly acknowledged and sought to eliminate the intangible effects of the discrimination.[140] The Supreme Court has been par-

135. See Army Clears Records of 11 Nisei Court-Martialed in 1944 Protest, Hokubei Mainichi, Sept. 8, 1983, at 1, col. 6.

136. Stephan v. United States, 133 F.2d 87, 90 (6th Cir.), cert. denied, 318 U.S. 781 (1943).

137. 18 U.S.C. § 2381 (1982).

138. Charge to Grand Jury, 30 F. Cas. 1024, 1025 (D. Mass. 1851)(No. 18269).

139. 18 U.S.C. § 2381 (1982).

140. See, e.g., Johnson v. Virginia, 373 U.S. 61 (1963)(per curiam)(courtrooms); New Orleans City Park Improvement Assoc. v. Detiege, 252 F.2d 122 (5th Cir. 1958), 358 U.S. 54 (1958) (per curiam)(parks); Browder v. Gayle, 142 F. Supp. 707, 710-12 (M.D. Ala. 1956), 352 U.S. 903 (1956)(per curiam)(buses).

ticularly sensitive to the social and psychological impact racial exclusion has upon the excluded. In the landmark case of *Brown v. Board of Education*,[141] the Supreme Court concluded that state-enforced segregation of Black children "solely because of their race generates a feeling of inferiority as to their status in the community that may affect their hearts and minds in a way unlikely ever to be undone."[142] Constitutional law scholar Laurence Tribe states that the most plausible justification for *Brown* is that "racial separation in our society conveys strong social stigma and perpetuates both the stereotypes of racial inferiority and the circumstances on which such stereotypes feed."[143]

In school desegregation cases, therefore, the minority children's interest in removing the odious stigma that results from state-enforced racial segregation has independent legal significance which courts seek to mitigate or eliminate. As a legally cognizable fact of life, this type of stigma arguably acquires the status of a legal disability; one's interest in removing it should generate a "case or controversy" within the meaning of article III of the Constitution.

E. The Japanese American Situation: Reputation for Loyalty Based on Race

One branded by the government as either disloyal or racially excludable suffers from a stigma which festers until all institutional approval is removed.[144] Both of these categories of social stigma are present in *Korematsu*. First, Korematsu was branded as potentially dangerous to the national security and was therefore suspected of disloyalty. Second, he was excluded by the government from the rest of society because of his race.

Today, this stigma continues to weigh upon the shoulders of the entire Japanese American community, because the controversy over whether the internment was justified remains alive.[145] The argument continues in the wake of

141. 347 U.S. 483 (1954).

142. Id. at 494.

143. L. Tribe, supra note 107, § 16-15 at 1020-21 (1978).

144. The most common social and psychological response of second generation Japanese Americans to their imprisonment is:

> [1] Attempting to deny or avoid the experience and refusing to acknowledge the significance of the losses . . . .
>
> [2] Losing faith in white America; maintaining a general distrust or hatred toward white society and choosing to associate only with Japanese Americans.
>
> [3] Turning aggressions inward, as rape victims do, by blaming themselves for something over which they had little control. Anger is internalized as feelings of guilt, shame, and racial inferiority; and energy is focused on attaining economic success in order to prove that one is not inferior . . . .
>
> [4] Identifying with the aggressor by refusing to associate with other Japanese Americans and proudly proclaiming ignorance of Japan, its language and culture . . . . This denial of who one was and how one looked bred ethnic hate and, ultimately, self-loathing.

CWRIC, supra note 1, at 299-300.

145. John J. McCloy, Assistant Secretary of War under President Franklin D. Roosevelt, continues to maintain that "military necessity" justified the incarceration. See, e.g., P. Irons, supra note 4, at 351-54; Repay U.S. Japanese?, New York Times, April 10, 1983, at E21, col. 1.

the report of the Commission on Wartime Relocation and Internment of Civilians, which concluded that the internment was due not to military necessity but to "race prejudice, wartime hysteria and a failure of political leadership."[146] It continues despite the documentary evidence uncovered by Korematsu's attorneys, which shows that the government, in prosecuting the wartime internment cases, had falsified evidence and suppressed relevant information in order to prevent the courts from rendering an informed decision.[147]

The inescapable fact is that the reputation of the Japanese American community remains at issue, and the injury persists. The stigma to be erased through the coram nobis cases has rarely been equalled and should generate a case or controversy sufficient to preclude dismissal on grounds of mootness.

CONCLUSION

Though often viewed as nothing more than a somber chapter in American history, the Japanese American internment remains a source of humiliation and frustration to its victims, many of whom died waiting for their government to confess its deliberate deception to the American people. The survivors and their descendants continue to witness the irony of the government's belligerent refusal to accept its responsibility to the ethnic community it left paralyzed and crippled. Three of the victims, unwilling to wait forever, brought the coram nobis lawsuits in an effort to vindicate their reputations and that of the Japanese American community in the courts. This Note has described the damage done to their reputation and articulated a basis for elevating social stigma to the status of a legally cognizable injury. The remedy for the community-wide injury can hardly be the mere vacating of Korematsu's, Yasui's, and Hirabayashi's convictions. Justice requires the courts to engage in fact finding which would expose the myth of disloyalty as the unethical scheme of high government officials. Only a community-wide remedy will relieve the psychic burdens and remove the stain on the reputation of Japanese Americans caused by this tragic episode in American history.

Notwithstanding the narrow focus of this Note, the full significance of these lawsuits transcends the interests of the Japanese American community. All Americans should be concerned, for at stake is the capacity of the Constitution to protect individuals, particularly those who are members of minority

---

He has stated, "There is, I submit, nothing whatever for which the country should atone." P. Irons at 354.

James Currieo, National Commander-in-Chief of the Veterans of Foreign Wars (VFW), declared there should "never be reparation paid or an apology offered" to former internees. He blamed the suffering by the Japanese Americans on the military government of World War II Japan. Statement by VFW Chief Condemned by Congressman, Pacific Citizen, July 22, 1983, at 1, col. 1. Currieo later apologized. VFW Nat'l Chief Meets with Nisei Posts, Pacific Citizen, Aug. 12, 1983, at 1, col. 1.

146. CWRIC, supra note 1, at 18.

147. See note 46 supra.

groups, in times of national crisis. At this broader level, the coram nobis cases have fallen short of their goal of vanquishing the legal precedent of the internment cases, for the courts have thus far refused to question the constitutionality of the evacuation and internment orders. Though modern legal principles minimize the possibility of another internment, minorities who feel secure in this "protection" may someday find their complacent delusion shattered.

MARC HIDEO IYEKI*

* The author gratefully acknowledges the assistance of the Asian American Legal Defense and Education Fund [AALDEF], for making available its extensive resources on the internment issue. The author also wishes to thank Moon Sam Sue, Class of 1984, New York University School of Law; Stan Mark, staff attorney at AALDEF; and Tehre Ikeda-Iyeki, for their constructive and perceptive comments.

Review Essay

At the Bar of History: Japanese Americans Versus the United States

Stanley I. Kutler

PETER IRONS, *Justice at War.* New York: Oxford University Press, 1983. Pp. xiii + 407. $18.95.

Writing recent American history presents numerous hazards and challenges. All historians bring some ideological, partisan, and emotional commitments to their subject, whether it be the Laws of Hammurabi or the constitutional and political crisis of Watergate. But the closer the historian's experience to the subject, the more the obstacles mount. Thus, the immediacy of Vietnam inevitably will influence historians of that conflict for decades to come.

Nostalgia and uncritical sentiment create further difficulties. Consider the transformation of Harry Truman from the president who left office in 1953, reviled, repudiated, and the object of wicked and often vicious humor, to his apotheosis as a sentimentalized neo-folk idol three decades later: In the presidential election of 1952, his own party sought to distance itself from him, while the opposition battered him for alleged incompetence. But in campaigning for election in 1984, the presidential candidates from both major parties competed for the mantle of Truman's rightful heir. Maybe Voltaire was right: History is a pack of tricks the living play on the dead. The critical, serious historian, with access to extensive documentation, must labor to find the historical truth, while properly accounting for both contemporary and subsequent images. Here, then, is another problem of writing recent history.

Finally, historians of the recent past must confront the mountains of data and documentation. Consider the federal government's twenty-five-year effort to deport Harry Bridges. The Immigration and Naturalization Service's (INS's) San Francisco office alone contains more than a quarter million documents—much of their content admittedly repetitious and insignificant—that must be sifted for the occasional nugget. What happened in the San Fran-

Stanley I. Kutler is E. Gordon Fox Professor of American Institutions and History, University of Wisconsin, Madison. B.A. 1956, Bowling Green State University; Ph.D. (history) 1960, Ohio State University. Russell Sage Resident in Law and Social Science, 1967–68.

cisco office, of course, constituted only one piece of the puzzle. The activities in Washington of the INS, the Federal Bureau of Investigation (FBI), the secretary of labor, and the attorney general; the proceedings in Congress; the litigation and the actions by judges and courts; and the politically motivated actions of President Roosevelt and his advisers—all have to be considered if we are to grapple with the scope and complexity of such an event.[1]

World War II offers a particularly vivid instance of the problems confronting historians. Truly a worldwide phenomenon, politically and militarily, the subject offers staggering challenges. The historian writing on military strategy cannot simply survey the materials of the Joint Chiefs of Staff. What about White House views? The ideas and maneuvers of the British, the Russians, the French, the Chinese? And the aims of adversaries in Berlin and Tokyo? What about the role of intelligence—was Ultra decisive, for example? How shall we measure the contributions of the generals, the fighting men, the civilian production? Obviously, all the questions have some bearing on the problem, and all must be in the historian's purview.

World War II is now at a critical juncture in the national consciousness. Glorified and romanticized, it runs the risk of being trivialized. Perhaps it was "the Good War," a phrase Studs Terkel significantly appropriated for his very successful oral history of the war.[2] Perhaps it offers a "glorious" saga in our history, as one reviewer noted in commenting on Ronald Spector's recent account of the war in the Pacific.[3] But glorious sagas are not without sordid chapters, and World War II is no exception.

The American response—or essentially lack of response—to the Nazis' destruction of European Jewry now stands revealed as a tragic scandal. The American government and responsible officials stand condemned as accomplices to the Holocaust. From Franklin D. Roosevelt down through the ranks in numerous departments and agencies, influential Americans who knew both German intentions and actions refused to take appropriate steps to thwart them and thus, however indirectly, facilitated the Nazis' crimes. Consider, for example, the consistent refusal to bomb Auschwitz or the rail lines that led to the crematoria. Assistant Secretary of War John J. McCloy contended that such action would require "the diversion of considerable air support essential to the success of our forces now engaged in decisive operations elsewhere." Meanwhile, American air squadrons passed over the death camp on numerous occasions and on one mission bombed German installations less than five miles away. McCloy also argued that an American bombing "might provoke even more vindictive action by the Germans." What could be more vindictive than Zyklon B?[4]

1. For some interesting reflections on the writing of recent and contemporary history, see Alan Brinkley, Writing the History of Contemporary America, Daedalus, Summer 1984, at 138–39. Stanley I. Kutler, The American Inquisition: Justice and Injustice in the Cold War ch. 5 (New York, 1982) discusses the Bridges case.

2. Studs Terkel, "The Good War": An Oral History of World War Two (New York, 1984).

3. Ronald H. Spector, Eagle Against the Sun: The American War with Japan (New York, 1984).

4. David S. Wyman, The Abandonment of the Jews: America and the Holocaust, 1941–1945, at 296 (New York, 1984).

The treatment of the Japanese Americans during the war offers another seamy, perhaps better known, chapter of World War II. The influences affecting the decision to evacuate Japanese Americans from the west coast and "relocate" them in detention centers converge at the familiar junction of law, politics, constitutionalism, and racism. The multifaceted story, with a lengthy cast of villains, is one that transcends the immediate events of the wartime years, for it continues to have a powerful effect on postwar society.

In early 1942, the United States government first limited the mobility of, then excluded, evacuated, and relocated, nearly 120,000 Japanese Americans from the west coast. Those actions and apparently the policy guiding them made no distinction between generations of Japanese Americans. The *issei,* immigrants who were barred by law from citizenship in the United States, and their children, the *nisei,* who were born in the United States and were citizens by virtue of the Fourteenth Amendment, were treated the same. President Franklin D. Roosevelt signed Executive Order No. 9066 in February 1942, authorizing the secretary of war to set up military areas from which he could exclude "any or all persons" (at 61-63, 63). The War Department, in turn, delegated that power to military commanders. The army's west coast commander quickly issued orders for exclusion, detention, and then removal of Japanese Americans to inland "relocation centers." After that, a new civilian agency, the War Relocation Authority, assumed control of the internees. "Relocation" was in fact incarceration. The action was without precedent in American history, excepting the nineteenth-century Indian removals.

The policy stemmed from a myriad of motives, including the insecurity of the army's west coast commander, the racism and hostility of the Pacific states' white population, bureaucratic ambitions, and the political advantages perceived by local, state, and federal officials. The affair involved a variety of officials and institutions, including high ranking military officers, heads and lower officials of the Department of Justice and the War Department, the FBI, the Supreme Court, and the president. Many of these officials knew at the time that the Japanese American community harbored very few disloyal persons; furthermore, knowledgeable parties in key agencies, such as the FBI and the Office of Naval Intelligence, long had been aware of those elements and knew that no military necessity existed to justify so Draconian a measure.

Historical treatments of the affair have unraveled the truths that were concealed during wartime and for nearly two decades afterward. The earliest scholarly studies, those arising from the Japanese American Evacuation and Resettlement Study, at Berkeley, emphatically disagreed on the responsibility for the evacuation. Morton Grodzins's *Americans Betrayed* blamed west coast political and economic interests and nativist groups.[5] But another study by Jacobus ten Broek and others laid the blame squarely on west coast mili-

5. Morton Grodzins, Americans Betrayed (Chicago, 1949).

tary commander General John L. DeWitt and his military advisers.[6] In a seldom-cited study, the army's own historical section published an account that disputed the "military necessity" argument and blamed the army's provost marshal, General Allen W. Gullion, and his assistant, Colonel Karl R. Bendetsen, for influencing the civilian leaders of the War Department. But without thorough access to all the archival materials, the study missed the eagerness of those civilian leaders to evacuate the Japanese Americans.[7]

Historian Roger Daniels's researches in the late 1960s and early 1970s documented the bureaucratic origins of the exclusion and evacuation orders, and he fleshed out the machinations of Gullion and Bendetsen, noting their influence on General DeWitt as well as on the civilian hierarchy in the War Department. In this research, set in the context of his earlier scholarship on the history of the Japanese in America, Daniels outlined the drives for personal aggrandizement as well as the political and racial motives that animated the policy.[8]

Daniels exposed a melancholy story of official chicanery and base political behavior. But for some, the scholarly effort did not adequately compensate for the personal losses or the constitutional scars. Nearly four decades after the event, questions and pressure from younger Japanese Americans forced the older, traditionally more accommodating leaders of the Japanese community to agitate for reparations and an official apology. In response, Congress in 1980 established the Commission on Wartime Relocation and Internment of Civilians to "review the facts and circumstances surrounding Executive Order Number 9066," and "to recommend appropriate remedies." The commission's report, prepared in 1982, offered few major revelations, but it confirmed and reinforced the scholarship that long had questioned the official explanations and rationalizations of military necessity.[9] In effect, the government sanctioned and legitimated the rewriting of its history and the condemnation of its policy.

Most recently, Peter Irons's account of the wartime conduct of cases that challenged the constitutionality of the policy has further exposed official misconduct. Using testimony and evidence secured by the Commission on Wartime Relocation, as well as documents he obtained under the Freedom of Information Act, Irons has retraced the preparations and strategies on all sides for the four test cases: *Hirabayashi v. United States, Yasui v. United*

6. Jacobus ten Broek, Edward N. Barnhart, & Floyd W. Matson, Prejudice, War, and the Constitution (Berkeley, Cal., 1954). The Berkeley project also produced accounts of the impact of the evacuation and relocation policy on the Japanese Americans: Dorothy S. Thomas & Richard S. Nishimoto, The Spoilage (Berkeley, Cal., 1946); Dorothy S. Thomas, The Salvage (Berkeley, Cal., 1952).

7. Stetson Conn, Rose C. Engleman, & Byron Fairchild, The United States Army in World War II: The Western Hemisphere: Guarding the United States and Its Outposts (Washington, D.C., 1964).

8. Roger Daniels, Concentration Camps U.S.A. (New York, 1972); *id.*, The Decision to Relocate the Japanese Americans (Philadelphia, 1975); *id.*, Concentration Camps, North America (Melbourne, Fla., 1981); *id.*, The Politics of Prejudice: The Anti-Japanese Movement in California and the Struggle for Japanese Exclusion (Berkeley, Cal., 1962).

9. Commission on Wartime Relocation and Internment of Civilians, Personal Justice Denied (Washington, D.C., 1982).

States, *Korematsu v. United States*, and *Ex parte Endo.*[10] Consequently, he has found some new villains, both inside and outside the government. But most significant, Irons has charged that in its defense of the policy on grounds of military necessity, the government deliberately withheld contradictory evidence from the Supreme Court.

Such wanton behavior, Irons contends, produced "a legal scandal without precedent in the history of American law" (at viii). Government attorneys certainly did not fully reveal many facts that would have damaged their case. But the dimensions of that scandal may be less imposing than Irons contends, particularly when we examine the full scope of the government's action.

In January 1944, Department of Justice (DOJ) lawyers Edward Ennis and John Burling received a copy of General DeWitt's *Final Report.* That the document was a classic of bureaucratic self-interest was evident in DeWitt's stressing the imminence of Japanese attacks and domestic sabotage. Predictably, the army document assailed Attorney General Francis Biddle and the DOJ for not sharing the same concerns. The report apparently was leaked to west coast newspapers, among them the Los Angeles *Times*. Ennis and Burling had cooperated (albeit reluctantly) with the War Department in 1942 when the policy was initiated, and they reacted bitterly toward the report and the publicity. The two lawyers, presumably in retaliation, prevailed on Biddle to have the FBI and the Federal Communications Commission (FCC) investigate DeWitt's espionage and sabotage allegations. The inquires demolished DeWitt's case. J. Edgar Hoover gained a measure of revenge on the army for its unwillingness to respect his bureaucratic turf when he told Biddle that no "espionage activity ashore or . . . any illicit shore-to-ship signaling, either by radio or lights," could be linked to any enemy attacks on shipping or land after Pearl Harbor. DeWitt's statements, the director concluded, were "false and erroneous." The FCC scoffed at the amateurism of DeWitt's counterintelligence people. Chairman James Fly flatly told Biddle that the FCC had "no evidence of any illicit radio signaling" on the west coast following Pearl Harbor (at 280–84).[11]

By April 1944, then, the DOJ had conclusive evidence that DeWitt's allegations owed more to imagination and hysteria than to fact. Furthermore, the refutation had come from Hoover, who by 1944 may have equaled the army in both credibility and intragovernmental clout. And at that stage of the proceedings, knowledge of that evidentiary counterattack could not have been a very tightly held secret.

Ennis and Burling had primary responsibility for preparing the government's briefs in the *Korematsu* and *Endo* cases. They decided on a curious

10. 320 U.S. 81 (1943); 320 U.S. 115 (1943); 323 U.S. 214 (1944); 323 U.S. 283 (1944).

11. Hoover had opposed the evacuation policy from the outset. In effect, its implementation implied that the FBI had not adequately done its job of maintaining proper surveillance in the Japanese American community.

strategy: on the one hand they would concede that army officials had "ample ground" to believe that a Japanese attack was imminent in the first months of 1942; yet they would also alert the Supreme Court to the unreliability of DeWitt's report. In the original draft of the brief prepared for Solicitor General Charles Fahy, Ennis and Burling provided a *footnote* stating that DeWitt's contentions conflicted with "information in the possession of the Department of Justice" (at 286). The positions flatly contradicted each other, but Irons passes over the contradiction, for he is intent on noting the decision to discredit DeWitt. Yet Ennis and Burlings's concession in the body of the brief was substantial.

Fahy and his assistant, Herbert Wechsler (formerly a clerk for Chief Justice Stone and for Justice Frankfurter), quickly moved to alter the footnote to a simple declaration that the DOJ differed with the army over the espionage and potential sabotage allegations. That vague admission still alarmed McCloy at the War Department and he instructed his deputy, Captain Adrian Fisher (a former Brandeis clerk), to negotiate the removal of any doubts of the army's position. Fisher, according to Burling (certainly a biased source), believed "that the Government would deal with sufficient honesty with the court if it would merely refrain from reciting the report without affirmatively flagging our criticism thereof" (at 289). Fisher may have been either a knave or a fool, but he deserves more than the reminiscences of an aggrieved party to condemn him.

The final word in the dispute belonged to Wechsler, who arbitrated between the feuding lawyers for Justice and War. The footnote he devised supposedly preserved a modicum of intellectual honesty for the DOJ, but in fact it did nothing to reveal the devastating reports from the FBI and the FCC. "We have specifically recited in this brief the facts relating to the justification for the evacuation," Wechsler wrote, "of which we ask the Court to take judicial notice; and we rely upon the Final Report only to the extent that it relates to such facts" (at 290–91).[12] The dramatic confrontation between two institu-

12. Fahy's behavior here contrasted sharply with that in a wartime treason case involving a citizen who had aided German saboteurs: Cramer v. United States, 325 U.S. 1 (1945). After a hearing in which the Court posed a number of historical questions on the treason clause, Fahy engaged Willard Hurst, then working in the Navy Department, to research the Court's queries. Hurst's answers, which later formed the basis of his landmark work on the history of treason, contained material contrary to the government's interest in the case. Apparently, some members of Fahy's staff were reluctant to submit such material to the justices, but according to Hurst's recollections, Fahy told them that the Court asked to be fully briefed on the history of the treason clause "and that's what we're going to give them" (Hurst, to the author, Jan. 21, 1985). Fahy apparently made his decision to use Hurst's findings shortly after his arguments in *Korematsu*. For Hurst's work, see The Law of Treason in the United States (Westport, Conn., 1971).

This discussion is not intended to excuse Fahy's failure to make a full disclosure to the Court. The rationalization whereby government lawyers are said to have primary responsibility to their client (i.e., the government) in effect argues against the notion of citizens as the ultimate sovereign. Edward Ennis might have had the last, and the most useful, word on moral and ethical behavior in this case. If there is a mini-hero in Irons's tale, it is Ennis, who fought an extended guerrilla war against his War Department counterparts. Yet, he consistently bowed to the decisions of his superiors, despite his serious misgivings. Four decades later, when Irons asked why he had not resigned instead of submitting, Ennis said: "Watergate hadn't happened yet" (at 351). For a discussion on making judgments out of historical context, see note 23 *infra*.

tional giants, staffed with a dazzling array of the "best and brightest" champions on both sides, ended unheroically—and with a substantial victory for McCloy and the War Department. The footnote conceded the issue of judicial notice of DeWitt's report, something it surely did not deserve, and Ennis and Burling signed the brief.

The dispute leaves us with some tattered reputations, to be sure; but was it a "legal scandal without precedent"? The government certainly waffled on the veracity of Dewitt's report. But the record suggests that despite Wechsler and Fahy's capitulation to the army, at least some of the Supreme Court justices realized or sensed the truth. Furthermore, we can question whether full disclosure would have changed the outcome of the *Korematsu* decision.

The Court was not oblivious to doubts regarding the *Final Report.* Charles Horsky, who argued an amicus brief for the American Civil Liberties Union (ACLU) in *Korematsu,* referred to the "extraordinary footnote" in the government's brief. Ennis and Burling had met with Horsky on several occasions and had discussed the FBI and FCC investigations. Horsky knew the facts; but it is unclear how specific he was in his oral arguments before the Court. He did, however, refer to the ample evidence in the public record, which was summarized in the brief filed by the Japanese American Citizens League (JACL), that severely impugned DeWitt's report—evidence that the justices also had before them. The lawyer later recalled that he "almost accused Fahy of misleading the Court." Fahy, reportedly a hot-tempered man, did not speak to the lawyer for another decade (at 314–15).

Justice Robert Jackson, in the conference room and in his subsequent *Korematsu* dissent, directly challenged DeWitt's contentions. He told his colleagues that they should not accept the report without some "inquiry into [its] reasonableness" (at 322). His dissent questioned both the accuracy and the reasonableness of DeWitt's "military necessity" claims, and he relied heavily on the JACL brief disputing the charges of espionage and sabotage (at 332). Jackson seemed aware of the truth; in any event, his opinion undermines Irons's assessment of the significance of the apparently unsuccessful cover-up.

Justice Frank Murphy's more famous dissent (asserting that the government's policy fell into the "ugly abyss of racism") directly assaulted the heart of DeWitt's case (at 335). Nothing in the record substantiated the charges, he said; furthermore, he found no evidence to demonstrate the inability or failure of the FBI and military intelligence agencies to keep "the espionage and sabotage situation well in hand" (at 336). Murphy, too, apparently sensed the truth. The former attorney general maintained very friendly ties with J. Edgar Hoover; it is difficult to believe that Hoover would have missed an opportunity to peddle his views, particularly given his contempt for DeWitt's intelligence staff (at 336).[13]

13. For the earliest discussion of Murphy's dissatisfaction with the government's policy, and the tensions within the Court, see Sidney Fine, Mr. Justice Murphy and the *Hirabayashi* Case, 33 Pac. Hist. Rev. 195 (1964). In the concluding volume of his biography of Murphy, Fine thoroughly explores the Court's

The circle of those who knew the facts regarding the *Final Report* actually was quite large. Lawyers in the Justice and War departments, in the solicitor general's office, and in the War Relocation Authority, as well as the plaintiffs' attorneys (who had been briefed by Ennis and Burling) knew. James Rowe, then in the DOJ but formerly a Roosevelt assistant, knew and probably informed people in the White House. Some of those involved in the various agencies and offices maintained close ties with the justices; indeed several had served as clerk to a justice. Given all the possibilities for judicial knowledge, it seems likely that some, perhaps most, of the justices realized the *Final Report*'s vulnerability. But in all probability, it made little difference.

Justices Murphy, Jackson, and Roberts had expressed differing levels of dissatisfaction when the Court upheld the army's curfew policy in the *Hirabayashi* case. Given the broader constitutional issues of exclusion and detention at stake in *Korematsu,* however, they chose to dissent. But the remaining justices were determined to uphold the government's policy, and the dissenters' basic criticisms barely troubled the majority. On a personal level, the latter understood and accepted the administration's political and military needs, as well as popular sentiments. Some of them—notably Black, Douglas, and Frankfurter—had close relationships with the president. Together with Chief Justice Stone and Justice Reed (and often Justice Jackson), they deferred to presidential war powers and military necessity arguments whenever asked to do so.[14]

Finally, it seems unlikely that the ambiguous, fuzzy Ennis-Burling footnote, as originally proposed, would have changed things. Submitting the FBI and FCC reports and disavowing DeWitt's report might have done so. But no one at any time proposed such public disclosure. That would have left the government with a different case, indeed with no case.

The real "scandal" may lie in the Court majority's maneuvers to assert constitutional scruples and yet avoid political embarrassment for the administration. Those justices found a practical route by linking the *Korematsu* and *Endo* cases. In the *Korematsu* decision the Court sustained the exclusion orders but on the same day in *Endo* rejected the continued incarceration of citizens whose loyalty had been established. Even the government's most loyal supporters on the bench could not tolerate *that* policy. But by that pair of decisions, the justices established an artificial distinction between exclusion and detention, whereas the military orders for exclusion clearly envisaged detention. Chief Justice Stone, in conference, particularly insisted on the fiction that *Korematsu* involved only an exclusion order (at 320–23). More important, the Court distinguished between exclusion as a military policy and internment as a civilian one. The civilian "do-gooders" in the

---

conflicts in the other cases as well: *id.*, Frank Murphy: The Washington Years ch. 18 (Ann Arbor, 1984). Justice Jackson, who also had been attorney general, did not appear to be as close to Hoover as Murphy was, but it is possible that the director also provided information to him.

14. See, e.g., Ex parte Quirin, 317 U.S. 1 (1942); Keegan v. United States, 325 U.S. 278 (1945); In re Yamashita, 327 U.S. 1 (1946).

War Relocation Authority (WRA) (some of whom indeed did good) offered an easier and more inviting political target for judges with nagging consciences about constitutionality.

Although he dissented in *Korematsu*, Jackson expressed misgivings and doubts about the Court's ability to judge the reasonableness of military actions. No such hesitation operated on his part or that of others in *Endo*. Fahy's arguments that the WRA operated in the "best interests" of the Japanese Americans left the justices unimpressed. They easily decided that the WRA had no authority, statutory or otherwise, to detain loyal citizens.[15] Meanwhile, Justice Black maintained the necessary fiction on the narrowest of grounds when he wrote in *Korematsu* that the defendant's "conviction was for remaining in a prohibited area, and we shall not here pass upon the legality of an internment the petitioner might have, but had not suffered" (at 327, 339).[16] But he had.

Irons's book raises the question of Franklin D. Roosevelt's role and responsibility. To ignore the president is to perform *Hamlet* without the prince of Denmark. Yet we have written precious few lines for FDR. His activities remain shadowy, largely, one suspects, because he preferred to have subordinates play the more prominent roles in this affair, and because aides and friends shielded him from contemporary criticism and involvement.

During arguments in the *Hirabayashi* case, for example, the ACLU attorneys sidestepped any challenge to Executive Order No. 9066, concentrating instead on DeWitt's orders as unreasonable and beyond the scope of the president's directive. The governing council of the ACLU, dominated by men with close ties to the Roosevelt administration, such as Roger Baldwin and Morris Ernst, deliberately sidestepped any direct challenge to the president and his civilian officers (e.g., at 108–12, 128–34).[17]

When Eugene V. Rostow published one of the first serious public criticisms of the policy in June 1945, he basically blamed the military and then excoriated the Supreme Court for applying the wrong standards of judicial review.[18] Such criticism established the basic liberal litany that deplored the Japanese American policy as excessive, yet exceptional to an otherwise exemplary wartime civil liberties record—a litany that also emphasized how the president had been victimized by bad advice from nervous generals and

15. The Commission on Wartime Relocation assailed Justice Douglas's majority opinon as unnecessarily "crabbed and confined." The *Endo* decision, the report concluded, "did not come with an air of generosity or largeness of spirit." Commission on Wartime Relocation, *supra* note 9, at 239.

16. The majority's decision in *Korematsu* offers a compelling validation for the use of institutional analysis in interpreting Supreme Court decisions. Clearly, in this case, Justice Black and his majority colleagues justified their action as deference to the judgment of "properly constituted military authorities" (323 U.S. at 223) and to the political judgments of the executive and legislative branches. See Neil K. Komesar, Taking Institutions Seriously: Introduction to a Strategy for Constitutional Analysis, 51 U. Chi. L. Rev. 366, 387–91 (1984).

17. Forty years later we learned that Morris Ernst regularly reported ACLU activities to J. Edgar Hoover. The ACLU's San Francisco office resisted the national council's pleas for restraint. Led by Executive Director Ernest Besig, it eagerly sought test cases to challenge the government's policy.

18. Eugene v. Rostow, The Japanese-American Cases—A Disaster, 54 Yale L.J. 489 (1945).

zealous, racist west coast politicians and citizens. Attorney General Biddle, who opposed the evacuation (albeit somewhat timidly), acknowledged that "the decision had been made by the President." But Biddle hastily added that Roosevelt described it as "a matter of military judgment."[19] Again, the implication is that Roosevelt succumbed exclusively to military considerations.

Thus we have a portrait of a president, pressured by military defeat abroad and fears of domestic insecurity, passively (but understandably) submitting to the demands of his military subordinates. But we now know that some of the president's high-ranking civilian officers supported an evacuation policy with enthusiasm. We also know too much about FDR to think of him as passive—unless passivity served some useful purpose. In this case, the president's silence apparently masked his consent and complicity. The policy offered numerous opportunities for political gain, from placating west coast antagonists of the Japanese Americans to satisfying a wider public seeking some measure of revenge for the humiliation of Pearl Harbor. In the meantime, military necessity, certainly an appealing rationalization in those perilous moments, minimized any constitutional or moral qualms.

Roosevelt was hardly oblivious to the rising chorus among west coast politicians and newspapers demanding evacuation of the Japanese Americans. A week before he issued Executive Order No. 9066, a delegation of Pacific Coast congressmen sent him a recommendation urging evacuation. Congressman Leland Ford of California probably was the group's most outspoken member. He later recalled how he had phoned Biddle's office "and told them to stop fucking around" on removal. He threatened to force the issue in Congress and to "give the bastards [presumably, administration officials] everything we could with both barrels."[20] Given the embarrassing news from the war fronts, Roosevelt could ill afford such a political assault on the home front.

Roger Daniels first called the president to account. Citing Roosevelt's longstanding antagonism toward Japan, he argued that it had nourished the president's political and racial animosity toward the Japanese Americans. Daniels also unearthed McCloy's memos summarizing telephone conversations between FDR and Stimson in which the president said that he considered evacuation militarily essential. The president added, according to McCloy, "Be as reasonable as you can." Daniels not only exposed the machinations of McCloy and such lower level *fonctionnaires* as Gullion and Bendetsen, but most significant, he charged that Roosevelt, as "the leader of the nation, was, in the final analysis, responsible."[21]

19. Francis Biddle, In Brief Authority 219 (New York, 1964).

20. Commission on Wartime Relocation, *supra* note 9, at 84.

21. Daniels, Concentration Camps U.S.A., *supra* note 8, at 65–73. FDR's racial views have not been systematically explored. We do know, e.g., that he asked the Smithsonian Institution to study the notion that the "evil-doing of the Japanese could be attributed to their less-developed skulls." Christopher Thorne, Allies of a Kind: The United States, Britain, and the War Against Japan, 1941–1945, at 725–30, 158–59 (New York, 1978). In 1936, the president suggested that all Japanese persons, citizen and non-

FDR's political concerns certainly dictated his behavior at some points—political concerns that prevailed over *both* constitutional arguments and new military recommendations. During a cabinet meeting in May 1944, Secretary of War Stimson discussed the possibility that the Supreme Court might soon declare the internment policy unconstitutional. Stimson, who was supposed to be politically neutral, told the president and his colleagues that he thought internment could be ended "without danger to defense considerations but doubted the wisdom of doing it at this time before the election." Roosevelt replied that Japanese Americans should be spread throughout the nation, rather than being simply returned to the west coast in large numbers (at 271-72).[22]

Interior Secretary Ickes, who had jurisdiction over the WRA, vigorously disagreed with the decision to delay. Ickes was under pressure from Interior Undersecretary Abe Fortas, who had visited the camps and had been appalled at the conditions, and Fortas's assistant, John Frank (a former clerk to Justice Black), both of whom believed that the internment policy was unconstitutional. But the president adamantly refused to reverse course, again citing political problems on the west coast.

Meanwhile, DeWitt's successor, General Delos M. Emmons (who had pursued tolerant policies toward the Japanese Americans in Hawaii), informed McCloy that the "military necessity" for exclusion no longer existed and proposed that a "substantial number" of Japanese families be allowed to return to their west coast homes. McCloy responded that the president had "put thumbs down on this scheme." Roosevelt and his political advisers, he added, did not want "to stir up the California situation." McCloy, however, had difficulty keeping *his* troops in line. General Charles H. Bonesteel succeeded Emmons at this time and reiterated the view that exclusion no longer was necessary. McCloy clearly was concerned about the president's political needs; equally important, he did not want to moot the pending *Korematsu* case and thereby deprive the government of any constitutional justification for its policy. Finally, McCloy ended Bonesteel's challenge by revealing the paramount political considerations. Policy had to be determined, he said, on "high political rather than military levels"; and after the forthcoming November elections, he concluded, "we shall have a greater opportunity for constructive plans" (at 271-76, 276).

In November Roosevelt defeated Dewey, and west coast congressmen allied to him were re-elected. Shortly afterward, the Court approved the exclusion and evacuation orders (in *Korematsu*) and within hours rejected the continuation of internment (in *Endo*). The scenario unfolded perfectly—from the administration's point of view. McCloy even had the last word.

---

citizen, who met any Japanese ships or officers in Hawaii should be identified and their names retained on a "special list of those who would be the first to be placed in a concentration camp in the event of trouble." Roosevelt memo, Aug. 10, 1936, submitted to the Commission on Wartime Relocation, *quoted in* N.Y. Times, Feb. 10, 1983. See John W. Dower's work on racial aspects of World War II in Asia, forthcoming in 1986.

22. Citing Stimson's diaries and Biddle's notes.

After only six months in his post, Bonesteel was transferred and McCloy allowed his successor to have the "honor" of rescinding the exclusion and internment orders. Politics does have its moments (at 276).

Finally, the inevitable question: "Can it happen again?" *Korematsu* has had no defenders; one searches the legal and scholarly literature in vain for supporting commentary. The Supreme Court avoids references to it as if it were plague ridden. Still, one wonders. The Vietnam War demonstrated anew the possibilities for a broad reading of the war powers. The future effectiveness of the War Powers Act of 1973 is problematical. Our historical experience demonstrates that in time of crisis, constitutional checks and balances often go unutilized as governmental institutions abdicate or delegate authority in order to marshal consensus, sometimes with results devastating to civil liberties. Witness the red scares accompanying the two world wars, and the World War II treatment of the Japanese Americans as an obvious case in point. Furthermore, who can doubt the persistence of racism that might again turn governmental policy against a feared and despised minority? And retribution—that other propelling force in the exclusion, evacuation, and internment of the Japanese Americans—certainly is ingrained in human nature.

Irons's book parallels his participation in a coram nobis brief asking the federal courts to vacate their rulings in *Korematsu* and the other three cases. Irons and his associates offered documentation from the Commission on Wartime Relocation's findings and from Irons's research for his book to argue that the courts had decided wrongly. Ten months after the brief was filed, the government capitulated and asked the court to set aside Korematsu's conviction. The maneuver gave Fred Korematsu and the Japanese American community some sense of retribution, but the government's move did not specifically repudiate the doctrine and viability of the Supreme Court's ruling.[23]

Korematsu is merely the form, not the substance, of the reality. Whatever we now do with the formal decision, we still must recognize that its animating

23. N.Y. Times, Oct. 5, 1983. Judge Marilyn H. Patel nevertheless granted the coram nobis petition, despite the government's opposition to the move. 586 F. Supp. 769 (N.D. Cal., 1984). The government decided not to appeal. A district court, of course, cannot effectively overrule a Supreme Court ruling; thus the *Korematsu* doctrine in effect remains intact. As of February 1985, the petitioners were appealing an adverse lower court ruling in the *Yasui* case. Most interesting, it seems, is a promised evidentiary hearing on the coram nobis materials in the *Hirabayashi* case, scheduled for June 1985. In preliminary arguments, the government denounced Judge Patel's *Korematsu* ruling.

Irons's book raises numerous questions about the dividing line between advocacy and writing history, a subject that requires lengthier examination than can be given here. Irons himself has raised the issue (at xii). Here I would state that historical judgments require sensitivity to context and attention to the full range of actors and issues. Briefs are designed to "win" a point: Both historians and advocates engage in selection, but the former must acknowledge the complexity and contradiction of the evidence. The historian inevitably raises and illuminates a wide spectrum of questions; briefs, however, focus on answers, indeed seek to avoid questions in the minds of readers. Aviam Soifer, in an otherwise favorable review, criticized Irons on this point. "[I]f his book has a significant flaw, it is that its boldest claims may have been tailored a bit to meet the burden Irons and his clients must satisfy in their *coram nobis* petitions." Aviam Soifer, Lawyers and Loyalty, 12 Revs. Am. Hist. 580 (1984).

ideas lurk as a brooding beast in our constitutional mélange. Consider the words of John J. McCloy: testifying in 1981 before the Commission on Wartime Relocation and Internment of Civilians, he staunchly defended the government's policy and, of course, his specific role. "There is, I submit, nothing whatever for which the country should atone." He reminded the commission of the fears—real or alleged—that pervaded American society following the Pearl Harbor attack. In an unguarded moment, he even acknowledged that the policy, in part, was adopted "in the way of retribution" for the Japanese raid (at 353).

It is most significant that McCloy deplored excessive "sensitivities" toward the Constitution. He scoffed at Edward Ennis as a man who saw "a civil liberties problem under every bedstead" (at 353). Wars and national emergencies required extraordinary action, McCloy said, and he urged the commission not to recommend measures that might restrict the government in similar future situations. He even conjured a scenario: "I don't mean to be alarmist," he stated, but went on to say he worried about a hostile government "within 90 miles right now of our shore." He then pointed to the large number of Cuban émigrés living those 90 miles away in south Florida and challenged the commission: "Wouldn't you think seriously about moving those people out if there was a raid there?" (at 353) The equally unrepentant Karl Bendetsen believed that "human nature" would have compelled Japanese Americans to assist invading Japanese troops; clearly, McCloy expected Cuban Americans to behave similarly (at 355).[24]

McCloy, of course, had a heavy stake in all this. Despite overwhelming contrary evidence, he still viewed Japanese Americans of 1941–42 as potential fifth columnists, and Cuban Americans apparently as no different. The "chairman of the American establishment," as Richard Rovere once dubbed him (at 351), seemed oblivious to the Cuban émigrés' passionate anticommunism and loyalty to the United States. Such misperceptions, such seemingly reasoned views, once endowed us with a legacy of constitutional abuse and social tragedy. When President Gerald Ford formally rescinded Executive Order No. 9066 in 1976, he called on the American people "to resolve that this kind of action shall never again be repeated." But "never again," like maintaining liberty, requires eternal vigilance. Clearly the legacy endures; whether we use it again remains to be seen.

24. Déjà vu. Compare the remarks of a federal court judge in Washington state in 1942: "How many [here] believe that if our enemies should manage to send a suicide squadron of parachutists to Puget Sound that the Enemy High Command would not hope for assistance from many such American-born Japanese?" *Ex Parte* Ventura, 44 F. Supp. 520 (W.D. Washington, 1942) (quoted in Irons at 113). Mary Ventura, a *nisei* married to a Filipino, had challenged the curfew, claiming that she was a loyal citizen. To this, the judge replied that if she were "as loyal and devoted as her petition avers she would be glad to conform to the precautions which Congress, the President, and the armed forces, deem requisite to preserve the Constitution."

[illegible] was forced [illegible] beaches [illegible] invasions [illegible] with later the [illegible] 1955 before the Commission on Wartime Relocation and Internment of Civilians, [illegible] the government's policies and [illegible] of [illegible]. [illegible] he reminded the [illegible] that the [illegible] provided [illegible] the [illegible] Pearl Harbor [illegible] the Navy [illegible] the Japanese [illegible].

It is most significant that McCloy declared [illegible] "Seriously [illegible]" the Constitution. He [illegible] as a man who saw [illegible] military matters [illegible] constitutional [illegible] conflict [illegible] similar [illegible] relocation. He [illegible] a central [illegible] mean to [illegible] however [illegible] Pearl Harbor [illegible] the [illegible] Japanese [illegible] was [illegible].

[illegible] the Japanese American [illegible] 1986 [illegible] loyalty to the United States. [illegible] "[illegible] in 1988 [illegible] American [illegible] never again [illegible] vigilance [illegible]."

[illegible]

FORGING A LEGEND: THE TREASON OF "TOKYO ROSE"*

STANLEY I. KUTLER**

Treason against the United States, shall consist only in levying War against them, or in adhering to their enemies, giving them Aid and Comfort. No Person shall be convicted of Treason unless on the Testimony of two Witnesses to the same overt Act, or on Confession in open Court.

United States Constitution
Article III, Section 3

"Se non è vero è ben trovato."
[If it is not true it is very well invented.]

Giordano Bruno

Willard Hurst's first major scholarly contribution appeared in a series of essays in 1945 on the law of treason. Twenty-five years later, the articles were collected and up-dated with post-World War II developments in a book, THE LAW OF TREASON IN THE UNITED STATES (1970). *The work reflects Hurst's assiduous research and his ever-probing speculative searches for implicit political and social values in the law. It stands as the definitive work on the subject. The following essay describes the treason case of "Tokyo Rose," who was convicted in 1949 for having broadcast for the enemy during World War II. The application of treason law in the case is, of course, intrinsically significant as a matter of legal doctrine. The story, however, involves far more than statute and case law. As such, it fulfills another of Hurst's themes: legal history must not be preoccupied with the work of judges and courts. In this case, the machinations of low-level government bureaucrats, the personal goals of extra-governmental interests, and the political ambitions and concerns of high officials—all in the context of the political and cultural milieu of the Cold War—operated to produce a decision that had little to do with abstract concerns of law or justice.*

---

* This essay is adapted from a forthcoming book on political justice in the Cold War, to be published in 1982.

** E. Gordon Fox Professor of American Institutions, University of Wisconsin, Madison.

INTRODUCTION

As American military forces poured into Japan at the close of World War II in late August 1945, they were inevitably accompanied by war correspondents eager for hard news or interesting features for a curious home audience. To be first with a story, to have that magical "scoop," was an occupational obsession. Clark Lee, a correspondent for International News Service, and Harry T. Brundidge, who worked for *Cosmopolitan* magazine, both part of the Hearst publishing empire, were in the first wave of American reporters. Both fashioned themselves in the swashbuckling Richard Harding Davis journalistic tradition—intrepid, daring, and arrogant—and both wrote in the same racy manner.

The arriving correspondents had a particular interest in finding "Tokyo Rose," an English-speaking woman who had broadcast from Radio Tokyo to Allied troops in the Pacific throughout much of the war. An air of mystery surrounded this woman. Some imagined her to be a beautiful Eurasian Mata Hari, others, an American turncoat. According to Lee, there was a "race" to find her because of the intense curiosity about her, a race that Lee described in dramatic tones, resembling, one observer noted, "a movie script, right out of M-G-M . . . [the part] could have been played by Clark Gable instead of Clark Lee."[1]

Lee and Brundidge won the race and found a twenty-nine year old Nisei. She was, as Lee admitted, "A pleasant-looking girl, but by no stretch of the imagination a siren." She identified herself as Iva Ikuko Toguri d'Aquino. Eager for an exclusive story, Lee and Brundidge hid Mrs. d'Aquino and her husband in the Imperial Hotel while they secured her confession that she was the "one and only Tokyo Rose."[2]

The two reporters claimed they had found Mrs. d'Aquino largely as a result of "tips" from her Radio Tokyo associates. She indicated that other women broadcasters could just as well be "Tokyo Rose," a name never actually used in the broadcasts, but she was apparently eager enough to give Lee and Brundidge the exclusive story they wanted. Lee carried a holstered gun during the interview. The reporters gave her a contract promising $2,000 upon publication of her story in *Cosmopolitan,* certainly not an insignificant amount in the desolated, war-ravaged Tokyo of September, 1945. The contract stated that Mrs. d'Aquino was "the one and original 'Tokyo Rose' " and that she had no female

1. C. LEE, ONE LAST LOOK AROUND 84-85 (1947).
2. *Id.*

assistants or substitutes. (Ironically, these are the only clearly provable lies d'Aquino told throughout her experience; as early investigations demonstrated, d'Aquino was only one of several female broadcasters for Radio Tokyo who were collectively denominated "Tokyo Rose" by Americans.)[3]

A few days later, however, d'Aquino refused the money and broke the contract, giving her story to *Yank* magazine instead. Lee and Brundidge persisted with subsequent stories. Ignoring evidence to the contrary, they certified the legend of the "one and only 'Tokyo Rose'"—a legend that, like many, blossomed from silly to dangerously exaggerated. Hollywood nourished the seed with an instant movie in 1946, appropriately entitled "Tokyo Rose" and described as "the story of the notorious Japanese Mata Hari," "the slinky female propagandist."[4]

Lee lost interest in the matter after 1947. Harry Brundidge, however, spent the next four years trying to give credence to that legend. Ultimately, he succeeded, with the aid of some fellow journalists, intimated government personnel, and his own suborning of perjury. In September, 1949, four years after the end of the war, Mrs. d'Aquino was convicted of one count of treason and sentenced to ten years in prison and a $10,000 fine.[5]

I. The Initial Investigations

A. *The Army: "Hands Off"*

Several days after the news stories of "Tokyo Rose" appeared, General MacArthur's Tokyo headquarters announced the impending arrest of American citizens suspected of treason-

3. *Id.* at 84-89, 118; *Record of Trial of Iva Ikuko Toguri d'Aquino, Testimony of Clark Lee* 7:478-53 (trial testimony hereinafter cited as *Record,* followed by relevant person and citations); transcript of Lee's notes of interview with d'Aquino, September 3, 1945, Justice Department File [hereinafter cited as *JDF*]. For comment on Lee's testimony, *see* San Francisco Chronicle, July 15, 1949 [hereinafter cited as *SFC*].

4. A copy of the aborted contract between d'Aquino and the correspondents, September 1, 1945, is in the Army Intelligence Files, Department of the Army, Fort George Meade, Maryland [hereinafter cited as *AIF*]. Mrs. d'Aquino's first formal affidavit to military authorities, detailing her negotiations with Lee and Brundidge, was on December 21, 1945, in *AIF*. Leslie Nakashima, a Japanese newspaperman, helped Lee and Brundidge find Radio Tokyo employees. For a statement that he did not find the "one and only 'Tokyo Rose,'" *see Deposition,* Wayne Collins Papers, in possession of Wayne Collins, Jr., San Francisco [hereinafter cited as *WCP*]. In Lee's trial testimony, he was uncertain whether Nakashima had told him he had found "a" or "the" Tokyo Rose. *Record, Lee,* 7:478-79. For wartime descriptions of "Tokyo Rose," *see, e.g.,* Time, April 10, 1944; Collier's, January 8, 1944.

5. *See* d'Aquino v. United States, 192 F.2d 338 (9th Cir. 1951), *cert. denied,* 343 U.S. 935 (1952).

ous activity during the war. First questioned on September 6, 1945, d'Aquino was arrested on October 16, five weeks after the warrant had been issued. She was held in Yokohama Prison until November 16, then transferred to Sugamo, near Tokyo, where she would remain for nearly a year. Until Christmas, 1945, she was held incommunicado. After that, she was granted twenty-minute monthly visits from her husband. d'Aquino and an American military policeman later testified that prison authorities treated her as a Japanese national rather than as an American. Most important, she was denied counsel and forbidden to write or receive letters.[6]

From October until April, 1946, army intelligence officers steadily interrogated d'Aquino. They readily elicited from her the basic story of her involvement, a story which never materially changed and which regularly received substantial corroboration. The basic facts concerned her trip to Japan in 1941; her entrapment by the war; her need for work; her recruitment for broadcasting; her admission of broadcasting essentially as a "disc jockey"; the names of other English-speaking women who had appeared on her program and on others; her denial that there was any such person as "Tokyo Rose"; her outspoken criticism of the war and support for the United States; her assistance to Allied prisoners of war; and her steadfast determination to adhere to American citizenship despite ample opportunities to renounce it and assume either Japanese nationality or the Portuguese citizenship available to her as a result of her marriage in April, 1945 to Felipe J. d'Aquino, a Japanese-Portuguese.[7]

The suspect identified two Allied prisoners of war who had recruited her for broadcasting. They were Major Charles Hughes Cousens, an Australian captured at Singapore, and Major Wallace E. (Ted) Ince, an American captured at Corregidor. Both had prior radio experience. In affidavits given to army investigators in late 1945 and early 1946, both men testified that the Japanese Army had ordered them to work for Radio Tokyo, that the army had ultimate command at the station, that they had enlisted d'Aquino, and that they had done so largely to undermine the effectiveness of the propaganda broadcasts. Cousens specifi-

---

6. *Record, Martin Pray,* 43:4712; *Record, Defendant,* 46:5182, 5206.

7. Memorandum, Special Agent 2260, September 10, 1945, Investigation Summary, March 14, 1946, Memorandum, Legal Section to Chief of Counter Intelligence (CIC), April 17, 1946, *AIF, supra* note 4.

cally stated that d'Aquino's lack of broadcasting knowledge, "combined with her masculine style and deep aggressive voice, . . . preclude[d] any possibility of her creating the homesick feeling which the Japanese Army were [sic] trying to foster." He denied that certain incriminating statements allegedly broadcast by "Tokyo Rose" had ever been given on d'Aquino's program.[8] Both Cousens and Ince testified that d'Aquino had smuggled gifts of food and supplies to Allied prisoners, along with truthful war news which she learned from her husband, who was employed at the Domei News Agency. Most important in explaining why d'Aquino was singled out from all the women employed as broadcasters, was the fact that Cousens and Ince kept d'Aquino alone informed of their tactics and real purposes, thus arousing the antagonism of other broadcasters.[9]

d'Aquino's statements were meticulously pursued and checked by army investigators. A comparison of her own affidavit of December 21, 1945, with the final report from the legal section of army counter-intelligence, discloses no significant factual discrepancies. The Army investigators even turned up additional evidence that buttressed d'Aquino's claims. The Foreign Broadcast Intelligence Service of the Federal Communications Commission had monitored Radio Tokyo broadcasts. They had determined that despite widespread rumors and legends of a charming entertainer named "Tokyo Rose," who allegedly had access to top secret American military plans, the name had been coined by Allied servicemen and referred to "at least two people" on different programs.[10]

The final army recommendations were somewhat ambiguous, yet it is clear the army investigators and legal advisors no longer desired jurisdiction. They found that the facts of the case fell within the definition of treason. Then followed a seemingly rare bureaucratic judgment. Another agency, obviously the Justice Department, would have to determine "whether the case should, as distinguished from could, be prosecuted."[11] Finally,

8. Charles Hughes Cousens, October 25, 28, November 10, 1945, *Id.*

9. *Id.*, Wallace E. Ince, January 8, 1946. For further description of d'Aquino's conflicts with the other Nisei, *see Record, Defendant*, 49:5484-85.

10. d'Aquino Affidavit, December 21, 1945, Counter Intelligence Memorandum, January 10, 1946, Foreign Broadcast Intelligence Service, "Radio Report on the Far East", *AIF, supra* note 4.

11. Memorandum, Legal Section to CIC, April 17, 1946, *AIF, supra* note 4. The report added that there was sufficient evidence to justify a charge of violation of a 1917 law regarding attempts to cause insubordination in the military forces. No subsequent document again discussed this potential charge.

the report concluded that since there were no offenses against military law, there was no need for the military to hold d'Aquino. The matter now belonged to civilian authorities. A recommendation followed a few days later that d'Aquino be released immediately. Nevertheless, she remained in Sugamo for six more months as the civilian bureaucracy insisted on an independent accounting.[12]

B. *The Justice Department and the FBI: "No prosecution unless more facts"*

The Criminal Division of the Justice Department had requested investigative assistance from the Federal Bureau of Investigation as early as October, 1945. At that time, FBI Director J. Edgar Hoover assigned Special Agent Frederick G. Tillman, then in Manila, to investigate the case.[13] Tillman appeared in Japan in early 1946 and secured access to the army materials. From this initial trip to Tokyo, through d'Aquino's trial in 1949, Tillman had an intimate connection with the case as investigator, witness, and, most important, as confidant and participant in Justice Department machinations. In April, 1946, d'Aquino gave Tillman a remarkably restrained twelve-page statement of her experiences. In it she matter-of-factly related her activities, offering neither an emotional confession of collaboration nor a heated defense of mitigating circumstances. This document, together with the army materials, provided a thorough record: subsequent investigators could not hide behind a veil of ignorance.[14]

1. THE TILLMAN REPORT: INNOCENT EXPLANATIONS

After relating the basic facts of her life through 1941, d'Aquino described for Tillman the events surrounding her departure to Japan. Shortly after she graduated from UCLA in June, 1941, her family received word that her mother's sister in Tokyo, Shizu Hattori, was ill and wanted Mrs. Toguri to visit her. Because of her own mother's illness, d'Aquino's unemployment, and her desire "to see Japan," the family designated the oldest daughter as its representative. Her father, Jun Toguri,

12. *AIF*, *supra* note 4, April 27, 29, 1946; Transmittal of Case Record, May 1, 1946.
13. Hoover to Tillman, October 11, 1945, *AIF*, *supra* note 4.
14. d'Aquino Statement to Tillman, April 30, 1946, State Department documents on Ikuko Toguri, *JDF*, *supra* note 3; American Consulate General documents, December 1, 1941, April 4, 1942, *AIF*, *supra* note 4.

made the travel arrangements, but, d'Aquino told Tillman, he was unable to get her a passport. She made no effort herself to get one.[15] She did, however, obtain a notarized identification certification.[16]

This simple, typewritten document produced in 1941 played havoc with the young woman's fate. On July 1, 1941, she identified herself as Ikuko Toguri and signed a statement entitled *"CERTIFICATE OF IDENTIFICATION TO FACILITATE RETURN TO THE UNITED STATES OF AMERICA."* Prepared by a Japanese-American notary public, the certificate actually was a deposition to the effect that Miss Toguri was born in Los Angeles on the Fourth of July, 1916; that she currently resided in the city; that she was "temporarily leaving" the United States on July 5 on the *Arabia Maru* for Yokohama; that she intended to visit her aunt who was "eagerly awaiting her" and that she intended to return to the United States in six months. The document was duly signed by Miss Toguri, witnessed and sealed by the notary public, and embellished with Toguri's photograph and finger prints.[17]

To subsequent generations of Americans who have traveled abroad, this piece of paper is hardly more than that; it strains credulity to imagine trying either to leave or re-enter the United States today under the authority of such a document. When Miss Toguri left the country, however, the document had substantial validity. Proof of citizenship was adequate for either leaving or re-entering at that time. A law of May 22, 1918, provided that citizens would need passports to enter or depart only in time of war or national emergency.[18]

President Franklin D. Roosevelt's proclamation of November 14, 1941—several months after Miss Toguri's departure—declared the existence of a national emergency; it specifically required the use of passports. Thus what had gone out did not necessarily have to come back in. Had Miss Toguri managed to get passage on a ship back to the United States between November 14 and the bombing of Pearl Harbor on December 7, there is no assurance she would have been admitted without a

15. Earlier, she stated to army interrogators that passports were not being issued at the time.

16. d'Aquino Statement to Tillman, April 30, 1946, *JDF, supra* note 3.

17. "Certificate of Identification," *JDF, supra* note 3.

18. An Act Making further provision for the allotment of officers, enlisted men, and civilian employees of the Army, and for other purposes, Pub. L. No. 66, 40 Stat. 384 (1917).

passport. With nothing more than this paper in hand, Toguri's fate, as of December 7, 1941, was for all purposes sealed for the duration of the war. Ostensibly a lifeline for a return to the United States, the certificate served instead as a hawser mooring her to Japan.

Toguri sailed from San Pedro on July 5 and arrived in Yokohama on July 24. She immediately went to her uncle's house in Tokyo, where she lived until June, 1942. She paid room and board and attended the Japanese Language and Culture School. In 1946, she claimed she moved out in 1942 because transportation to the school had become too expensive. At her trial in 1949, she contended her family's fears of the secret police *(kempeitai)* prompted the move.[19]

In August, 1941, Toguri received her residence permit from the Japanese police. The police instructed her to register at the American Consulate office in the Embassy. On the advice of a consular official, she submitted an application for a passport, along with her birth certificate. According to her 1946 recollection, she made periodic inquiries about the passport until October. She then was told the office would contact her when it received word from Washington.[20]

It is not clear whether she understood the implication of the President's November, 1941 proclamation of national emergency as it related to her passport situation. In any event, she "became nervous" upon hearing of the Hull-Nomura-Kurusu negotiations concerning outstanding differences between the United States and Japan. She called her father in Los Angeles in late November, but Jun Toguri could not estimate the danger for his daughter, advising her simply to see as much of Japan as she could. Mr. Toguri soon afterward perceived the danger, for on December 1 he cabled her to hurry home. She learned a ship was scheduled to depart for the United States on the 2nd and immediately went to the American Embassy for clearance. She had as yet no passport and her birth certificate had been sent to Washington. The Embassy gave her a letter confirming the passport application and the accompanying birth certificate. For some reason, she was told to secure a certificate of attendance from the language school in order to facilitate re-entry into the

19. d'Aquino Statement to Tillman, April 30, 1946, State Department documents on Ikuko Toguri, *JDF, supra* note 3; *Record, Defendant,* 46: 5122-69.

20. d'Aquino statement to Tillman, April 30, 1946, State Department documents on Ikuko Toguri, *JDF, supra* note 3; American Consulate General documents, December 1, 1941, April 4, 1942, *AIF, supra* note 4.

United States. When her uncle attempted to buy a ship ticket, he was told his niece would also need clearance from the Finance Ministry in order to take back any money she had brought with her. That would take three or four days—alas, no clearance, no ticket, no passage. Less than a week later, the war came. With it, her bureaucratic tangles increased by geometric bounds.[21]

On December 8, Toguri appeared at the U.S. Embassy to advise the Vice-Consul that she did not possess dual citizenship, and offered a certified copy of her family's census register *(koseki)* recognizing her renunciation of Japanese nationality. In the meantime, she worked as a part-time typist at the language school. In March, 1942, she appeared at the Swiss Legation and filed an application for evacuation. She was advised that without a passport, it was unlikely she could be sent home on the first repatriation ship. By April, the American Consul in Yokohama determined that Toguri's citizenship had not been proven. In September, the Swiss notified her of a second ship which would provide free passage to Goa, Portuguese India. Passage from there to New York, however, required $425, payable in advance or upon arrival. Her own resources were exhausted; her uncle had no funds, and there was no way to contact her parents who, unknown to her, had already been evacuated from Los Angeles to Arizona. She owned real estate worth $2,000 in Los Angeles, but was apparently uncertain whether her holdings could be converted to cash. Her uncle, she claimed, advised her to stay. Accordingly, on September 2, she filed a document with the Swiss Legation stating: "I hereby wish to express my wish to remain in Japan for the present and hereby withdraw my request to be evacuated." The following day, she notified the Japanese police of her decision. According to her recollections, the police notified her she would be treated as a foreigner and she would have to renew her residence permit every six months. She could not travel outside Tokyo without a permit. The police, she recalled in 1946, visited her at irregular intervals, suggesting she become a Japanese citizen. Otherwise, she was "never jailed or ill-treated by the Japanese police."[22]

In July, 1942, Toguri began work at the Domei News Agency *(Domei Tsushin Sha)*, monitoring English-language

21. d'Aquino Statement to Tillman, April 30, 1946, State Department documents on Ikuko Toguri, *JDF*, *supra* note 3; American Consulate General documents, December 1, 1941, April 4, 1942, *AIF*, *supra* note 4.

22. d'Aquino Statement to Tillman, April 30, 1946, State Department documents on Ikuko Toguri, *JDF*, *supra* note 3.

broadcasts. She did not take verbatim notes but prepared rough translations of the news broadcasts into Japanese. Her training at the language school had been successful. She worked approximately five hours per day for 130 yen per month. In 1946, she stated she did "not know what disposition was made of my work and I made no comment upon, or evaluation of, the news I received." She worked in this capacity until December, 1943.[23]

In August, 1943, Toguri also found typing work in the business office of Radio Tokyo through contacts with Ed Kuroishi, a Nisei employed there. For this she received 100 yen per month. Her duties largely consisted of typing material for broadcast—"lists of Allied army personnel and the like"—in English.[24]

Toguri's broadcasting career began in November, 1943 when George Nakamoto Mitsushio asked her to audition for a planned "entertainment" program to be directed at Allied soldiers in the South Pacific. Mitsushio in turn introduced her to the two Allied prisoners of war already working at the station, Cousens and Ince. Cousens described the projected program as consisting of messages from prisoners of war, recorded music, and news from the American home front as well as the war areas. He added that he would write the scripts and she would only introduce the music. In 1946, she claimed that Cousens chose her because he thought she had "a Yankee personality." After the voice test, Cousens said he would coach her in using a "cheerful voice."[25]

In d'Aquino's 1946 statement to Tillman, she said she accepted the position because she thought she could entertain American soldiers. She offered no hint of coercion or duress: "No pressure was put on me by Cousens, Ince, the Japanese or other persons to force me to take the job and no one threatened me if I did not take the job or continue in it."[26] During her trial in 1949, however, she and other defense witnesses testified that the Mitsushio and Cousens "requests" were tantamount to army orders which one did not disobey. She also emphasized that she feared the army and the secret police and therefore continued to broadcast.[27] Her work began in mid-November of 1943. Cousens wrote her scripts until he became ill in the summer of 1944. Af-

23. d'Aquino Statement to Tillman, April 30, 1946, State Department documents on Ikuko Toguri, *JDF, supra* note 3.
24. *Id.*
25. *Id.*
26. d'Aquino Statement to Tillman, April 30, 1946, *AIF, supra* note 3.
27. *Record, Defendant,* 46:5122-69.

ter that, she wrote scripts herself, using Cousens' work "as a guide," with the help of Norman Reyes, a captured Philippine army officer and fellow broadcaster.[28]

The program, entitled "Zero Hour," was broadcast daily, except Sunday, from 6:00 to 7:15 p.m. The typical format consisted of prisoner of war messages, music introduced by Toguri, American home front news read by Reyes or Ince, more music, general news usually read by Ince, more music, and news commentaries by Charles Yoshii, a Nisei. Around Christmas 1943, Cousens and Ince told Toguri that they were trying to make the program "as entertaining as possible rather than propaganda." They also said there was "a double meaning" in their scripts and that they wanted to increase the POW messages. Although "they never directly or positively told me that they were trying to defeat the Japanese purpose of the program," Toguri also stated that Cousens instructed her to laugh when she referred to herself as "the enemy."[29]

Throughout the broadcasts, Toguri was introduced as "Orphan Ann," "Orphan Annie," "your favorite enemy, Ann," and "your favorite playmate and enemy, Ann." The names had evolved from "Ann" which was shorthand for "Announcer" on the script forms. She identified for Tillman at least three other women who appeared on the Zero Hour: Ruth Hayakawa, whom Toguri believed had once lived in Los Angeles; Mary Ishii, an Eurasian; and Miyeko Furuya, the wife of fellow employee Ken Oki. She admitted she never heard them broadcast and was not familiar with their scripts.[30]

Life in wartime Japan continued to be marginal. Toguri received no extra salary as a result of her new career, although every employee at the station received a raise to 150 yen per month in August, 1944 because of increased living costs. In December, 1943, however, she resigned her Domei job to work for the Danish Minister, Lars Tillitse, as a clerk-typist. She worked for him until his departure in July, 1945 and received 150 yen per month, 20 more than she had earned with Domei. At the same time, however, she further complicated her life by marrying Felipe Jairus d'Aquino, a Japanese-Portuguese with Portuguese citizenship. He was employed at Radio Tokyo and also

---

28. d'Aquino Statement to Tillman, April 30, 1946, State Department documents on Ikuko Toguri, *JDF, supra* note 4.

29. *Id.*

30. *Id.*

was a linotype operator. They were married in the Jesuit church at Sofia University in Tokyo and registered the marriage with the Portuguese Consul. She later claimed that the Consul gave her a certificate of Portuguese citizenship, but she continued to register as an American citizen with Japanese officials.[31]

The conclusion of the statement to Tillman in 1946 is heavily laden with contradiction and ambiguity, perhaps reflecting the tension between an interrogator anxious to establish some culpability and a suspect equally determined to proclaim her innocence. She admitted the program itself was "propaganda" designed to lower Allied troop morale and at the same time to demonstrate Japanese "sportsmanship" by sending POW messages. Her purpose in participating, she insisted, "was to give the program a double meaning and thus reduce its effectiveness as a propaganda medium." Cousens' remarks about his ulterior aims "almost convinced" her she was defeating the Japanese purposes. Finally, she acknowledged "all of their programs were propaganda," but maintained she did not believe she was working "against the interests of the United States."[32]

2. THE FIRST ROUND OF ADMINISTRATIVE DECISIONMAKING

The statement was forwarded to Washington where Theron L. Caudle, Assistant Attorney General, Criminal Division, turned it over to Nathan T. Elliff, Chief of the Internal Security Section, for evaluation and recommendation. Elliff's response was prompt and straightforward: prosecution for treason was "not warranted."[33]

Elliff's memorandum began with a forlorn attempt to destroy an important misconception: there was no "Tokyo Rose," he said. The name never had been used by Radio Tokyo or by the Japanese; rather it had been popularized by American soldiers and applied to any of a number of female radio announcers. Subsequent investigations and depositions identified ten English-speaking broadcasters at Radio Tokyo. There were

31. *Id.*

32. d'Aquino Statement to Tillman, April 30, 1946, State Department documents on Ikuko Toguri, *JDF, supra* note 3; American Consulate General documents, December 1, 1941, April 4, 1942, *AIF, supra* note 4. The FBI was fully aware of the suspect's attempts to leave Japan; see Report of Confidential National Defense Informant S-90 (n.d.), *AIF, supra* note 4. *Record, Defendant,* 46:5122-69, offers Mrs. d'Aquino's formal trial testimony on the background of her story. For her explanation of her fears, *see Record, Defendant,* 49:5502-05.

33. Elliff to Caudle, May 15, 1946, *JDF, supra* note 3.

still others in Manila, Batavia, and Bangkok. Elliff then reviewed the peculiar circumstances that trapped Toguri in Japan after Pearl Harbor and her need for work. Although she admitted broadcasting, Elliff found decisive the unanimous claims of Cousens, Ince, and Reyes that she merely introduced music. In his report to Caudle, Elliff attached script portions that substantiated her simple role. Although he recommended that the case be dropped, he suggested the Department await a report from James Carter, the United States Attorney in Los Angeles who had been considering the case for possible prosecution.[34]

Throughout the summer of 1946, Caudle prodded Carter for an opinion on prosecution. By then, the War Department was pressing Justice as to the future of the occupation army's famous prisoner. Carter finally replied on September 13 with a terse telegram: further investigations, he said, had "not strengthen[ed]" the case; he and his staff found the "evidence inadequate"; and he recommended that a "treason prosecution be declined."[35]

Following Carter's recommendation, Elliff wrote again to Caudle on September 19, agreeing with the California prosecutor. He again suggested the matter be closed and the suspect released. Although Elliff acknowledged that d'Aquino was an American citizen who had broadcast for the enemy, presumptive treason was not for him. "[T]he available scripts and the testimony of the majority of witnesses indicate that her broadcasts were innocuous, and could not be considered giving aid and comfort to the enemy." Elliff qualified his conclusion by adding that additional information might offer cause for reconsideration. Perhaps that was a typical bureaucratic hedge, but in this case it was prophetic. Caudle agreed with Elliff's report and asked him to prepare a separate memorandum for the attorney general regarding the findings and recommended "that the file be closed."[36]

Caudle relayed the formal recommendation to Attorney General Tom Clark on September 24, 1946. He reiterated the decision as he had given it to Elliff, but in two brief paragraphs he summed up the evidence, the weight of which never substantially changed during the next two years, although the interpre-

34. *Id.*

35. Caudle to Carter, August 27, 1946, September 13, 1946, Carter to Attorney General Clark, September 13, 1946, Carter to Caudle, June 19, 1946, *JDF, supra* note 3.

36. Elliff to Caudle, September 19, 1946, Caudle to Elliff, n.d., *JDF, supra* note 3.

tation altered. Caudle acknowledged that Radio Tokyo never identified *a* "Tokyo Rose"; that there were several women announcers who were indiscriminately given that name by American troops; that Toguri's activity "consisted of nothing more than the announcing of musical selections"; that her co-workers corroborated this fact, except for "two or three" who vaguely charged she had made anti-United States remarks; that recording cylinders, scripts, and transcripts from monitoring by the Federal Communications Commission substantiated her role as a disc jockey; and finally, there apparently was a so-called "Tokyo Rose" broadcast which occurred prior to Toguri's employment by the station in November, 1943. On the 26th, Attorney General Clark's assistant acknowledged receipt of Caudle's report and replied: "We agree with you—no prosecution unless more facts."[37]

The Director of the FBI had to have his separate accounting. Three days after Caudle sent his "final" report to Clark, the Director queried him "for an opinion as to prosecution." By the time Caudle replied, the War Department had been advised on October 1 to release Mrs. d'Aquino. The grinding wheels of bureaucracy delayed a release order from the War Department to MacArthur's headquarters and thence to the Sugamo prison commander until October 23. Two days later, d'Aquino left Sugamo a free woman—at least for the moment. The decision not to prosecute was played in a low key, for the Department allowed Carter to make the announcement in Los Angeles, and it merited little more than a brief news item in the *New York Times* on October 22.[38]

II. Enter Public Opinion: Arbiters of Patriotism, the Media, and the Justice Department

Upon her release from Sugamo, d'Aguino chose not to return immediately to the United States. The reasons are not clear: her subsequent pregnancy or simple fear of a hostile reception may have played a part in her remaining in Japan. One year later, she initiated efforts to come home. A desire to have her child on American soil may have motivated her. In any

37. Caudle to Clark, September 24, 1946, Clark to Caudle, September 26, 1946, *JDF*, *supra* note 3.

38. Caudle to Hoover, October 4, 1946, General Headquarters, U.S. Army, Pacific to Commandant, Sugamo Prison, October 23, 1946, *JDF*, *supra* note 3; New York Times, October 22, 1946.

event, her passport application in the fall of 1947 triggered renewed, and hostile, interest in her case.

On October 20, 1947, the State Department's Passport Division queried the Justice Department regarding d'Aquino's status. Four days later, T. Vincent Quinn, Assistant Attorney General, offered no objection to the passport request, replying that the evidence in the case had not warranted prosecution.[39] The case truly seemed closed. Word leaked, however, that "Tokyo Rose" had applied for a passport, bringing a swift and predictable response. Veterans' organizations and West Coast nativist groups vociferously criticized issuing a passport and called instead for prosecution.[40]

In November, the national commander of the American Legion demanded a treason trial. A few months later, the Legion's Executive Committee joined in protesting "the lack of interest [in] and the lenient treatment of war criminals." The Chairman of the Native Sons of the Golden West Americanism Committee protested to the Justice Department, J. Edgar Hoover, and Secretary of State George C. Marshall, among others. His letter noted that "from all reports, her mission was one of lower and vicious propaganda," indicating that the decision to drop the case had not been accompanied by adequate public explanation.[41] Had the Justice Department made clear the insufficiency of evidence and, more important, the flimsiness of the charges, such protests might have been easily countered.

Instead, a number of key actors in the Department responded to the winds of public opinion. Within a month, the correspondent representing the Native Sons of the Golden West was informed that d'Aquino would not receive a passport and that the investigation had been renewed. Attorney General Clark's assistant, Peyton Ford, replied more elaborately to a congressman who had passed on a local Legion protest. Ford pointed out that at least six women had been identified on the Radio Tokyo broadcasts, but only d'Aquino was American-born. He also noted she had been investigated for two years, but the Department had not been able to secure the required two witnesses to obtain a treason conviction. Ford had all flanks secured: the investigation would continue and the government

39. Quinn to Secretary of State, October 24, 1947, *JDF, supra* note 3.

40. National Legionnaire 3 (Nov., 1947).

41. *Id.* at 2 (May, 1948); Edward L. Meyer to Tom Clark, *et al.*, November 13, 1947, *JDF, supra* note 3.

would present the case to a grand jury as soon as the "necessary proof" could be obtained. In other words, legal technicalities, it now appeared, rather than the lack of evidence and the insignificance of d'Aquino's action, prevented prosecution. Ford noted that she was not permitted to return to the United States in the meantime.[42]

In a press release dated December 3, 1947, the Justice Department indicated its sensitivity to the growing pressures for action. The Department mentioned its ongoing investigation, although admitting difficulty in securing evidence to fulfill the two-witness requirement for a treason conviction. The Department promised, however, that if such evidence could be obtained, the case would be presented to a grand jury promptly. Accordingly, there was an appeal for persons who witnessed the broadcasts or recognized her voice to contact the FBI. The release also noted that d'Aquino's passport would not be issued while the investigation progressed.[43]

Patriotic and veterans' protests continued to bombard Justice. But apparently Attorney General Clark was decisively impressed by Walter Winchell's public demands for action. Winchell was then at the peak of his power and influence. Millions read his daily column and heard his weekly radio newscasts. His appeal was seductive. Blending petty gossip of celebrities with a political primitivism befitting the style of a neighborhood saloon, Winchell both satiated mass desires and articulated a simple grassroots political mood of cynicism toward government. Before and during World War II, Winchell had been militantly anti-fascist and pro-Roosevelt. In the postwar period, he swung sharply to the right, probably not much differently from many ordinary citizens, the mainstay of his constituency. A monumental dose of vanity and a swollen ego animated Winchell. His column periodically chided or maligned public officials who failed to respond to his concerns.

Sometime after d'Aquino applied for her passport in late 1947, Winchell lent his public support to the opposition. His

---

42. Eldred L. Meyer to Quinn, December 22, 1947; Congressman Franklin J. Maloney to Clark, February 5, 1948; Ford to Maloney, February 20, 1948, *JDF*, *supra* note 3. Ford's statement that d'Aquino was the only American-born woman announcer on Radio Tokyo was wrong, of course. He knew better from the evidence; furthermore, Clark Lee's book, published some months earlier, severely criticized the broadcasts of Mrs. Genevieve Faville ("Mother") Topping, an American missionary then over eighty years old. C. Lee, One Last Look Around, *supra* note 1, at 84.

43. Press Release, December 3, 1947, *JDF*, *supra* note 3.

sentiments were clear, and presented in his usual breathless, racy style: "If 'Tokyo Rose' can be brought to this country and tried for treason, o.k.," Winchell said in one column. "Otherwise," he continued, "we're for letting her stay forever in Japan. And if protests are what the State Department needs to keep her there, here's one. We imagine plenty more will be forthcoming."[44] Winchell had the wrong agency in mind, but the message was clear: if enough protests could be made, such pressure would keep d'Aquino in Japan or force the Justice Department to prosecute. A protest Winchell received from a Gold Star mother in early 1948 regarding Mrs. d'Aquino's request for a passport probably sparked his involvement. He broadcast the complaint and called for Tom Clark to consider the matter carefully. Shortly thereafter Winchell heard from Harry Brundidge, or possibly from J. Edgar Hoover, who had been in touch with Brundidge, that the former war correspondent had d'Aquino's "confession" which she had given to himself and Lee.[45] The so-called confession, however, consisted of Brundidge's story as developed from Lee's notes, and unsigned, had little validity. According to Brundidge, Winchell and other columnists were giving him "a kicking around" and a "shaming"—which only made Brundidge more determined to bring d'Aquino to trial.[46]

Attorney General Clark was sensitive enough to Winchell's activity that he dispatched U.S. Attorney James Carter from Los Angeles as an emissary to explain the facts to the columnist. Carter, of course, was no stranger in the matter, for he had recommended in 1946 that prosecution be dropped.[47] He still held that opinion in late 1947, but his report to Clark of his visit to Winchell, at the Twentieth Century Fox film studios, had the obsequious flavor of the temporal lord's retainer trying to appease the spiritual lord of the realm.[48]

Notwithstanding the flat, narrative, diplomatic tones of

---

44. Undated clipping, c. November 1947, *JDF*, *supra* note 3.

45. *See* Introduction, *supra*.

46. Wayne Collins to Lyndon B. Johnson, November 4, 1968, *WCP*, *supra* note 4. This was a formal request for a presidential pardon. Nashville Tennessean, May 2, 1948. Brundidge claimed the announcement that d'Aquino would not be prosecuted made him "indignant" and he thereupon contacted Hoover offering to get the necessary evidence in Japan. But according to Brundidge's account, this was in December 1947; the announcement regarding the dropping of the investigation had been made by Carter over a year earlier. What obviously prompted his renewed interest was the passport application and/or the possibility that Winchell was getting mileage out of a story that Brundidge regarded as peculiarly his.

47. *See* text accompanying note 35 *supra*.

48. Carter to Clark, December 5, 1947, *JDF*, *supra* note 3.

Carter's report, there is a surrealistic quality to the account. Carter persistently adhered to the facts of the case during the interview with Winchell to demonstrate to the columnist the weaknesses of a potential prosecution. Winchell, on the other hand, recited a litany of comments and charges as to the Department's duties and obligations, berated Clark for his personal ingratitude toward him, and complained about the failure to mention Winchell's role in re-kindling interest in the case. In particular, Winchell expressed considerable pique that he "had been plugging" Clark for attorney general and "had never received proper thanks." Moreover, he was miffed because Clark had not told him in advance he would receive the appointment—in other words, Winchell missed a "scoop," a not inconsiderable event in his special world. While Winchell admitted that many of d'Aquino's broadcasts were innocuous, he characteristically added that ex-soldiers who offered to testify that she was a morale builder "were probably communists."[49]

Carter told Winchell he would not recommend prosecution unless there were a stronger case. He believed a trial probably would result in a dismissal or an acquittal; Winchell agreed that that would be worse than no prosecution. As a final observation, Carter perceptively reported to Clark that "somewhere along the line his [Winchell's] pride had been injured," and that a personal conversation between Clark and Winchell "could go a long way towards smoothing things out." There is no reply from Clark to Carter in the files, but there is a notation on the letter that Clark's assistant, Peyton Ford, sent a *"PERSONAL AND CONFIDENTIAL"* note to the U.S. Attorney.[50]

It is tempting to dismiss Carter's report and its subject as trivial—but for the fact that this pressure and fear of public opprobrium drove Clark enthusiastically to endorse subordinate recommendations for prosecution and explicitly countermand every opinion that recommended dropping the case.[51] Winchell was indeed a powerful figure and could not be treated lightly. Furthermore, harrassing the attorney general and the Department to "do something about" people such as d'Aquino fit into a

49. *Id.*

50. *Id.* Carter soon successfully prosecuted another Nisei treason case, involving brutality toward U.S. prisoners of war. United States v. Tomoya Kawakita, 96 F. Supp. 824 (S.D. Cal. 1950). Kawakita's death sentence was commuted by Eisenhower and he was pardoned by Kennedy. Carter subsequently was appointed to the Ninth Circuit Court of Appeals.

51. *See* text accompanying note 77 *infra*.

larger movement decrying the administration's "softness" and lack of concern toward disloyalty.[52] In short, the political credibility of Clark and the administration made prosecution expedient—and d'Aquino expendable.

Carter's efforts to appease Winchell proved fruitless. So too did a similar attempt by one of Winchell's friends, Hollywood producer and night club operator Earl Carroll. Carroll visited Tokyo in March, 1948 in preparation for a film. While there, he arranged to meet d'Aquino at Army G-2 headquarters. He asked if she had ever called herself "Tokyo Rose"; predictably, she denied the canard. Carroll thereupon informed her that Major General Charles A. Willoughby, MacArthur's intelligence chief, had told him d'Aquino's "trial days are over" and as far as the army was concerned, her case was closed. Carroll had very close contact with Willoughby. After Carroll returned home in April, Willoughby sent him a summary of the dossier on d'Aquino, adding that he could attribute the information to an "authoritative" source.[53] Carroll was aware of Winchell's role in delaying her passport clearance and he told d'Aquino to write a personal letter to the columnist explaining her case. Clearly persuaded that the affair was frivolous, Carroll mailed the letter upon his return, but he received an enigmatic reply. "I have confidence in the administration of United States justice," Winchell wrote, "and she will get a fair trial in court." At the end of May, Carroll passed the message to d'Aquino, adding that he was bewildered by Winchell's response in the light of Willoughby's statements. He promised to meet personally with Winchell in New York in June. On his way east to attend the Republican National Convention, however, Carroll was killed in a plane crash.[54]

---

52. *See* text accompanying note 98 *infra.*

53. Carroll to d'Aquino, April 20, 1948, *WCP, supra* note 4 (reprinted in John Juji Hada, The Indictment and Trial of Iva Ikuko Toguri d'Aquino, M.A. Thesis, University of San Francisco, 1973) [hereinafter cited as Hada]; Carroll to Willoughby, April 20, 1948, G-2 Memorandum, May 4, 1948, Willoughby to Carroll, May 4[?], 1948, *AIF, supra* note 4. Willoughby's disdain for the civilian case is expressed in a note to his aide. Willoughby to Bratton, March 3, 1949, *AIF, supra* note 4.

54. d'Aquino to Winchell, April 14, 1948, Winchell to Carroll, May 14, 1948, Carroll to d'Aquino, May 27, 1948, *WCP, supra* note 4 (also reprinted in Hada, *supra* note 53). Interestingly, in his May 27 letter, Carroll used the salutation, "Dear Rose."

III. The Justice Department's Decision to Prosecute: A Study in Administrative Politics

A. The Evidence

1. Early Evaluation

Concurrently with Carter's "negotiations" with Winchell, the Criminal Division of the Justice Department was preparing its first report of the re-opened investigation. Caudle's successor as head of the Criminal Division, T. Vincent Quinn, forwarded the first lengthy evaluation to Peyton Ford, Attorney General Clark's assistant, on December 12, 1947. Based on comparison with other documents and in light of Quinn's general reticence in the case, there is a strong possibility that John B. Hogan, an attorney in the Department's Internal Security Section, actually wrote the memorandum. It summarized the suspect's background and her activities in Tokyo, offered excerpts from her broadcasts, mentioned possible witnesses against her in both Japan and the United States, and then presented a brief recommendation for further investigation.[55]

Despite firm evidence on the matter, d'Aquino's recruitment by Radio Tokyo was handled a bit gingerly.[56] This account merely states that Cousens and Ince wanted "a person who would not report them to the Japanese authorities and who would not give the program a sentimental touch." The memorandum then summarizes the basic evidence—which consisted simply of portions of two scripts in which "Ann" of Radio Tokyo introduced music to Allied soldiers. The script dialogue blended idle chatter with the introductions. The following samples as reconstructed from the scripts (with no intended ellipses) are characteristic:

FEB. 22 . . . .

> ANN: Hello there Enemies . . . how's tricks? This is Ann of Radio Tokyo, and we're just going to begin our regular programme of music . . . news and the zero hours for our Friends . . . I mean, our enemies! . . . in Australia and the South Pacific . . . so be on your guard, and mind the children don't hear! . . . All set? . . . O.K. here's the first blow at your morale . . . the Boston Pops . . . playing 'STRIKE UP THE BAND' . . . .

55. Quinn to Ford, December 12, 1947, *JDF*, *supra* note 3.
56. *See* text accompanying note 25 *supra*.

'STRIKE UP THE BAND'

> ANN: How's that for a start? . . . well now listen to me make a subtle attack on the Orphans of the South Pacific. Sergeant! . . . where the Hell's that Orphan Choir? . . . Oh, there you are Boys . . . this is Ann here! . . . How about singing for me to-night? . . . You won't alright you thankless wretches, I'll entertain myself and you go play with the mosquitoes . . . thank you Mr. Payne . . . when you're ready!
> 'THE LOVE PARADE'

APRIL 10 . . . .

> ANN: Thank you kindly Sir, she said! . . . Hello Everybody, this is Ann of Radio Tokyo with our regular programme for our Friends in Australia and the South Pacific. It's pretty music to-night, guaranteed to take your mind off the present, and to make you forget for a while 'the haunting spectres of the might-have-been'! . . . Yes, its a quotation, but I don't know who wrote it, do You? . . . think it over, and here's some music to oil the wheels . . . Franz Lehar's 'Gipsy Love' . . . played as a Concert Waltz by Nat Shilkret and Orch . . .
>
> 'GIPSY LOVE' . . . CONCERT WALTZ . . .
>
> ANN: Like that? well be good and we'll have an even better one directly, in the meanwhile here's an old smoothy for you, Savino's 'A Study in Blue' . . . please to listening!
>
> 'A STUDY IN BLUE' . . .
>
> ANN: This is Radio Tokyo's special programme for listeners in Australia and my Boneheads in the South Pacific. Right now I'm lulling their senses before I creep up and annihilate them with my nail file . . . but don't tell anybody! . . . Now here's the next waltz I promised you, Victor Herbert's 'Kiss Me Again' . . . you heard me!
>
> 'KISS ME AGAIN' . . .[57]

The other evidence cited was d'Aquino's own admission to army investigators and to FBI Agent Tillman that she had worked as a broadcaster. The memorandum also noted the contract between d'Aquino and the Hearst reporters, Lee and

---

57. Quinn to Ford, December 12, 1947, *JDF*, *supra* note 3.

Brundidge, for an exclusive *Cosmopolitan* story detailing her activities as the "one and original 'Tokyo Rose.'" The memorandum, however, acknowledged she never used that name and that there were other female broadcasters on Radio Tokyo. Citation of the contract implied that it constituted reliable evidence of an admission of culpability. The fact that the contract was a nullity was ignored.[58]

In conclusion, the memorandum stated that d'Aquino's voluntary acceptance of the position, her performance, and the Japanese military's approval of her duties demonstrated an adherence to the enemy. Moreover, her broadcasting lent "aid and comfort" to the enemy to such an extent that prosecution could be considered. The basic recommendation was that the various witnesses, largely employees and technicians at Radio Tokyo, be extensively "reinterviewed" to determine whether they heard or saw d'Aquino on the programs. Until this was accomplished, the case should not go to the grand jury. Moreover, if the new interviews still failed to produce the necessary witnesses, "then further consideration can be given as to the advisability of presenting the case to a grand jury on the evidence that is or may become available in the United States." Ford did not receive this document until January 9, 1948; apparently, the holiday season delayed things. In any event, Ford told Quinn to proceed with the case: "All available lead[s] should, of course, be developed," he wrote. "There are several loose ends."[59]

2. SEARCH FOR WITNESSES

The "loose ends" had to be tied up in Japan. Accordingly, on March 11, 1948, Clark directed that John Hogan go to Tokyo—with none other than Harry T. Brundidge, the reporter with a longstanding stake in the story. From his initial contact with d'Aquino in September 1945, through the trial in 1949, Brundidge had a consuming passion for her case. From his trip in March 1948 until the trial began, he is particularly omnipresent. Not only did he travel in 1948 to Japan, but he returned a year later to join FBI Agent Tillman in further investigations. Significantly, the Japanese he talked to believed he served in an official capacity—an impression he rather easily conveyed. Whatever his role, official or unofficial, Brundidge carried on his

58. *Id.*

59. Quinn to Ford, January 9, 1948, *JDF*, *supra* note 3; Ford to Quinn, n.d., *JDF*, *supra* note 3.

peculiar crusade. Whether it was the crusade of a true-believing zealot or an ambitious man is perhaps immaterial. No doubt his personal involvement, which brought him close to Attorney General Clark, and the articles he published after he returned with Hogan, pushed the decision ever closer to prosecution. The oddity that someone so familiar with the case never appeared at trial as a witness is probably due to the government's reluctance to call a man who, by the time of the trial, had according to all indications suborned perjury.[60]

On March 20, 1948, Brundidge located a prospective witness, Hirom Yagi, then employed by the Japan Travel Bureau. Yagi not only was willing to cooperate himself but offered to find other witnesses. For example, Hogan later claimed that Yagi located Leslie Nakashima, a former wartime Domei News Agency employee who worked in 1948 for the Associated Press in Tokyo. Nakashima, however, denied ever having seen or heard d'Aquino broadcast and stuck to his story despite "prolonged cross-examination" by both Hogan and Brundidge; the latter were convinced he was lying. Nakashima certainly was no surprise find for Brundidge. Nakashima had brought d'Aquino to Brundidge and Lee in 1945 and witnessed the abortive contract.[61]

Another witness, identifying himself only as a friend of Yagi's, offered exactly the same story as Yagi but refused to testify publicly. Brundidge and Yagi surmised that the friend feared reprisals from American military authorities. Yagi promised to "persuade" his friend to testify in the future. Because of Brundidge's own fear—imaginary or real—of the army people, he insisted that prospective witnesses be interviewed in his hotel room rather than in the office provided by the Civilian Intelligence Service of Army G-2. Actually, General Willoughby's aide in G-2 provided Hogan with complete cooperation, including access to army files, although he denied Brundidge the same privilege. Hogan accommodated his friend's fears by refusing to identify any of the witnesses to army authorities.[62] Nine months later, the murkiness surrounding Yagi and his friend would be clarified.[63]

Hogan and Brundidge interviewed d'Aquino on March 26,

---

60. *See* text accompanying notes 91-93 *infra*. For Brundidge's suborned perjury, *see* text accompanying notes 93-99 *infra*.

61. *See* note 4 *supra*.

62. Bratton to Willoughby, March 22, 1948, C. A. W. to Willoughby, April 8, 1948, *AIF*, *supra* note 4.

63. *See* text accompanying notes 94-98 *infra*.

1948. Confronted with Clark Lee's notes of their September 1945 interview, she stated that they were accurate to the best of her recollection, but she refused to sign a statement that a story Brundidge had prepared from those notes was correct, insisting there were too many discrepancies. She also was unable to identify any scripts shown to her as ones she actually had broadcast. Most important, she refused to admit she had broadcast any intelligence material.[64]

3. INITIAL PROSECUTION STRATEGY AND THE INTRUSION OF AN OLD FRIEND

Hogan and Brundidge returned to Washington on April 5 and met with the attorney general the next day. Clark asked for another review of the case and a recommendation. In his initial summary, dated April 12, 1948, Hogan attached Yagi's statement to his own.[65]

Hogan did not prepare his recommendations until April 20. The April 12 memorandum, sent to Quinn, had been promptly forwarded to Ford and the attorney general. Clark did not need a final report, for on the covering page, carrying a date of April 15, Clark noted: "Let[']s expedite consideration. Do we need anything further?"[66]

Clark's views only reinforced Hogan's apparent eagerness. Hogan's April 20 recommendations flatly asserted that the case should be brought to a grand jury. His only concession was that the name "Tokyo Rose" should never be mentioned in the indictment or in court. He acknowledged the name had not been used by Radio Tokyo and added there was "an unanswered question" as to whether any other English-speaking women had broadcast. For a man who supposedly had reviewed the files carefully, he had conveniently overlooked some well-established evidence.[67]

In his recommendations, Hogan dealt primarily with the problem of locating two witnesses to the overt act of broadcast-

64. Hogan to Quinn, April 12, 1948, *WCP*, *supra* note 4.

65. *Id.* That document is not in the current files.

66. *Id.*; Quinn to Ford, April 13, 1948, Ford notation, April 15, 1948, *JDF*, *supra* note 3. *See also* Bratton to Willoughby, April 1, 1948, *AIF*, *supra* note 4; Nakashima Deposition, *WCP*, *supra* note 4. Nakashima had witnessed the September 1945 contract with Mrs. d'Aquino. The defendant confirmed her conflicting view of Lee's notes and Brundidge's story. *Record*, *Defendant*, 47:5219-20.

67. Hogan to Quinn, April 20, 1948, *JDF*, *supra* note 3. *See* text accompanying notes 6-12 *supra*, for a discussion of the overlooked evidence.

ing, and the extent of harm of d'Aquino's broadcasts. He admitted that thus far he had not turned up two people who had observed or heard a single broadcast. At this time, Hogan had hopes that two people in Los Angeles who knew the suspect before the war, and who recognized her voice on radio, might have heard the same broadcasts. He remained optimistic, moreover, that Yagi and his still-unnamed friend would come forth. Hogan's primary resources, however, were Cousens, Ince, and Reyes, d'Aquino's former colleagues at Radio Tokyo. Obviously, he was not interested in any mitigating or ulterior purposes that the Allied prisoners may have had; he simply wanted their testimony that d'Aquino had broadcast for the enemy.[68]

As for the extent of harm, Hogan believed a sufficient number of ex-soldiers would testify that their morale had been lowered by the broadcasts. If not, he still insisted the government was duty-bound to prosecute on the theory "that the mere act of broadcasting popular music and indulging in friendly chatter was intended to and did hold an audience to which the Japanese could broadcast propaganda." This conduct, he concluded, met the constitutional requirement of "aid and comfort" to the enemy. He nowhere intimated that d'Aquino herself had broadcast propaganda. Finally, Hogan recommended that she be brought to the United States in custody. Fearful that anti-Japanese sentiment would preclude a fair trial on the West Coast, he suggested she be flown to the United States via the Atlantic and be tried somewhere in the East.[69]

Once again, there was prompt action on Hogan's recommendation. Quinn sent it to Clark on April 26, concurring in the suggestion that "Tokyo Rose" (ignoring Hogan's injunction against that label!) be brought to the east coast. Two days later, Clark pushed a bit more, telling Quinn: "OK—let[']s expedite prosecution if it[']s decided to proceed." In turn, Quinn told one of the lawyers in his division to "Go to work."[70]

While Hogan was writing his reports, Brundidge had his own enterprise going. Upon his return he prepared a ten-part series on "Tokyo Rose" for the *Nashville Tennessean,* which was published in May. The articles largely covered familiar ma-

68. *Id.* Hogan suggested the number of Japanese witnesses be kept to a minimum to avoid appearances that the prosecution had resulted from Japanese nationals' hatred of Nisei.

69. *Id.*

70. Quinn to Clark, April 26, 1948, Clark to Quinn, April 26, 1948, Quinn to Whearty, April 28, 1948, *JDF, supra* note 3.

terial, but from Brundidge's point of view, d'Aquino's recent remarks in her interview with Brundidge and Hogan amounted to a "confession." One article was headlined, "Rose Confesses Tokyo Broadcasts." The stories even had an advance "tip" in Drew Pearson's radio broadcast of April 22 in which he explicitly congratulated Tom Clark and Stilliman Evans, publisher of the *Tennessean,* for "co-operating in getting this confession."[71]

Understandably, Brundidge's Justice Department friends were not pleased. Indeed, they betrayed a tentativeness and uncertainty quite contrary to their internal recommendations. Quinn sent Clark a copy of the Pearson story and complained that Brundidge was "going pretty far" in his stories. "We don't think too much of this case from a prosecution standpoint, but Harry is blowing it up," he added. Clark responded it might be best to "ask him to hold up as it would affect [the] case." Soon afterward, J. Edgar Hoover sent Clark and Quinn some copies of the Brundidge stories. Clearly, Brundidge's stories unnerved high officials in the Department. Peyton Ford wanted "to put a clamp on Harry's mouth or typewriter." He feared calling him, knowing Brundidge would "blow it into a story thus giving him a leg . . . of credence." Clark personally wrote to publisher Evans, an old political friend: "The boys are a little disturbed on the Tokyo Rose matter. It would be helpful if no publicity came out until after action was taken. I will let you know."[72]

Despite the apparent whole-hearted endorsement of Hogan's recommendations, nagging doubts persisted. Stacked against earlier investigations and recommendations, Hogan's letter indeed seemed a flimsy reed. MacArthur's army investigators, no strangers to political decision-making under the color of law, already had favored d'Aquino's release; Carter's supplemental investigation in behalf of the Justice Department had concurred in that finding; and the Department's higher bureaucracy had supported both decisions. Therefore, following Hogan's memorandum, Quinn turned to the Department's resident treason expert in May 1948 for further examination of the files and of Hogan's recommendation.

---

71. Nashville Tennessean, May 2, 1948, *JDF, supra* note 3. Brundidge's articles appeared in the *Tennessean* on May 2, 5, 9, 10, 12, 14, 19, 20, 21, 23, 1948.

72. Quinn to Clark, May 21, 1948, Clark to Quinn, n.d., Hoover to Clark, June 11, 1948, Quinn to Clark, May 28, 1948, Quinn to Whearty, June 3, 1948, Ford to Whearty, n.d., Clark to Evans, June 3, 1948, *JDF, supra* note 3.

B. *"Prosecute It Vigorously"*

1. AN INTERNAL HITCH . . .

Tom DeWolfe had been with the Justice Department since 1927. Most important, he had spent considerable time as a member of the prosecution team in the treason trials of Douglas Chandler and Robert H. Best, both of whom were convicted for giving aid and comfort to the German government by broadcasting propaganda during World War II.[73] Neither the crime of treason nor the special province of broadcasting as incidental to the crime was unfamiliar terrain to DeWolfe.

DeWolfe prepared a six-page report for Raymond P. Whearty, Quinn's top assistant in the Criminal Division. It is best read as a master's dissection of a novice's, Hogan's, memorandum. DeWolfe's report was about evenly divided between a statement of known facts and his own recommendations. After reciting the familiar story of d'Aquino's trip to Japan in 1941 and her entrapment by the war, DeWolfe concentrated on her work as a broadcaster from November 1943 until the end of the war. He noted that three prisoners of war—the Australian, Cousens; the American, Ince; and the Filipino, Reyes—had charge of the radio program, "Zero Hour." All had been extensively investigated and cleared by their governments of any charges of treasonous activity. Because of their connection to the program and to d'Aquino, the three obviously had to be prime witnesses. But for whose benefit? DeWolfe knew they would testify that d'Aquino's broadcasts had neither military nor propaganda value. They also would testify that d'Aquino personally endeavored to see that no propagandistic material was inserted in the scripts. Furthermore, these witnesses would testify that they selected her because she was the only available woman, "white or Nisei, whom they could trust not to betray" their efforts to sabotage any Japanese attempts to use the program for propaganda.[74] DeWolfe also knew Cousens and Ince would claim they selected her because her "masculine voice" would not have much appeal to Allied soldiers.[75]

---

73. Chandler v. United States, 171 F.2d 921 (1st Cir. 1948), *cert. denied*, 336 U.S. 918 (1949); Best v. United States, 184 F.2d 131 (1st Cir. 1950), *cert. denied*, 340 U.S. 939 (1951).

74. This incidentally offers some light on the animosity between d'Aquino and the other Nisei at Radio Tokyo who rather easily identified her as "Tokyo Rose" and later testified against her at the trial.

75. DeWolfe to Whearty, May 25, 1948, *JDF*, *supra* note 3.

DeWolfe's argument was simple: with such prosecution witnesses, the defense would need none. The three Allied military men, he noted, lived in a rather luxurious Japanese hotel during the war and had no more police surveillance than the suspect. "They," DeWolfe argued, "seem just as much, or more, culpable than she"—and they, including the American, Ince, had been cleared. Finally, as to the overt acts of broadcasting certain scripts, DeWolfe gave the evidence short shrift: "The scripts of her programs seem totally innocuous and might be said to have little, if any, entertainment value." Incidentally, DeWolfe contemptuously dismissed the "evidence" gained by correspondents Lee and Brundidge. Their methods, including the offer of a cash payment for the statement, made the "confession" of doubtful propriety and a jury would have to determine whether it was given of her own free will.[76]

DeWolfe's summary of the case was damning enough; his recommendations were equally forthright: *"THERE IS INSUFFICIENT EVIDENCE TO MAKE OUT A PRIMA FACIE CASE."* He grounded his opinion on the Supreme Court's recent decision in *Cramer v. United States*,[77] ruling that to support a treason conviction the accused's overt acts must be accompanied by an *intention* to betray. DeWolfe flatly stated that the government could not make a case able to withstand a motion for an instructed verdict for acquittal. In the Boston trials of Chandler and Best which he handled, DeWolfe contended that the evidence had clearly demonstrated the pro-German and anti-American sentiments of the defendants, that their scripts clearly intended to dissuade Americans from supporting the war, and that they clearly broadcast military information on Allied losses to impede the American war effort by lowering morale. By comparison, the evidence in the present case, DeWolfe declared, "is the direct antithesis of that available and utilized in the Boston litigation." DeWolfe clearly favored dropping the case; if, however, the Department found it expedient to present the evidence to a grand jury, then he recommended that a no true bill be sought.[78]

2. . . . AND HOW IT WAS OVERCOME

On May 27, Quinn passed DeWolfe's memo directly to Attorney General Clark. The covering form had a notation to the

---

76. *Id.*
77. 325 U.S. 1 (1945).
78. DeWolfe to Whearty, May 25, 1948, *JDF*, *supra* note 3.

effect that the attorney general might want to see the document "in the light of all the publicity" recently given to the case. Clark's response was almost immediate and must have jolted the bureaucracy from Quinn to Whearty to DeWolfe: "Prosecute it vigorously," Clark instructed.[79]

In retrospect, Clark's decision was more than a veto of DeWolfe's memorandum. In fact, it reaffirmed a policy decision he had been fostering over the previous five months. From that standpoint, it seems clear that DeWolfe had been brought in as a final reviewer, probably anticipating his role as prosecutor. His negative recommendation, however, was too late to stop the policymaking process that ground inexorably toward a decision to prosecute.

Yet Quinn, and apparently Clark, needed something to counter DeWolfe's devastating critique. Quinn suggested they turn to a third party. They chose Frederick Bernays Wiener, who was then in the Solicitor General's office and had participated in the German broadcasting cases, as had DeWolfe. In July 1948, Wiener recommended that d'Aquino and Major Ince should both be presented to a grand jury. He admitted that the cases were "not open and shut . . . in which conviction can be guaranteed," but their innocence was not clear enough to justify dropping their cases.[80]

Wiener believed d'Aquino's defense would be based on lack of treasonable intent, while Ince's would rely on duress. He considered both insubstantial. The subjects had assisted an enemy in furthering war aims, the very essence of "aid and comfort." However innocuous their broadcasts appeared, the fact that the Japanese continued to use them indicated the enemy believed they advanced their war aims. Wiener said that the *Cramer* case's requirement of treasonous intent could be met instead through the presumption that the defendants intended "the natural and probable consequences" of their actions. Significantly, Wiener adamantly insisted that if d'Aquino were tried, then Ince certainly had to be prosecuted, for he had recruited her. Wiener was disinclined to believe Ince had acted under duress; his claim, Wiener said, had "a fishy smell," but that was a jury's task to determine. Long standing treason law recognized mitigating circumstances if the subject acted "for fear of Death, and

---

79. Quinn to Clark, May 27, 1948, Clark to Quinn, May 28, 1948, *JDF*, *supra* note 3.

80. Wiener to Quinn, July 6, 1948, *JDF*, *supra* note 3.

while the Party is under Actual Force."[81]

Wiener believed there had been too much concern to prove individual, "segmented" overt acts. The fact that d'Aquino admittedly had broadcast for a continuous period was sufficient, a fact corroborated by ample witnesses. He suggested adding a few Japanese policy officials as witnesses to establish the importance of the radio program in the larger scheme of psychological warfare. Wiener believed the Supreme Court's recent decision in the *Haupt v. United States* treason case unduly created an "artificiality" to the segmented theory. "That sort of thing is legal hocus-pocus at its sorriest," he stated, "and I am far from convinced that even the law of treason requires it."[82]

Finally, Wiener thought it best not to return d'Aquino to the east coast. He feared the "possibly adverse overtones" resulting from the government's discretionary power to pick the venue. A trial in a district distant from her home and family would create hardships that would reflect unfavorably on the government.[83]

Wiener's summary ended the intra-Departmental hesitation. Less than three weeks later, Quinn received a letter from the United States District Attorney in the Canal Zone describing how he would proceed in presenting the d'Aquino case to the grand jury.[84] That man was none other than Tom DeWolfe. Although he had disagreed with Wiener's views in discussions with John Hogan and Raymond Whearty, DeWolfe ultimately succumbed to superior directives—and, of course, served them. DeWolfe found himself in the classic bureaucratic crunch. As late as January 1949—after he had secured an indictment—he informed the Department that he had reviewed all treason investigations in the San Francisco area. These included Ince, d'Aquino, Mark Streeter, and John Provoo—all suspected of collaboration with the Japanese. He evaluated the importance of their actions and the comparative evidence in the following order: 1) Provoo; 2) Streeter; 3) Toguri (d'Aquino); 4) Ince. Of the four, however, only d'Aquino ever was successfully prosecuted

81. *Id.* On English law, *see* FOSTER, CROWN LAW, *216. *See* Cramer v. United States, 325 U.S. 1, for a discussion of intent. The definitive study of the historical growth of treason law in England and the United States is in J. W. HURST, THE LAW OF TREASON IN THE UNITED STATES: SELECTED ESSAYS (1970). *See* Chapter 6 for post-World War II developments.

82. 330 U.S. 631 (1947).

83. Wiener to Quinn, July 6, 1948, *JDF, supra* note 3.

84. DeWolfe to Quinn, July 26, 1948.

and convicted.[85] DeWolfe continued to follow higher orders, however.

With the decision made to seek an indictment, the way was at last open for d'Aquino's return—not, of course, in a manner of her choosing. In contrast to the casual treatment of her departure and attempts to return in 1941, the United States government now displayed extraordinary concern about her voyage home in 1948.

IV. Return and Indictment

A. *What Forum? The Politics of Venue*

d'Aquino returned to the United States for indictment on the transport ship, *General H. F. Hodges,* accompanied by army guards. She departed Yokohama on September 15, 1948, and arrived in San Francisco on the 25th, where a familiar figure, FBI Special Agent Fred Tillman, took custody of the prisoner. The route home was curious, almost leisurely, like a lengthy holiday cruise. From Japan, the ship sailed to Naha, Okinawa and then to Inchon (Ginsen), Korea to discharge troops. From there, the *Hodges* sailed directly to California, avoiding an obvious resting or refueling stop in Hawaii. This was a carefully calculated maneuver to avoid berthing in any American port prior to landing in California. For if the ship had stopped in Hawaii, or any American territory in the Pacific, jurisdiction would lie in the local federal court. Attorney General Clark had made his preference clear to Justice Department subordinates: the case was to be presented in the Northern District of California. Apparently, the attorney general found no comfort or security in risking the probability of a polyglot jury in Hawaii, or anywhere off the mainland.[86]

The Justice Department's dealings with Secretary of the Army Kenneth Royall to arrange passage home offer a model for bureaucratic dissembling. Assistant Attorney General Alexander

---

85. *Id.* DeWolfe to Whearty, August 23, 1948, DeWolfe to Campbell, January 12, 1949, *JDF, supra* note 3. Provoo was tried in New York in 1954 and convicted for treason as a result of aiding the enemy while a prisoner of war. The conviction later was reversed by the Circuit Court and the government dropped the case. United States v. Provoo, 124 F. Supp. 185 (S.D.N.Y. 1954); United States v. Provoo, 215 F.2d 531 (2d Cir. 1954). Hogan successfully argued that Ince's case not be joined with d'Aquino's. Hogan to Whearty, August 20, 1948, *JDF, supra* note 3.

86. "Receipt of Prisoner," September 25, 1948, signed by Tillman, Hogan to Quinn, July 27, 1948, *JDF, supra* note 3. Given then-current Nisei enthusiasm to demonstrate their loyalty, Clark's fear may have been misplaced.

Campbell, Quinn's successor in the Criminal Division, told Royall that Justice would be very anxious to know if there were any planned diversion in the ship's route back to the United States. He frankly admitted the intention to seek indictment in San Francisco, and noted that a prior stop in any territory might vest jurisdiction in a territorial court. But he explained that the Department's desire to avoid the territories was based on "budgetary and financial restrictions, and for other reasons." Campbell requested confirmation from Royall that the ship would not make a call in the Territories of Alaska or Hawaii. In the meantime, Royall was to hold this seemingly innocuous request in strictest confidence.[87]

The decision to try the case in San Francisco did not meet with universal approval. The presidents of both the California Federation for Civic Unity and its San Francisco Council asked Tom Clark to consider the possible adverse affects upon racial feelings in the area. They emphasized that they "unquestionably assume[d]" there were ample grounds for prosecution, but expressed concern that the trial would hamper efforts to "weld together" the different, loyal racial elements in the community. Campbell merely replied that the action was "foreclosed by statutory provisions" for venue.[88]

B. *Indictment—And Its Dubious Aftermath*

Whatever public pressures the Department experienced for prosecuting d'Aquino, the fact is that Tom DeWolfe had no easy time persuading the grand jury in San Francisco to return a true bill. He nonetheless succeeded in doing so in October, 1948. "It was necessary for me to practically make a fourth of July speech in order to obtain an indictment," he told Whearty (Quinn's aide) on November 12. Still, two jurors voted against the govern-

87. Campbell to Royall, August 13, 1948, *JDF, supra* note 3. Campbell noted that d'Aquino had a husband and that she was "believed to be the mother of a minor child." Persecutors, it would seem, should know simple biographical facts: d'Aquino's child had been stillborn in January 1948. In any event, Campbell said that Justice would not defray the costs of transporting the husband and child. But Campbell was not without heart: he told John Hogan to interview d'Aquino on arrival, "but don't delay so that they can say we used a hose." Campbell to Hogan, August 2, 1948, *JDF, supra* note 3. d'Aquino's lawyer and the ACLU, however, bitterly protested his inability to see his client during the FBI's initial interrogations in San Francisco. Collins to Tom Clark, *et al.*, September 27, 1948, Ernest Besig, Director, ACLU to Clark, September 27, 1948, DeWolfe to Campbell, September 28, 1948, *JDF, supra* note 3.

88. Isaac Pacht to Clark, August 25, 1948, Ralph Reynolds to Clark, August 24, 1948, Campbell to Pacht, September 8, 1948, *JDF, supra* note 3.

ment. The same day, in a more formal message to Campbell (by then head of the Criminal Division), DeWolfe reported he personally presented the case "in a rather forceful manner." Apparently the grand jury was disturbed that the government was asking for an indictment against d'Aquino, without similar action against Ince. As a result, DeWolfe told the jurors Ince's case would be presented "in the immediate future" after further investigation in the Orient. (It never was.) DeWolfe pressed both Whearty and Campbell to apprise Attorney General Clark and his assistant Ford of the situation so that they, in turn, could secure J. Edgar Hoover's cooperation for post-indictment investigation.[89]

Initially, the FBI Director balked at further participation. Agent Tillman had done his work in 1946 and early 1948, and apparently Hoover believed that sufficient. Or perhaps Hoover, always eager to chalk up victories, was reluctant to involve his agency in a potentially losing cause. Ultimately, however, he consented and Fred Tillman returned to Japan on December 28 to secure more witnesses.[90]

1. WITNESSES: TWO, ONE, ZERO

It is impossible to know exactly what had troubled the grand jury besides the Ince matter. Perhaps the jurors detected a touch of incredibility surrounding one of the key Japanese witnesses, Hirom Yagi. Yagi, it will be recalled, was discovered and interviewed by Hogan and Brundidge during their trip to Tokyo in March, 1948.[91] In his grand jury testimony, Yagi claimed that, while in the company of a friend, he witnessed d'Aquino doing one broadcast. That "friend" was the same person Brundidge had hidden in his Tokyo hotel room because the man allegedly feared the military authorities. Yagi's evasiveness before the grand jury, however, definitely disturbed the FBI, who questioned him "many times" while he was in San Francisco, trying to learn the identity of his "friend." By then Yagi must have been terrified, for he consulted with Brundidge, who then was living in Monterey, and identified his friend to the FBI as one Toshikatsu Kodaira.[92]

---

89. DeWolfe to Whearty, November 12, 1948, DeWolfe to Campbell, November 12, 1948, *JDF*, *supra* note 3.

90. *Id.*

91. *See* text accompanying notes 60-61 *supra*, for Hogan's and Brundidge's March 1948 interview of Yagi.

92. FBI Report, San Francisco (Yagi) October 5, 1948.

The FBI agents informed DeWolfe and Hogan, who was also in San Francisco, of this development. The Justice attorneys asked the Army Counter-Intelligence Corps (CIC) agent who had escorted the Japanese witnesses back to Japan to conduct further investigations in Japan and to locate and interrogate Kodaira. The subsequent CIC report was devastating and sent shock waves throughout the bureaucracy that had developed the case. Kodaira "emphatically" denied having attended any broadcasts with Yagi. Yagi stuck to his original story until November 5 when the CIC man threatened to bring Yagi and Kodaira together to resolve the conflict. Thereupon, the investigator reported

> Yagi begged [him] . . . not to do so, stating, 'I will tell you the truth this time'. Yagi then advised, 'My friend, Harry T. Brundidge, came to Japan in March or April 1948. He asked me to go to the United States of America as a witness against Toguri. I told him I never seen Toguri broadcasting. He stated, "If you tell the story to Mr. Hogan then you will make a trip to the United States and we will have a nice time together." At this point of the interview, Yagi seemed to gain his composure. In response to a question . . . whether he would make a sworn statement to the above, he replied, 'Yes, I want to tell the truth now even though it gets me in lots of trouble.' Yagi continued, 'The statement I signed (that is the statement given to Hogan in Tokyo in March or April 1948) was not the truth.'[93]

Assistant Attorney General Campbell duly reported all this to Attorney General Clark on December 2. The attorney general's assistant, Ford, took the all-important first step in a cover-up that ultimately tarnished the whole affair. The report had a handwritten notation stating, "I would run this darn carefully. Confront Brundidge with it, etc."[94]

2. RE-ENTER THE "DISCOVERERS" OF "TOKYO ROSE"

Less than a fortnight later, Campbell sent a rather formal letter to Brundidge at his New York residence, asking him to come to Washington to talk about "certain matters" relating to the indictment. The letter was written "For the Attorney General." Brundidge was on vacation at the time and did not receive the letter until the 23rd. He thereupon called his old traveling companion Hogan, demanding to know what was happening.

---

93. Campbell to Clark, December 2, 1948, *JDF*, *supra* note 3.
94. *Id.*

Hogan informed his superiors he replied to Brundidge that he "was unable to tell him." Brundidge refused to come to Washington at his own expense and told Hogan to inform Campbell that the Department should send a representative to see him in New York—preferably John Hogan. Brundidge, however, finally appeared in Washington on January 5, 1949, and conferred with Campbell, Hugh Fisher, a special assistant to Clark, and Dean Schedler, Director of Public Information for the Department and an old acquaintance. He categorically denied he had prompted Yagi's testimony. He could offer no explanation for the witness's recantation except to say that he probably had "been stricken with fear" since his return to Japan. Clearly, there now were serious doubts about Yagi and Brundidge among some people in the Department. Hogan at least still knew the proper course for the agency. On Campbell's report to Clark of his interview with Brundidge, Ford noted to the attorney general that "We can still make the case according to my information from Hogan."[95]

The chummy relationship between Hogan and Brundidge continued through the spring of 1949. Hogan told the reporter of a postponement in the trial and that the government still intended to subpoena him and keep him on the witness list. In May, Brundidge called Hogan from New York to offer confidential information he believed the government should know. Apparently Brundidge's former colleague, Clark Lee, was outraged at the persecution of d'Aquino and had prepared a series of six articles for his employer, International News Service. (In his 1947 book, Lee already had referred to Mrs. d'Aquino as a "fall guy.") According to Brundidge's information, the articles "included a most bitter and vitriolic denunciation of the Government for having even considered a prosecution . . . in this case." The Chief Editor of INS refused to publish the articles because they were so hostile. Brundidge drew the obvious conclusion: Lee would be "extremely sympathetic" to the defendant if called as a witness by the government. Curiously, Hogan prepared this information only for "The Files" and apparently did not relay it to his superiors.[96]

Whatever doubts existed in the Department about the exis-

95. Campbell to Brundidge, December 14, 1948, Hogan to William Foley, December 24, 1948, Campbell to Clark, January 5, 1949, *JDF*, *supra* note 3.

96. Hogan to Brundidge, April 10, 1949, *JDF*, *supra* note 3; C. Lee, One Last Look Around, *supra* note 1, at 85; Hogan to The Files, May 26, 1949, *JDF*, *supra* note 3.

tence and nature of the Yagi-Brundidge link apparently were resolved with a report from Criminal Division Attorney Noel E. Story, who was in Japan with FBI Agent Tillman in the spring of 1949. On May 27, Story informed Campbell and DeWolfe that Theodore Tamba, one of d'Aquino's lawyers who was in Japan at the time gathering depositions and interviewing prospective witnesses, had found Yagi and confronted him with the charge of perjury before the grand jury. Yagi admitted the perjury to Tamba and then went to Tillman to offer a complete confession. Tamba also located Kodaira, from whom Story and Tillman then secured a deposition confirming Yagi's duplicity. Tamba desperately tried to get a deposition to this effect from Yagi himself, but Story advised the Japanese he did not have to do so and that it could be used against him in subsequent proceedings. According to Story, Yagi then left Tokyo and could not be found. Story also reported that Brundidge represented himself to Kodaira as an official representative of the Justice Department. Because of the "serious nature" of the allegations against Brundidge, Story recommended that the Department launch a "vigorous investigation" of his actions "with a view of possible prosecution action."[97]

Brundidge had powerful friends, and Justice Department officials realized that any prosecution of Brundidge would "completely destroy any chance of a conviction" of d'Aquino, as Campbell admitted to Clark on June 8. Not only was there a final wrap put on the cover-up, but it was done with a disingenuous twist. Campbell told Clark it would be unwise to prosecute Brundidge for there was only a slight chance of convicting a white man in California on the testimony of two Japanese.[98] Brundidge's case was not only dropped for the present, but there would be no future action—for if d'Aquino were convicted, prosecution of Brundidge would reveal his machinations in securing d'Aquino's prosecution and conviction, casting doubt on the evidence and likely constituting grounds for reversal or, at least, appeal. If she were acquitted, the same revelations from a

---

97. Story was fully aware of Tamba's questioning of Yagi; army censors regularly monitored Tamba's calls to Wayne Collins, d'Aquino's chief lawyer in San Francisco.

98. Story to Campbell, May 27, 1949, *JDF, supra* note 3. The Civil Censorship Detachment of the Occupation Army monitored Tamba's and Felipe d'Aquino's phone calls to Collins. Given Story's knowledge of Tamba's movements, possibly all of Tamba's calls were tapped. *See* Transcripts, June ? 1949, December 15, 1948, *AIF, supra* note 4. Tillman to DeWolfe, May 20, 1949, *JDF, supra* note 3, reports on the FBI agent's observations of Tamba.

trial of Brundidge could render the Department vulnerable for having undertaken d'Aquino's prosecution, and subject it to charges of covering-up an obstruction of justice.[99]

Better yet for Brundidge, he would not have to testify in d'Aquino's public trial. The government would not call him to the stand, and the defense could not: the common law rule then in effect required one to vouch for the veracity and truthfulness of one's own witness. Harry Brundidge sat in the proverbial catbird seat.

V. The Trial: Setting and Historical Perspective

The Cold War summer and autumn of 1949 marked a particularly grim moment for the United States. The "fall of China" and the Soviet Union's explosion of an atomic device occurred in rapid succession in August and September. Both events were viewed from within the United States as calamities facilitated by the connivance and disloyalty of Americans who aided and abetted those ominous, sinister developments. That perception had profound internal consequences in the United States. The external events confirmed rising fears of international communism, fears that already had led to increased persecution of alleged subversives and spies, as well as an increased demand for patriotic conformity. The ever-fragile supporting props for diversity and civil liberty collapsed as national governmental institutions and the media successfully peddled the idea of a domestic "Red Menace."

Resorting to legal institutions and processes legitimates dubious public policy. Thus the period is noteworthy for more than "Red-baiting," whether by the president, other executive branch officers, or congressional committees. From 1949 through the next half decade, the federal government initiated a wave of criminal trials involving alleged subversives and spies. The charges were criminal, but political purposes heavily freighted the prosecutions. The lengthy trial of the Communist party leadership for violating the Smith Act heads the list. The summer of 1949 also brought Alger Hiss's first trial for perjury. More sweeping, and more indicative of the times, were the new, not-

99. Campbell to Clark, June 8, 1949, *JDF, supra* note 3. The U.S. attorney in San Francisco apparently was uninformed as to the Washington decisions, as several weeks later he informed his superiors of his intention to interview Brundidge to get a complete explanation. He never did, of course. Frank Hennessy to Campbell, June 22, 1949, *JDF, supra* note 3. Also see Hennessy to Campbell, September 2, 1949, *JDF, supra* note 3.

so-subtle demands for loyalty from ordinary citizens. The mania for loyalty, officially sanctioned since 1947 by Executive Order 9835 (which established the Federal Employee Loyalty Program), spread throughout the country. Varied individuals, however notorious or ordinary, were required publicly to attest to their past, present, and future loyalty. The Board of Regents of the University of California implemented a loyalty oath program for professors in 1949, provoking bitter institutional divisions that persisted for a generation.

The world was choosing-up sides, and individuals, like nations, could not avoid involvement. The times dictated a premium on conformity and loyalty, however adverse and trying to one's personal situation. This was the cold reality that Iva Ikuko Toguri d'Aquino confronted in her treason trial from July to September 1949. Her alleged crimes had been committed during World War II. The immediate popular and official concerns for loyalty, however, coupled with a merciless, publicity-seeking pursuit by a few self-chosen persecutors, explain Mrs. d'Aquino's indictment, trial, and conviction far more than do the distant events of 1941-1945. These factors were both aided and nurtured by bureaucratic inertia, timidity, and, at times, chicanery.

The war with Japan was four years past; it might as well have been four light years. Japan's place as the linchpin of the United States' East Asian defense perimeter was established, and economic reconstruction, with American aid, was in full bloom. Just before the d'Aquino trial began, the *San Francisco Chronicle* carried a long feature on changes in Japan and American plans for the area. The article particularly emphasized that reconstruction would accelerate in order to make Japan "the workshop of Asia." Equally symbolic, perhaps, was the Naval Academy's appointment at this time of its first nisei, a young man who had spent the war years in the Minidoka Relocation Center in Idaho.[100]

The d'Aquino trial opened on July 5, 1949, with Federal District Judge Michael J. Roche presiding. It began amidst fanfare and sensationalism appropriate to the public's perception of the notorious "Tokyo Rose," but the trial pales in contrast to the drama of the case's covert side from 1945. The adversaries were experienced and skillful. Tom DeWolfe was a veteran prosecutor, and he was on familiar terrain as a result of his exper-

---

100. *SFC*, *supra* note 3, July 3, 8, 1949.

iences with the Nazi broadcasters. Defense lawyer Wayne Collins also had extensive trial experience, and he was familiar with the force of governmental power, having taken an active role in most of the major cases involving the rights of Japanese-Americans since 1942. Until his death in 1974, he devoted much of his time and practice to serving these clients in a myriad of cases and issues. The d'Aquino trial lasted nearly three months, but it was mostly uneventful and tedious, save for Collins' desperate, unavailing attempts to offer depositions discrediting governmental witnesses and to prove the taint of perjury. Judge Roche excluded as hearsay Kodaira's deposition denying that he and Yagi had attended any broadcasts together, but Collins managed to get a statement from Tillman on the record that Yagi had confessed to him he had been bribed to come to San Francisco and give false testimony to the grand jury.[101] Yagi, of course, having thoroughly recanted his original statements as given to the grand jury, was never called during the trial.

The government prosecuted the case in San Francisco to insure an all-white jury, and it secured one with little difficulty. The prosecution used only seven of twenty peremptory challenges, disqualifying six Blacks and one Chinese, an army veteran who had fought in the China-Burma-India theater. Following prevailing fashion, the government segregated witnesses during the trial: whites in one room, Orientals and Blacks in another. The jurors consisted of middle-class professionals and housewives, equally divided between men and women.[102] They were not sequestered during the fifty-two days of testimony. The long parade of witnesses they heard offered conflicting views on the nature of d'Aquino's role as broadcaster, with prosecution witnesses portraying her as an active participant in Radio Tokyo's propaganda purposes and those for the defense emphasizing her working under duress and as a ready accomplice in attempts to undermine the Japanese Army's propaganda goals.

The indictment voted by the grand jury the previous October alleged eight overt acts of treason that involved the defendant's broadcasting for the enemy. Overt Act #6 quickly emerged as the focus of the government's case, with its charge

101. *Record, Tillman,* 16:1597-99; Hennessy to Campbell, September 2, 1949, *JDF, supra* note 3. In the concluding arguments, one of the defense lawyers raised the matter of bribery and delicately suggested Brundidge's involvement. The issue, however, necessarily was restrained by the boundaries of the trial testimony. *Record, Arguments,* 2:251-56.

102. *SFC, supra* note 3, July 6, August 2, 1949.

that the defendant had broadcast information "concerning the loss of ships" during October 1944. As distinguished from Japanese or Allied soldier witnesses who claimed to have *heard* these remarks, two key witnesses testified they *saw* d'Aquino deliver the broadcast. George Nakamato Mitsushio was chief of the Overseas Section of Radio Tokyo and Kenkichi Oki was production supervisor at the station. Both men were California Nisei who had gone to Japan in 1940 and subsequently renounced their American citizenship. Both claimed to have witnessed the broadcast cited in Act #6, and Mitsushio testified he directed her to make the remarks. Mitsushio quoted d'Aquino as saying: "Now you have lost all your ships. You really are orphans of the Pacific. How do you think you will ever get home?" And Oki: "Now you fellows have lost all your ships. You really are orphans of the Pacific. Now how do you think you will ever get home?" These statements concerned the heavy American naval losses caused by a typhoon following a decisive victory over the Japanese in the Second Battle of the Philippine Sea. d'Aquino emphatically denied the testimony, but the government had its required two witnesses. The jury had to decide who was telling the truth.[103]

The jurors began their deliberations on the afternoon of September 26. From statements issued after the verdict, and from the foreman's memoir, we know the jury wrangled bitterly for nearly four days. There was "a general opinion" that the overt acts had been committed, but the problem was whether there was sufficient proof of the defendant's intent to betray the United States. Specifically, they divided 9-3 on overt Act #6. On several occasions, the jurors returned to the courtroom, telling Judge Roche they were "hopelessly deadlocked," but the judge refused to dismiss them. In the late afternoon of the 29th, the jurors asked for clarification of the judge's instructions. Roche had instructed the jury that when overt acts were judged "in the light of related events, [they] may turn out to be acts which were not of aid and comfort to the enemy." The holdouts wanted clarification on this point in order to use the evidence of d'Aquino's assistance to prisoners of war as demonstrating a lack

---

103. *Record, Mitsushio,* 10:919, 11:971; *Record, Oki,* 9:672; *Record, Defendant,* 49:5434-36; 5513-15. The Chicago Tribune published an article on March 22, 1976 claiming that one of the chief accusers in the trial had admitted giving false testimony. Several months later, however, both Oki and Mitsushio refused to retract any of their testimony or to submit an affidavit in support of d'Aquino's pardon request. Yasuhiro Fujita (Tokyo lawyer) to Wayne Collins, Jr., October 14, 1976, *WCP, supra* note 4.

of intent to commit treason. Roche refused to elaborate on his point, insisting the jury must give "uniform consideration" to all of his instructions. According to Foreman John Mann, Roche's position and pressure for a verdict led to the capitulation of the holdouts. In the early evening of the 29th the jury delivered its verdict, finding d'Aquino guilty of one count of treason. The decision shocked not only the defense, but the reporters covering the trial who had voted among themselves by a 9-1 count that the jury would acquit.[104]

Judge Roche imposed a sentence of ten years' imprisonment and a $10,000 fine on October 6, 1949. Speaking for his *client*, Wayne Collins said that "her own conscience is clear. She is sorry because she wishes she could say the Government's witnesses' consciences were also clear."[105] Neither Collins nor d'Aquino had the information to question the consciences of the bureaucrats who had determined her fate.

Appeals to the Ninth Circuit Court of Appeals and to the Supreme Court failed,[106] and the defendant was sent to her new prison, the Federal Women's Reformatory at Alderson, West Virginia. She was released in January 1956—the first convicted traitor to return to freedom—and then went to Chicago where her father had established a successful business following the war. There were talks of deportation—but to where?[107] The treason conviction ironically confirmed d'Aquino's persistent, almost pathetic, claim of American citizenship. This was not the end of d'Aquino's encounters with bureaucracy. Her attorneys made several attempts to secure a pardon after her release from prison, but to no avail. In fact, the Department counter-attacked by pursuing d'Aquino for full payment of the $10,000 fine, the final demand coming in 1971 from Attorney General John Mitchell.[108] The papers requesting a presidential pardon worked

---

104. Mann Memoir, Collection of Antonio Montanari, Jr., San Francisco; *SFC, supra* note 3, September 30, 1949. In 1976, Mann said he was surprised at the severity of the sentence. Furthermore, he stated he regarded Roche's final comments as virtually directing the verdict. Obviously remorseful, Mann added: "In my idea of justice, . . . the prosecution failed miserably to convict her." *SFC, supra* note 3, February 16, 1976.

105. *SFC, supra* note 3, October 7, 1949.

106. Iva Ikuko Toguri d'Aquino v. United States, 192 F.2d 338 (9th Cir. 1951), *cert. denied*, 343 U.S. 935 (1952). Rehearings denied, 343 U.S. 958 (1952); 345 U.S. 931 (1953). The Circuit Court found at least twelve errors in the trial court, but excused them as non-prejudicial.

107. The treason conviction carried a loss of citizenship rights, but being native-born, d'Aquino could not be deported.

108. On the attempts to collect the fine, *see* J. Walter Yeagley to Edward Hanrahan, May 9, 1966, *JDF, supra* note 3; Chicago Tribune, March 23, 1971.

through the bureaucratic labyrinth in the White House and the Justice Department through successive administrations. Finally, on his last day in office, President Gerald R. Ford, notoriously open-handed with pardons, pardoned Iva Ikuko Toguri d'Aquino.

AFTERWARD

Iva Ikuko Toguri d'Aquino was a pathetic, isolated, relatively insignificant individual who classically confronted political justice. She symbolically served the government's goals and purposes. Her acknowledged acts of broadcasting for the enemy took on a legendary mystique that heightened her importance far beyond the innocuous substance of her activities. Similarly, she served the professional and political needs of a few relentless lay prosecutors who successfully mounted enough pressure to enlist the cooperation of both ambitious and timid bureaucrats. d'Aquino was a symbolic sacrifice for their energies and fears; furthermore, they used her to symbolize the stringent, politically expedient meaning of loyalty. She was a pawn, and nothing more, from 1941 to 1949. In her desperate plea to Walter Winchell in 1948, d'Aquino perceptively captured the essence of her situation. "Under normal circumstances I should be called a 'bad girl' but there were so many complicated cobwebs . . . I . . . had to find a way to survive," she told Winchell. "I . . . had little chance in choosing a way to survive."[109] She was speaking of her Japanese experiences, of course; she had no idea of how well that explained her Kafkaesque experiences after 1945.

109. d'Aquino to Winchell, April 14, 1948, *WCP*, *supra* note 4.

AMERASIA 5:2 (1978), 69–94

The Pardoning of "Tokyo Rose": A Report on the Restoration of American Citizenship to Iva Ikuko Toguri

CLIFFORD I. UYEDA

On 19 July 1977, on his last full day in office, President Gerald R. Ford signed a full and unconditional pardon for Iva Ikuko Toguri d'Aquino, wiping the record clean. It was the first time in United States history that such a pardon had been granted in a treason conviction.

The detailed account of Iva Toguri's experience in Japan and of the trial are contained in the booklet Iva Toguri (d'Aquino): Victim of a Legend, *published by the Japanese American Citizens League's Committee for Iva Toguri (1975). For this reason, these details are not repeated in this present paper. Rather this report deals with the events leading to and the people involved in obtaining a presidential pardon for Iva Toguri.*

This report is also a story which chronicles the indomitable loyalty of a young woman who maintained her love for the United States in the face of incredible pressures. It is a story of justice denied in a trial in which the government's witnesses were ordered to perjure themselves. It is a story of the faith that sustained Iva Toguri despite a quarter century of harassment and persecution. Finally, it is a story of decent and fair-minded people who belatedly became aware of their culpability in neglecting her when she needed help, and set out

CLIFFORD I. UYEDA served as chairman of the JACL National Committee for Iva Toguri, 1975–77. He is currently the national president of the Japanese American Citizens League, 1978–80.

This account is a condensation of an earlier report. The original, along with other pertinent documents, is housed in the Bancroft Library, University of California, Berkeley.

finally on the almost impossible mission of vindicating her.

Few know all the details of this story. I learned many of them through involvement in the campaign to obtain a presidential pardon for Mrs. d'Aquino. This presentation is a condensation of an earlier report (October 1977).

There are many lessons to be gained from experience such as this, but only if the review is completely honest and candid. This I have attempted to do.

Formation of the Campaign Committee

Sometime during the summer of 1973 I happened to hear that John Hada, who had retired from the United States Army with the rank of lieutenant colonel, had recently written a thesis titled "The Indictment and Trial of Iva Ikuko Toguri d'Aquino—'Tokyo Rose' " for his master's degree in history at the University of San Francisco.

Out of curiosity, I telephoned Hada and invited him to speak at one of the monthly meetings of the San Francisco Center for Japanese American Studies. He consented, and in September 1973 I heard the story of "Tokyo Rose" for the first time.

I was fascinated, and troubled. During the following year-and-a-half I visited the San Francisco Public Library evenings after work. There I read the day by day coverage of the Tokyo Rose trial as reported in the *San Francisco Chronicle* and the *Examiner*. Later I reviewed the *Pacific Citizen*, the weekly publication of the Japanese American Citizens League (JACL), and sought out other published materials on Tokyo Rose. By then the enormity of the injustice perpetrated against this single unfortunate woman had become overwhelming.

There had been little sympathy or support from the general public for Iva Toguri during her 1949 trial. Even the Nisei remained aloof. To some 30,000 Nisei veterans of the war just recently ended, going to the support of someone charged with crimes against the very country they fought to defend was inconceivable. To other Nisei recently released from mass detention camps where they had been imprisoned during the war, their own problems of survival overshadowed all else. Besides, all the information they had about the case of Tokyo Rose came from the same sources that informed the rest of America, the press and radio which, especially in California, were openly hostile toward the Japanese and Japanese Americans.

Most Japanese Americans at the time preferred not to hear about Tokyo Rose, and over the years they largely forgot about her. To John Hada belongs a major credit for reawakening among them the interest and the sense of outrage at the injustice of her case that led eventually to her pardon.

One wonders what brought about the change in Nisei attitude toward Iva Toguri. Maturity, economic security, improved social acceptance into the mainstream of American life were important; but time was the greatest factor. Thirty years after the end of the war, anti-Japanese sentiment in the United States was minimal, the civil rights movement was a popular cause; and especially after the Vietnam War and Watergate scandal, Americans were not reluctant to look into and point out mistakes made by their own government.

John Hada was the speaker at the installation dinner of the San Francisco JACL chapter in March 1975. At the end of the program I went up to John Hada and said, "Let's find out if the JACL is doing anything about this case." John and I approached David Ushio, the JACL National Executive Director.

"Dave, what is JACL doing for Iva Toguri d'Aquino?" I asked.

"Nothing," David replied.

I said to John: "Let's form a committee to see what we can do. You be the chairman, because you know more about the case than anyone else."

John replied that he was very busy.

"I'll be your co-chairman," I replied, and that was the beginning.

The first meeting of the committee was held at my apartment in San Francisco on the night of 2 April 1975. Nine were present from over a dozen invited.

The committee was never any particular group of individuals who worked from the beginning of the campaign to the end. Different individuals joined at different phases of the campaign. In retrospect, the relatively smooth functioning of the committee was due to this free flow of members in and out of the committee. Those who felt they could contribute were comfortable in their functions within the committee doing what they did best.

At the meeting of April 2, the obvious first question was, "What can we do?"

Since the Supreme Court had reviewed the case and had

affirmed the judgment, legal avenues seemed pretty well closed. The one remaining possibility seemed to be a presidential pardon.

The committee was not formed at the direction of the National JACL Headquarters, as some had intimated later, nor in response to a resolution passed at the National JACL Convention in Portland, Oregon in the summer of 1974 which offered the League's aid to Iva Toguri if she desired. The existence of the resolution was completely unknown to me, and it was not brought to my attention until after the committee had been formed.

Soon after the first committee meeting I went to see attorney Wayne Merrill Collins, the son of the chief counsel for Iva Toguri during the 1949 trial. I found him to be a lean six-footer. He occupied the same office and desk used by his father on the thirteenth floor of the Mills Tower Building in the financial district of San Francisco. Japanese scrolls and woodblock prints adorned the walls. In his second year of law practice, the young Collins had known Iva Toguri almost as another member of the family and was thoroughly familiar with the details of the case.

"Several of us interested in the case of Iva Toguri have gotten together," I explained to Collins. "We would like to explore the possibility of what can be done for her at this time."

I sensed that he was skeptical.

"I would like to mobilize the Japanese Americans behind the movement to correct the injustice done to Iva Toguri," I continued. "To do so I believe the JACL can be a useful vehicle."

JACL has never been a respected name in the Collins family. As long as young Collins could remember, his father used the term "Jackals" in referring to JACL members. He was furious at the JACL organization for not opposing the war time expulsion from the West Coast more vigorously.

"I know how your father felt about JACL," I said. "JACL, however, also reflects the changes of the time. It is organized, and its members can be valuable in a national campaign."

Wayne Collins did not express his own feelings toward the JACL. I had assumed them to be somewhat similar to that of his father's.

"Have you seen Olshausen's article published in 1952?"

Wayne Collins said. George Olshausen was the only survivor of the three attorneys who had represented Iva Toguri in the 1949 trial.

"No, I haven't."

"It's called 'Folklore and Justice,' and was published in an obscure journal called *City lights*." He also referred me to the 1964 article by Dean Lipton, "Did we commit the wrong Tokyo Rose?" in *Nexus Magazine*.

I was delighted with the references he was giving me.

"I would like to write to Iva Toguri, informing her of what we want to do," I said.

Then with a handshake and my promise to keep him informed of our activities I left his office. I was satisfied with the first meeting with Wayne Collins.

Meanwhile, the committee obtained a loan of $1,000 from the Northern California-Western Nevada JACL District Council to begin. This was good news. The bad news was that John Hada was forced to drop out after the first committee meeting. His new job took him to Tokyo.

Documenting the "Tokyo Rose" Case

Throughout the campaign, Federal District Court Judge Robert Takasugi of Los Angeles was a voice at the other end of the telephone line. We did not have an occasion to meet, but his sound and friendly advice sustained us during the early uncertain months of the committee's existence.

Judge Takasugi's suggestion was to go the "pardon route," and mobilize public support for the campaign. The three essentials he saw were: 1) a comprehensive narrative of the case directed toward lay consumption, 2) legal briefs and the appeal document, and 3) the backing of prominent legislators and persons in positions of influence.

Attorney Theodore Tamba had filed a petition for executive clemency on 7 June 1954. There was no response. Attorney Wayne Mortimer Collins filed a second petition for presidential pardon on 4 November 1968. The second petition was also denied. Another petition was not immediately filed since an appeal for presidential pardon can be filed only once during a two-year period.

The advice we received from Judge Takasugi reinforced

what the committee had discussed must be done. The reeducation of the American public about "Tokyo Rose," we all felt, was the primary task. Our tool would be a comprehensive narrative of the case.

Mention Tokyo Rose to anyone, and the first reaction was "traitor." Not only the American public, but most Japanese Americans as well were totally ignorant of the true story behind the legend of Tokyo Rose.

At the committee's third meeting on 15 May 1975, Raymond Okamura and Mrs. Isami Arifuku Waugh volunteered to come up with the draft of a comprehensive narrative of the "Tokyo Rose" case. They were in possession of considerable material on "Tokyo Rose." Using their personal notes together with the more recently compiled information, each began writing different sections of the draft. "We began from scratch," said Ray Okamura.

A month later at the committee meeting on June 19, copies of the thirty-eight-page draft were distributed. The main criticism of the first draft was that the writing was abrasive. We wanted the document to be objective, convincing and appealing, not argumentive. Reader sensitivity and concerns had to be considered.

Committee writing is one of the most nerve wracking experiences. Every word and phrase in the style and presentation had to be weighed by different standards and personalities. We wanted the document to be logical and factual. At times solutions seemed impossible; but compromises finally won the day.

During the summer of 1975 four to five members of the committee met frequently to continue work on the manuscript. Disagreements and suggestions for changes were never taken personally. At the end of these working sessions there was always a feeling of accomplishment and satisfaction. Slightly over four months after the task had begun, the sixth and final draft was completed on September 24.

When the eye-catching orange and black booklets finally appeared, committee members who had labored long and hard on the drafts were elated. It looked like a good product. If it had not been for the patience and dedication in particular of Raymond Okamura, Isami Waugh and Yasuo Abiko, the manuscript could never have been completed. Frequent correspondence with Bill Hosokawa of the *Denver Post* also provided us with many valuable critical comments.

The Role of the Collins Family

Immediately after the first meeting with Wayne Collins I wrote to Iva Toguri, telling her of the committee's hopes of winning a pardon for her with the help of public and congressional support. Iva's reply was polite and reserved. She apparently had consulted with Collins. She thanked me for the interest, but did not hold out any great hopes for success.

From the very onset we let both Wayne and Iva know that their wishes would be followed explicitly. Iva indicated clearly that she wanted a low profile campaign. We could understand. Iva had a reflex fear of the media, which thirty years ago played a major role in her indictment and conviction. To give them a second chance to crucify her was unthinkable to Iva. Her fears had to be respected, but this posed a difficult problem because of the need to build up public support, which could only be done with publicity.

The committee felt Iva must be vindicated by the American people, not just pardoned secretly by the President. Her story had to be told to the people, and these people had to put pressure on the President. It was a delicate maneuver; how to reeducate the American public with minimal publicity.

We knew, of course, that the mood of the country had changed drastically since 1949. We felt confident that if Iva's story was presented carefully, factually, and nonabrasively, Americans would listen and react positively.

In November of 1975, the newly printed booklet, *Iva Toguri (d'Aquino): Victim of a Legend*, was forwarded to a few selected individuals. Response from this group was most gratifying. As Christmas 1975 approached, booklets were being mailed to the hundreds who were requesting copies.

Iva Toguri was to later tell us that whenever an article about Tokyo Rose appeared in the press, she would be subjected to unmerciful crank calls. Crackpots had disturbed her so much that she stopped opening mail when she could not identify the sender. Years later, she was told that some of the envelopes contained money for her. She had burned them all.

My first meeting with Iva Toguri was not until 6 August 1975 in San Francisco. She was with the younger Wayne Collins, who towered nearly a foot above her. She walked briskly and stood very erect. Wayne introduced me to her, and she extended her hand. It was a firm grip.

Iva thanked us for the campaign we were waging in her behalf, then turned the conversation away from herself. She spoke endearingly of the late Wayne Mortimer Collins. She even seemed to admonish some Nisei for not recognizing nor appreciating to the full the personal benefactions they had received from this great libertarian.

"There are Nisei who owe their success if not their lives to Mr. Collins," she said. "They wouldn't be where they are now if it were not for Mr. Collins."

She was obviously very close to the Collins family with whom she lived when she returned to San Francisco in May 1956 to defend herself in a deportation hearing. The younger Wayne Collins was then just eleven years of age.

"Mr. Collins used to say to me so often, 'Look Iva, at least you've led an eventful life full of excitements, though mostly painful. The rest of us lead a dull and monotonous life.'"

She did not speak one word about her own ordeal. She was crusading for the memory of the elder Collins.

When she and the young Collins left, the impression I received after this first meeting was that Iva Toguri was a dynamic individual with incredible strength of character. Coming face to face with her, I finally could understand how she withstood the harsh injustices of so many years. She had not been broken by the ordeal.

Who was Wayne Mortimer Collins who had impressed Iva so profoundly? Few individuals have done more for Japanese Americans than he. Acting alone when others failed to join his crusade for many good causes, his fierce devotion to principle was without equal.

The elder Collins was born in Sacramento, California, on 23 November 1899. His family moved to Oakland just prior to the San Francisco earthquake of 1906. From 1908 he resided in San Francisco. He was one of the seven original founders of the Northern California American Civil Liberties Union in 1934, and he served as a director for many years.

Wherever he sensed injustice, he was instantly on the scene to oppose it. In the *Korematsu* case he challeged the constitutionality of the government's indiscriminate expulsion and incarceration of Japanese Americans on the basis of race. Almost single-handedly he took up the cause of the renunciants, the discarded citizens, not only to prevent their being expatriated to Japan but to recover their citizenship which had been renounced under governmental duress. Using the threat of a full scale media exposé of atrocities perpetrated on American citizens, he was instrumental in obtaining the release of those imprisoned in the Tule Lake stockade during the summer of 1944.

After the outbreak of World War II hundreds of Japanese Peruvians were brought to the United States at our instigation to be used as hostages to be exchanged for Americans stranded in Japanese-occupied areas. At the end of the war, Peru refused to let these prisoners return home even though many were Peruvian citizens. All were scheduled for removal to Japan, a country many never had seen.

Wayne Collins, through legal maneuvers, prevented Attorney General Tom Clark from deporting these Peruvians of Japanese descent. Eventually some 300 of the Japanese Peruvians rescued by Collins chose to remain in the United States with many becoming U.S. citizens under the 1952 U.S. Naturalization Law.

In 1948 when Iva Toguri d'Aquino was brought to San Francisco for trial, the Toguri family searched for a lawyer. Many attorneys turned them down. Wayne Collins heard of the case and offered to represent her without fee. Even though Iva was convicted Wayne Collins was convinced of her innocence, and he sought to vindicate her for over twenty-five years until his death in 1974.

A Meeting with Iva Toguri

The Committee had spent most of its first nine months in preparing the information booklet. Now came the task of gathering support from individuals and organizations. This meant not only talking about the case but also talking about Iva Toguri the person, and most committee members had not even met her.

Just before Christmas 1975, Iva wrote that she was planning to come to San Francisco in early January and would like to meet with the committee. "Tokyo Rose" had been an American legend for over thirty-five years. Few had met her, even fewer had had the opportunity to speak to her. This was the first time, since her conviction twenty-six-and-a-half years earlier, that she would be talking about herself before a group.

It was the largest committee turnout to date. There were fourteen in all present. To most committee members this was the first meeting not only with Iva Toguri but also with Wayne Merrill Collins. The atmosphere was relaxed and cordial. Soon everyone was seated in a circle, all facing Iva.

"I really appreciate what you are doing for me," Iva began. "After so many years I didn't think any movement like this would actually happen." She hesitated a moment, then continued, "When first informed of the campaign I didn't know what to think of it. Would it again be like in the past?"

We knew she was referring to past attempts in her behalf which disappeared like burst bubbles. Now she was afraid to hope for much.

"What was your reaction when you learned that the JACL had offered you help?" one member asked. This referred to the JACL resolution offering the League's aid, passed at the Portland convention in July 1974.

Iva answered, "I said to myself, 'Why now, after 30 years?' I didn't know what it meant."

Iva was correct, because after the single notification of the resolution, nothing came of it.

"How did you manage to keep going all these years?" It was a question foremost in all our minds.

"I shut everything out. I set up a mental block about everything that happened to me." Her lips became ruler-straight. "Now, I often draw a blank when I try to recall the details."

"I might say something wrong because I just don't recall clearly, and I would be accused of lying," she said. Iva told us that she had not reviewed the details of her trial since 1949.

Iva recalled her arrest on 17 October 1945, and her subsequent imprisonment in Japan. After an exhaustive investigation by the Army and the FBI, Iva was cleared and freed on 25 October 1946. By simply remaining quietly in Japan after her release from Sugamo, Iva could have escaped prosecution. She had, however, considered herself only as an American, never a citizen of any other country. She therefore decided to return home.

When the news of her intended return was seized by the

American media and blown into a full scale campaign against her, Iva was rearrested in 26 August 1948. "If the trial was necessary to clear my name, I welcomed the opportunity," she told the committee members.

"Were you bitter after the trial?" someone asked meekly.

Iva's answer was immediate. "All through the war I clung to my American citizenship. I believed in America, that justice could be obtained in the U.S. court." She hesitated a moment. "I was disappointed."

Iva glanced for an instant at Wayne, then to all the others in the room. "When the truth could not be established in the court of law, I just said, 'What's the use!' I gave up."

"What are your reactions to the present campaign?"

"I'm grateful, of course, But I'm also worried. I'm not alone anymore. I'm responsible for the livelihood of numerous employees and their families."

Before the evening was over the meeting was more like an informal tea-party. People were talking in different groups with conversation turning to other topics. Iva appeared relaxed. Iva was no longer just a name to the committee members present; she was their friend with whom they sensed a deep emotional bond. The result was reinforcement and reconfirmation of our belief in the incredible injustice of the case and the strengthening of our motivation to work toward the committee's goal.

The Media Responds

Despite Iva Toguri's reflex revulsion for the media, the committee felt that public reeducation about the case was essential. How to reconcile the committee's belief with Iva's feelings posed a dilemma.

The very first public article to result from the booklet was a *Denver Post* editorial dated 8 January 1976. It stated: "It Is Time for Clemency."

After consultation with Iva and Wayne Collins we obtained permission to seek publicity in the well known and established newspapers and magazines in order to build a foundation of credibility. The information booklet was sent to each of the one hundred daily newspapers and weekly magazines with a circulation of over 200,000.

In February and early March of 1976, over a dozen sympathetic articles and editorials appeared. A big break in the campaign came on the morning of 22 March 1976. Iva called to say: "There's a huge crowd gathered outside my store. Reporters and cameramen are all over the place. I'm not stepping out of my office." Then she explained that the *Chicago Tribune* had published a story by their correspondent in Tokyo, Ronald Yates, under the headline: "Tokyo Rose's Accusers Claim U.S. Forced Them to Lie." The two former witnesses interviewed were Kenkichi Oki and George Mitsushio of Tokyo. The article went on to say:

"We had no choice," said one of the witnesses whose damaging testimony helped send Iva Toguri d'Aquino to prison for six years and two months. "U.S. Occupation Army police came and told me I had no choice but to testify against Iva or else.

"We were told that if we didn't cooperate, Uncle Sam might arrange a trial for us too. . . . So we cooperated and we did what we were told and now many of us have a guilty conscience because of it."

Iva was still afraid of the media, afraid that whatever she said might come out twisted and unrecognizable and she would have no recourse, as had happened thirty years ago.

Bill Kurtis, now anchorman of Chicago CBS-TV News, had been Iva's friend for many years. Back in 1968 he went to see Iva on a routine news assignment. He saw Iva briefly but was intrigued with her story and followed up his investigative work by going out to San Francisco to meet with Wayne Mortimer Collins and Ted Tamba.

Iva saw in Bill Kurtis a sympathetic friend who seemed to understand her, a person who had the interest and took the trouble to seek her attorneys in San Francisco. Unlike the other media people she had met, Bill Kurtis was thorough. Kurtis had read the trial transcipt at the Federal Archives and Records Center in San Bruno, California. He strongly believed in Iva's innocence and the injustice of the conviction.

On this day, March 22, Iva called Kurtis for help. He came to the store immediately to reassure her. That evening on the "Walter Cronkite Evening News," the public finally saw and heard Iva Toguri.

In April, Kurtis arranged to have Iva filmed on CBS's "60 Minutes," and the segment was shown on 20 June 1976. Interviewer Morley Safer was marvelous, according to Iva, in putting her at ease. His gentle approach enabled Iva to go through the interview with confidence.

Said Morley Safer, "It was not Iva on trial, it was Tokyo Rose."

During the interview Morley Safer said, "You could have either become a Japanese citizen or be interned."

Iva's voice was deliberate. "I said intern me. Then they said, 'No, it would cost money to feed you three times a day.'"

Commented Morley Safer, "Iva Toguri was functioning basically as a disc jockey. Where did the legend of Tokyo Rose come from? Iva was one of dozens of women announcers broadcasting not only from Tokyo but from at least nine different Japanese strongholds in the Pacific. All these voices were regarded as one voice, and the Americans nicknamed it Tokyo Rose."

Safer continued, "When the war ended, American reporters were eager to find a legend. But they were looking for someone who did not exist, Tokyo Rose, the creature of their own invention."

Attorney George S. Guysi of Oklahoma City, who back in 1945–46 was the counter-intelligence officer in charge of investigating Iva Toguri in Tokyo, appeared on the same program, and states, "The State Department abandoned her." Concerning Iva's broadcasts, he said, "She wasn't asked if she wanted to do it. She was told to do it."

"In 1948," Safer commented, "with no new evidence the Justice Department reopened the case. Iva was rearrested in Tokyo, and shipped back to San Francisco."

Iva said, "I suppose if they got someone and got the job over with they were all satisfied."

The voice of John Mann, foreman of the jury that convicted Iva, was next heard, although he refused to be filmed: "There was a great deal of anti-Japanese prejudice existing throughout the country, especially here in California. And that had some effect on the jury. Of that I'm quite certain."

Mann's voice went on: "There have been very few months since the trial that I did not think of her and think that she was not guilty. And I am rather sorry that I did not stick to my guns."

Morley Safer seemed surprised at Iva's lack of obvious rancour. "Do you have any bitterness toward this country?"

"I don't think bitterness is the word. I am kind of disappointed that I had to go through what I did because of some untruths, and that they couldn't separate the myth from facts."

Safer continued, "There's terrible irony that on this bicentennial year your birthday is the Fourth of July."

Iva's voice trailed away softly, "Isn't it? It really is. It is."

Morley Safer ended the interview with a statement: "This year a pardon application is being filed in her behalf. A presidential pardon, we remind you, does not bestow innocence on the pardoned person. Nor can it restore the fine nor the years spent in prison. But in Iva's case it would give back something of great value to her, her American citizenship."

The impact of "60 Minutes" was tremendous. A week later not one adverse comment was heard on the follow-up program. Editorials across the nation came to her support. Letters poured in. Two weeks later on July 4, the country's two-hundreth and also Iva's sixtieth birthday, Iva mentioned over the phone that her store looked like a florist shop. Brilliant colored carnations, gladiolus, birds-of-paradise, and potted plants of all shapes and sizes were delivered to her store. Some came from those she knew, but many were from total strangers. "I can't believe it," she said. "It's a 180-degree turn from twenty-seven years ago."

Mobilization of Political and Organizational Support

The growing success of the publicity campaign created concerns among many regarding the actual filing of the petition for presidential pardon. "Why not strike while the iron is hot?" they reasoned. What happens if President Ford is not elected? Would we have to begin all over again?

This is where the younger Collins steered the course with both insight and foresight. Wayne explained thus:

— We must consider the political consequences of any actions we take.

— Hesitancy to act on a controversial pardon must be expected prior to the November election.

— Regardless of the administration in power, letters on the case received by the White House would remain in the files of the Pardon Attorney. Thus, those letters written would not be lost due to a change of administration.

— For us, it is perfectly possible to place again in motion considerable publicity surrounding the case on the date upon which we file the pardon application. Our best course would be to file it immediately after the election when the pressure on President Ford would be off.

All those involved in the campaign were informed of Wayne's strategy. It was realistic, and was to prove wise.

Sample petitions for public signatures and sample resolutions were mailed to over a hundred JACL chapters, and to other interested individuals and organizations.

California State Assemblyman S. Floyd Mori (D-Pleasanton) became deeply interested in the case. He introduced a resolution supporting pardon for Iva Toguri, co-sponsored by fifty-seven Assemblymen and Senators. On 24 June 1976, this resolution passed the State Assembly 60-0, and the State Senate 22-0. The California State Legislature had provided a major boost to the campaign.

The San Francisco Board of Supervisors, led by president Quentin L. Kopp, adopted a supporting resolution on 6 July 1976, with Mayor George R. Moscone endorsing. The city of San Francisco, where the trial was held twenty-seven years earlier, became the first city to officially come to the support of Iva Toguri.

Honolulu followed suit on July 14, Los Angeles on August 16, and San Jose on September 28. The Los Angeles resolution had a special meaning to Iva because Los Angeles was not only her place of birth but because the present resolution also rescinded the 8 December 1947 City Council resolution which opposed Iva Toguri's return home to America.

Some JACL members wanted to follow up these successes by asking every politician running for office to take a stand on the issue. The danger was obvious. The committee cautioned that without knowledge of the case, the natural reaction of politicians would be to support the legend which had been established so firmly in the public mind.

Instructions were sent to supporters to make sure that candidates had adequate and accurate information before confronting them with a question about Iva in public.

Support for Iva's pardon by former American soldiers was an important part of the campaign. Congressman (now Senator) Spark M. Matsunaga from Hawaii, a decorated veteran of World War II, was the leading advocate of the campaign in Congress. He spoke to his colleagues, spoke in the House chamber, and wrote numerous letters to the heads of veterans organizations.

Some 33,000 Japanese Americans had served in the U.S. armed forces during World War II, 6,000 of them in the Pacific theater. In California alone there are fourteen Nisei VFW (Veterans of Foreign Wars) Memorial Posts and two Military Intelligence Service (MIS) Associations. Each was sent an informational packet and a request for support. At first there was only silence.

The VFW Post (Willard Anderson Post #2471) in The Dalles, Oregon came to Iva's support. We were not sure how many of its members were Japanese Americans, or how this particular post became interested in the campaign. However, none came from the California Nisei VFW Posts. We repeated our requests.

Meanwhile, we heard from the 41st Infantry Division Association. The 41st was a National Guard division from Oregon, Washington, Idaho and Montana. Called to active duty in September 1940, it was the first American division to be sent overseas. This division fought its way up from Australia through numerous Pacific islands to the Philippines, and was one of the first units sent to Japan for occupation duty.

The 41st Infantry Division Association has 3,000 active members, and has held conventions annually since 1948. At its annual meeting in Dearborn, Michigan on 8–9 July 1976, more than 500 members of the 41st unanimously agreed to put the Association on record as being in favor of an immediate presidential pardon for Iva Toguri. In its association publication, *The Jungleer,* it declared: "The case is considered a travesty on justice and a blot on our postwar behavior."

Another veterans groups, the surviving crewmen of an aircraft carrier, *U.S.S. Gambier Bay*, sunk by the Japanese during the Battle of Leyte Gulf on 25 October 1944, made a gallant effort to pass a resolution supporting Iva Toguri's pardon. It failed only because of opposition from members who were still unaware of the true story behind the legend. However, the fact that so many came to Iva's support was an enormous comfort to us and to Iva.

Still, we were unable to get the endorsement of the Nisei 442nd at its 1976 summer reunion in Chicago. These delegates had gathered primarily for fellowship and felt that they could not make decisions as representatives of their home organizations. Reluctance to speak for others and concerns about the reactions of others, however, appeared to be the unspoken reasons. Although we were disappointed, it was to the credit of Tom Hibino of the JACL Midwest Regional office in Chicago that no attempts were made to force the issue.

The 442nd moved to Washington, D.C. for a banquet. There was a suggestion that a resolution could still possibly be passed at this gathering. If there was anything we had learned from the campaign it was that no attempts should ever be made to ramrod a statement of support through an organization when there is reluctance for any reason. We redoubled our efforts because we knew there were many Nisei veterans supporting us.

The breakthrough came when Nisei VFW Memorial Posts of Monterey Peninsula, Sacramento and San Jose (all in California) publicly announced their support during October and November 1976.

Many Nisei veterans had personally contributed to the committee's campaign fund. Yet these same veterans had remained silent when the support issue was brought up at their local meetings. Apparently it was "going public" that deterred them.

Filing the Pardon Petition

On 5 December 1975, the Committee requested a special recognition for Iva Toguri and Wayne M. Collins to be presented at the 1976 JACL National Convention in June 1976. Non-cooperation by the JACL National Headquarters came as a complete surprise and was a mystery to the committee.

The proposal therefore was taken directly before the National Council at the convention, and two hours before the scheduled recognition luncheon the proposal was unanimously adopted. Obviously the recipients were not present. The citations the committee had prepared many months earlier were accepted by proxies.

It was Wayne Collins' idea to announce the filing of the official petition for presidential pardon at a press conference to be held at Seventh and Mission Streets, San Francisco, the site of the Federal District Court where twenty-seven years earlier Iva Toguri had been convicted.

The day selected was Wednesday, November 17. By 10 A.M. sunlight was breaking through the fog. Media representatives completely filled the area near the southwest entrance to the building. Iva Toguri and Wayne Collins arrived by cab at 10:10 A.M. as arranged, and Iva answered questions put to her by the reporters.

"Why are you filing for pardon again after all these years?" someone asked.

"Because I want to recover my American citizenship," she said. "America is my home. It will always be my home. American citizenship is a deep and valuable thing to me. You don't fully realize its value until you lose it."

"How did you manage to survive all these years as a convicted person if you are innocent?"

"Because my conscience is clear. I can sleep nights."

"Do you really think President Ford will pardon you?"

"I have hopes. Times have changed. If the trial was held today there is no way I would be convicted."

Iva appeared nervous and grim, but she was managing well. At the conclusion of the press conference Postmaster Lim P. Lee accepted the official petition to be mailed to the Pardon Attorney Lawrence M. Traylor at the Justice Department in Washington, D.C.

On the same day, the American Veterans Committee (AVC) notified the JACL National Headquarters that on November 13, its national board meeting had passed a resolution "requesting the President of the United States to grant pardon to Iva Toguri so that her American citizenship may be restored."

The press conference made the front page of the evening *San Francisco Examiner*, and was carried on the major networks' evening news. With the exception of CBS (both national and local) and KQED (PBS local), however, the TV networks still presented Iva Toguri as "the infamous Tokyo Rose of World War II years." This was particularly true in the coverage of both NBC's nightly news of November 17 and its "Today Show" on November 18.

Mrs. Constance Yasuda of Bloomington, Indiana was among those who protested the NBC report. David Brinkley replied to her in a letter that said in part: "We offered no grounds for Mrs. Toguri's (sic) appeal for a pardon because there were no grounds, beyond the fact that she simply wants a pardon . . . It is simply a woman who after all these years wants to wind up her life a full-fledged citizen with all her rights restored. That is about all there is to it, and that was what we said."

Katherine Pinkham Harris, former correspondent who covered the 1949 trial, was another who wrote to Brinkley protesting the distorted report. Don McGaffin of KING-TV, Seattle, Washington, called stating that they had received outraged reactions at Brinkley from viewers of the NBC report. McGaffin requested materials for preparing a program to counter the Brinkley presentation.

The committee had wanted to let the American people know that Iva Toguri was "The Woman Who Was Not Tokyo Rose," which is the title used by Dr. S. I. Hayakawa in his syndicated newspaper column of 20 March 1976.

Seeing the distorted news reports did not create a sense of failure or depression among committee members. We knew this could happen. Emotionally, we were prepared for this. It just gave us first hand experience on how news reporting can be distorted.

The President Acts

Senator-elect Samuel I. Hayakawa informed me early in December that he had discussed the Iva Toguri case with President Ford. The matter was to be kept confidential, he said. About a week later Iva Toguri mentioned that nothing had happened following the two previous filings, but this time FBI agents came to see her less than a month after the petition had been filed. They also visited each person she had named as a character reference in her pardon application. One of these individuals told her that the FBI agent had said he was under instruction to complete and file his report by mid-December.

The year-end holiday season is a traditional time for pardons. We waited anxiously. But Christmas came and went and there was no word. Those JACL members in Washington were mainly identified with the Democratic party. "I don't think President Ford will do anything," the JACL Washington representative reported. "We're concentrating our efforts with the incoming Carter administration. The Republicans are busy preparing to leave Washington."

As 1976 came to a close our Washington contacts were silent except for S. I. Hayakawa who called to say he was still in touch with the White House on the matter.

During the last week of the Ford administration, I kept the radio on for any possible news concerning Iva's pardon. On Monday, January 17, Iva sounded resigned over the phone.

Apparently it wasn't going to happen, although this time her hopes had been raised. She was remaining at home by the radio. She was too nervous to work at the store.

A few minutes after 3 P.M. that day, KCBS Radio suddenly announced that President Ford intended to pardon Iva Toguri. Immediately, I telephoned KCBS for details. The source of the report was a *Washington Post* article to be published the next morning, January 18. However, the day was over in Washington. Nothing further could be learned.

Excited but uncertain, we retired for the night. At 1:30 A.M. the phone rang. It was our Tokyo member on the committee, Ms. Kamisaka, calling from Japan to ask if it was true that the President had decided to pardon Iva. I told her what I knew.

Next morning I called Wayne Horiuchi, JACL Washington representative. Yes, he had heard the news also. He said he had called S. I. Hayakawa who wondered whether this could be "a deliberate leak by the Justice Department to test the public's reaction to the pardon." I thought not. It would seem too late for testing public reactions. Only two full days remained for the outgoing administration. I sent a telegram to President Ford stating we hoped the report we had heard on the radio about his granting a pardon to Iva Toguri was true.

Wayne Collins finally got in touch with the pardon attorney, Lawrence M. Traylor. Wayne was told the Justice Department had forwarded a recommendation to the White House, but Traylor could not say what that recommendation was.

I spoke to Wayne Horiuchi in Washington again. He had also gotten in touch with Traylor. "He told me, 'Don't worry'," Horiuchi said.

I felt then that we had won. Why would a pardon attorney say, "Don't worry," to a Japanese American unless it was a favorable recommendation?

All day Tuesday (January 18) the phone rang requesting clarification of the radio and TV reports of the pardon. However, no official confirmation came from Washington.

Imre Horvath, producer of CBS's "60 Minutes," was in frequent contact with Iva Toguri, transmitting reports he had been receiving from his contact at the White House. He told Iva that her petition was definitely on the President's desk.

By nightfall we could imagine how Iva must be feeling. Tomorrow was President Ford's last full day in office.

Ever since the first radio news, Iva Toguri had remained at

home to wait for confirmation. When nothing was said at the 2 P.M. press briefing in Washington on January 19, Iva decided to report to work at the store. It was a good try, she figured, but seemed like another failure. The legend of Tokyo Rose was too powerful even for the President of the United States to destroy.

The back room at Iva's gift shop was like a command post. All lines were in constant use. Bill Kurtis of CBS-TV in Chicago was there, in constant touch with Washington. Tom Hibino, Director of JACL Midwest Regional office in Chicago, was also covering the phone, answering the stream of incoming calls.

I was on the line speaking to Iva when she said, "Wait a minute. Wait a minute. Hold on. Bill Kurtis just said he got through to the Justice Department."

I squeezed the phone to my ear. I could hear several voices in the distance.

"Really?" I could hear Iva saying.

She returned to the phone. "It's true," she said. She was hardly audible. She continued, "Bill just spoke to Deputy Pardon Attorney David Stephenson. He confirmed that President Ford had signed a full and unconditional pardon."

All I could think to say was: "I'm glad. Congratulations." I could hear shouts in the background. "I'll get back to you later," I said, and hung up.

It was lunch time in San Francisco. I was alone in a reading room on the first floor of the JACL National Headquarters building, the room the committee was permitted to use. The area was completely separated from the offices in the building.

When I hung up the phone I walked over to a chair and sat down. I linked my fingers of both hands behind my neck and closed my eyes to gather my thoughts. So it had finally happened. Thirty years of waiting was over for Iva Toguri. Nothing could restore nor change those thirty miserable years, but at last the President had recognized the injustice of the ordeal she had undergone. The dignity with which she suffered in silence, the indestructible loyalty she held toward the United States was remarkable.

If Wayne Mortimer Collins and Theodore Tamba were still alive, I thought; how they would have rejoiced this day. In particular, Collins, who became Iva Toguri's protector when none other would touch her case, remained her champion for over quarter of a century after the trial, completely convinced of her innocence. His death came two and a half years too early. The consolation is that it was his son, Wayne Merrill Collins, who carried on his father's work to a successful conclusion.

I was glad it was President Ford who pardoned her, because he had been forced to give the matter thorough deliberation. The President did not need further criticism of his pardon decisions. The unfavorable public reaction to his pardoning of Richard Nixon was still present. But once Iva Toguri was separated from the legend of Tokyo Rose, I felt there could be no doubt in anyone's mind of the enormous injustice committed against this woman.

The official notification, affixed with the seal of the Department of Justice, was signed by Attorney General Edward H. Levi, stating that "this action is the act of the President being performed at his direction." It stated that President Gerald R. Ford "has this day granted unto Iva Ikuko Toguri d'Aquino a full pardon," and had empowered the Attorney General "to sign this grant of executive clemency."

According to Wayne Collins this was the first time in United States history that a full and unconditional pardon had been granted where the conviction was for "treason."

In Chicago Iva told reporters: "After all these years it's hard for me to believe that the pardon is really true." She also said she was grateful to President Ford and to all those who worked in her behalf during the pardon campaign.

There are so many people to thank in the campaign to obtain the pardon for Iva Toguri. Foremost are the late Wayne Mortimer Collins and his son Wayne Merrill Collins. Since his father's death in July 1974, the younger Collins has represented Iva Toguri. Two generations of Collins attorneys have represented Iva for twenty-nine years without charge.

The committee members labored for nearly two years with the help of many other dedicated volunteers. Many JACL members from throughout the country gave us the support we desperately needed. Once the true story behind the legend of Tokyo Rose was revealed, the media support was phenomenal; and this triggered grass roots support from Americans all across the nation.

Governor George R. Ariyoshi of Hawaii was the first elected public official to endorse the campaign. Dr. S. I. Hayakawa was the first prominent mainland figure to publicly state his support. Congressman Spark M. Matsunaga's eloquent voice was the first from the Congress to urge pardon and the restoration of her United States citizenship. Many, many others came to her support, and to each of them we are most grateful.

The political reality of the Washington scene must be considered. The White House was occupied by a Republican administration, and to get attention it helps to have a member of that party on your side. Senator S. I. Hayakawa, one of the most talked-about new Republicans in Washington, played a crucial role in the pardoning of Iva Toguri. On 23 January 1977, I finally received permission to release his role to the public.

When the President was bidding farewell to his fellow Republicans at the White House on the morning of January 19, he told Senator Hayakawa that he had signed Iva's pardon. The Senator, of course, was not at liberty to reveal the news, but within a few hours the Department of Justice announced the pardon.

Antonio (Tony) Montanari, Jr., a free-lance television producer, became interested in the case back in 1972. From the Justice Department files, Tony obtained interoffice memos concerning the case. He was most helpful in supplying our committee with important documents, and we are deeply grateful.

There was support from Japan during the campaign. We were fortunate in having two prominent Japanese writers on our committee. Their efforts resulted in wide awareness of the case among the Japanese people, who had known even less about "Tokyo Rose" than the Americans.

One of the writers was Mrs. Masayo Umezawa Duus who wrote a book based on her research into Iva Toguri's story. It was published in December 1976 under the title, *Tokyo Rose: 30 Year Stigma as Traitor.*

The other, Ms. Fuyuko Kamisaka began writing about the case in July 1976. She obtained signatures from Seiji Ozawa of the Boston Symphony Orchestra, Nobel prize winner Leo Esaki, and many other prominent Japanese on a petition sent to President Ford. She also served as a chief liaison person for Iva Toguri with her former Japanese acquaintances in Japan.

Why was there interest in Japan concerning the pardon for Iva Toguri? Although Iva was imprisoned by American authorities, episodes for which she was convicted occurred at the order of the Japanese wartime military government. The Japanese, therefore, felt some responsibility for her predicament.

On February 4 I received a phone call from the JACL National Headquarters saying there was talk about "recognizing" about half a dozen persons for their contributions toward the successful campaign to have Iva Toguri pardoned.

I was furious. It seemed JACL headquarters still had not understood the purpose of the campaign, that it still looked upon it as an opportunity to blow its own horn.

The only two persons that should be recognized are Iva Toguri and the elder Collins, I insisted. Iva for her loyalty to the United States in spite of what the country did to her; and Collins for devoted support for a cause he and his son never doubted.

Where are you going to draw the line among the vast number of supporters who worked so hard for the campaign? The pardon campaign was not undertaken to seek publicity for the JACL, the committee, or its members. I felt selecting a few committee members for "recognition" was an affront to the many whose roles were also significant. How can JACL pat themselves on the back for doing something that should have been done many, many years ago? We should be asking forgiveness for past timidity and inaction.

The final meeting of the committee was held on 7 February 1977, at a no-host dinner meeting at a restaurant in San Francisco. Both Iva Toguri and Wayne Collins were present.

Now that the objective for which the committee was formed had been achieved, we heard complaints about the inadequacy of the pardon. Some held that the government should make some form of restitution.

When the idea was broached to Iva, she said she would like to be allowed to enjoy her new status for a while. "Let me bask in the sun," she said. "It's such a pleasant feeling." A sense of relief and spiritual contentment has finally been achieved, after years of suffering in silence. It would be cruel and heartless, indeed, to thrust Iva back into the midst of another controversy, especially if it is to satisfy someone else's desires. Wayne Collins stated that he had no comments to add at this time. It was, therefore, suggested that the committee be declared inactive. There was no dissent.

Obviously, the cause for which so many gave their time and efforts so generously has to have an universal appeal, in this instance, the righting of an injustice so long overlooked. But the real satisfaction is in knowing that we have made an individual happy again. The greater the injustice, the more difficult it becomes to admit it and to correct it. An individual stands helpless before the gigantic figure of a nation which, in this case, was the perpetrator of the injustice.

For a nation it is easy to forget the past and let the victims suffer unnoticed in the shadow. That we were able to prick the nation's conscience and arouse it enough to persuade the President to act was the result of the concern of millions of reawakened Americans.

We are grateful for each and every committee member's dedication to the task we set out to accomplish nearly two years ago. We appreciate Iva Toguri for permitting us to organize and campaign in her behalf. Without Wayne M. Collins' understanding and expert counseling the campaign could not have succeeded as it did. Our thanks go also to others across the nation and overseas who supported Iva Toguri's cause.

*(*ed. note:* Raymond Okamura's research on the Iva Toguri case can also be found in the article, "Iva Ikuko Toguri: Victim of an American Fantasy," published in *Counterpoint: Perspectives on Asian America*, 1976.)

Acknowledgments

Daniels, Roger. "The Decisions to Relocate the North American Japanese: Another Look." *Pacific Historical Review* 51 (1982): 71–7. Reprinted with the permission of the University of California Press. Courtesy of Yale University Sterling Memorial Library.

Dembitz, Nanette. "Racial Discrimination and the Military Judgment: The Supreme Court's Korematsu and Endo Decisions." *Columbia Law Review* 45 (1945): 175–239. Courtesy of Yale University Law Library.

Fine, Sidney. "Mr. Justice Murphy and the Hirabayashi Case." *Pacific Historical Review* 33 (1964): 195–209. Reprinted with the permission of the University of California Press. Courtesy of Yale University Sterling Memorial Library.

Freeman, Harrop A. "*Genesis, Exodus,* and *Leviticus*: Genealogy, Evacuation, and Law." *Cornell Law Quarterly* 28 (1942–43): *414–58*. Courtesy of Yale University Law Library.

Gotanda, Neil. " 'Other Non-Whites' in American Legal History: A Review of *Justice at War*." *Columbia Law Review* 85 (1985): 1186–92. Reprinted with the permission of Columbia University in the City of New York. Courtesy of *Columbia Law Review*.

Irons, Peter. "Fancy Dancing in the Marble Palace." *Constitutional Commentary* 3 (1986): 35–45. Reprinted with the permission of Constitutional Commentary Inc. Courtesy of *Constitutional Commentary*.

Morris, Arval A. "Justice, War, and the Japanese-American Evacuation and Internment." *Washington Law Review* 59 (1984): 843–62. Reprinted with the permission of the Washington Law Review Association. Courtesy of Yale University Law Library.

Nash, Philip Tajitsu. Book Review. "Moving for Redress." *Yale Law Journal* 94 (1985): 743–55. Reprinted by permission of

The Yale Law Journal Company and Fred B. Rothman & Company. Courtesy of Yale University Law Library.

Rostow, Eugene V. "The Japanese American Cases—A Disaster." *Yale Law Journal* 54 (1944–45): 489–533. Courtesy of Yale University Law Library.

Takahata, Sandra. "The Case of *Korematsu v. United States:* Could It Be Justified Today?" *University of Hawaii Law Review* 6 (1984): 109–75. Reprinted with the permission of the *University of Hawaii Law Review*. Courtesy of Yale University Law Library.

tenBroek, Jacobus. "Wartime Power of the Military Over Citizen Civilians Within the Country." *California Law Review* 41 (1953): 167–208. Courtesy of Yale University Law Library.

Christgau, John. "Collins versus the World: The Fight to Restore Citizenship to Japanese American Renunciants of World War II." *Pacific Historical Review* 54 (1985): 1–31. Reprinted with the permission of the University of California Press. Courtesy of Yale University Sterling Memorial Library.

Daniels, Roger. "Japanese Relocation and Redress in North America: A Comparative View." *Pacific Historian* 26 (1982): 2–13. Reprinted with the permission of the University of the Pacific. Courtesy of *Pacific Historian*.

Daniels, Roger. "Redress Achieved, 1983–1990." In Roger Daniels, Sandra C. Taylor and Harry H.L. Kitano, eds., *Japanese Americans: From Relocation to Redress*. (University of Washington Press, 2nd ed. 1991): 219–23. Reprinted with the permission of the University of Utah Press. Courtesy of Yale University Law Library.

Iyeki, Marc Hideo. "The Japanese American Coram Nobis Cases: Exposing the Myth of Disloyalty." *New York University Review of Law & Social Change* 13 (1984–85): 199–221. Reprinted with the permission of the New York University School of Law. Courtesy of the *New York University Review of Law & Social Change*.

Kutler, Stanley I. "At the Bar of History: Japanese Americans Versus the United States." *American Bar Foundation Research Journal* 1985 (1985): 361–73. Courtesy of Yale University Law Library.

Kutler, Stanley I. "Forging a Legend: The Treason of 'Tokyo Rose.'" *Wisconsin Law Review* 1980 (1980): 1341–82. Reprinted with permission of the Wisconsin Law Review. Copyright 1980, The University of Wisconsin. Courtesy of Yale University Law Library.

Uyeda, Clifford I. "The Pardoning of 'Tokyo Rose': A Report on the Restoration of American Citizenship to Iva Ikuko Toguri." *Amerasia Journal* 5 (1978): 69–93. Reprinted with the permission of the *Amerasia Journal*. Courtesy of Yale University Sterling Memorial Library.